Industrial Pressure, Level, and Density Measurement

Second Edition

Industrial Pressure, Level, and Density Measurement

Second Edition

Donald R. Gillum

ISA™

Industrial Pressure, Level, and Density Measurement, Second Edition

67 Alexander Drive
P.O. Box 12277
Research Triangle Park, NC 27709

Printed in the United States of America.
10 9 8 7 6 5 4 3 2

ISBN: 978-1-934394-34-2

Notice
The information presented in this publication is for the general education of the reader. Because neither the author nor the publisher have any control over the use of the information by the reader, both the author and the publisher disclaim any and all liability of any kind arising out of such use. The reader is expected to exercise sound professional judgment in using any of the information presented in a particular application. Additionally, neither the author nor the publisher have investigated or considered the affect of any patents on the ability of the reader to use any of the information in a particular application. The reader is responsible for reviewing any possible patents that may affect any particular use of the information presented.
Any references to commercial products in the work are cited as examples only. Neither the author nor the publisher endorses any referenced commercial product. Any trademarks or trade names referenced belong to the respective owner of the mark or name. Neither the author nor the publisher makes any representation regarding the availability of any referenced commercial product at any time. The manufacturer's instructions on use of any commercial product must be followed at all times, even if in conflict with the information in this publication.

Library of Congress Cataloging-in-Publication Data

Gillum, Donald R.
Industrial pressure, level, and density measurement / Donald R.
Gillum. -- 2nd ed.
p. cm.
ISBN 978-1-934394-34-2
1. Pressure transducers. 2. Level indicators. 3.
Liquids--Density--Measurement. I. Title.
TJ223.T7G48 2009
681'.2--dc22

2008044970

Any success that I have or may ever achieve I owe to my family and close colleagues. First of all, I owe much to my Mother and Father who always had confidence in me and encouraged me to better myself when I was happy to settle for much less. My wife, Barbara has always supported me in all endeavors. My children, Darcy and Jeff were understanding and patient when I was trying to build a career and depriving them of time we could have spent together. Darcy and Jeff have given much support, lending their efforts in many ways including research and desktop publishing. My grandchildren, Lisa, Katie, Sarah, Samuel, Daniel and great-grandchildren Joseph and Hannah have also been patient. The little ones have been very understanding when they ask "Pa, is your book finished so you can play with us?"

The Instrumentation Faculty and Staff at Texas State Technical College -TSTC have been very helpful in technical support and research.

Thanks to Susan Colwell and the ISA staff for their patient help in this project. Much appreciation goes to the reviewers, especially Jerry Gerlich who offered many valuable recommendations.

—Donald R. Gillum

TABLE OF CONTENTS

Appendixes

List of Illustrations

List of Tables

PREFACE

This book demonstrates the principles of pressure, level, and density measurement and how they are applied. The basic theory and technology of each measurement system are discussed. Because each application has its own requirements and measurement demands, it is important to consider the most suitable scheme for a given situation. This involves studying the means by which a particular measurement system is applied to a process and selecting an instrument that will yield the most satisfactory results. Selection necessitates consideration of maintenance and calibration requirements and procedures in addition to accuracy, reliability, and cost. No measurement system can be expected to fulfill all requirements exactly, and trade-offs of advantages and shortcomings generally must be made, particularly in the selection of level instruments.

Chapter 1 presents an introduction to the science of measurement; Chapters 2 and 3 deal with pressure and differential pressure measurement and applications; Chapter 4 addresses transmitters and transmission systems; Chapter 5 introduces level measurement theory and visual measurement techniques; Chapters 6 and 7 cover various level measurement principles; Chapter 8 discusses density measurement; Chapter 9 discusses the principles and applications of hydrostatic tank gaging; and Chapter 10 explores various methods of instrument selection and typical applications.

No attempt has been made to discuss all of the many applications dealing with the measurement of these variables or to list all of the manufacturers of specific equipment or systems. The most common transducers/transmitters and their general operating principles are discussed with the hope that an application engineer or technician will be able to design, install, and implement a measurement system for pressure, level, or density.

The material presented takes an application-oriented approach to measurement and is fundamental in nature. The reader should have a background in general physics, applied algebra or math, and basic instrumentation as well as a desire to learn more about measurement and control.

ACKNOWLEDGMENTS

The author expresses gratitude to the following companies for their cooperation in granting permission for the reproduction of several drawings as well as the use of various technical articles and materials.

ABB Kent-Taylor, Incorporated

Ametek, Incorporated

Arjay Engineering, Ltd.

Ashcroft-Heise, Incorporated

Beamex, Inc

Bindicator

CRC Press

Clark-Reliance Corporation

Crystal Engineering Corporation

Delta Controls

Dover Company (Quartzdyne)

Dresser-Masoneilan

Emerson Process Management (Fisher-Rosemount, Micro Motion)

Flow Control Magazine

Fluke Corporation

Gauging Systems Incorporated

HACH Company (Great Lakes Instruments)

Haenni Instruments, Incorporated

HART Communication Foundation

Invensys Process Systems (Foxboro)

ISA

ISE-Magtech

KROHNE, Incorporated

Magnetrol International

Oil & Gas Journal

Ohmart/VEGA Corporation

Ruska Instrument Corporation

Scott Fetzer Company (Meriam Process Technologies)

The University of Texas Petroleum Extension Service

3-D Instruments, LLC

Toshiba

Transmation, Incorporated

Validyne Engineering Corporation

Yokogawa Corporation

About the Author

Don Gillum's industrial background includes ten years as an instrument/analyzer engineering technician and application engineer at what is now Lyondell Petrochemicals. Before that he was a process operator at that plant and a chemical operator at a gaseous diffusion nuclear facility. He has worked at Texas State Technical College for 36 years and held positions as Master Instructor, Assistant Professor, Department Chair, and Division Director. In these jobs he has been actively involved in all facets of student recruitment, curriculum development, course development, laboratory design, instruction, industrial training and continuing education.

Don has developed and taught courses for ISA since 1982, was instrumental in the purchase and development of the original training center in downtown Raleigh. He has taught courses at the training institute and on location since that time and has written various articles for technical magazines and training manuals.

On the volunteer side, Don has served two terms on ISA's Executive Board as Vice-President of District VII and Vice-President of the Education Department—now PDD. Presently he is Director of PDD serving in the area of certification and credentialing, has worked in the development of CCST, NICET, is a program evaluator for ABET, past commissioner of the Technical Accreditation Commission (TAC of ABET) and is now a member of the ABET Board of Directors. Don is a recipient of ISA's Golden Achievement Award. He holds a BS from the University of Houston and is a Professional Engineer in Control System Engineering.

1

Introduction to Measurements

In our advanced industrial society, it is readily understood that measurement is of vital interest to the populace as a whole, just as it has been to every civilization throughout recorded history. The many efforts to measure time, weight, length, pressure and so forth by various cultures separated by vast distances reveal the universal human desire for known quantities and the ability of man to develop different solutions to common problems.

Little evidence exists to reveal the actual practices and methodologies used to accomplish many ancient human feats. However, it is well understood that various achievements in earlier times required sophisticated measurement techniques. The pyramids were built with such close tolerances that even if the force required to transport the gigantic building stones and elevate them to great heights is discounted, the precise measurements certainly inspire awe. Other classic achievements of earlier civilizations—Stonehenge, developments by the Aztecs, and irrigation techniques used in Egypt thousands of years ago—leave little doubt that the Industrial Revolution was not the beginning of measurement technology; however, many modem measurement methods are directly related to developments in the late 1800s.

Making measurements is generally considered to be a preliminary procedure for a greater result. It is a means to an end—usually the control of a variable, a scientific experiment or evaluation, data obtained for research or data maintained to discover historical trends.

Probably more than any other single item, the steam engine enabled the concentration of industrial facilities, the predecessor of present manufacturing and process installations. The first prominent industrial application of feedback control is generally considered to be the flyball governor used to control the speed of steam engines.

In early industrial facilities, it became apparent that many variables needed to be measured. The first crude instruments were simple pointer displays located at the process. Common applications where observation of a parameter needed to be remote from the actual point of measurement included hydraulic impulse lines filled with the process fluid and connected to pressure gages mounted in a panel for local indication. The gages were calibrated in units corresponding to the measured variable.

As the size and complexity of process plants increased, the need for transporting measurement signals through greater distances, isolated from the process material, became apparent. Transmitters, as they later came to be called, were developed for this purpose. The sensor or input transducer of a transmitter is often a pressure element connected to the process. A change in process condition causes a response in the form of motion or force at the free end of the element. This response is used as an input to the output transducer, which generates a scaled signal that is transmitted to a receiver instrument. The receiver then converts the transmitted signal into a measurement.

While this evolution in measurement increased the operator's knowledge of the process, part of the problem was yet to be solved. Periodic adjustments to components for load changes still required human intervention through the manipulation of valves in the process. Many times these valves were in hazardous locations, distributed throughout the process unit (which was growing in size), separated from the central control station and from each other by considerable distances, and located in inaccessible areas. Another development driven by necessity was an air actuator or diaphragm powered by compressed air to position the valves. A spring was used to oppose the force generated by the diaphragm, causing the valve to return to the open or closed position as required for safety. Being able to observe the process conditions of several control points at a central location and to remotely position a valve to maintain a process value helped to establish the principle of centralized control.

To establish this centralized control, manual regulators were used to enable an operator to position valves from the control station. A few operators could then monitor and control entire units or sections of large plants from a central control station. From this constant monitoring and regulation, operators soon learned the techniques of feedback control. As a process condition started to deviate from a control point, the valve position was exaggerated; an overcorrection was made to anticipate a condition based on the amount and rate of change. If the original adjustment failed to return the process to the control point, increments of valve movement would be continued over a period of time until the deviation was eliminated. This continuous type of control by operator intervention enhanced control quality and was a considerable improvement over previous methods.

During times of repetitive load changes in processes with interactive control loops, operators were unable to keep up with the demands on the system. In practice, a single valve can be manipulated with little effort or deep mental concentration; adjusting several valves where functions are interdependent becomes quite challenging, however. The answer, in the form of automatic feedback control, was not long in coming. Although field-mounted automatic controllers had been used prior to the development of control centers, controllers were developed to receive a signal from a field measurement transmitter, compare the process value with a reference point established by the operator, and generate a signal that was sent back to the valve for corrective action. Algorithms developed for these early pneumatic controllers were basic control equations that emulated the manual control provided by operators and established the concept and theory used in modem proportional, integral and derivative (PID) control technology. Although limited to distances of a few thousand feet, pneumatic transmitters and controllers were the workhorses of control instrumentation for years; electronic measuring instrumentation was slow to be developed. Pneumatic transmitters were not replaced in great numbers until the advent of solid-state and microelectronic devices. Development was first concerned with controllers, control rooms, and other auxiliary equipment.

The importance of measurement is often overlooked because the end result of a complicated control system is considered to be the function of control. Regardless of the simplicity or complexity of a control application or research project, it soon becomes apparent that a measurement must be made. When measuring instruments are unavailable, or those available are inadequate to provide reliable results with the desired accuracy and precision, other measuring devices must be developed or existing devices improved. The development of measuring systems has largely been governed by the currently available technology. Like control instruments and auxiliary devices, measuring instruments have progressed through various development stages—from mechanical to pneumatic to electronic—vacuum tubes, to transistors, to integrated circuits, to microprocessor-based intelligent transmitters. While this development was brought about by generic components and devices available at any point in time, much was a result of regulative and economic demands.

In the late 1960s and early 1970s, Environmental Protection Agency (EPA) stipulations required record keeping and data-reporting functions that stretched the limits of many analytical and process-measuring instruments in use at that time. More accurate and reliable measuring means were developed as a matter of necessity. The evolution of microprocessor control systems, full distributed control systems (DCSs) and single-loop digital controllers enabled more accurate control to closer tolerances.

Many control system revamps were justified by increased product quality, economy of resources and energy efficiency. The engineers who installed DCS technology and failed to upgrade measuring systems that were of a previous generation soon realized that the axiom "control is no better or accurate than measurement" always rings true.

While microprocessor-based measurements using digital transmitters were initially developed for enhanced flexibility in configuration, the improved accuracy and stability resulting from digital circuitry soon became primary selling points. Totally digital signal generation, manipulation and transmission are advantages being fully realized with digital transmitters; the general compatibility of digital communication between transmitters, receiving instruments and other components is common. Digital-to-analog and analog-to-digital conversion for 4–20mA DC signal utilization is still used in many applications, but this technology is declining.

The lack of a fieldbus standard resulted in manufacturers providing input-output (I/O) processors to provide a mix-and-match capability between various types and brands of instruments. Highway Addressable Remote Transmitter (HART) technology, a hybrid between analog and digital communication, was considered by some to be a short-lived technology, to last until a true digital technology could be developed. HART has staying power, however, and its use is still expanding. FOUNDATION™ Fieldbus technology was developed and the bus wars that impeded the growth of all-digital control systems are no more. Complete compatibility and interoperability are provided by FOUNDATION Fieldbus, and an added bonus is the first field control system (FCS); control is moving back to the field. Wireless technology is expanding and finding near-plant and off-plant applications.

Despite the cumbersome beginning of the total integration of digital measuring and control devices, analog electronic transmitters are giving way to digital transmitters as digital devices are moving into the field. Although generally not considered a viable option for most new installations, pneumatic transmitters are still used in a few applications: local measurement in some remote locations, hazardous atmospheres and other isolated instances. Those still in use are most often replaced with digital transmitters.

The advantages of digital instruments in terms of configuration and programming flexibility, economics, excellent performance specifications and networking are leading to the total integration of measurement and control. Bidirectional communication between field and control-room components over a data highway is state-of-the-art technology. Microprocessor-based intelligence built into field devices, both valves and transmitters, use software to provide nearly any desired control strategy by the initiation and intervention of numerous measured variables. Intelligent field transmitters communicating with digital controllers communicating with valves over a digital highway are commonplace. The era of total integration of digital components in a measurement and control system is here.

EXERCISES

1.1 List the stages of development or evolution of control systems, from total manual control to the latest technology for automatic control.

1.2 What was the first industrial application of feedback control?

1.3 List two advantages of microprocessor-based measuring instruments. What are these instruments called?

1.4 Describe an application where pneumatic transmitters may be preferred over electronic transmitters.

1.5 What does the term "total integration" imply with respect to measurement and control instrumentation?

2

Pressure Measurement and Calibration Principles

2.1 INTRODUCTION

Industry's current emphasis on improving control quality sometimes implies that the importance of measurement is overlooked. However, measurement is the first requisite of any control scheme. Lord Kelvin once summarized the significance of measurement science: "If you can measure that of which you speak and can express it by a number, you must know something of the subject. But, if you can not measure it, your knowledge is meager and unsatisfactory—measurement is the basis of all knowledge." If you cannot measure, you cannot control.

2.2 FLUID PROPERTIES RELATING TO PRESSURE MEASUREMENT

Fluids and Pressure

Pressure is a fundamental measurement from which other variables can be inferred. Pressure values rank with those of voltage and temperature in defining the energy (primarily potential) or state of matter. Temperature is the potential for doing thermodynamic work, voltage is the potential for doing electrical work, and pressure is the potential for doing fluidic work. The importance of pressure measurement is demonstrated by the need for transmitting signals for indication and control, inferring fluid flow in pipes, and using filled thermal systems in some temperature applications. Liquid levels in tanks and other vessels can also be inferred from pressure quantities.

Pressure can be best understood through Pascal's law, which describes the behavior of fluids at rest. According to this law, a change in the pressure of an enclosed fluid is conveyed undiminished to every part of the fluid and to the surface areas of its container.

In this discussion, the term "fluid" refers to both liquids and gases. Both occupy the container in which they are placed; however, a liquid, if it does not completely fill the container, will present a free liquid surface, whereas a gas will always fill the volume of its container. When a gas is confined in a container, molecules of the gas strike the container walls. These collisions result in a force, exerted against the surface area of the container. Pressure is equal to the force applied to an object (here, the walls of the container) divided by the area that is perpendicular to the force. The relationship between pressure, force, and area is expressed as:

$$P = F/A \tag{2.1}$$

where P is pressure, F is force, and A is area. In other words, pressure is equal to force per unit area.

For a fluid at rest, the pressure exerted by the fluid at any point will be perpendicular to the boundary point. In addition, whenever an external pressure is applied to any confined fluid at rest, the pressure is increased at every point in the fluid by the amount of the external pressure. The practical consequences of Pascal's law are apparent in hydraulic presses and jacks, hydraulic brakes, and pressure instruments used for measurement and calibration.

To understand the significance of Pascal's law, consider the hydraulic press shown in Figure 2–1. A force applied to the small area of piston 1 is distributed equally throughout the system and is applied to the large area of piston 2. Small forces exerted on the small piston can cause large forces on the large piston. The following relationship exists in the hydraulic device because the pressure at every point is equal:

$$P_1 = P_2 \tag{2.2}$$

Combining Equations 2.1 and 2.2 leads to the following relationships:

$$P_1 = F_1/A_1$$

$$P_2 = F_2/A_2$$

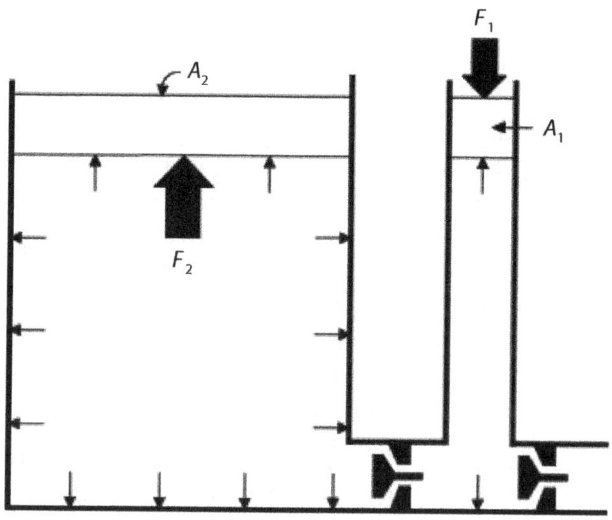

Figure 2–1 Hydraulic Press

$$F_1/A_1 = F_2/A_2$$

$$F_2 = (A_2/A_1)F_1$$

EXAMPLE 2-1

Problem: For a hydraulic press with the following specifications, find F_2:

$$A_1 = 0.5 \text{ in}^2$$

$$A_2 = 4 \text{ in}^2$$

$$F_1 = 150 \text{ psi}$$

Solution: By rearranging Equation 2.2 and solving for F_2,

$$F_2 = \left(\frac{4\text{in}^2}{0.5\text{in}^2}\right)(150\text{psi}) = 1200\text{psi}$$

Pressure Units

Pressure is caused by two forces: gravity and compression. The pressure exerted by a volume of liquid is proportional to the vertical height, the mass density of the liquid, and the value of gravity at the local point (Figure 2–2). Work performed on a volume of contained fluid compresses the fluid, causing the pressure to rise in direct proportion to the work performed. Likewise, a compressed fluid is capable of doing work at the expense of pressure release (Figure 2–3). Also, in liquids the gravitational effects can be a primary source of pressure, while in gases the work effects predominate. Both effects are always present to some degree in the physics of fluids and are significant considerations in the science of pressure measurement.

The Système International d'Unités (a standard international system, abbreviated SI) recognizes length, mass and time as the fundamental units applicable to pressure. An intermediate unit is force expressed as mass × length or mass × acceleration. Time is a universal unit in all unit systems. Mass, length, time, pressure, and force can be expressed in a variety of forms. Details are given in Table 2–1.

Table 2–1 is by no means exhaustive in terms of practical units. Every trade and occupation that deals with pressure has a favorite unit and is reluctant to change for the sake of standardization. It is best, then, to get used to the idea of multiple units and to keep conversion tables handy (see Tables 2–2 and 2–3).

The standard unit of pressure in the English or foot-pound-second (FPS) system is pounds per square inch (psi), where the force is pounds and the area is square inches. In the meter-kilogram-second (MKS) system of measurement, pressure is expressed as kilograms per square meter (kg/m^2). In SI units, pressure is measured as newtons per square meter (N/m^2) or Pascals (Pa).

Gage and Absolute Pressure

We live in an environment of about 1 atm pressure produced by the effect of gravity on the air molecules that surround the earth. This pressure is an unnoticeable, permanent condition and thus is a natural reference to which higher or lower pressures can be compared. This situation is particularly useful since many pressure-measuring instruments actually measure differential pressure. These devices, such as the familiar Bourdon tube, yield an output that is proportional to the

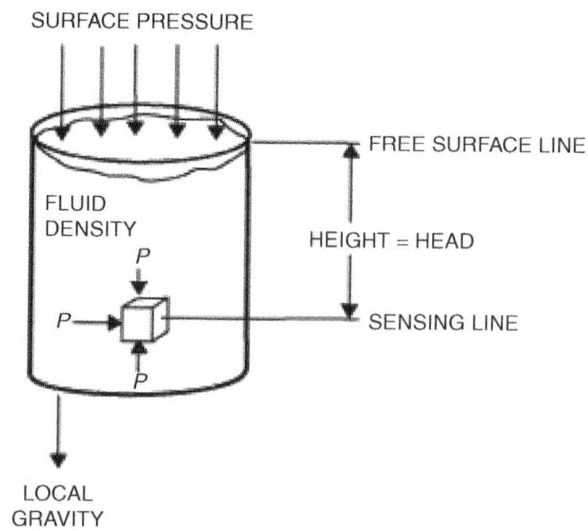

Figure 2–2 Pressure Due to Gravity (*Courtesy of Transmation Inc.*)

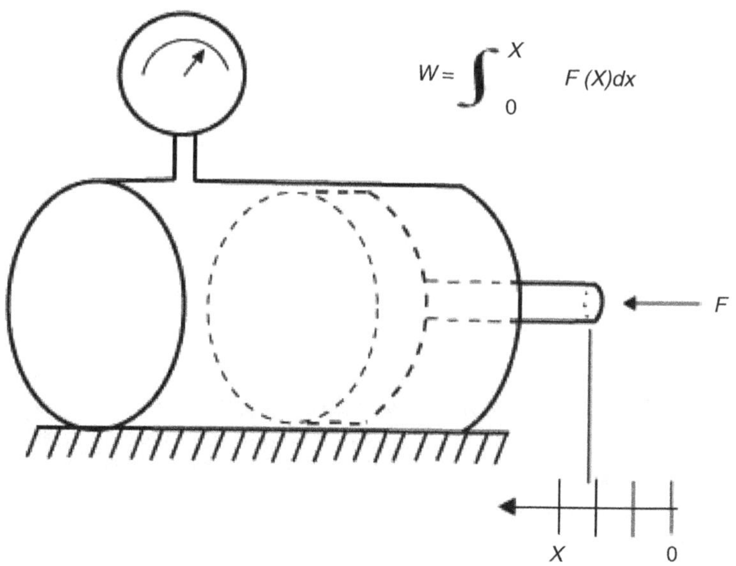

Figure 2–3 Pressure Due to Applied Work

applied pressure minus the value of the environmental pressure surrounding the instrument. The pressure indicated by any instrument referenced to the local atmosphere is called gage pressure; the corresponding units, such as psi, are postscripted with the letter "g". Thus, psig means pounds per square inch gage pressure [Ref. 1].

Table 2–1 Some Fundamental Units of Measurement (*Courtesy of Meriam Instrument Division, Scott Fetzer Company, and Transmation Inc.*)

Physical Measure	System of Units		
	SI	**Metric**	**English**
Length	meter (m)	meter (m) centimeter (cm) millimeter (mm)	foot (ft) inch (in.)
Mass	kilogram (kg)	kilogram (kg) gram (g) metric slug	pound (lb) ounce (oz) slug
Time	second (s)	second (s)	second (s)
Force	newton (N) dyne	newton (N) kilogram froce (kgf) gram force (gf)	pound force (lbf) ounce force (ozf)
Pressure	N/m^2	kg/cm^2 $dyne/cm^2$ cm H_2O mm Hg bar atmosphere (atm) pascal (Pa) kilopascal (kPa)	lbf/in^2 = psi in. H_2O in. Hg atmosphere (atm)

Pressure values that are lower than atmospheric are referred to as vacuum or negative gage pressure. If, by diligent pumping, all free molecules are removed from a volume, then there is no longer an agent within the volume that could exert a force on any area. A perfect vacuum exists, and the pressure is true or absolute zero. The value of this pressure is independent of the pressure in the atmosphere surrounding the evacuated vessel and hence is the true zero point on any pressure scale, regardless of unit.

An absolute pressure scale is of major significance in many engineering and scientific endeavors because it eliminates the need to measure and perform calculations for the pressure of the local atmosphere. If the local atmosphere does affect a pressure measurement, then its conversion to absolute terms by adding the locally measured value of atmospheric pressure (as is done with a barometer) yields a number—that is, an absolute pressure—which can be compared with similar measure-

Table 2–2 Pressure Unit Conversion Factors
(*Courtesy of Meriam Instrument Division, Scott Fetzer Company*)[a]

1 in. H_2O equals	**1 in. of Red Oil equals**
0.0360 lb/in^2	0.0297 lb/in^2
0.5760 oz/in^2	0.4752 oz/in^2
0.0737 in. Hg	0.0607 in. Hg
0.3373 in. No. 3 fluid	0.2781 in. No. 3 fluid
1.2131 in. Red Oil	0.8243 in. H_2O
1 ft. H_2O equals	**1 in. No. 3 fluid equals**
0.4320 lb/in^2	0.1068 lb/in^2
6.9120 oz/in^2	1.7088 oz./in^2
0.8844 in. Hg	0.2184 in. Hg
4.0476 in. No.3 fluid	2.9643 in. H_2O
14.5572 in. Red Oil	3.5961 in. Red Oil
62.208 lb/ft.^{2}	
1 in. Hg equals	**1 oz/in^2 equals**
0.4892 lb/in^2	0.1228 in. Hg
7.8272 oz/in^2	0.5849 in. No. 3 fluid
4.5782 in. No. 3 fluid	2.1034 in. Red Oil
13.5712 in. H_2O	1.7336 in. H_2O
1.1309 ft. H_2O	
16.4636 in. Red Oil	
1 lb/in^2 equals	**Other Conversions**
2.0441 in. Hg	1 kg/cm^2 = 14.22 psi
27.7417 in. H_2O	1 kg/cm = 98.067 kPa
9.3586 in. No. 3 fluid	1 kPa = 0.1450 psi
2.3118 ft. H_2O	
33.6542 in. Red Oil	

a. All fluids at a temperature of 71.6°F (22°C).

ments made in different places and at different times without the varying atmospheric pressure further entering the comparisons.

Pressure measurements can be referenced to atmospheric pressure, which is 14.696 psi at sea level, or to zero pressure, which is a vacuum (no positive pressure is expended). When referenced to atmospheric

Table 2–3 Pressure Conversion Factors (*Courtesy of Transmation Inc.*)

Pascal + N/sq. m	dyne/ sq. cm	bar	kg/sq. cm	torr = mm Hg at 0°C	Atmosphere (atm)	psi (lb/sq. in.)	in. H_2O at 0°C
1	10.000,00	0.000,01	0.000,010,197,16	0.007,500,671	0.000,009,869,304	0.000,145,037,7	0.004,014,630
0.100,000,0	1	0.000,001	0.000,001,019,716	0.000,750,067,1	.000,000,986,930,4	0.000,014,503,77	0.000,401,463,0
100,000.0	1,000,000.	1	1.019,716	750.067,1	0.986,930,4	14.503,77	401.463,0
98,066.50	980,655.0	0.980,655,0	1	735.564,5	0.967,848,1	14.223,34	393.700,8
133.321,4	1,333.214	0.001,333,214	0.001,359,500	1	0.001,315,789	0.019,336,64	0.535,236,2
101,324.3	1,013,243.	1.013,243	1.033,220	760.000	1	14.695,84	406.779,5
6,894.759	68,947.59	0.068,947,59	0.070,306.96	51.715,31	0.068,046,45	1	27.679,90
249.089,0	2,490.890	0.002,490,890	0.002,540,000	1.868,334	0.002,458,334	0.036,127,29	1

pressure, the unit is gage pressure and is designated psig, and when referenced to absolute zero pressure or a vacuum, the term psia (absolute pressure) is used. Most pressure gages, transmitters and other pressure-measuring devices indicate a zero reading when the measuring point is exposed to atmosphere. This is 0 psig. However, measuring devices designed to produce readings referenced to absolute pressure indicate a reading of 0 psia when a perfect vacuum is applied to the measuring point. Such devices indicate a reading of 14.7 when the measuring point is exposed to atmospheric pressure, because a pressure of 14.7 psi (at sea level) is applied to the input. The operation of such instruments is explained later in the text. Figure 2–4 helps to explain the relationship between absolute and gage pressure, which can be expressed as

$$\text{psia} = \text{psig} + \text{atmospheric pressure } (P_{atm}) \tag{2.3}$$

when expressing values above atmospheric pressure.

It should be noted that in Equation 2.3 the relationship between absolute and gage pressure is a function of the local atmospheric pressure. A change in atmospheric pressure will cause a change in absolute pressure. Thus, a change in atmospheric pressure would cause a change in an absolute pressure-measuring instrument but not in a gage pressure-measuring instrument. This is true because atmospheric pressure represents a variable point on the absolute pressure scale. Pressure measurements are referenced to standard conditions that can be mutually established by consenting parties but which are usually expressed as atmospheric pressure (1 atm) at sea level and 60°F.

EXAMPLE 2-2

Problem: A pressure gage that measures psig indicates a reading of 40. The local atmospheric pressure is 14.6 psi. What is the absolute pressure corresponding to the psig reading?

Solution: From Equation 2.3,

$$\text{psia} = 40 + 14.6 = 54.6$$

EXAMPLE 2-3

Problem: A psia pressure instrument indicates a reading of 35.5. The atmospheric pressure is 14.8 psi. Calculate the corresponding gage pressure.

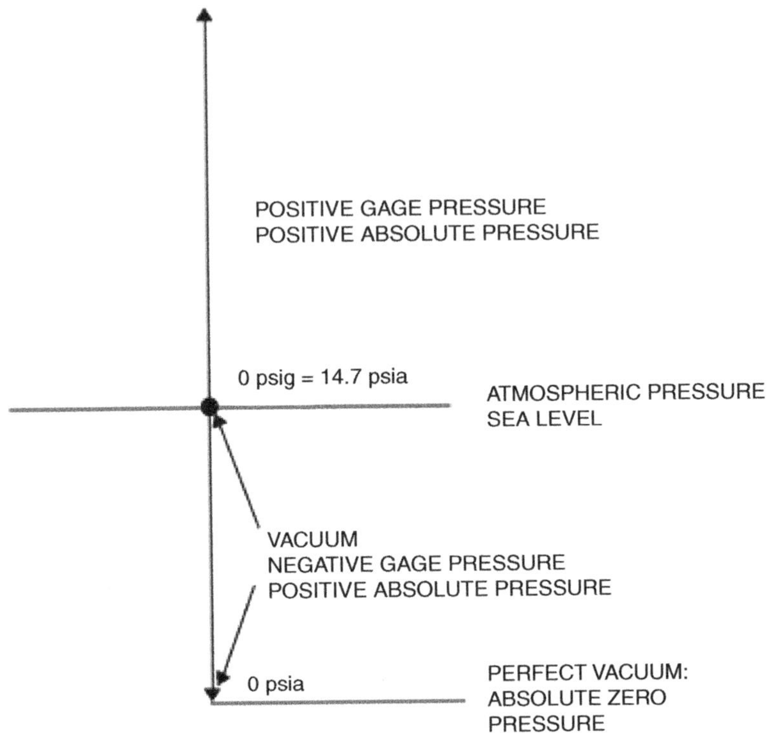

Figure 2–4 Relationship between Absolute and Gage Pressure

Solution: From Equation 2.3 (solving for psig),

$$\text{psig} = 35.5 - 14.8 = 20.7$$

EXAMPLE 2-4

Problem: A vacuum gage reads a value of 11.5 psig. The atmospheric pressure is 14.7 psia. Find the absolute pressure.

Solution: When dealing with pressures below atmospheric pressure, psia is equal to atmospheric pressure minus psig:

$$\text{psia} = 14.7 - 11.5 = 3.2$$

Manometric Principles

Because pressure is a fundamental parameter of a system, its measurement is extremely important. Wide ranges of this measurement with variable degrees of accuracy are required. The manometer is a useful instrument for measuring pressure.

To understand the operating principle of a manometer, consider the illustration shown in Figure 2–5. The weight of 1 ft^3 of water at 60°F is 62.342 lb. This weight is exerted over the surface area of 1 ft^2 or 144 in^2. The total pressure on the surface area of 1 ft^2 is

$$\frac{62.342\,lb}{144\,in^2} = \frac{0.433\,lb}{1\,in^2}$$

This pressure is the result of a 12-in. column of water on 1 in^2. A 1-in. column of water exerts a pressure of

$$\frac{0.433\,psi}{12\,in.} = \frac{0.036\,psi}{1\,in.}$$

One cubic inch of water at 60°F weighs 0.036 lb, and since this force is exerted over a surface area of 1 in^2, a very important relationship for pressure measurement has been established: A pressure of 0.036 psi will cause a column of water to be raised 1 in. A 12-in. column of water exerts a pressure of

$$0.036\frac{psi}{in.}(12\ in.) = 0.43275\,psi$$

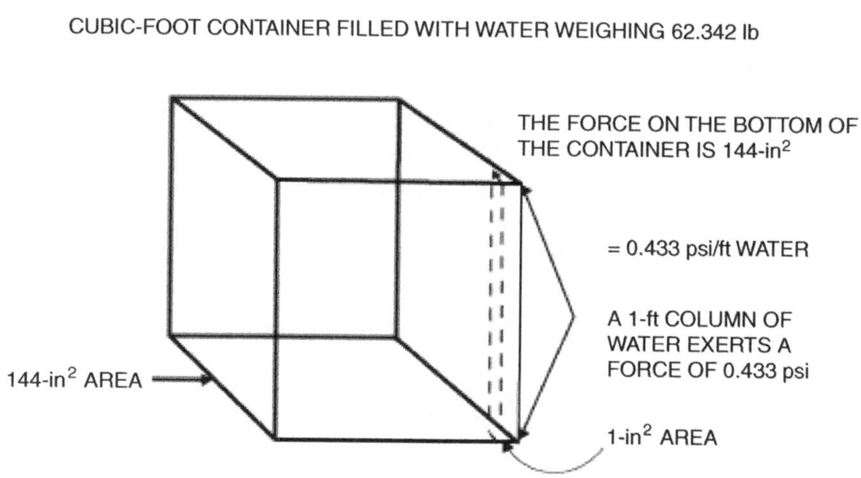

Figure 2–5 Relationship between Pressure and a Column of Liquid.

A pressure of 0.433 psi will cause a column of water to be raised 1 ft. From the foregoing discussion, it can be seen that the following relationships exist:

$$\text{1-in. wc} = 0.036 \text{ psi} \tag{2.4}$$

$$\text{1-ft wc} = 0.433 \text{ psi} \tag{2.5}$$

$$1 \text{ psi} = \text{2.31-ft wc} = \text{27.7-in. wc} \tag{2.6}$$

where wc is water column. (This designation is often not written as it is understood in such discussions of hydrostatic head pressure measurement.) Figure 2–6 illustrates the relationship between pressure and the corresponding displacement of a column of liquid.

Equations 2.4 to 2.6 suggest a means by which pressure can be determined by measuring the height of a column of liquid. A manometer usually has two transparent tubes and two liquid surfaces (some designs have one tube; more on that later). Pressure applied to the surface in one tube causes an elevation of the liquid surface in the other tube. The amount of elevation is read from a linear scale, and that reading, by the use of appropriate equations, is converted into pressure. Often, however, the scale is calibrated to read directly in pressure units. It can be seen in Figure 2–7 that a positive pressure in the measurement tube will cause an elevation of level in the readout tube. A negative pressure ($P < P_{atm}$) applied to the measurement tube will cause a decrease of level in the readout tube. This is caused by the atmospheric pressure in the readout tube. With equal pressure applied to each, the levels in both tubes are equal.

In general, an unknown pressure is applied to one tube and a reference pressure is applied to the other tube. The difference between the known pressure and the unknown pressure is balanced by the weight per unit area of the displaced manometer liquid. Water is the fluid most commonly used for manometers. However, practically any fluid can be used as long as its specific gravity or density is known. The relationship between pressure and liquid displacement is

$$P = hm\left(0.036\frac{\text{psi}}{\text{in.}}\right)(SG) \tag{2.7}$$

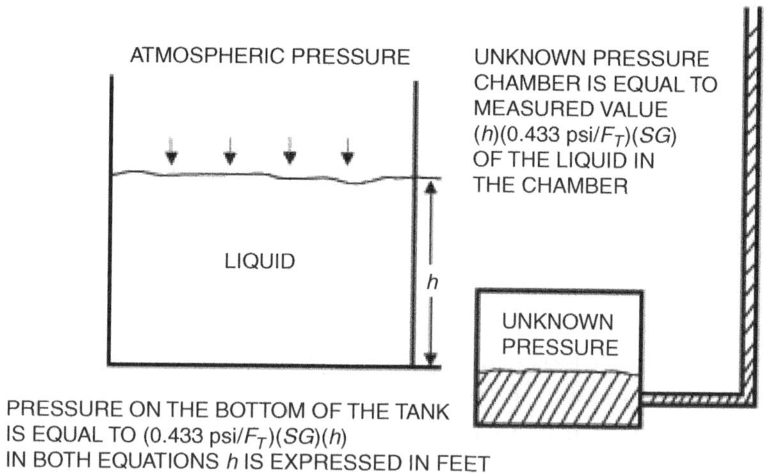

Figure 2–6 Basic Hydrostatic Head Pressure Measurement Concept

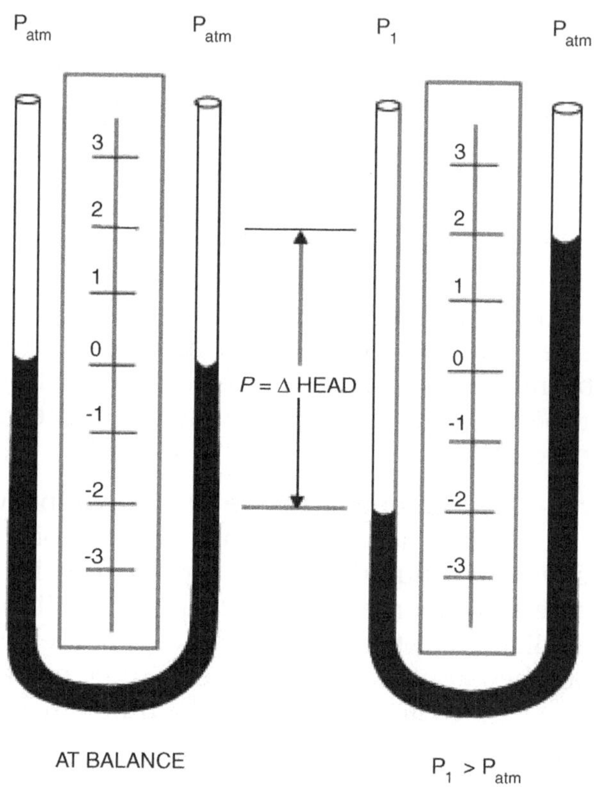

Figure 2–7 U-Tube Manometers (*Courtesy of Transmation Inc.*)

where P is pressure (psi), hm is the amount of displaced liquid (height measured in inches), SG is the specific gravity of the liquid, and 0.036 is the pressure caused by a 1-in. displacement of a column of water.

Manometers can provide very accurate pressure measurements and are often used as calibration standards. The measured pressure range of most manometers is usually between a few inches and about 30 psi, depending on the physical length of the tubes and the specific gravity of the fill fluid.

An important consideration in the use of manometers concerns the capillary effects of the liquid. The shape of the interface between two fluids at rest—air and the liquid in the manometer—depends on the relative specific gravity of the fluids and the cohesive and adhesive forces between the fluids and the walls of the tubes. The liquid surface at the interface, called the meniscus, is concave upward for water/air/glass combinations and downward for mercury/air/glass combinations. For water/air/glass applications the adhesive forces dominate, resulting in an elevation of the water level at the glass surface. Cohesive forces dominate in mercury/air/glass applications, and the meniscus is concave downward. The mercury level, therefore, is depressed by capillary action. The liquid-level displacement caused by capillary action can result in significant amounts of error if proper precautions are not taken. With the same fluid in both legs of a U-tube manometer, the capillary action in one tube counterbalances that in the other tube and can be neglected. In reservoir or well-type manometers, as shown in Figure 2–8, the most effective precaution is the use of large tubes. The tubes for manometers used as laboratory standards should be greater than 0.375 inches in diameter. Also, to eliminate error caused by meniscus effects, the reading should be taken at the center of the tube. Figure 2–9 shows the meniscus effects in water and mercury manometers.

Reservoir or well-type manometers have only one tube, with the other liquid surface being in a larger chamber or reservoir. As pressure is applied to the well, the liquid level in the tube is displaced upward. This must also be accompanied by a decrease in the liquid level in the well. When the diameter of the well is large compared with the diameter of the tube, the fall of level in the well can be neglected and only the displacement of the elevated level in the tube is read.

In effect, the difference in level of two fluid columns is still being measured, but one column level remains almost the same. Of course,

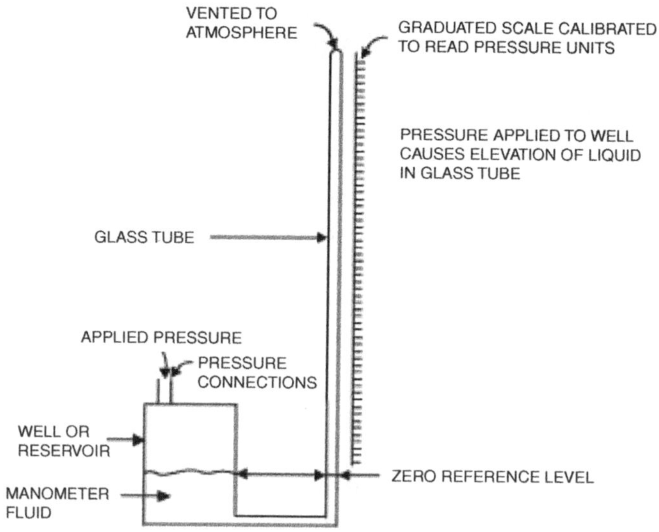

Figure 2–8 Well Manometer

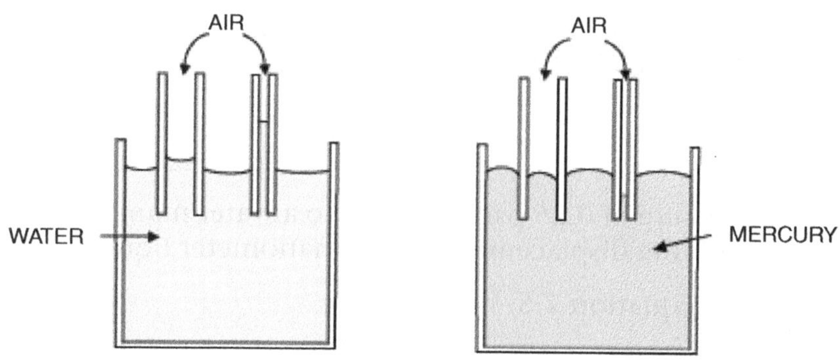

Figure 2–9 Meniscus Effects in Water and Mercury Manometers

there is some change in level on the well leg of the U (called "well-drop"), but this is corrected by the graduations on the scale. Thus, the scale of a well-type manometer, although it may be graduated in terms such as inches or tenths, will not measure exact linear inches. Instead, the scale incorporates a precise correction factor that accounts for the drop in level on the well side of the U. A typical factor is 0.9975.

Well-type manometers offer many advantages in terms of convenience and versatility. The most important improvement over the basic U-type design is the direct-reading single indicating column. This means that the pressure, vacuum, flow or liquid level reading is taken directly from a single indicating column, rather than by a measure of the difference between two column levels.

The scale used to read the displacement of the manometer liquid can be calibrated to correspond to any pressure value. Tables 2–2 and 2–3 and Equations 2.3 to 2.6 establish this principle. Manometers used in most instrument applications, however, usually are calibrated to be read in inches of water, inches of mercury or pounds per square inch. Using mercury and the appropriate conversion factors, laboratory manometers can be used to measure up to about 550 in. of water or about 20 psi. The following examples will help to illustrate this point.

EXAMPLE 2-5

Problem: A pressure applied to a manometer causes a water displacement of 34.5 in. Find the pressure in psi.

Solution: From Equation 2.3,

$$\frac{0.036\text{psi}}{\text{in.}}(34.5\text{in.}) = 1.244\text{psi}$$

EXAMPLE 2-6

Problem: A pressure of 0.26 psi is applied to a water manometer. What is the corresponding displacement of the manometer liquid?

Solution: From Equation 2.5,

$$27.7\frac{\text{in.}}{\text{psi}}(0.26\text{psi}) = 7.20\text{in.}$$

EXAMPLE 2-7

Problem: A pressure of 20 psi is applied to a mercury manometer. The specific gravity of mercury is 13.619. Determine the displacement of the manometer liquid.

Solution: From Equation 2.6 (solving for *hm*),

$$hm = \frac{20\text{psi}}{\left(0.036\frac{\text{psi}}{\text{in.}}\right)(13.619)} = 40.72\text{in.}$$

EXAMPLE 2-8

Problem: A mercury manometer is 30 in. high. What is the maximum pressure in psi that can be measured?

Solution: From Equation 2.6,

$$P = (30\text{in.})\left[0.03606\frac{\text{psi}}{\text{in.}}\right](13.619) = 14.733\text{psi}$$

EXAMPLE 2-9

Problem: One leg of a U-tube water manometer is connected to a vacuum pump and all of the air is evacuated. The other leg is vented to the atmosphere. The water-level displacement is 33.00 ft. Determine the atmospheric pressure.

Solution: From Equation 2.4,

$$\left(0.433\frac{\text{psi}}{\text{ft}}\right)(33.00\text{ft}) = 14.280\text{psi}$$

Although water is most commonly used in manometers, other fluids are used as well. Common trade names for fluids used in industrial manometers are No.1, No.2 and No.3 fluids and Red Oil. Care must be taken to use the proper fluid for the appropriate manometer scale.

It is apparent that manometric principles can be used for level measurement. However, in measurement applications manometers are utilized primarily as secondary calibration standards for the shop calibration of instruments.

When the maximum operating pressure of a manometer has been exceeded, the fluid is blown out of the top. This presents a health hazard if mercury is the fluid being used, and thus care must be exercised when using such manometers. Special traps can be used to catch liquid blown from a manometer, but not all hazards can be eliminated. For this reason, mercury manometers have been replaced by other types of measurement standards in many facilities.

The pressure measurement made with a manometer is always referenced to a standard pressure (most commonly, atmospheric pressure) on one leg. The pressure applied to the other leg is referenced to atmosphere and is gage pressure. If the pressure on a reference leg is evacuated to form a vacuum, the pressure measurement is referenced to a vacuum and is absolute pressure. A barometer is a special type of manometer with a vacuum in one leg and atmospheric pressure in the other leg. The mercury-level deflection varies with atmospheric pressure changes.

Several arrangements have been designed to achieve greater accuracy and sensitivity in manometers. The inclined manometer, a variation of the common manometer that consists simply of a tube inclined toward the horizontal at some angle, is used to measure low pressures (Figure 2–10). The inclined tube has an open end; the pressure being measured is applied to a reservoir to which the other end of the tube is connected. Because the angle of inclination is fixed, an exact relationship exists between the liquid movement along the tube and the vertical displacement. A scale is usually placed along the tube, and the vertical displacement is read directly from the scale. The vertical displacement is related to the liquid travel by:

$$h = h' \sin \alpha \tag{2.8}$$

where h is the vertical displacement of the manometer liquid, h' is the distance traveled by the liquid up the tube, and α is the angle of inclination to the horizontal. The expanded scale of an inclined manometer means quite simply that more graduations for an inch of vertical height are possible and that measurements of low pressure can be made with more ease and accuracy.

The increased sensitivity and readable accuracy of the inclined manometer become more apparent when one considers that a 6-in. range (range is always expressed in terms of a vertical column height) can be expanded by an inclined-tube manometer to cover a 48-in. long scale. Imagine the reading accuracy in this case, where 8 in. of scale is equivalent to 1 in. of range. An instrument of this type readily lends itself to a scale directly graduated in hundredths of an inch. Using water as an indicating fluid, this would make graduations a full tenth of a linear inch apart equal to 0.0005 psig. Because the angle α is referenced to the horizontal plane, it is very important that the inclined manometer be leveled. Failure to do so can create substantial errors in measurement.

The hydrostatic measurement principle is based on the force exerted by a column of fluid of known weight (or density); therefore, the specific gravity of the indicating fluid is an important factor in determining the pressure range of the manometer. Readings for water can

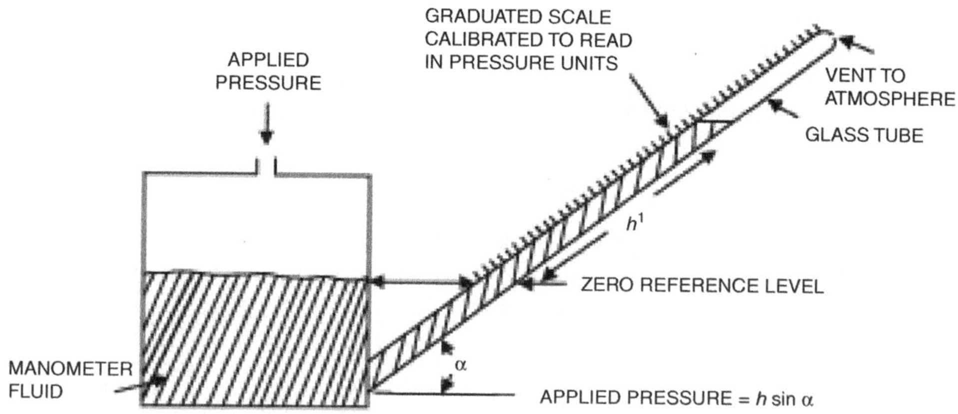

Figure 2–10 Inclined Manometer

readily be expressed in terms of inches of water or some other linear measure. If a fluid three times as heavy as water is used, at the same pressure the movement of fluid in the indicating tube will be one-third as great, and, for a given instrument range (vertical linear height), there will be three times the pressure range. If the indicating fluid has a specific gravity lower than that of water, it will move a proportionately greater distance, and the pressure range will be less than when using water. For a given instrument range, the pressure range can be expanded by the use of a fluid with a higher specific gravity and reduced with a fluid of lower specific gravity.

Inches of water and inches of mercury are widely accepted units of pressure measurement. However, scales can be graduated in any other units desired, such as pounds, centimeters of water, miles per hour, knots airspeed, flow in gallons per minute or any other variable inferred from a pressure or differential pressure measurement. Scales in terms of inches of water using mercury as the indicating fluid were once quite common. Such scales were used when readings were desired in inches of water over a range far exceeding the practical linear scale length required when using a lighter fluid.

The accuracy of a manometer depends on the use of an indicating fluid with certain clearly defined characteristics. For this reason, specially designed fill fluids and water, being readily obtainable and quite stable, have been long recognized as standards.

Density

The most important single factor is the density value of the indicating fluid. A manometer, no matter how carefully constructed, can be no more accurate than the known accuracy of the indicating fluid density.

Viscosity

The viscosity of a fluid governs its flow characteristics. It is relatively unimportant in vertical instruments, but is more important in inclined manometers.

Temperature

All fluids change density with variations in temperature. Since the most important single factor in instrument accuracy is density, it naturally follows that temperature must be a consideration. For extreme accuracy in the laboratory, room temperature can be controlled or the mean temperature of the indicating fluid can be determined and proper corrections applied to the scale reading. The latter method is most common.

Standard Gravity and Density

Instrument range is selected on the basis of (1) the overall pressure, vacuum, or pressure differential range to be measured, (2) the reading accuracy desired, and (3) convenience in handling the instrument. To give exact meaning to column height as a system of pressure measurement, standard conditions must be defined for gravity and for fluid density. When the actual measurement conditions are other than standard (as they almost always are), corrections must be applied to the actual manometer reading to reduce it to standard conditions. Only then can the fluid column height be related to a true pressure measurement.

The standard acceleration of gravity has been almost universally accepted as 980.665 cm/s^2. This is approximately equal to the acceleration of gravity at sea level at 45° latitude. It is not important how the value was established; it is only necessary that a particular value be chosen and accepted as standard.

By almost universal agreement, the standard density of mercury is defined as 13.5951g/cm^3. This is very close to the actual density (by best determinations) of pure mercury of naturally occurring isotopic proportions at a temperature of 0°C.

For water, there is no such unanimity of acceptance of a standard density. Some scientific and industrial usages have favored 1.000000 g/cm^3; this is a hypothetical value, slightly above the maximum density of pure water. General industrial usage more often accepts the following value for the density of water at 20°C: 0.998207 g/cm^3. This value is used by ISA as an adopted standard.

Other standard usages exist. For example, hydraulic engineering uses the pressure measurement "head of water" in feet, which is based on a water density of 62.4 lb/ft^3 (under standard gravity). This is the density of water at approximately 53°F. The American Gas Association uses a 60°F density standard; 15°C has been used by other authorities.

To avoid misunderstanding or uncertainty where precise measurements are involved, it is recommended that any pressure measurement in terms of water column height be qualified with the standard density temperature on which the measurement is based. For example,

xx.xx inches of water 4°C (1.000000 g/cm^3)

or

xx.xx inches of water 20°C (0.998207 g/cm^3)

It is also suggested, pending universal acceptance of a standard, that whenever the above qualification is not made, the measurement should be assumed to be based on a standard water density of 0.998207 g/cm^3 (20°C) [Ref. 2].

Conditions of Manometer Use

The environmental conditions surrounding the manometer at the time of actual pressure measurement are henceforth called working conditions to distinguish them from standard conditions of temperature and gravity.

Local gravity must be determined by a surveyor by calculation. A gravimetric survey will give results of very high accuracy. Determination by calculation is always an approximation because of unique local abnormalities, but it often gives results of sufficient accuracy for practical purposes.

As we have seen, temperature affects fluid density. To determine true working fluid density, the temperature measurement must be exact, and the temperature must be uniform throughout the fluid. Reliable temperature-measuring equipment must therefore be used, and there must be assurance that the point of measurement truly indicates the actual fluid temperature.

Height measurement is generally by means of an engraved scale. Because of the coefficient of thermal expansion of the scale material, temperature will affect the scale length and this must be considered in determining the true column height.

The use of a fluid manometer to determine pressure involves a true determination of fluid column height, fluid density, and local gravity. This pressure can be expressed as a column height only if density and gravity determinations are used to correct the true height reading to a corresponding theoretical value under predefined standard conditions of density and gravity. Fluid manometers provide a very accurate means of pressure measurement, but because of the health hazards involved with mercury, limited portability and inconvenience in usage, fluid manometer applications have declined in recent years and electronic manometers are gaining in popularity. Because of their simplicity, reliability and accuracy, fluid manometers are still used to a limited extent as laboratory calibration standards.

Calibration Principles

Nearly all process-measuring instruments require validation checks to ascertain the degree to which a measured value agrees with an actual or true value. This is accomplished through calibration procedures that conform to set policies in compliance with adopted standards. Calibration is performed to verify the output(s) of a device corresponding to a series of values of the quantity which the device is to measure, receive or transmit. Data so obtained are used to

- Determine the location at which scale graduations are to be placed.
- Adjust the output to bring it to the desired value within a specified tolerance.

- Ascertain the error by comparing the device output reading against a standard.

Performance statements are used to establish a criterion of exact true measurement values and the degree to which the measurement of an instrument under test (IUT) conforms to the true value. The most common measures of performance are percent of full scale (FS) and percentage of measurement or rate. There is a significant difference between these performance classifications. Manufacturers should clearly state the criterion used and the user should know what the criterion means. The following examples and figures aid in the understanding of this concept.

EXAMPLE 2-10

Problem: Determine which specification expresses greater accuracy: 0.5% FS or 1% of measurement.

Solution: It seems that the full-scale specification represents greater accuracy; however, at lower values of measurement the full-scale specification results in greater error. The value at which both specifications are equal is:

$$(\text{\% of measurement})\ (\text{actual measurement}) = (\text{\% FS})\ (\text{full-scale measurement})$$

By solving for the actual measurement and substituting known quantities:

$$\text{Actual measurement} = (0.5\%/1\%)\ (\text{full scale measurement})$$

The errors are equal at 50% of full scale. The full-scale specification expresses a higher degree of accuracy from 50 to 100% of scale, while the measurement specification is superior over the lower portion of the range. It should be pointed out that when the measurement is above 50% of measurement scale, the maximum difference between the specifications is 0.5% of measurement (at 100% of scale), whereas, for example, at 10% of scale the difference is 9.5% of measurement. Accordingly, if the measurement is always between 50 and 100% of scale, the full-scale specification is better. If the measurement will vary through the entire measurement range, the percent of measurement is superior.

EXAMPLE 2-11

Problem: Compare the errors of a pressure instrument with a measuring range of 0 to 100 psi and stated accuracies of ±0.5% FS and ±1% of measurement.

Solution: At a value of 100 psi, the errors will be:

$$100 \text{ psi} \pm 0.5\%(100) = 100 \text{ psi} \pm 0.5 \text{ psi}$$

$$100 \text{ psi} \pm 1\%(100) = 100 \text{ psi} \pm 1 \text{ psi}$$

At a value of 50 psi, the errors will be:

$$50 \text{ psi} \pm 0.5\%(100) = 50 \text{ psi} \pm 0.5 \text{ psi}$$

$$50 \text{ psi} \pm 1\%(50) = 50 \text{ psi} \pm 0.5 \text{ psi}$$

At a value of 10 psi, the error will be:

$$10 \text{ psi} \pm 0.5\%(100) = 10 \text{ psi} \pm 0.5 \text{ psi}$$

$$10 \text{ psi} \pm 1\%(10) = 10 \text{ psi} \pm 0.1 \text{ psi}$$

EXAMPLE 2-12

Problem: The accuracy of a pressure instrument is expressed as ±0.025% of indicated value from 25 to 100% FS and 0.00625% FS below 25% FS. For a range of 0 to 100 psi, which statement expresses the greater accuracy?

Solution:

At 25 psi, the accuracy is: $25 \pm [(0.025)\ (25)] = 25 \pm 0.63$ psi.

At 100 psi, the accuracy is: $100 \pm [(0.025)\ (100)] = 100 \pm 0.25$ psi.

At 20 psi, the accuracy is: $20 \pm [(0.00625)\ (100)] = 20 \pm 0.625$ psi.

At 5 psi, the accuracy is: $5 \pm [(0.00625)\ (5)] = 5 \pm 0.03$ psi.

It is important to note that some manufacturers may specify measuring instruments with performance statements at defined operating conditions for a stated accuracy which can be much greater than the results achieved over the instrument's operating range. It may be difficult to simulate exact design conditions in an industrial measuring environment, especially for specific applications. Thus, performance stated as a function of one set of conditions can be quite different from that obtained under actual industrial conditions.

Calibration Terms

The terms used to define the prescribed performance of an instrument are included in Appendix A. They will be briefly reviewed here to explain calibration principles and procedures.

Precision and *accuracy* are two terms that because of redundancy and misrepresentation of correct usage, have lost much of their significance in defining the performance criteria of measuring instruments. Accuracy, as defined in Appendix A, is the degree of conformity of an indicated value to an accepted standard value. Precision defines instrument performance as a result of the care that is used in the design and manufacture of the instrument. This term has little exact meaning, however. For example, two different wrist watches can cost $800 and $15, yet each is commonly called a precision instrument. A term that more closely defines the quality of a device with regard to precision is repeatability, which is the ability of an instrument to reproduce the same measurement each time the same set of conditions is repeated. This does not imply that the measurement is correct, but rather that the measurement is the same each time.

Figure 2–11 illustrates the relationship between repeatability and accuracy. For many process applications, repeatability is a greater consideration than accuracy. Accuracy is obviously important, but often control is more important than the absolute accuracy of the measurement of a variable.

Of course, the first prerequisite to any control scheme is measurement. However, a repeatable measurement that is not necessarily accurate means that a variable can be maintained at a value that may not be accurate. Repeatable measurement reproduces the same value, though not necessarily accurately. Non-repeatable measurement can have an accuracy as defined by performance statements. An accurate but non-repeatable measurement could result in a poorer control quality. In summary,

- Poor repeatability means poor accuracy.
- Good accuracy means good repeatability.
- Good repeatability does not necessarily mean good accuracy.

EXAMPLE 2-13

Problem: Two pressure instruments with a range of 0 to 100 psi are measuring a process value of 50.0 psi. The accuracy of both devices is ±1% FS. Data obtained from five measurements are listed in the following table; determine which instrument has a greater degree of repeatability, using Figure 2–11 as an example to explain the relationship between repeatability and accuracy.

Measurement	Value, psi	
	Instrument A	Instrument B
1	49.9	49.9
2	49.7	49.6
3	50.1	50.4
4	49.8	49.7
5	50.2	50.5

Note: The values in the table do not correspond to the values in Figure 2–11.

Solution: From the examples shown, and the principle illustrated in Figure 2–11, it can be seen that instrument A is more repeatable. The measurements from both instruments are within the tolerance expressed by the ±1% accuracy stated (±1 psi).

Hysteresis is a term that describes an instrument's ability to reproduce a measurement with respect to a change in the direction of the measurement. A *dead band* will occur when a measuring instrument is insensitive to changes in measurement that are smaller than that required for the instrument's response. Figure 2–12 illustrates dead band and hysteresis.

Linearity is the closeness with which an instrument calibration curve approximates a straight line. It is usually measured as nonlinearity and expressed as linearity. Appendix A defines the terms associated with linearity: independent, terminal-based and zero-based.

The terms hysteresis, dead band and linearity are introduced here and will be reviewed later in the chapter.

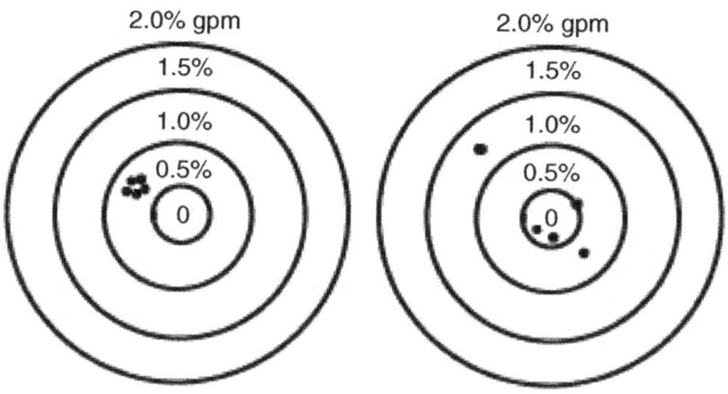

DEVICE 1	MEASUREMENT	DEVICE 2
0.8 gpm	1	–0.5 gpm
0.9	2	–0.3
0.7	3	–0.1
0.8	4	–1.1
0.7	5	–0.8

Figure 2–11 Examples of Repeatability and Accuracy

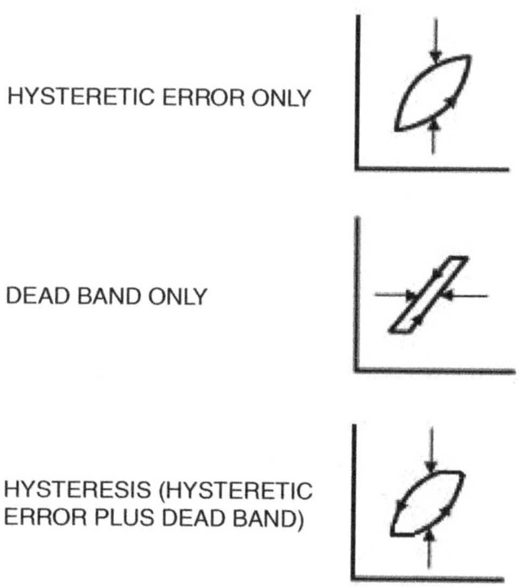

Figure 2–12 Dead Band and Hysteresis

EXAMPLE 2-14

Problem: Two pressure instruments with ranges of 0 to 100 psi produce the following data. Determine which measurement is more linear.

Actual Value, psi	Measured Value, psi	
	Instrument A	Instrument B
0	0.6	0
20	20.3	19.3
40	40.6	40.3
60	60.4	60.5
80	80.3	79.3
100	100.5	99.4

Solution: As shown in Figure 2–13, graphical representation of the data on an absolute scale does not present a clear picture of the instruments' response. A normalized graph shows that instrument A is more linear than instrument B.

Turndown expresses the ratio of the maximum measurement of an instrument within the stated accuracy, usually full scale, to the minimum measurement within the stated accuracy.

EXAMPLE 2-15

Problem: Calculate the turndown of a pressure gage with a range of 0 to 100 that can measure 5 to 100 psi within the stated accuracy.

Solution: The maximum and minimum accurate measurements are 100 and 5 psi, respectively. The turndown is 100/5, or 20:1.

A composite accuracy statement for an instrument is a measure of the combined effects of repeatability, linearity and accuracy. In evaluating the overall performance of an instrument, it is important to realize that no one term can express the true performance of an instrument's response. As has been explained, an instrument can be accurate but not repeatable and repeatable but not linear.

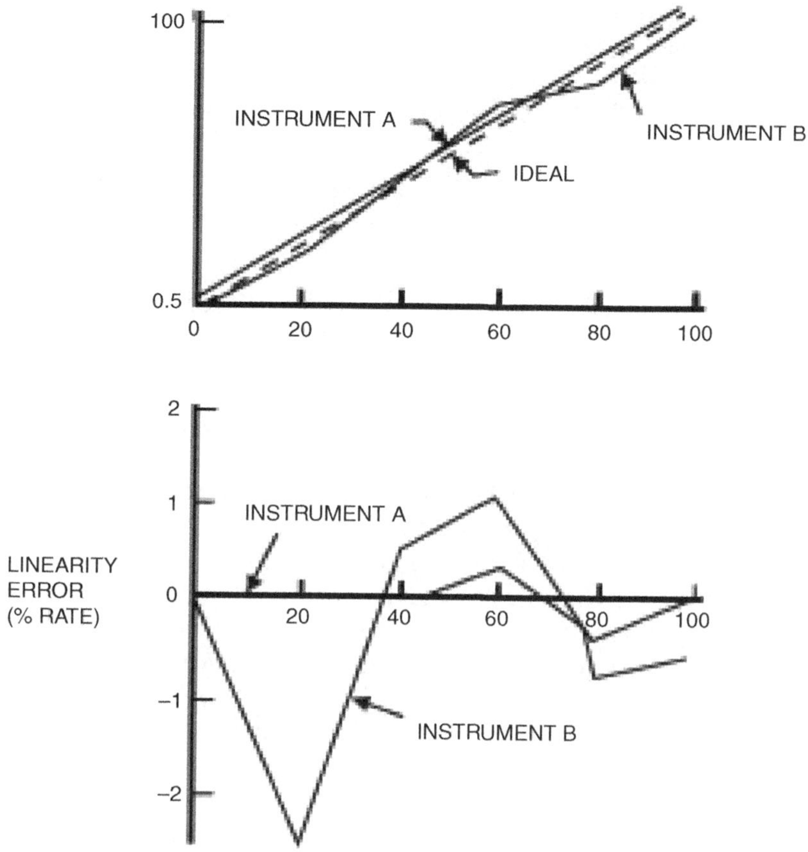

Figure 2–13 Linearity Example

Calibration Standards

Many companies have adopted a model for quality assurance in production and installation as a guideline for selecting and maintaining instruments that affect product quality. The International Organization for Standardization (ISO) has developed the ISO 9000–9004 series of quality management and quality assurance standards. ISO 9001/2000 specifies requirements for a quality management system where an organization

1. needs to demonstrate its ability to consistently provide a product that meets customer and applicable regulatory requirements, and

2. aims to enhance customer satisfaction through the effective application of the system, including processes for continual improvement of the system and the assurance of conformity to customer and applicable regulatory requirements.

All requirements of ISO 9001/2000 are generic and are intended to be applicable to all organizations, regardless of type, size and product provided.

Where exclusions are made, claims of conformity to this international standard are not acceptable unless these exclusions are limited to requirements within clause 7, and such exclusions do not affect the organization's ability, or responsibility, to provide a product that meets customer and applicable regulatory requirements [Ref. 3].

Quality systems requirements with respect to management responsibility, contract review, document control, purchasing, product identification and traceability, process control, inspection and testing, control of nonconforming product, corrective action, handling, storage, packaging, delivery, quality records, internal quality audits, training and statistical techniques are topics that must be considered and addressed for compliance with the standard. More specific for actual calibration purposes, however, is section 4.10, titled "Inspection, Measuring and Test Equipment":

> *The supplier shall control, calibrate and maintain inspection, measuring and test equipment, whether owned by the supplier, on loan, or provided by the purchaser, to demonstrate the conformance of product to the specified requirements. Equipment shall be used in a manner which ensures that measurement uncertainty is known and is consistent with the required measurement capability.*
>
> *The supplier shall:*
>
> *a. identify the measurements to be made, the accuracy required, and select the appropriate inspection, measuring and test equipment;*
>
> *b. identify, calibrate and adjust all inspection, measuring and test equipment and devices that can affect product quality at prescribed intervals, or prior to use, against certified equipment having a known valid relationship to nationally recognized standards—where no such standards exist, the basis used for calibration shall be documented;*

c. establish, document and maintain calibration procedures, including details of equipment type, identification number, location, frequency of checks, check method, acceptance criteria and the action to be taken when results are unsatisfactory;

d. ensure that the inspection, measuring and test equipment is capable of the accuracy and precision necessary;

e. identify inspection, measuring and test equipment with a suitable indicator or approved identification record to show the calibration status;

f. maintain calibration record for inspection, measuring and test equipment;

g. assess and document the validity of previous inspection and test results when inspection, measuring and test equipment is found to be out of calibration;

h. ensure that the environmental conditions are suitable for the calibrations, inspections, measurements and tests being carried out;

i. ensure that the handling, preservation and storage of inspection, measuring and test equipment is such that the accuracy and fitness for use is maintained;

j. safeguard inspection, measuring, and test facilities, including both test hardware and test software, from adjustments which would invalidate the calibration setting.

Where test hardware (e.g., jigs, fixtures, templates, patterns) or test software is used as suitable forms of inspection, they shall be checked to prove that they are capable of verifying the acceptability of product prior to release for use during production and installation and shall be rechecked at prescribed intervals. The supplier shall establish the extent and frequency of such checks and shall maintain records as evidence of control. Measurement design data shall be made available, when required by the purchaser or his representative, for verification that it is functionally adequate.

2.3 STANDARD INSTRUMENTS FOR CALIBRATION

Instruments used as calibration standards are called test instruments. Test instruments against which the accuracy of all other calibration devices is compared are called primary standards and are maintained by the National Institute of Science and Technology (NIST), formerly the National Bureau of Standards (NBS). When referring to the accuracy of measuring instruments, it is generally understood that a measured value is compared with an accepted primary standard. It is inconvenient, if not impossible, to actually calibrate process monitoring and test measurement instruments to a primary standard. However, it is generally desirable to express a measurement with absolute accuracy that is traceable to a primary standard. Secondary standards serve this purpose.

Secondary standards are test instruments whose accuracy is directly traceable to primary standards. Many calibration laboratories have secondary standards that are used in the calibration of process monitors. These instruments are relatively expensive, and secondary calibration laboratories are usually maintained in environmentally controlled conditions. To reduce the use of and thereby maintain the validity of secondary standards many industrial process calibrations are made with shop standards. A shop standard, at least as the term is used here, is a loosely defined type of standard that has been calibrated to a secondary standard. It is also called a transfer standard.

Secondary standards are available and employed in many industrial instrument laboratories, but often are used only to check the accuracy of other instruments used in day-to-day calibration exercises. This helps to maintain the accuracy of secondary standards by reducing the amount of operation and usage to which they are subjected.

It should be pointed out that shop standards as defined here can be just as accurate as secondary or even primary standards. The type of standard for which an instrument is used is not entirely related to the precision of the instrument used. For example, an instrument may be of a high quality and precision, but if it has not been certified with accuracy traceable to a primary standard, it is not a secondary standard. Of course, the particular calibration situation will determine the type of standard to be used for the application, but the accuracy of shop standards is sufficient for most process measurement applications. A good precision test instrument is very accurate and can be used as a shop standard for calibration. When procedures require conformity to a set of standards which dictate that test instruments have accuracy traceable to a primary or national standard, secondary standard instruments are used. A hierarchy of calibration standards is shown in Figure 2–14.

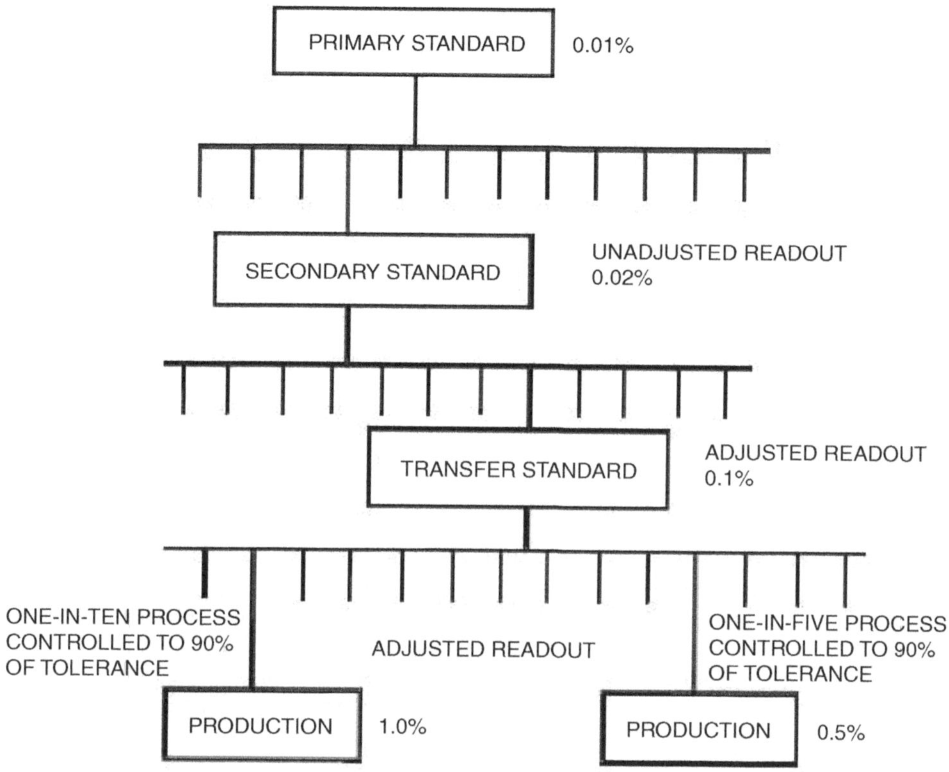

Figure 2–14 Hierarchy of Calibration Standards

Calibration Checks

Conformance with ISO 9001/2000 stipulates that calibration procedures be established and documented. Calibration procedures for individual instruments are therefore detailed, and tabulated data and corresponding records are maintained. While specific calibration procedures are not mandated, the procedures that follow will aid in understanding the general concepts involved in adopting established procedures such as those listed in section 5 of ISA-51.1-1979 – (R1993) – Process Instrumentation Terminology.

The calibration of an instrument consists of simulating and measuring the input to the IUT, which is the measured variable, and observing the corresponding response or output. This output may be a scaled electronic or pneumatic value in the case of transmitters or a mechanical indication for pressure gages and other indicating instruments. An arrangement for instrument calibration is shown in Figure 2–15. The

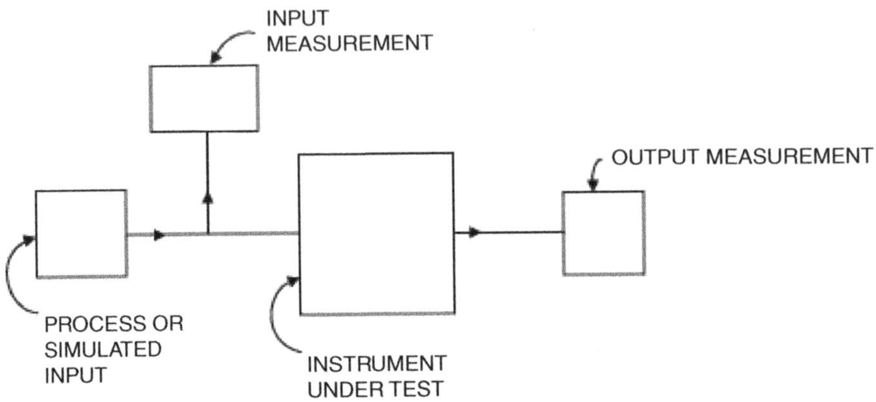

Figure 2–15 Block Diagram of a Calibration Arrangement

number of test points to determine the desired performance characteristic of an IUT should be distributed over the entire range and should include points within 10% of the lower and upper range values. The location and number of test points, at least five and preferably more, should be consistent with the degree of exactness desired and the characteristic being evaluated.

The calibration and test procedures given here can be used for the following determinations, which are defined and expanded on below:

- Measured accuracy
- Dead band
- Drift point
- Hysteresis
- Linearity, independent
- Linearity, terminal-based
- Linearity, zero-based
- Repeatability
- Reproducibility

The error rating of the test instrument should be one-tenth or less than that of the IUT, but in no case should be greater than one-third the allowed tolerance. The IUT and associated test equipment must be allowed to stabilize under steady-state operating conditions. Any operating conditions that would influence the test should be observed and recorded; when reference operating conditions are required for performance characteristics, they must be maintained as such.

Prior to obtaining calibration data, the IUT should be exercised by a number of full-range traverses in each direction. Tapping or vibrating the IUT is not permissible unless the performance characteristic under observation requires such action.

A calibration cycle is obtained by following the conditions and preconditions described. Output values for each established input are observed and recorded for one full-range traverse in each direction, starting near the midrange value. The input should be applied at such a rate as not to overshoot the intended input values; the final input must be approached from the same direction as the initial input.

In order to relate performance characteristic values such as accuracy, linearity, and dead band, equivalent units should be used. A good practice is to express all values as percent of measurement or percent of full scale. A calibration curve can then be prepared as a deviation plot. For this, the difference between each observed output and the actual input or desired output is determined.

EXAMPLE 2-16

Problem: For the following data obtained from a calibration cycle, determine the deviation at each point expressed as a percent of output span. The instrument's input range is 0 to 200 psi, and the output range is 4 to 20 mA. For an accuracy of ±1% FS, determine whether the response of the instrument is within the stated tolerance.

Input, psig	**Output, mA**
0	4.10
50	7.95
100	11.88
150	16.20
200	19.85

Solution: The span of the instrument's output is 16 mA (20 – 4 mA). The percent deviation is

$$\frac{0.1}{16} = +0.625\%$$

$$\frac{0.05}{16} = -0.312\%$$

$$\frac{0.12}{16} = -0.750\%$$

$$\frac{0.2}{16} = +1.25\%$$

$$\frac{0.15}{16} = -0.938\%$$

The instrument is not within the stated tolerance at 150 psi, or 75% of the input range.

Once the data obtained from a calibration cycle have been tabulated and the deviation calculated, a calibration curve or deviation plot can be prepared. The deviation is expressed as a percent of ideal output span and is plotted against the input span. Values are shown as positive or negative, depending on whether the observed output is greater or less than the true output value.

EXAMPLE 2-17

Problem: Prepare a calibration curve for the data in Example 2-16.

Solution: The deviation plot is shown in Figure 2–16.

Definition of Terms

The following terms will help to explain the calibration process and how to evaluate the performance of the IUT.

Accuracy, Measured

Measured accuracy can be determined from the deviation values of a number of calibration cycles. It is the greatest positive and negative deviation of the recorded values (from both an upscale and a downscale output traverse) from the reference or zero deviation line. Measured accuracy can be expressed as a plus or minus percentage of

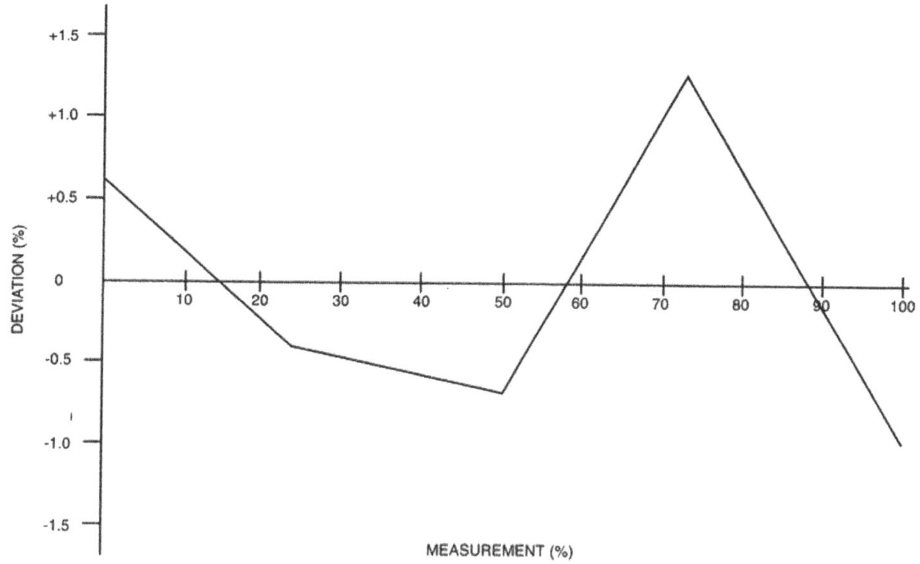

Figure 2–16 Deviation Plot for Example 2-17

output span. Example: The measured accuracy is +0.26% to –0.32% of output span.

Band, Dead

Maintain test conditions and precondition the test device as mentioned and proceed as follows:

1. Slowly vary (increase or decrease) the input to the device being tested until a detectable output change is observed.
2. Observe the input value.
3. Slowly vary the input in the opposite direction (decrease or increase) until a detectable output change is observed.
4. Observe the input value.

The increment through which the input signal is varied (differences between steps 2 and 4) is the dead band. It is determined from a number of cycles (steps 1 to 4). The maximum value is reported. The dead band should be determined at a number of points to make certain that the maximum dead band has been observed. Dead band can be expressed as a percentage of input span. Example: The dead band is 0.10% of input span.

Drift, Point

Maintain test conditions and precondition the test device as previously explained; proceed as follows:

1. Adjust the input to the desired value without overshoot and record the output value.

> Note: The test device should be permitted to warm up (if required) before recording the initial output value.

2. Maintain a fixed input signal and fixed operating conditions for the duration of the test.
3. At the end of the specified time interval, observe and record the output value.

In evaluating the results of this test, it is presumed that dead band is either negligible or of such a nature that it will not affect the value of point drift.

Point drift is the maximum change in recorded output value observed during the test period. It is expressed in percent of output span for a specified time period. Example: The point drift is 0.1% of output span for a 24-hr test.

Hysteresis

Hysteresis results from the inelastic quality of an element or device. Its effect is combined with the effect of dead band. The sum of the two effects can be determined directly from the deviation values of a number of test cycles and is the maximum difference between corresponding upscale and downscale outputs for any single test cycle. Hysteresis is then determined by subtracting the value of dead band from the corresponding value of hysteresis plus dead band for a given input. The maximum difference is reported and can be expressed as a percent of output span. Example: The hysteresis is 0.12% of output span.

Linearity, Independent

Independent linearity can be determined directly from the calibration curve using the following procedure:

1. Plot a deviation curve that is the average of corresponding upscale and downscale output readings.
2. Draw a straight line through the average deviation curve so as to minimize the maximum deviation. It is not necessary that the straight line be horizontal or pass through the end points of the average deviation curve.

Independent linearity is the maximum deviation between the average deviation curve and the straight line. It is determined from the deviation plots of a number of calibration cycles. It is measured in terms of independent nonlinearity as a plus or minus percent of output span. Example: The independent linearity is ±0.8% of output span.

> Note: The average deviation curve is based on the average of corresponding upscale and downscale readings. This permits observation of independent linearity without regard for dead band or hysteresis. This concept assumes that if no hysteresis or dead band were present, the deviation curve would be a single line midway between the upscale and downscale curves.

Linearity, Terminal-Based

Terminal-based linearity can be determined directly from the calibration curve using the following procedure:

1. Plot a deviation curve that is the average of corresponding upscale and downscale output readings.
2. Draw a straight line so that it coincides with the average deviation curve at the upper-range value and the lower-range value.

Terminal-based linearity is the maximum deviation between the average deviation curve and the straight line. It is determined from the deviation plots of a number of calibration cycles. It is measured in terms of terminal-based nonlinearity as a plus and minus percent of ideal output span. Example: The terminal-based linearity is ±0.3% of output span.

> Note: The average deviation curve is based on the average of corresponding upscale and downscale readings. This permits observation of terminal-based linearity independent of dead band or hysteresis. This concept assumes that if no hysteresis or dead band were present, the deviation curve would be a single line midway between the upscale and downscale readings.

Linearity, Zero-Based

Zero-based linearity can be determined directly from the calibration curve using the following procedure:

1. Plot a deviation curve that is the average of corresponding upscale and downscale output readings.
2. Draw a straight line so that it coincides with the average deviation curve at the lower-range value (zero) and minimizes the maximum deviation.

Zero-based linearity is the maximum deviation between the average deviation curve and the straight line. It is determined from the deviation plots of a number of calibration cycles and is measured in terms of zero-based linearity as a plus or minus percent of ideal output span. Example: The zero-based linearity is ±0.4% of output span.

> Note: The average deviation curve is based on the average of corresponding upscale and downscale readings. This permits observation of zero-based linearity independent of dead band or hysteresis. This concept assumes that if no hysteresis or dead band were present, the deviation curve would be a single line midway between the upscale and downscale readings.

Repeatability

Repeatability can be determined directly from the deviation values of a number of calibration cycles. It is the closeness of agreement among a number of consecutive measurements of the output for the same value of input approached from the same direction. Fixed operating conditions must be maintained.

Observe the maximum difference in percent deviation for all values of output, considering upscale and downscale curves separately. Report the maximum value from either the upscale or the downscale curves.

Repeatability is the maximum difference in percent deviation observed above and is expressed as a percent of output span. Example: The repeatability is 0.05% of output span.

Reproducibility

1. Perform a number of calibration cycles as described in the calibration procedures.
2. Prepare a calibration curve based on the maximum difference between all upscale and downscale readings for each input observed. The deviation values are determined from the number of calibration cycles performed for step 1.
3. Maintain the test device in its regular operating condition, energized and with an input signal applied.
4. At the end of the specified time, repeat steps 1 and 2.

The test operating conditions may vary over the time interval between measurements, provided they stay within the normal operating conditions of the test device. Tests under step 4 must be performed under the same operating conditions as existed for the initial tests.

Reproducibility is the maximum difference between recorded output values (both upscale and downscale) for a given input value. Considering all input values observed, the maximum difference is reported. The difference is expressed as a percent of output span per specified time interval. Example: The reproducibility is 0.2% of output span for a 30 day test [Ref. 4].

Calibration Procedure Summary

Although the exact procedures for calibration of a specific pressure or differential-pressure instrument will vary in accordance with application, instrument type, and so forth, general procedures with the aid of a manufacturer's manual will enable a technician to calibrate almost any device. Calibration of an instrument consists of simulating and measuring the measured variable and observing the corresponding instrument response or output (see Figure 2–15). The output may be a scaled electric or pneumatic signal in the case of transmitters and receivers or a mechanical indication for pressure gages and other indicating instruments. The observed output, called the measured value of the instrument, is compared with the true value, or the instrument response when no correction is needed, and appropriate calibration adjustments are made to make the measured value equal to the true value.

Calibration Adjustments

Calibration adjustments are provided by the instrument manufacturer to change the instrument's response to an input signal. In pressure-measuring instruments, this is normally a magnitude function, as the phase shift or timing relationship between input and output is generally insignificant.

Zero Adjustment

Zero adjustment is normally the first adjustment made in instrument calibration. This adjustment shifts the instrument reading up or down scale by an equal amount at every point on the measurement scale. Figure 2–17 illustrates the effect of zero adjustment. It can be seen that the error caused by zero is constant at all points. Correction for zero error can usually be made at any point on the scale as this adjustment will change the reading the same amount at every point. Zero adjustment on mechanical instruments is done by slipping a link on a hub, a pointer on a shaft or by some other means of establishing a starting or reference point for measurement. Although this adjustment can be made at any scale value, it normally is made at the zero point or at the nominal point of operation. It should be emphasized that the actual point of adjustment does not matter as long as the span (defined below) is checked after each zero adjustment.

Span Adjustment

Span, sometimes referred to as multiplication or gain, determines the relationship between input and output by establishing the amount that the instrument acts on the input signal. As can be seen from the calibration curve in Figure 2–18, the error caused by span increases as

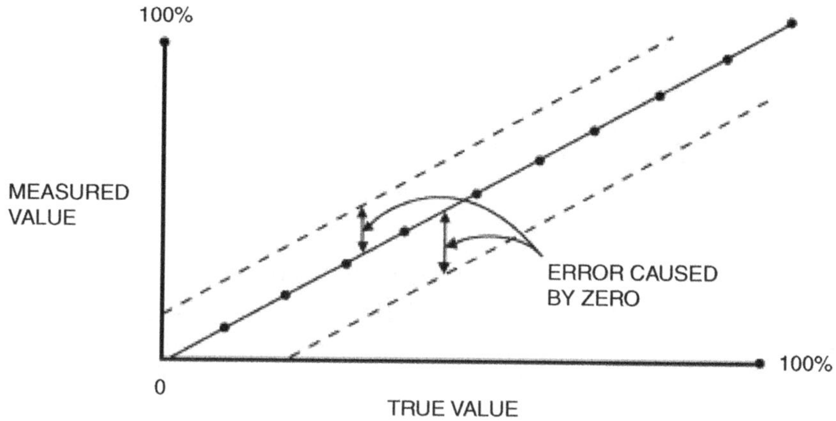

Figure 2–17 Zero Error

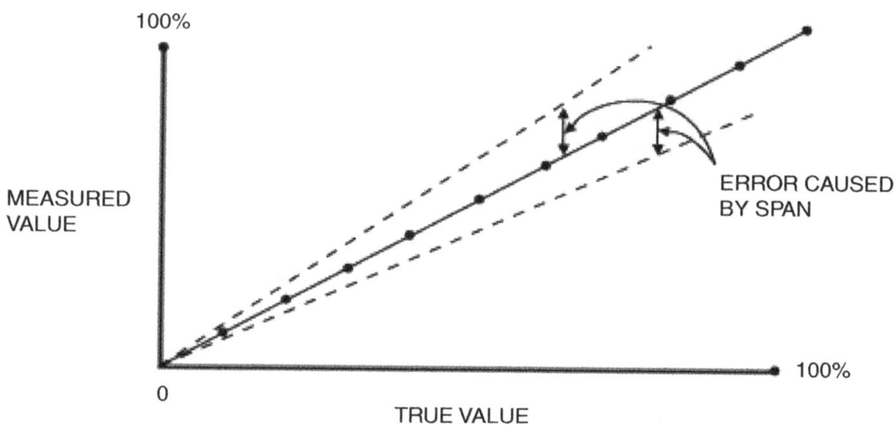

Figure 2–18 Span Error

the magnitude of input increases. Once a span adjustment is made, the instrument input must be changed in order to observe the effect of this adjustment. Also, if the span is too great, i.e., the gain is too high, the measured value will reach 100% before the true value (which is the measured input to the instrument) reaches 100%. Conversely, when the span is too small (the gain is too low), the applied input to the instrument will reach 100% before the instrument output reaches 100%.

Span (gain) adjustments in electrical instruments are accomplished by varying an electrical component to change the circuit gain of an amplifier. In mechanical instruments, span adjustments are usually made by changing the mechanical advantages of a lever system that relates the output response to an input response. A span adjustment will sometimes affect a zero reading, so a zero check and, if necessary, a zero adjustment should follow a span adjustment. This is especially true for mechanical devices.

The interaction between span and zero adjustment can complicate calibration unless special attention is paid to the procedure. For example, assume that after a zero adjustment a span correction is made to cause an instrument to read a full-scale value when a 100% measurement is applied. When the input is returned to a zero value, it is found that another zero adjustment is needed. The need for this adjustment is caused by the zero shift resulting from the span adjustment. To help remedy this time-consuming problem, an over-compensation in span adjustment (as found in the vendor's instructions) can be made and then the instrument set to the correct value by the zero adjustment. This can reduce the time required to calibrate an instrument with an interactive zero and span.

Nonlinearity Adjustment

This adjustment is necessary to compensate for the degree to which the span or amplification of the instrument is not constant. As shown in Figure 2–19, an instrument can respond properly at 0 and 100% or at two other points on the scale, but be in error at a third point. This error, which is sometimes referred to as angularity, is caused by improper angles between the drive and driven lever in a four-bar mechanism and the link that connects the two. Electrical amplifiers can be nonlinear, but errors caused by nonlinearity are most prominent in mechanical devices and are usually associated with pneumatic instruments.

Sample Calibration Procedure

The block diagram in Figure 2–15 illustrates a typical calibration arrangement for a pressure instrument. The components were described and discussed earlier. The process simulator is normally a pressure supply and regulator. To measure pressure from a few inches of water to several hundred psi, the input measuring device is a preci-

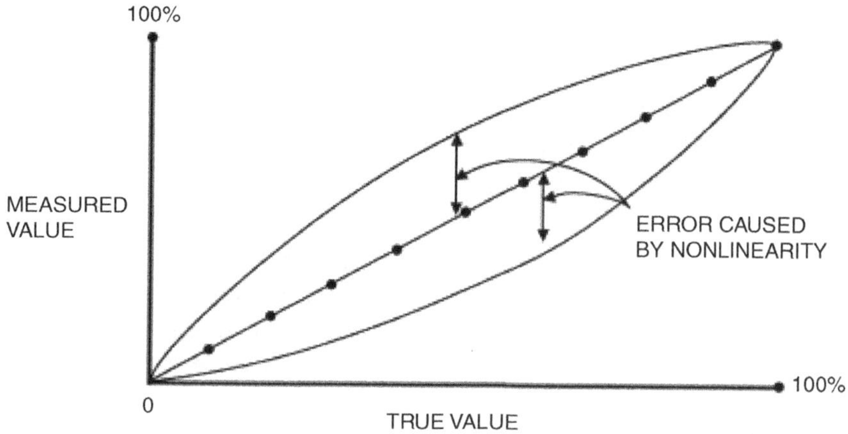

Figure 2–19 Nonlinear Error

sion pressure gage or manometer. The output measuring device is a precision pressure gage or current device to measure 4 to 20 mA, 10 to 50 mA, 3 to 15 psi, or some other output signal.

To calibrate a pressure or differential-pressure instrument, the instrument under test, the simulator and the test instruments are connected as shown in Figure 2–15 and in accordance with the manufacturer's recommended procedures for calibration hookup. With 0% of measured value applied to the instrument input, the output is made to be zero by the zero adjustment. An input signal of 100% instrument value is applied and measured with the input test instrument, and the output is observed. If the output is less than a value of 100%, the instrument span is increased. If the output is greater than 100%, the instrument span is decreased. After the span adjustment is made, the input is returned to a zero value and the instrument output is checked for a zero value and, if necessary, a zero adjustment is made. These procedures are repeated until instrument accuracy is obtained for the 0 and 100% measuring points. Some procedures recommend or specify 10% and 90% for the lower and upper calibration points. For process pressure gauges, calibration checks between 25% and 75% may be sufficient.

Once an instrument has been calibrated for zero and span, the accuracy has been verified at these two points only. This situation is illustrated in Figure 2–19. As is shown, considerable error can exist, however, for measured values that lie on the calibration span between the 0 and 100% points. Unless calibration checks are made at other points, the nonlinear condition shown will not be revealed. For this reason, calibration checks at 0, 25, 50, 75 and 100% values are normally included in calibration procedures. More or fewer calibration checkpoints may be desired, depending on the particular situation, but the midpoint accuracy should be verified. When nonlinearity error occurs, adjustments must be made to correct for it.

As has been mentioned, error caused by instrument nonlinearity is most prominent in mechanical and pneumatic instruments, where a series of links and levers are used to transfer the motion of a pressure element to the signal-generating mechanism or readout device. When the same type of arrangement is used in electrical instruments to generate movement of the movable coil of a linear variable differential transformer (LVDT), for example, nonlinearity can also exist, although it is less common. Linearity adjustments or procedures will not be available for many such instruments.

Some causes of nonlinearity may be worn parts, deformed pressure elements, tolerances in manufacturing, process coating of sensors, improper alignment of components during repair, or component aging. Repair and parts replacement may be required to eliminate errors. Where nonlinearity is caused by improper adjustments, it can be corrected by a procedure known as link and lever calibration.

The procedure just given for instrument calibration is valid for any calibration situation but is probably not applicable to more modern digital/microprocessor-based instruments. Such instruments are very accurate, repeatable, and have few or no moving parts; therefore, there is little chance of error introduced by friction, hysteresis, component wear, and so on. Calibration is much less involved and is usually performed less frequently than with mechanical or analog electronic devices. Calibration checks should, however, be performed after maintenance or parts replacement. Maintenance or parts replacement is seldom done, with the possible exception of replacing a primary element or electronics module in a pressure or differential pressure instrument. When this is done, a procedure called sensor trim or characterization can be performed as recommended by the manufacturer. Some manufacturers of modern microprocessor based smart devices claim that calibration checks need only be done once every three to five years. It is important to remember that often their test data is based on controlled conditions and the owner would do well to consider the process and environmental conditions that the device is exposed to when determining a calibration schedule.

The terms lower range value (LRV) and upper range value (URV) are often used in calibration discussions and calibration procedures. LRV is the pressure applied to the transmitter for a zero output value and URV is the pressure applied for a 100% output value. For differential pressure applications, the LRV is the net pressure difference in the high and low side pressure for zero output while URV is the net pressure for 100% output. Span, as has been used in this discussion, is the difference between URV and LRV. These terms are particularly useful in suppression and elevation applications.

Sensor trim requires an accurate pressure input and adds additional compensation that adjusts the position of the factory characterization curve to optimize performance over a given pressure or differential pressure range. It should be pointed out that since sensor trim adjusts the position of the factory characterization curve, it is possible to degrade the performance of the transmitter. If the trim adjustment is not performed properly, inaccurate instrument performance can result. (Most devices will allow factory defaults to be restored.) One of the overriding considerations in trim adjustment is the required accuracy of the test instruments used. In earlier times calibration procedures dictated that the test instruments should be at least ten times as accurate as the IUT. As field instruments became more accurate it became increasingly difficult to acquire test instruments many times more accurate than the IUTs. The accuracy ratio dropped to about five with transmitter composite accuracy on the order of 0.1% of rate. Many times the test instruments are not a great deal more accurate. For this reason, care and caution should be exercised in making sensor trim adjustments. This topic will be discussed in greater detail in Chapter 4.

Link and Lever Calibration

A four-bar lever system is used by some pressure instruments. The motion of the sensing element is conveyed to the pen arm or indicator by an arrangement similar to that shown in Figure 2–20. This lever system, a double rocker, has the following exact ratios:

$$\frac{\text{Angular rotation of } AB}{\text{Angular rotation of } DC} = \frac{DF}{AE} \tag{2.9}$$

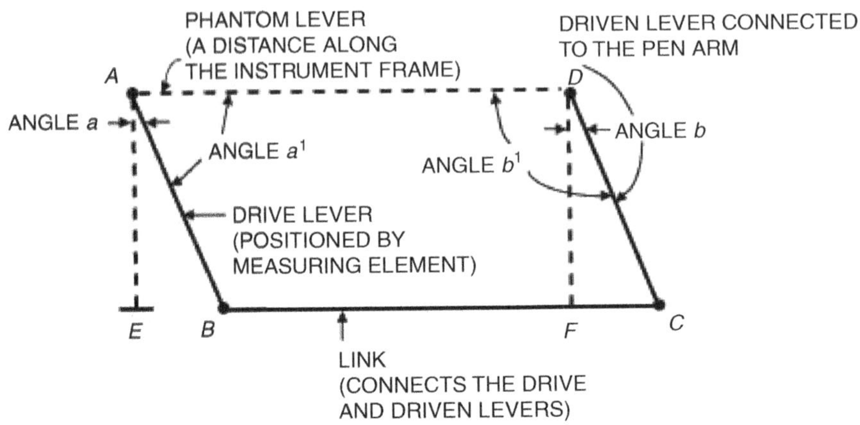

Figure 2–20 Four-Bar Lever System

When multiplication is not required, or when the movement of the measuring element is sufficient to produce enough pen motion for adequate resolution, *AD* = *BC* and *AB* = *DC*. In this case, angularity is not a problem. The geometric configuration of the four-bar system is a parallelogram. Angles *A* and *B* are equal, and the rotation of *DC* is equal to that of *AB*. This situation is seldom the case, however, because in most applications utilizing pressure measuring elements, multiplication of the angular rotation is necessary to increase the relatively small movement of the measuring element to the larger movement required for the pen or indicator. This multiplication is achieved by adjusting the ratio of *AB* and *DC*. In most instruments the pen or indicator (driven lever) is fixed, and the length of the drive lever (the pressure measuring element) is adjusted. This adjustment usually is made by a knurled knob attached to a multiplication screw. The multiplication is increased by increasing the drive-lever length and decreased by decreasing the drive-lever length.

When multiplication is introduced into the system by changing the drive lever length and changing the ratio of *AB/DC* in Equation 2.7, angularity always results. Angularity, as shown in Figure 2–21, causes a nonlinearity error, is detrimental to instrument operation and accuracy, and must be resolved. If it is allowed to exist, equal increments of element travel will produce unequal increments of indicator or pen travel. To overcome the angularity problem, the length of the connecting link is changed. This increases the multiplication over one half of the measurement range while decreasing the multiplication over the other half of the measurement range.

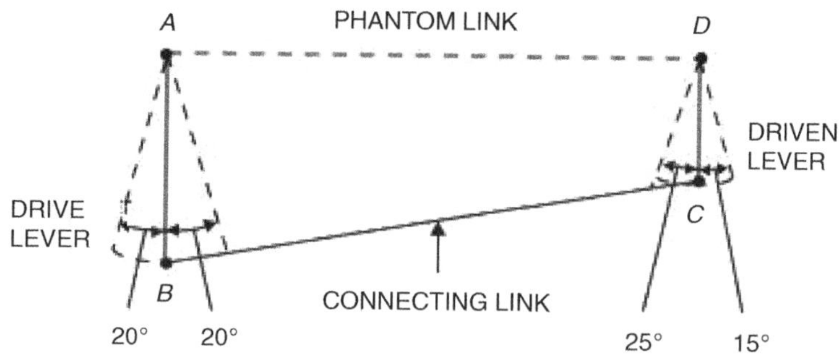

Figure 2–21 Angularity in a Four-Bar System

Instrument calibration should always include a check for linearity problems, and corrective action should be taken when necessary. The proper amount of adjustment to correct for angularity can be difficult if a hit-or-miss approach is used; therefore, the following calibration procedure is recommended.

Once the instrument under test has been checked for proper zero and span adjustment, the measured value is adjusted to a value of 50% FS. The connecting link and/or both the drive and driven levers should be adjusted so that angle a' = angle b' = 90° (Figure 2–20). This is usually done by loosening a holding screw and slipping either the drive or driven lever on its shaft to make the desired right angle between the connecting link and both levers.

Instrument accuracy is then checked for at least five points on the scale. If the instrument response is nonlinear, caused by angularity problems, the length of the connecting link is incorrect. With a measured value of 50% FS applied to the instrument, the error of the measured value is noted by the instrument response. The length of the link is then adjusted until the error is increased by a factor of approximately 5% in the direction of the error. This procedure establishes the proper angle between the connecting link and lever. As one adjustment often affects another in mechanical instruments, checks should be made for zero and span error, and adjustments made if necessary. The calibration procedure should be repeated until the instrument accuracy is within the desired limits.

If these established procedures fail to produce accurate results within the tolerances listed in the instrument instruction or specification manual, the instrument should be examined for worn or damaged components. A worn or damaged instrument should be repaired and the calibration procedures repeated.

Test Instruments for Pressure Calibration

The calibration of pressure instruments is one of the most common types of calibration because of the fundamental applications of pressure measurement. This section deals with conventional test instruments and calibration procedures. HART and FOUNDATION Fieldbus will be covered in Chapter 4.

Calibration equipment is relatively inexpensive, easy to use, and available in most instrument laboratories for the process industries. This can include even secondary laboratories. It should be noted, however, that the wide range of pressure measurements, from a few inches of water to hundreds or thousands of psi, can require test instruments with more than a million to one turndown. Several different test instruments may therefore be needed. This number will increase when the calibration of vacuum or compound ranges are considered. As will be seen in the next section, deadweight testers (DWTs) can be used successfully in wide-range applications.

One of two procedures can be used for instrument calibration. The instrument can either be calibrated in the field at the location of use or taken to a calibration shop. Zero checks can usually be made in the field, but complete calibration requiring range checks is normally accomplished in a calibration laboratory or instrument shop.

As previously discussed, calibration requires a measured input for the instrument; the instrument response or output is compared with the input. The input or simulated process for pressure and differential-pressure instruments is usually supplied by a variable-pressure regulator connected to an air supply source. Most differential pressure applications involve pressure measurements of full-scale ranges less than 20 psig. Thus, the process can be simulated by air supply sources used in calibration shops, which are sometimes 100 psig or higher. For the calibration of pressure instruments with ranges above available air supply pressure, cylinders of compressed gas with pressures as high as 2000 psig can be used for process simulation. Deadweight testers (DWTs) also can be used for this purpose. The test instrument used will depend on the specific application, but the choice will largely be a function of the calibration required and the calibration site (shop or field).

Deadweight Testers

A deadweight tester (also called a piston pressure gage or pressure balance) is often regarded as an absolute instrument because of its principle of operation. An absolute instrument is one capable of measuring a quantity in the fundamental units of mass, length, and time. Certain types of DWTs fall into this category.

Deadweight testers are similar to the hydraulic press shown in Figure 2–1 in terms of construction and theory of operation. Such testers are used to provide both process simulation and measurement for calibrating pressure instruments with ranges from zero to a few psig up to several hundred or a few thousand psig. A schematic diagram of a DWT is shown in Figure 2–22.

Deadweight testers and their matched set of weights are sold as a unit. The weights have been matched to the specific piston to compensate for manufacturing tolerances. It is important that weights not be mixed and matched among deadweight testers. It is also critical that the weights be protected from corrosion and damage.

This type of calibration device is filled with a lightweight oil or hydraulic fluid. A plunger or piston connected to a screw and hand-wheel is used to reduce the fluid volume when the handwheel is turned to screw the piston farther into the chamber (or as in Figure 2–22, a pump may be used). This reduction in volume results in an increase in pressure. Since the fluid is nearly incompressible, a slight reduction in volume results in a relatively high increase in pressure which is applied to both chambers. The force caused by the weights and piston is equal to the mass of the weights plus the mass of the piston multiplied by the acceleration of gravity. When the downward force from the weight of the calibrated weights plus the weight of the piston equals the upward force of the pressure acting against the surface area of the small piston or pump, the piston attached to the weight cable will float within the cylinder. The pressure required to cause the piston and calibration weights to float can be determined by markings on the weights, which are calibrated in pressure units.

The following principles apply:

- Upward force = (pressure)(piston area)
- Downward force = (mass)(acceleration due to gravity) = weight
- At balance (piston floating), upward force = downward force
- Pressure = weight divided by piston area

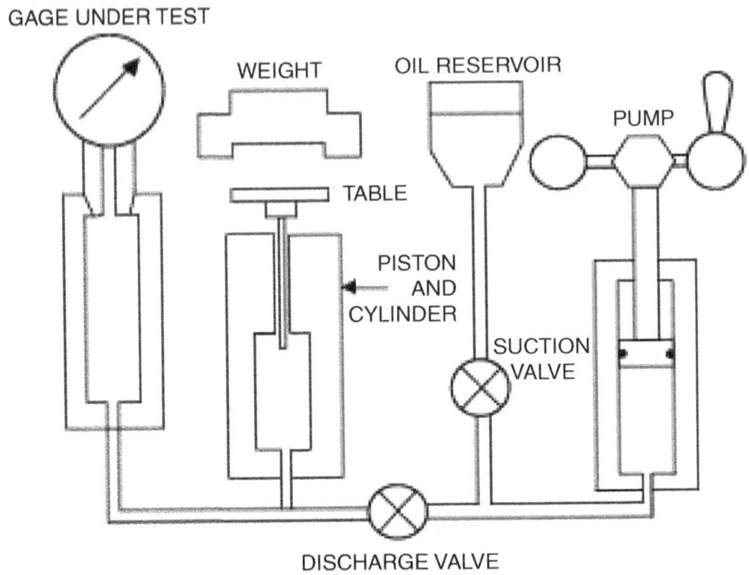

Figure 2–22 Deadweight Tester (*Courtesy of Ruska Instrument Corporation*)

The accuracy of DWTs is dependent upon the accuracies of the piston areas and of the mass of the weights and piston. For piston instruments, the overall accuracy can be ±0.1% of measurement for working standards, down to ±0.003% of measurement for primary standards. This primary standard accuracy requires compensation for several factors that include:

- Local gravity
- Air buoyancy
- Temperature
- Air density
- Fluid surface tension
- Piston and cylinder change in volume under high operating pressure

Each of these effects will be discussed in more detail later. The calculation of these effects will usually be accomplished by a computer program supplied by the manufacturer. Compensation for local gravity is a major factor since it directly affects the value of the mass of the weights times gravity (that is, the weight of the weights).

Pressure ranges covered by DWTs are from about 5 psi to 100,000 psi, but this wide range will require more than one test instrument because of limited turndown capabilities. The weight of the piston and carrier assembly determines the minimum measurable pressure as this is the minimum weight that can float.

Absolute pressure calibrations are performed using a tester with the weights and piston in an enclosure that has been evacuated to a nearly perfect vacuum.

It is common for standard DWTs to have two interchangeable piston and cylinder assemblies with different piston areas and weight ranges to increase the turndown of the tester.

Table 2–4 will help in the explanation of the increase in turndown capabilities of a DWT with two selectable piston and carrier assemblies.

Table 2–4 Turndown Capabilities of a Dual Piston Deadweight Tester

Total weight of piston and carrier (lbs)	Area of piston (in^2)		Maximum measurable pressure (psi)
	.01	.1	
0.5		X	5
1.0		X	10
2.0		X	20
10.0		X	100
110		X	1100
0.5	X		50
1.0	X		100
2.0	X		200
10.0	X		1000
110	X		11,000

From the data in the table, several observations should be considered.

- In each listing, the maximum pressure that can be measured is the weight divided by the piston area:

$$P = F/A$$

- The turndown using the large piston is

$$(1{,}100) \div (5) = 220$$

- The turndown using the small piston is

$$(11{,}000) \div (50) = 220$$

- The combined turndown is

$$(11{,}000) \div (5) = 2200$$

- The turndown using each piston is very good.
- The combined turndown is excellent.

The weights used in this application would be a combination of the following:

- Four weights at 0.5 pounds
- Four weights at 2.0 pounds
- One weight at 9.5 pounds
- Nine weights at 10.0 pounds

Advantages of DWTs include

- Excellent turndown
- Excellent accuracy
- Ease of operation
- Low maintenance

Limitations of DWTs include [Ref. 5]

- Possible overshooting of calibration points as weights are added.
- Limited to shop use for most applications.
- Limited incremental pressure steps.
- Excessively viscous or dirty fill fluid can cause errors and poor operation.
- Limited volume supplied by the fill fluid chamber may preclude the successful use of DWTs in some applications.

Deadweight testers are excellent transfer standards for calibrating pressure gages and transmitters. They are also used as secondary standards where extreme accuracy is required. Their use is normally limited to relatively small-volume pressure elements.

When using DWTs it is important to have all fluid connections tight to avoid loss of the fill fluid. Also, the instrument being calibrated should be free of process fluids, which could contaminate the fill fluid. Care should be exercised so that the piston and cylinder pressure does not exceed the pressure value of the calibrated weights. If it does, the piston will be blown out of the cylinder, resulting in possible operator injury or the loss of fill fluid.

An advantage of DWTs is that they can be used for both measurement and process simulation. When pressure supply sources and regulators are used for process simulation, test instruments must be employed for the simulation.

Figure 2–23 illustrates the three most common types of cylinder arrangements used in piston pressure gages (DWTs). When the simple cylinder shown in Figure 2–23(A) is subjected to an increase in pressure, the fluid, exerting a relatively large total force normal to the surface of confinement, expands the cylinder and thus increases its area. A pressure drop appears across the cylinder wall near point *x*, resulting in an elastic dilation of the cylinder bore. Measurement of pressure with a piston gage is subject to uncertainties resulting from effects other than those of elastic distortion. However, elastic distortion will be discussed first, since this characteristic is largely responsible for the various designs that have been developed.

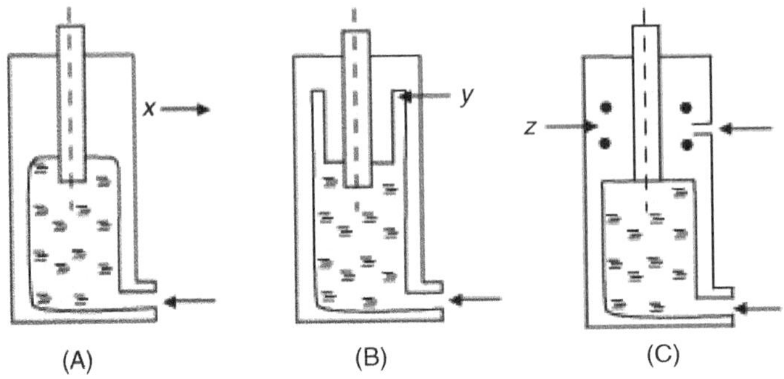

Figure 2–23 Cylinder Arrangements of Piston Gages
(Courtesy of Ruska Instrument Corporation)

It can be shown that the effective area of the piston and cylinder assembly is the mean of the individual areas of the piston and of the cylinder; therefore, as the pressure is increased, the cylinder expands and the effective area becomes greater. The rate of increase is usually (but not always) a linear function of the applied pressure. The piston also suffers distortion from the end-loading effects and from the pressure of the fluid, but to a much smaller extent than the cylinder. It is evident, then, that the simple cylinder of Figure 2–23(A) would be inadequate for a primary DWT unless some means of predicting the change in area was available.

The increase in the effective area of the simple cylinder is also accompanied by an increase in leakage of fluid past the piston. Indeed, the leakage becomes so great at some pressures that insufficient floating time can be maintained for a proper pressure measurement.

In Figure 2–23(B), the fluid under pressure is allowed to surround the body of the cylinder. The pressure drop occurs across the cylinder wall near the top of the cylinder, at *y*, but in a direction opposite to that of the simple cylinder in Figure 2–23(A). In consequence, the elastic distortion is directed toward the piston, tending to decrease the area of the cylinder.

The change in area with changing pressure places a limit on the usefulness of cylinder (B) as a primary instrument. However, some benefit results from the use of cylinder (B) in the construction of a DWT because higher pressures can be attained without a loss-in-float time. A small sacrifice is made in float time at lower pressures because the total clearance between piston and cylinder at low pressure must necessarily be greater for cylinder (B) than for cylinder (A).

In the controlled-clearance design of Figure 2–23(C), the cylinder is surrounded by a jacket to which a secondary fluid pressure system is connected. Adjustment of the secondary, or jacket, pressure permits the operator to change the clearance between the cylinder and piston at will. A series of observations involving piston sink rates at various jacket pressures leads to empirical determination of the effective area of the assembly.

Piston gages having very high resolutions can be made by using regular cylinders. The distortion coefficients of such gages can be determined by direct comparison with a controlled-clearance gage. Most piston gages have some elastic distortion, but some, used in very low pressures, have only small coefficients, and correction for distortion can sometimes be neglected.

Measurement processes proposed for high accuracy are disturbed by limitations in equipment performance, by small changes in the environment, and by operational procedures. The disturbances can be reduced to a degree by controlling the environment and by developing skill in the operation of the apparatus. Some of the disturbances are difficult to control; it is easier to observe their magnitudes and apply corrections for their effects.

The factors that disturb a pressure-measurement process when conducted with a piston pressure gage are described below. It is important that the operator become acquainted with these factors and become accustomed to recognizing their presence. The success of the measurement will depend on the degree to which control has been maintained and on the completeness with which corrections have been applied for the disturbances.

Factors that affect the performance of the piston gage and the measurement process are

- Elastic distortion of the piston and cylinder
- Temperature of the piston and cylinder
- Effects of local gravity on the masses
- Buoyant effect of the atmosphere upon the masses
- Hydraulic and gaseous pressure gradients within the apparatus
- Surface tension effects of the liquids

Elastic Distortion of the Cylinder

As has been mentioned, as pressure is increased within a piston gage, the resulting stress produces a temporary deformation of the cylinder. The net effect is a change in the effective area of the piston. If the change in the area is a linear function of the applied pressure, the relationship can be described by

$$A_e = A_0(1 + bp) \tag{2.10}$$

where A_e is the effective area at a pressure P, A_0 is the area of the piston-cylinder assembly at a reference pressure level and b is a coefficient of elastic distortion that is determined experimentally. The value of b is the fractional change in area per unit of pressure.

Temperature

Deadweight piston gages are temperature sensitive and thus must be corrected to a common temperature datum. Variations in the indicated pressure resulting from changes in temperature arise from the expected change in effective area of the piston. Treatment, therefore, is a straightforward application of the thermal coefficients of the materials of the piston and cylinder. By substituting the difference between the working temperature and the reference temperature and the thermal coefficient of area expansion in Equation 2.7, the area corresponding to the new temperature can be found:

$$A_{0(t)} = A_{0(Ref.\ t)}(1 + C\,dt) \tag{2.11}$$

where $A_{0(t)}$ is the area corrected to the working temperature, $A_{0\ (Ref.\ t)}$ is the area of the piston at 0 psig and at the selected reference temperature, C is the coefficient of superficial expansion as indicated in the test report and *dt* is the difference between the working temperature and the reference temperature. The magnitude of error resulting from a temperature change of 5°C for a tungsten carbide piston in an alloy-steel cylinder is approximately 0.008%. For work of high precision, gage temperatures are read to the nearest 0.1°C.

Effects of Gravity

The confusion and ambiguity that have resulted from the careless use of the terms mass, weight, and force are under severe and just attack by those who recognize the consequences of such practice. The scientific community has long been careful in the use of terms involving units and dimensions. On the threshold of international agreement regarding a unified system of weights and measures, it is important that, regardless of how inconvenient or mysterious these new units appear, we quickly become acquainted with them and contribute our best effort toward a rapid and successful transition to the new system. The new units and their definitions leave little room for ambiguity. An understanding of the relationships between SI units will help correct old habits.

All objects are attracted toward the earth by the action of gravity. This attraction is significant because it provides a way to apply accurately determined forces to the piston of a pressure gage. The force can be evaluated if the mass of the object to be placed on the gage piston is known. The quantity is expressed as

$$F = kMg \tag{2.12}$$

where g is the local acceleration due to gravity, M is the mass of the object, and k is a constant whose value depends on the units of F, M, and g:

$k = 1$ for F in newtons, M in kilograms, and g in meters per second squared

$k = \dfrac{1}{980.665}$ for F in kilograms force, M in kilograms, and g in centimeters per second squared

$k = \dfrac{1}{980.665}$ for F in pounds force, M in pounds mass, and g in centimeters per second squared

Buoyant Effect of Air

According to Archimedes' principle, the weight of a body in a fluid is diminished by an amount equal to the weight of the fluid displaced. The weight of an object (in air) that has had its weight corrected for the effects of local gravity is actually less than the corrected value indicates. The reduction in weight is equal to the weight of the quantity of air displaced by the object or the volume of the object multiplied by the density of the air. However, the volume of an irregularly shaped object is difficult to compute from direct measurement. Buoyancy corrections are usually made by using the density of the material from which the object is made. If the value of mass is reported in units of apparent mass versus brass standards rather than of true mass, the density of the brass standards must be used. Apparent mass is described as the value the mass appears to have as determined in air having a density of 0.0012 g/mL, against brass standards of a density of 8.4 g/cm^3, whose coefficient of cubic expansion is 5.4×10^{-5}/°C, and whose value is based on true mass in vacuum.

Although the trend is toward the use of true mass in favor of apparent mass, there is a small advantage in the use of the latter. When making calculations for air buoyancy from values of apparent mass, it is unnecessary to know the density of the mass. If objects of different densities are included in the calculation, it is unnecessary to distinguish the difference in the calculations. This advantage is obtained at a small sacrifice in accuracy and is probably unjustified when considering the confusion that is likely to occur if it becomes necessary to alternate in the use of the two systems.

A satisfactory approximation of the force on a piston that is produced by the load is given by

$$F = M_a 1 - \left(\frac{r_a}{r_b}\right) \text{kg} \tag{2.13}$$

where M_a is the mass of the load reported as "apparent mass versus brass standards," r_a is the density of the air, and r_b is the density of brass (8.4 g/cm^3).

Reference Plane of Measurement

The measurement of pressure is strongly linked to gravitational effects on the medium. Whether in a system containing a gas or a liquid, gravitational forces produce pressure gradients that are significant and must be evaluated. Fluid pressure gradients and buoyant forces on the piston of a pressure balance require the assignment of a definite position at which the relation $P = F/A$ exists. It is common practice to associate this position directly with the piston as the datum to which all measurements made with that piston are referenced. This is called the reference plane of measurement, and its location is determined from the dimensions of the piston. If the submerged portion of the piston is of uniform cross section, the reference plane is found to lie conveniently at the lower extremity. If, however, the portion of the piston submerged is not uniform, the reference plane is chosen at a point where the piston, with its volume unchanged, would terminate were its diameter uniform.

When a pressure for the deadweight gage is calculated, the value obtained is valid at the reference plane. The pressure at any other plane in the system can be obtained by multiplying the distance of the other plane from the reference plane by the pressure gradient and then adding (or subtracting) this value to that observed at the piston reference plane.

For good work, a pressure gage should be provided with a fiducial mark for associating the reference of the piston with other planes of interest within a system. Not only does the mark serve to establish fixed values of pressure differences through a system, but it also indicates a position of the piston with respect to the cylinder at which calibration and subsequent use should be conducted. If the piston is tapered, it is important to maintain a uniform float position for both calibration and use.

In normal operation, the system is pressurized until the piston is in a floating position slightly above the fiducial or index mark. After a short period, the piston and its load will sink to the line, at which time the conditions within the system are stable. If there is a question as to the error that may be produced by accepting a float position that is too high or too low, the error will be equivalent to a fluid head of the same height as the error in the float position. This statement assumes, of course, that the piston is uniform in area over this length.

Effects of Liquid Surface Tension

One of the smaller disturbances that affect the performance of a piston gage is that resulting from the surface tension of the liquid. The strong meniscus that is formed around the piston as it enters the cylinder is visible evidence of a force acting on the surfaces. Numerically, the force on the piston that results from surface tension is

$$F_{st} = gC \tag{2.14}$$

where g is the surface tension of the liquid in dynes per centimeter or pounds force per inch, and C is the circumference of piston in centimeters or inches.

Conditions Favorable for a Measurement

A pressure measurement is no better than its beginning. All pressure measurements are made with respect to something. When a value of pressure is expressed, it is implied that the difference between two pressure levels is the value stated. In order to determine the difference between two pressures, each of the pressures must be measured. Furthermore, if a level of confidence is stipulated for the expressed value of pressure, the confidence figure must include the errors of each of the pressure measurements. This problem is not unique to pressure measurement and is brought to attention here to impress, by repetition, the importance of proper zero measurement at the start.

Errors in establishing a starting-point zero at the beginning of the test arise principally from uncertain oil heads in various parts of the equipment. In general, the vertical dimensions of a hydraulic calibrating system, such as would normally be connected to a laboratory deadweight gage, are small; therefore, the total head error is relatively small. If the pressure to be measured or generated is large, the small starting error may possibly be neglected. But if pressures in the low ranges are expected to be measured with high accuracies, small head errors at the beginning are quite significant.

When using the piston gage as a standard of pressure for the calibration of elastic-type sensors in the low ranges, a datum must be established for the sensor. The sensor usually has a mechanical or electrical adjustment for setting the zero. Some adjustments may have a considerable range, enabling the device to indicate zero (although a rather large bias may be present at the time the zero is adjusted).

If the manufacturer does not stipulate a pressure reference plane for the sensor, one must be chosen. The choice may be arbitrary or may be based on obvious details of sensor geometry: All pressures must thereafter be referenced to this datum.

When calibrating gages or transducers with gas, using a diaphragm barrier between the oil and gas portions of the system, the readout mechanism of the diaphragm device must be adjusted to zero when no pressure exists across the diaphragm. The diaphragm terminates the liquid portion of the system; thus, it becomes the interface of the oil and gas media. Head corrections are computed for the difference in height of the piston gage reference and interface diaphragm for the oil and from the diaphragm to the transducer for the gas system. If the gas system is opened to atmosphere and the liquid system is also opened at a point of equal height of the diaphragm, the pressure across the diaphragm will be zero and the readout mechanism can be adjusted to zero. After the transducer is also carefully adjusted to zero, the gas system can be closed and the calibration begun.

One source of error at low pressures is the presence of air in the system. If a quantity of air is present in the vertical section of a connecting tube, the assumed head correction will be in error. If air migrates to the deadweight gage, the reference plane may be shifted due to upset of the buoyancy of the oil on the piston.

A quantitative measurement of the amount of air in the system can be made by noting the number of turns on the hand pump handle required to cause a response in the IUT. The deadweight gage must have a weight on the piston to make the test effective. When no air exists in the system, the IUT will respond almost immediately as the pump handle is rotated slowly.

Establishing the Pressure

Assuming that the deadweight tester is to be used in the calibration of a pressure-measuring device, and assuming also that the device has been properly tested and exercised prior to calibration, the selection of weights is placed on the DWT according to the results of the calculations. The pressure can be raised until the deadweight gage is floating in a rather high position.

For the first few moments, the weights on the DWT may be observed to fall rather rapidly—more so than usual. The unusual descent rate is observed for three or more reasons. (1) The packing of the valves, pump plunger, and o-ring seals is compressible to some extent and reluctant to be squeezed into the packing gland. The movement of the packing is not instantaneous. (2) Any air in the system is forced into solution by the increased pressure. Dissolution of the gas causes a reduction in the volume of the pressurized fluids, which is manifested by an apparent increase in fall rate. (3) The greatest contribution to the increased sink rate is the contraction in fluid volume from the transfer of the heat to the apparatus. The increase in pressure is accompanied by a sharp rise in temperature of the oil and also a rather large volume expansion. As the vessels of the system absorb the heat, the oil contracts and thus causes an apparent increase in leakage in the DWT.

The effects of the three causes are additive and serve to indicate an apparent high rate of leakage. The sink rate can be measured with a scale and a watch. If measurements are plotted for each minute interval for several minutes, the curve will drop sharply the first few minutes and then level off to a constant value. After raising the pressure to values as high as 600 to 700 atms, 7 to 8 min are required for the thermal effects to die out [Ref. 6].

Pneumatic Deadweight Tester

The pneumatic DWT is similar in operation to the hydraulic DWT. It can be used for pressure calibration in the low inches of water range. An accurate calibration pressure is produced by establishing an equilibrium between the air pressure on the underside of a ball against weights of known mass on the upper side of the ball. The pneumatic tester illustrated in Figure 2–24 shows that a precision ball is floated within a tapered steel nozzle. A flow regulator produces pressure under the ball, lifting it in the tapered annulus until equilibrium is established—at which time the ball floats. The vented flow equals the fixed flow from the supply regulator, and the output pressure applied to the IUT is equal to the force caused by the weights on the weight carrier. During operation the ball is centered by a dynamic film of air, prohibiting physical contact between the ball and the nozzle.

As weights are changed on the weight carrier, the ball is positioned up or down, affecting the airflow. This change in flow is detected by the regulator, which adjusts the pressure under the ball to return the system to equilibrium and establish a new output pressure. The output

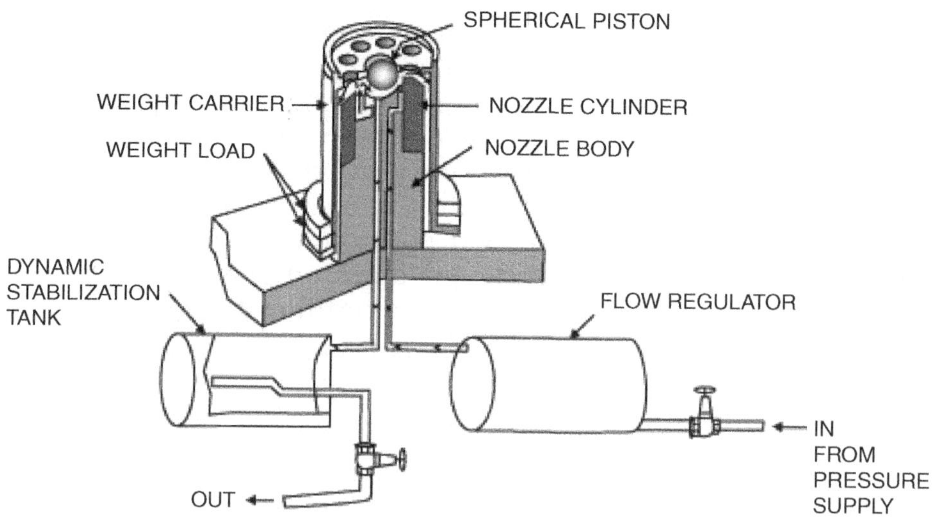

Figure 2–24 Pneumatic Deadweight Tester
(*Courtesy of Ashcroft-Heise*)

pressure regulation is then in accordance with, and proportional to, the force caused by the weight of the carrier acting on the top of the ball. Flow regulation is unaffected by the supply pressure as long as it is maintained above a minimum threshold value. Accuracy is a function of measurement and varies from 0.15 to 0.05% of reading based on weights calibrated for international standard gravity. Measurement ranges are related to the mass of the weights. Ranges are available from 4 inches of water to 30 psi up to 100 to 1500 psi.

Test Gages

Although production-type pressure gages used in process applications are accurate enough for most process measurements, they normally are not used as calibration standards. The accuracy of production gages may be no better than 1 to 3% overall. Gages specially manufactured with greater precision—called test gages—are used for calibration purposes.

Considerably more expensive than production-quality gages, test gages are designed and constructed for greater accuracy: With overall accuracy (including hysteresis and linearity) as great as 0.1% FS, these gages can be used in some calibration applications. Such gages have been designed with mirrored scales to reduce errors caused by parallax, jeweled bearings to reduce errors caused by friction, and scale sizes up to 16 in. to improve resolution. Compound scales for pounds per square inch, inches of water, inches of mercury, feet of seawater, and so on, are available. Vacuum and compound ranges are available with absolute pressure or gage pressure scales. Some test gages have bimetallic bars forming the link between the pressure element and the gear sector for temperature compensation. Changes in ambient temperature result in changes in the angular position and effective length of the gear segment slide. Specially designed carrying cases prevent transport damage during field use. Test gages certified as secondary standards are available.

Pressure Comparators

The pressure comparator shown in Figure 2–25 combines the operation of deadweight testers with the precision and accuracy of digital test gages for pressure calibration. This device offers the advantage of being fast, portable, relatively low in cost, easy to operate and, depending on the test gages, can be used in the low to several thousand psi range with as low as 0.1 inch of water resolution. This device is recommended primarily for gage calibration but can also be used for calibrating pressure transmitters and setting pressure switches.

For calibration operation, a suitable test gage is connected to one port and the IUT is connected to the other port. For ease and speed of operation, quick disconnect fittings are available that do not require thread lubricant and can operate with finger tight assembly.

Manometers

Because of their accuracy and simplicity, liquid manometers are used in both field and shop calibration applications. As they have no mechanical moving parts, they require no calibration, and excessive use causes no loss of dependability or accuracy. Well manometers are primarily used for shop calibration of process instruments and shop standards. Because shop calibration is inconvenient for some process instruments portable manometers for field calibration are available.

Many applications, such as furnace draft and low differential pressures in air and gas ducts, require the accurate measurement of low pressures. For such measurements, as has been earlier described, the manometer is arranged with the tube inclined. This provides an expanded scale and can allow 12 in. of scale length to represent 1 in. of

Figure 2–25 A Pressure Comparator Test Instrument
(*Courtesy of Crystal Engineering*)

vertical height. With scale subdivisions to 0.01 in. of liquid height and water as the fill fluid, the equivalent pressure of 0.00360 psi per division can be read.

Three types of pressure measurements are common in process applications and in the calibration of process instruments: (1) positive or gage pressures—those above atmospheric; (2) negative pressures or vacuums—those below atmospheric; and (3) differential pressure—the difference between the two pressures. Differential pressures are measured by connecting one leg to each of the two pressures. The higher pressure causes the fluid level to be depressed, while the lower pressure causes the fluid level to be elevated. The true differential pressure is measured by the difference in height of the fluid level in the two legs.

Sometimes it is convenient to use one manometer to measure two different pressures. Such is the case in the calibration of a pneumatic differential-pressure instrument. A dual-tube manometer can be used for this purpose. The instrument input range, which is normally in inches of water, is connected to one tube for measurement; the instrument output pressure, which is normally psig, is connected to the second tube and read on an associated scale. When using water as the fill fluid, a 10-in. fluid height will measure 0.360 psi. Another tube using mercury as the fill fluid will measure 4.892 psi for a 10-in. level displacement. The range of the two measurements is 13.57:1, which happens to be the ratio of the specific gravities of mercury and water.

Another principle common to manometers is the use of expanded and contracted scales to increase both the resolution of scale readings and the range of pressure that can be measured. Most laboratory manometers for shop calibration are 35 or 40 in. long. When the fill fluid is water, the maximum pressure that can be measured is equal to the tube length read in inches of water. It is often desirable, however, to measure a pressure greater than 40 in. of water with a 40-in. manometer. In this case, a fill fluid with a higher specific gravity can be used, with the scale calibrated to read inches of water. For example, when mercury is used as the fill fluid and the scale is divided into 13.57 divisions per inch, each division represents 1 in. of water; 13.57 in. of water pressure will cause a 1-in. displacement of the mercury level. Similar combinations of fill fluid and scale markings can be used to measure many different pressure units.

Table 2–5 lists several available scale graduations and fill fluid specific gravities. Both of these factors must be considered in manometer measurement. These data normally are indicated on the scale. For example, the left scale of the B/C duplex manometer would specify: using mercury, read inches of water. The right scale likewise would state: using mercury, read pounds and tenths.

Electronic manometers have largely replaced glass tube manometers for low pressure applications. The term inches of water, which was derived from the use of glass tube devices, is still commonly used for low pressure applications even with electronic manometers. These manometers measure pressure in inches of water to a few thousand psi. Such a manometer is shown in Figure 2–26.

Table 2–5 Manometer Scale Graduations
(*Data courtesy of Scott Fetzer Company*)

A: Linear inches and tenths B: Inches water pressure using mercury calibrated dry C: Pounds and tenths using mercury G: Inches and tenths water pressure using Meriam 827 Red Oil H: Inches and tenths water pressure using Meriam 175 Blue Fluid I: Inches and fifths water pressure using Meriam 295 No. 3 Fluid
A/C: Duplex Left: Linear inches and tenths Right: Pounds and tenths using mercury
B/C: Duplex Left: Inches of water pressure using mercury Right: Pounds and tenths using mercury
A/B: Duplex Left: Linear inches and tenths Right: Inches water pressure using mercury calibrated dry
A/F: Duplex Left: Linear inches and tenths Right: Linear centimeters and millimeters
Fluids available, nominal specific gravity Instrument mercury, 13.54 1000 Green Concentrate, 1.00 827 Red Oil, 0.827 100 Unity Oil, 1.00 295 Red Fluid, 2.95 175 Blue Fluid, 1.75

EXAMPLE 2-18

Problem: A manometer using mercury reads pounds per square inch. Assume the nominal specific gravity of the mercury to be 13.54. Find the linear scale length representing 1 psi of pressure.

Solution: From Equation 2.5, 1 psi = 27.7 in. of water. Assuming the nominal specific gravity of mercury to be 13.54, the scale length is 27.7/13.54 in. = 2.045 in.

A General-Purpose Multifunction Documenting Process Calibrator

Throughout this chapter on calibration, information has been given relating to calibration procedures and various test instruments that can be used for this purpose. Conventional calibration usually consists

Figure 2–26 Electronic Manometer (*Courtesy of Scott Fetzer Company*)

of providing an input to the instrument under test (IUT), measuring that input, measuring the corresponding output, logging the values on a data table, and making a calibration plot to determine the validity of the instrument response and the need for calibration adjustment. This procedure requires a source, measurement of the source value and measurement of the output. The accuracy of the test instrument should be several times greater than that of the IUT. With the measurement accuracy of process instruments sometimes better than 0.1% of rate, the test instrument must be very accurate.

A generation of electronic, microprocessor process calibrators are now available that can be used to check and calibrate many components used in a process control loop. They are battery powered, hand-held instruments that measure and source electric and physical parameters and many are HART compatible. They can be used to troubleshoot, calibrate, and verify, and some can even document work progress and recordkeeping on instruments. Some of the features available with these calibrators are

- Analog display of measurements
- Setup options for different applications

- Thermocouple input jacks for thermocouple measurement with internal temperature compensation and external temperature reference
- Ability to log several thousand data points
- Serial computer interface for uploading/downloading tasks, lists and results
- Automatic calibration procedures for transmitters and limit switches
- Split screen display to simultaneously show measured and source values
- Transmitter mode enabling the configuration of emulated results from a process instrument
- Built in calculator with square root functions
- Accessible registers containing measure and source values
- Adjustable damping for inputs to be measured with a display indicator of damped status
- Auto ranging, display in engineering units, percent of scale, damping and user selected custom units
- Measurement of a secondary value and value units
- Display of measurements in square law inputs or custom units
- Capture and display of minimum and maximum values

A special feature on some process calibrators is to provide a 4 – 20 mA source that is normally supplied by a 4 – 20 mA transmitter. Also, these calibrators can supply a 250 ohm resistor and voltage source to provide a loop current.

Many of the functions available can include

- DC volts
- AC volts
- Frequency
- Resistance
- DC current

- Continuity
- Thermocouple
    - Types E, N, J, K, T, B, R, S, C, L, or U
- RTD (numbers in parenthesis are for identifying the RTD)
    - 100 ohm platinum (3926)
    - 100 ohm platinum (3851)
    - 120 ohm nickel (672)
    - 200 ohm platinum (385)
    - 500 ohm platinum (385)
    - 1,000 ohm platinum (385)
    - 10 ohm copper (427)
    - 100 ohm platinum (3916)
- Pressure with the use of a factory supplied transducer. Some calibrators can be equipped with two pressure transducers for measurement of differential pressure. For sourcing pressure applications it is often necessary to use a pressure pump, a squeeze bulb or another pressure source.

The accuracy of process calibrators can be as good as

- DC volts 0.025% of rate with 1×10^{-6} volt resolution on the lowest range
- DC current 0.01% of rate with 1×10^{-6} amp resolution on the lowest range

There may be ambient temperature limits for the above specifications. A general range is around 18°C to 28°C. For this and all specifications, the manufacturer's data should be checked.

A multifunction documenting process calibrator is shown in Figure 2–27.

CONCLUSION

The accuracy of an instrument is verified by calibration; some calibration procedures and details have been presented in this chapter. Although the success of calibration depends on various factors, the accu-

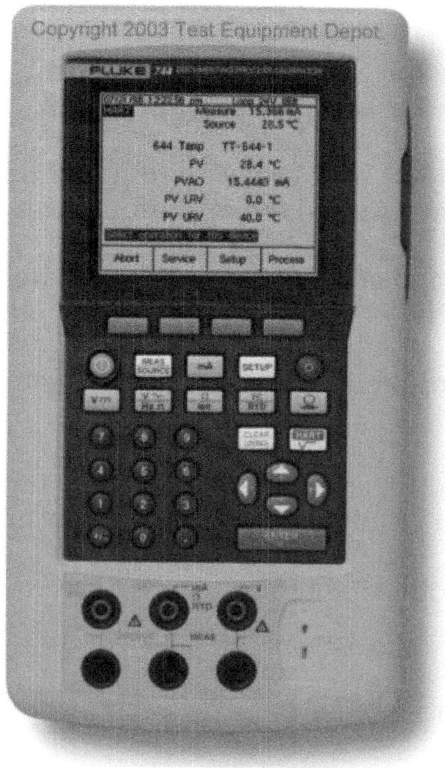

Figure 2–27 A Multifunction Documenting Process Calibrator
(*Courtesy of Fluke*)

racy and reliability of test instruments play an important role in establishing performance criteria. For this reason, test instruments that are used as accuracy standards must themselves be checked on a routine basis. Because the accuracy of deadweight gages can be expressed as natural fundamental relationships, they are often used to establish the highest order of accuracy in pressure measurement. However, procedures dictate that even these instruments must have accuracy validation checks, and they are often checked against other deadweight gages. Although this procedure will normally be conducted in a secondary standard laboratory and require efforts generally beyond the capabilities of most calibration laboratories, it is important to realize the careful and tedious operations involved in calibrating a deadweight gage. Appendix B outlines the procedure.

EXERCISES

2.1 Calculate the force on a 10 × 12 ft wall caused by a differential pressure of 0.5 psi on the sides of the wall.

2.2 State Pascal's law with respect to fluids at rest and pressure. How does this relate to the operation of a hydraulic press?

2.3 Define gage and absolute pressure and state the relationship between the two.

2.4 Express atmospheric pressure in feet of water and inches of mercury.

2.5 Explain the difference between accuracy statements expressed as a function of full-scale values and measured values.

2.6 Define the following terms:

- accuracy
- precision
- hysteresis
- linearity
- dead band

2.7 State the purpose of calibration and outline a general calibration procedure.

2.8 List three calibration adjustments.

2.9 What is a primary standard? A secondary standard?

2.10 Is there a difference between a test instrument and a standard production instrument? List any distinguishing characteristics.

REFERENCES

1. Transmation Instrument Company, Inc., "Teach Tips 11, 12."

2. Schwien Engineering Corporation, "Manometer Principles," Course No. 969.
3. ISO 9001
4. ISA-51.1-1979 - (R1993) - Process Instrumentation Terminology
5. "The Squeeze is Pressure," Keith Cheatle, *InTech*, June 2006.
6. "Types of Piston Pressure Gages," Bulletin Q-0573-8B and "Procedures for Calibrating a Dead Weight Tester," Static Pressure Measurement Training Manual, Ruska Instrument Corporation.

3

Pressure Transducers and Pressure Gages

3.1 INTRODUCTION

Pressure measurement for indication and control has been classified by measurement span, measurement range, the type of application, and other principles involved. A primary consideration in the measurement of any variable is the process of converting a characteristic of the measurand (the parameter being measured) to a response that can reveal its value. This almost always requires the utilization of transducers that convert one form of energy into another form of energy, e.g., pressure into an electrical signal or pressure into mechanical motion. This second form of energy is of a more useful type that can represent a measurement value.

3.2 PRESSURE TRANSDUCERS

The terms transducer, sensor, and transmitter are sometimes used interchangeably and at other times have completely different connotations. The most basic definition of a transducer is that it converts one form of energy (pressure or mechanical) to another form (often electrical, but may be mechanical). Early applications in the late 1940s and 1950s that encouraged the development of pressure transducers involved measurements for hydraulic systems, engines, and other related equipment in the process industry. Before the development of solid-state electronics, signal-conditioning circuits for transmission systems would have required bulky electronic amplifiers; therefore, the signal was usually in the low-level millivolt range produced directly by a transducer. This device then converted the pressure to a proportional electrical signal.

As industrial facilities expanded and transmission distances increased, low-level signals from transducers were attenuated by the I^2R drops in the transmission line. The need for amplification became apparent. Transmitters were developed that could condition the signal generated by a transducer, enabling transmission over longer distances. Although the distinction between a transducer and a transmitter may not be exact, the difference seems clear: A transmitter generates a scaled signal strong enough for transmission over fairly long distances. A transducer converts the process measurement to a signal capable of conditioning, amplification, and scaling by the transmitter. When the signal-conditioning circuitry is considered to be a transducer, a pressure transmitter would have two transducers: one for the input—pressure to resistance or voltage, for example, and one for the output—resistance or voltage to milliamps.

The distinction between a transmitter and a transducer has become vague; the difference seems simple, however. A transmitter generates and outputs a signal scaled to represent an exact quantity of a variable to be measured. The signal from the transmitter is usually 4 to 20 mA for electrical service and 3 to 15 psi for pneumatic applications. Special types of signal conditioning or modification are now possible by the use of microprocessor-based circuitry housed within the transmitter, which can be rated for use in flammable or hazardous areas.

3.3 PRESSURE ELEMENTS

The most important part of a pressure instrument, and sometimes the instrument's most distinguishing characteristic, is the pressure element. Pressure elements perform the function of converting pressure into mechanical motion. They are constructed such that an applied pressure causes a deformation in physical shape which produces a distortion that results in a motion or force at a free end. This output or indication is then used to generate an intelligent signal or reading that is proportional to the applied pressure. Many different types of elements are presently used and are classified by design, function, and application. Figure 3–1 shows a block diagram of a pressure element.

Bourdon Pressure Elements

Pressure elements represent a broad classification of transducers used in pressure applications. Most mechanical pressure measurements rely on a pressure acting on the surface area inside the element to produce a force that causes a mechanical deflection. The deflection, a mechan-

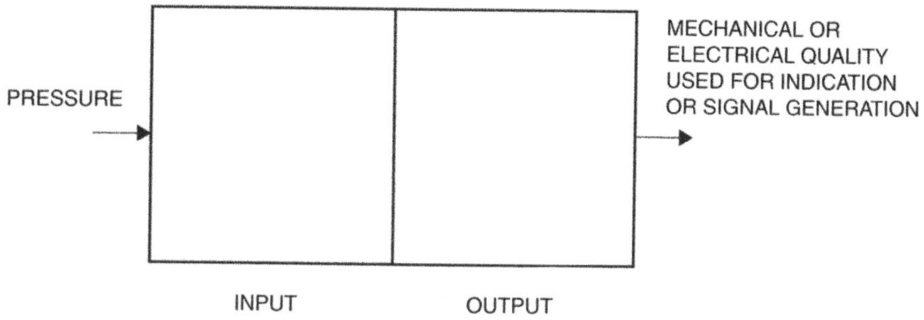

Figure 3–1 Block Diagram of a Pressure Transducer

ical motion, is used to generate an electrical or pneumatic output and sometimes simply a mechanical linear or rotary indication. The most common type of pressure element is the Bourdon tube, which is available in many forms, shapes, and sizes. Bourdon tubes are made of thin, pliable, springy metal.

The metal is formed to make a flattened tube with an open end to which the pressure is applied; the other end is closed and is usually free to move or flex when the pressure causes a deformation. The most common form of Bourdon tube pressure element is the C-tube.

Bourdon C-Tubes

Figure 3–2 shows a Bourdon C-tube used in a direct-reading gage; the tube usually has a 270° arc. The process pressure is applied to the fixed and open end of the tube, while the tip or closed end is free to move. Because of the difference in inside and outside radii, the measured pressure is acting on different areas, creating a differential force on the tube surfaces that causes the tube to straighten. The resulting tip motion is nonlinear because less motion results from each increment of additional pressure. When used in pressure gages, this nonlinear motion must be converted to a linear angular indication. A geared rack-and-pinion sector achieves this purpose. A number of important characteristics of C-tube materials are listed in Table 3–1.

Care must be taken so that the applied pressure is not sufficient to cause the tube stress to exceed the elastic limits of the tube, because that level of stress will result in permanent deformation and thus permanent tube damage. Some tubes have overrange stops to protect them against this type of damage. The C-tube is inexpensive, can be

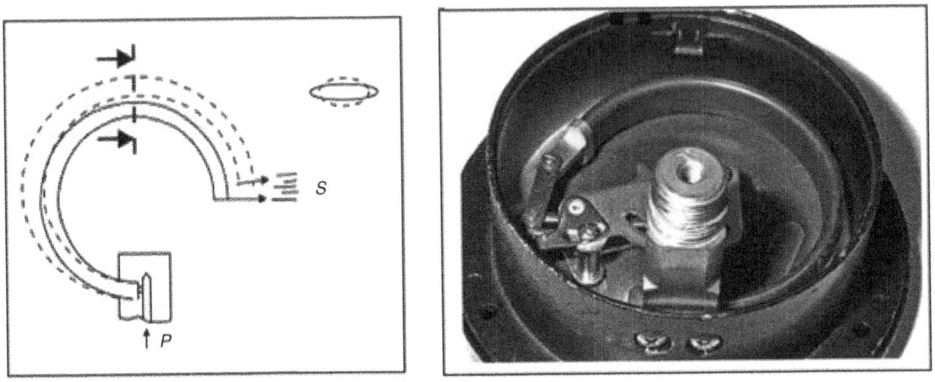

Figure 3–2 Bourdon C-Tube

Table 3–1 Characteristics of C-Tube Materials
(Courtesy of CRC Press)

Tube Material	Corrosion	Spring Rate	Temperature Coefficient	Hysteresis	Maximum Pressure (psig)
Phosphorus bronze	Poor	Fair	Poor	Fair	800
Beryllium copper	Poor	Good	Poor	Good	5,000
316 Stainless steel	Good	Poor	Poor	Poor	10,000
403 Stainless steel	Good	Poor	Poor	Poor	20,000
Ni-SpanC®	Good	Good	Good	Good	12,000
K-Monel®	Good	Poor	Poor	Poor	20,000

made to be accurate, and is rugged and reliable. A disadvantage of the C-tube is the limited amount of travel at the free end compared with other types of pressure elements. This movement must be amplified in some applications (discussed later). C-tubes are used in most pressure gages and in many pressure transmitters. They are not used to a great extent, however, in receiver instruments.

Spiral Pressure Elements

The free-end travel and force of the C-tube are insufficient to operate many motion balance devices. Motion balance describes the means by which the motion of the pressure element moves the sector, mechanically transmitting a rotary motion to the pinion assembly and indicator. The spiral element shown in Figure 3–3 can be considered to function as a series of C-tubes connected end to end. When pressure is applied, the element tends to uncoil and, by producing greater movement without amplification, yields greater accuracy and sensitivity. Less hysteresis and friction result because of the reduced need for additional links and levers. In general, the same types of materials are used for spiral element construction as for C-tubes.

Spiral pressure transducers produce more tip travel and higher torque capability at the free end than C-tubes. They are thus more capable of use as direct-drive readout devices. Instruments that utilize these types of transducers can respond to lower pressure and thus are more sensitive, partly because of the reduced friction involved in the movement. The operating range of spiral and helical elements is from a few inches of water to a few hundred pounds per square inch. These transducers are used primarily in pressure transmitters, pneumatic controllers and receiver instruments. Their use in pressure gages is limited.

Helical Pressure Elements

The helical elements shown in Figure 3–4 produce more free-end travel and torque than spiral elements. Other advantages include a higher range of operating pressure with sufficient overrange protection. The operating range is affected by coil diameter, wall thickness, the number of coils used, and the material of construction, which is the same as may be used in C-tube elements (Table 3–1). Applications of helical elements include direct reading for the following: indicators, recorders, controllers, switches, transmitters, and differential instruments when pressures are applied to both sides of the elements.

Bellows

Bellows (Figure 3–5) are made of springy metal formed in the shape of thin-wall tubes. The tube is then shaped to form deep convolutions, which enable the element to expand in much the same manner as an accordion. The bellows are sealed on the free end; pressure is applied to an opening on the stationary end. Pressure applied to the inside acts on

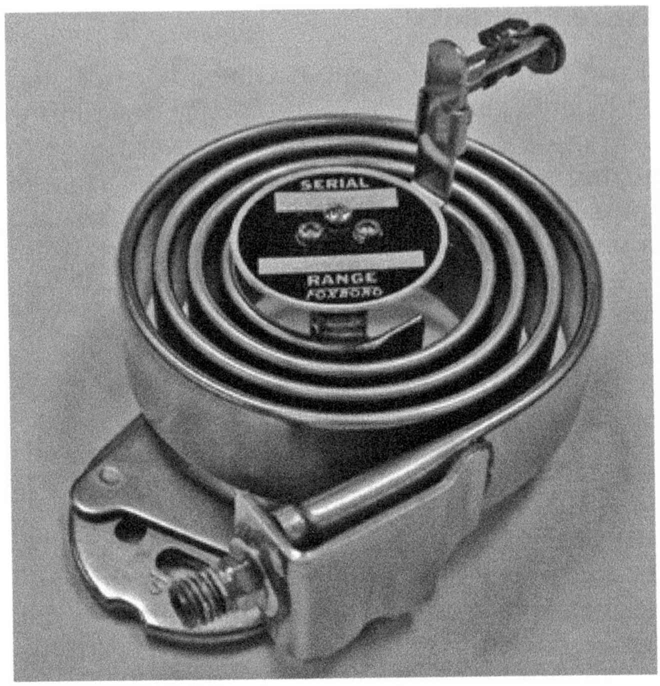

Figure 3–3 Spiral Pressure Element

Figure 3–4 Helical Pressure Elements

the inner surface, producing a force that causes the bellows to expand in length, thus producing movement at the free end. Because of the relatively large surface area compared to physical size, the movement-to-pressure ratio is greater than for the devices previously discussed. Bellows are more sensitive, very accurate, and used primarily for lower-pressure applications found in receiver instruments.

Figure 3–5 Bellows-Type Pressure Element

Diaphragms

Pressure sensors that depend on the deflection of an element have been used for many years. However, the overall performance of diaphragms has been improved, with less elastic hysteresis, friction, and drift and more linear response. They generally consist of a flexible disk with either flat or concentric corrugations constructed of beryllium copper, Ni-Span C®, Inconel®, stainless steel, or in some cases quartz. Sometimes two diaphragms are fused together to form a capsule. The capsule can be evacuated to form a vacuum and used as a reference for absolute measurement. Capsules and diaphragms can range in sizes from 1 to 6 in., with the larger sizes used in low-pressure applications.

Diaphragms used for very low pressure measurement ranges are usually made of soft, pliable material, either metallic or fibrous (Figure 3–6). This type of pressure element is sensitive to very little pressure and can be used to measure full-range pressure values as low as 5 to 10 inches of water. The increased sensitivity of diaphragm pressure elements is attributable to the large surface area of the element. It should be remembered from previous discussions that force is equal to pressure times area ($F = PA$). The force exerted by a pressure

Figure 3–6 Diaphragm Pressure Element

element is therefore directly proportional to the area of the device over which the pressure is exerted. For this reason, high pressure can be measured by pressure elements with relatively small surface areas. Such pressure elements are characterized by the small amount of area exposed to the process pressure. Diaphragms, on other hand, regardless of physical configuration, can be characterized by the relatively large surface area exposed to the pressure to be measured. The most extensive use of diaphragm pressure elements is in pressure gages and pressure transmitters.

The pressure elements discussed in this section convert the applied pressure acting over the surface area of the element to create a force or movement and are the primary transducers in pressure measuring devices. A secondary transducer converts the force or motion of the primary transducer to an indication when used in pressure indicators and a scaled signal for transmission in the case of transmitters.

Pressure Gages

Although unsophisticated in nature and used only for local indications, industrial-quality pressure gages are employed in more pressure measurement applications than any other device. Industrial pressure gages range in size, usually specified by case diameter, from about 1-1/2 in. to 12 in. Most, however, are 2-1/2 in., 4 in., or 6 in. They are used to measure pressures from a few inches of water to several thousand pounds per square inch. This wide application range is made possible by the specific physical design of the C-tube and in general the pressure element. By altering the area over which the process pressure is applied, the ratio of pressure to movement or force can be changed. For example, pressure for pressure, a wide thin tube with a relatively large area and less resistance to movement will produce more motion at the free end than one with a smaller surface area, and can be used for lower range applications. It should be noted, however, that such devices are more susceptible to damage caused by overrange pressure and should be equipped with overrange protection. Pressure elements with reduced area and more rigid material can be used for higher range applications.

The pressure gage shown in Figure 3–2 utilizes a C-tube pressure element as the transducer. The more detailed illustration in Figure 3–7 aids in understanding the operation of a C-tube gage. Pressure applied to the open end of the C-tube causes a mechanical deformation of the physical structure—the tube expands. With one end fixed so that no movement can occur, the tube tries to straighten, causing a movement at the free end. This movement is transferred through the adjustable link and sector to the pin and gear, causing the gear to rotate. A pointer attached to the gear by a pointer shaft rotates. The end of the pointer traverses an arc along a calibrated circular scale (Figure 3–8). The relative position of the rotating pointer on the scale is a direct result of the amount of deformation of the tube caused by the applied pressure.

Pressure gages and most pressure instruments used for local indication are of the motion balance type. From a steady-state condition that exists when the measured pressure is constant and no movement of the pointer occurs, a change in pressure will cause the free end of the pressure element to move. This changes the position of the sector and pointer, the motion of which will continue until the spring tension on the C-tube is at equilibrium with the force caused by the pressure inside the tube.

Motion balance describes the means by which the motion of the pressure element (primary transducer) moves the sector (secondary transducer), mechanically transmitting a rotary motion to the pinion assembly and indicator. The output motion from the pressure element is usually small, a few degrees for angular movement or limited to a few tenths of an inch for linear-movement devices. This is normally sufficient for most local indicators or recorders. However, when considerable friction or load is imposed on the primary transducer by the secondary transducer, the accuracy and sensitivity of the instrument are compromised. For sustained accuracy and precision, the

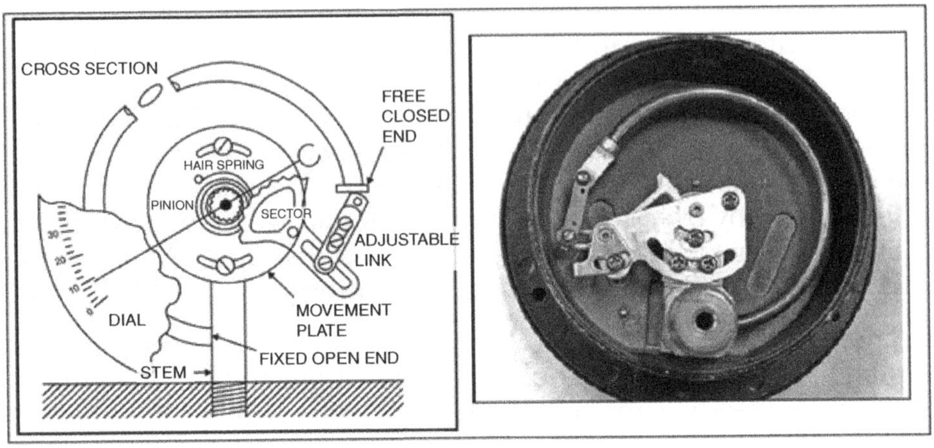

Figure 3–7 Bourdon Pressure Gage

Figure 3–8 Pressure Gage

secondary transducer must have linear response with limited hysteresis and friction effects. Overall accuracy of motion balance devices can be on the order of 0.05 to 0.1% FS (full scale), but is more commonly in the range of 0.1 to 0.5%.

The relationship between the pressure applied to the pressure element and the position of the pointer is established during the calibration of the gage. This general procedure consists of causing the output of the device (the position of the pointer) to correspond to an input. The input is the pressure on the tube. Calibration procedures were covered in Chapter 2.

Other Designs

The Bourdon C-tube type of gage described in the previous discussion is the oldest and most common technology used in the manufacture of pressure gages. However, other designs are gaining in prominence. The gage illustrated in Figure 3–9 is a direct-drive device that utilizes a helically wound Bourdon tube. The tube is permanently bonded to a calibration clip positioned to provide an arc of 270° at full-scale deflection. The pointer shaft is also permanently bonded to the clip, and the pointer is a press-fit assembly on the shaft. The rotation of the pointer shaft on synthetic jeweled bearings is the only motion in the gage. The direct-drive principle eliminates the major cause of wear and gage failure. The absence of springs, gears, sectors, links, and levers reduces the problem of nonlinearity and reduces calibration requirements.

The direct-drive mechanism also enables the gage to operate under harsh environments of mechanical vibrations and high pulsations. The specified accuracy of this type of gage is listed in three basic grades: 0.25% test gage, 0.5% test gage, and process gages for 0.5% midscale and 1% full-scale accuracy; overrange protection to 150% of full scale without quality degradation is provided. The sensitivity and repeatability of the gage are stated to be ±0.025% of full scale. With the shaft and pointer as the only moving parts, the reduction in calibration drift caused by wear eliminates the need for regular recalibration. The lack of levers and connecting links results in built-in linearity, eliminating the need for link and lever calibration procedures and reducing error caused by hysteresis. Zero adjustments are made by rotating the dial by a screw adjustment on the front of the gage.

The smaller size of the helix sensor can make it more susceptible to plugging if the process is not relatively clean. Most of these gages come with a built-in filter. In severe process conditions, an isolating diaphragm can be used.

Other specifications of this direct-drive helically wound Bourdon tube gage include pressure ranges from 0 to 30 psi to 0 to 20,000 psi in sizes of 2-1/2, 4-1/2, 6, 8-1/2, and 12 in. dial sizes and compound ranges from 30 in. Hg to 30 psi, to 30 in. Hg to 300 psi. Standard connection sizes and mounting configurations are available.

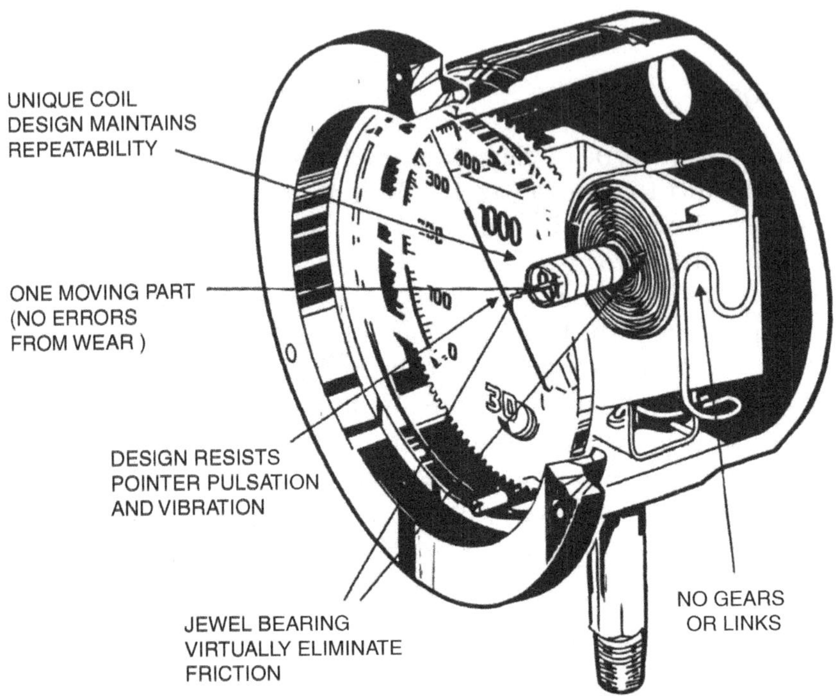

Figure 3–9 Direct Drive Helical Bourdon (*Courtesy of 3-D Instruments, Inc.*)

Another type of gage uses a capsule diaphragm pressure element, constructed from two thin, concentrically corrugated plates. The pressure to be measured is applied to the space between the two metal plates. Force on the plates resulting from the applied pressure causes them to flex or bulge inward or outward, depending on the relationship between the pressure on the measure side and on the reference side of the capsule element. When atmospheric pressure is on the outside of the element, measured pressure is referenced to atmospheric pressure and the pressure scale is gage pressure. If the area within the volume chamber is at a vacuum (a volume with the absence of pressure), however, the measured pressure is referenced to this vacuum and is absolute pressure.

Because of the relatively large surface area of the element exposed to the measured pressure, a relatively small amount of pressure is needed to create a force to move the measuring element and central pointer.

This type of gage can be used for pressure ranges from about 0.5 in. H_2O to a few pounds per square inch. A pressure gage using a diaphragm element is shown in Figure 3–10.

Zero Shift

The fixed-end tube used in most pressure gages is usually fused, soldered, or welded to a hollow pipe with a standard pipe thread that is used to attach the gage to the process. The pipe thread is generally 1/4 in. or 1/2 in. male National Pipe Thread (NPT). Since the gage is used for local indication only, it must be mounted at an accessible location. This sometimes presents a problem. For example, if pressure is being measured at the top of a steam drum, the pressure gage will usually be located near ground level so that the operator will not have to ascend stairs or a ladder to the top of the drum to read the value on the gage. The steam in the line connecting the gage to the drum, being exposed to ambient temperatures, will condense to water. The water will then cause an additional pressure to be exerted on the gage. This, you will recall, is pressure caused by hydrostatic head.

Figure 3–10 Diaphragm Element in Pressure Gage

To eliminate the error in measurement caused by the pressure exerted by the column of water, the reference of measurement (or zero reference point) is shifted. The gage is adjusted so that it will read zero pressure with the pressure caused by the column of liquid applied to the gage. The point of zero pressure measurement is thereby lowered an amount equal to the hydrostatic head pressure caused by the column of water in the pipe connecting the gage to the steam drum.

The mechanical pressure gages described are very reliable and have served the industry well for many years, but the limited accuracy can result in limited applications. The pressure gage shown in Figure 3–11 is one of the latest innovations in pressure indication. The digital display is protected by a hard polycarbonate lens, the circuitry is mounted in elastomer for shock resistance, and the case and sensor are stainless steel which offers protection to the circuits from electromagnetic interference. This gage has excellent overrange protection without changing calibration. The most striking advantage of the digital gage, however, is the high accuracy of ±0.01% of reading over the upper 80% of scale and ±0.02% of full scale over the lower 20% of scale. Calibrated full range values can be set from 15 psi to 10,000 psi. A five digit display offers excellent resolution. It is battery-powered with 1,500 hours battery life and is designed for intrinsically safe applications.

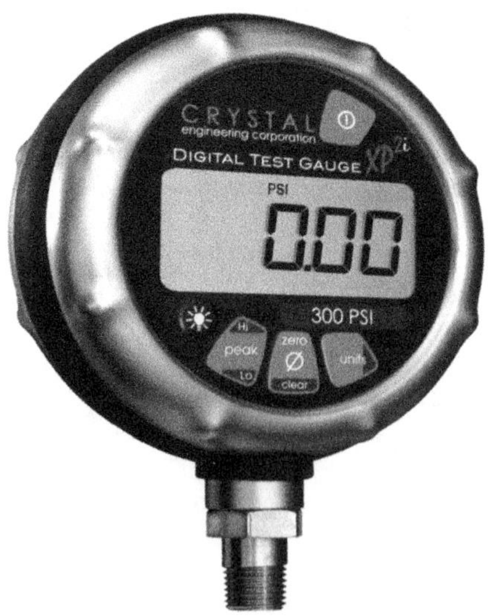

Figure 3–11 A Digital Pressure Gage (*Courtesy of Crystal Engineering Corporation*)

EXAMPLE 3-1

Problem: A pressure gage is mounted 30 ft below the point on a steam drum where a pressure measurement is required. Calculate the amount of zero shift required to compensate for the error caused by the 30-ft hydrostatic head.

Solution: From Equation 2.4,

$$1 \text{ ft } H_2O = 0.433 \text{ psi}$$

The pressure caused by a 30-ft column of water is

$$0.433 \text{ psi/ft } (30 \text{ ft.}) = 12.98 \text{ psi}$$

Rounding off to the least number of significant digits, the pressure is 13 psi.

To accommodate such a zero shift, most pressure-measuring devices are equipped with zero suppression or elevation kits. The terms suppression and elevation refer to a corrective offset in instrument zero to compensate for the zero offset caused by the instrument process connection. When a pressure gage is mounted as in Example 3-1, the static pressure caused by the hydrostatic head of the condensate in the connecting (impulse) line causes the zero reference point of the gage to be elevated above actual zero. This amount of pressure (12.98 psig in Example 3-1) must be suppressed so that the instrument reads zero pressure with the hydrostatic head of the impulse line applied. Any pressure indication on the gage is then caused by the pressure in the steam drum.

Zero elevation is required to raise the instrument zero point when process fluid in connecting lines causes the instrument reading to go below zero. Such would be the case when hydrostatic head pressure is applied to the low side of a differential-pressure instrument. Suppression and elevation are covered in more detail later in the chapter.

It should be noted that the amount of zero offset, suppression, or elevation required is not only a function of the vertical displacement between the measuring device and the process, but also depends on the specific gravity of the process fluid. This is pointed out by Equation 2.7.

Vibration and Pulsation

Certain mounting considerations and precautions should be taken into account to prevent damage to, and excessive wear on, the pressure-measuring device. It is sometimes necessary to mount pressure gages at points where excessive mechanical vibration exists. This vibration, when transferred to the pressure gage, can cause damage to the instrument. Special types of vibration dampers are available and should be used in such applications.

When the process temperature at the point of pressure measurement is high enough to cause damage to the gage, heat sinks or insulating devices should be used.

When the process pressure at the point of measurement is pulsating, such as would occur at the discharge of a positive-displacement pump, pulsation dampers should be used. Pulsation dampers (or snubbers, as they are sometimes called) are small orifices or flow restrictors placed at the point of process inlet to the pressure gage. This prevents the pressure gage from responding to sudden changes in the process pressure. The gage reading then represents an average value instead of an instantaneous value. This average value is normally sufficient, because the sudden changes of pressure are soon equalized by the resistive and capacitive elements of the process.

When using pulsation dampers, care must be taken not to damp the pressure changes to such a point that the desired process pressure measurement cannot be achieved. Snubbers, having small passages, may be susceptible to plugging.

When desired or required, damping and isolation can be used, but performance curtailment is to be expected. For pressure ranges of 30 psi and lower, 0.5 to 2% accuracy should be added to the gage specifications. For pressure ranges of 30 psi and above, isolators have little effect on gage accuracy.

Gages with direct drive helix sensors are more immune to damage caused by pressure pulsations.

As we have seen, because of their mechanical arrangement the pressure-to-motion transduction of energy supplied by most gages causes them to be susceptible to damage and excessive wear under severe to extreme service conditions. Such conditions include vibration, pulsating pressure, and harsh media. Table 3–2 lists measurement applications for various industries, with notes for specific recommendations.

Table 3–2 Application Notes for Pressure Gages
(*Courtesy of 3-D Instruments*)

Key Application	Notes
Power Generation Industry Power plants often offer an excellent potential for gage applications. These plants have their own small instrument shops and many gage problems. The *Directory of Electric Utilities* published by McGraw-Hill lists all power plants by city, size, and type. In nuclear plants, it is very expensive to replace gages inside the containment boundary. Gages used inside this boundary may require special rating and approval.	
Process gages	
Chemical additive pumps for water treatment	Chemical additive pumps are usually reciprocating at up to 120 strokes/min.
Compressors	
Fly-ash recirculating pumps (coal)	Fly-ash systems have a notorious "hammer" that destroys gages.
High-pressure boiler water feed pumps	High pressure feed pumps are know for vibration and pulsation.
Pulverizers—bail mill (coal)	
Raw water intake pumps	
Turbine gland seal	Turbines, often running at 3600 rpm, create significant vibration.
Turbine inlet/output	
Water reclamation pumps	
Test Gages	
Calibration in instrument shop	
Field instrument calibration	
Heart rate (efficiency) testing	
Hydrostatic testing group (nuclear)	Hydrostatic testing group often uses 1/2% test gages in large quantities.
Accessories	
Hand pump with test gages	Hand pump with test gages used for field instrument testing.
Isolators	Internal isolator often used in fly-ash recovery systems, with No. 6 oil and with river water. Also provides inexpensive "freeze protection."

Table 3–2 Application Notes for Pressure Gages (*Courtesy of 3-D Instruments*) (Continued)

Key Application	Notes
Maximum pressure pointers	Maximum pressure pointer provides indication for relief valve check and calibration.
Petroleum and Chemical Processing Industries Petroleum refining, petrochemical and chemical processing applications often involve corrosive streams and high temperatures. Safety is usually a major concern. Most plants also have their own powerhouse and water and wastewater treatment facilities.	
Process gages	
Additive pumps	
ammonia environment	Stainless steel and inconel have excellent corrosion resistance.
Blending systems	
Compressors	Multistage compressors have surge problems as well as high vibration.
Filing and packaging operations	Rapid value closures cause significant pressure cycles.
Meeting pumps	
Near quarter-turn valves or solenoid valves	
Process stream lines	
Product Pumps	Pumps often punish gages on the discharge side.
Receiver gages for pneumatic control system	True zero reading on receiver gages indicates when system is shut down.
Recirculating pumps	
Stream-driven equipment	
Test gages	
Calibration in instrument shop	
Field instrument calibration	Field calibration checks versus bench operation saves time.
Meter testing	
Pressure switch testing/setting	

Table 3–2 Application Notes for Pressure Gages (*Courtesy of 3-D Instruments*) (Continued)

Key Application	Notes
Reid vapor test	
Relief valve testing	
Accessories	
Hand pump with test gages (and carrying cases)	Hand pumps with test gages used to check static element in differential-pressure flow instrumentation. Hand pumps and test gages also used to check and set relief valves—often with a maximum pressure indicator in the gage.
Isolators	Internal isolators are less expensive than standard C-shape gages with diaphragm seals—particularly useful in suspensions, light slurries, high-viscosity fluids, and corrosive service. Also provides inexpensive "freeze protection."
Oil and Gas Production Industry The oil and gas production industry often requires high-pressure gages of quality materials of construction to handle corrosive environments, and gages that can handle pulsation or vibration while maintaining repeatability. At the well head and in sour gas service, the National Association of Corrosion Engineers (NACE) specifies materials of construction for instrumentation.	
Process gages	
Christmas-tree well-head installations	Stainless steel case is often used in offshore applications.
Compressors	
Fire protection systems	
Pressure reduction chokes	
Product pumps	
Receiver gages for pneumatic control system	
Salt water disposal pump discharge	

Table 3–2 Application Notes for Pressure Gages (*Courtesy of 3-D Instruments*) (Continued)

Key Application	Notes
Water/steam injection pumps	Sour oil and gas (i.e., sulfur content) attacks steel and copper. All stainless and inconel construction is useful under these conditions. NACE certification is also available where required.
Well pumps	
Test gages	
Calibration in instrument shop	
Meter testing	
Reid vapor test	
Well-head testing and measurement	Precision gage is accepted in lieu of a deadweight tester by some regulatory agencies.
	High-pressure ranges (e.g., 20,000 psi) are sometimes needed.
Accessories	
Hand pumps with test gages	Hand pumps with test gages used to check static element in differential-pressure flow instrumentation. Hand pumps and test gages also used to check and set relief valves. Internal isolators are often used in corrosive service and are less expensive than standard gages with diaphragm seals.
Natural Gas Transmission and Distribution Industry	
Process	
Compressors (gas plants)	Suction, discharge, and recycle pressures on compressors.
Cooling water pumps	
Distribution line pressure	
Expander drier discharge	
Odorant injection pump	

Table 3–2 Application Notes for Pressure Gages (*Courtesy of 3-D Instruments*) (Continued)

Key Application	Notes
Propane storage cavern (brine and propane pumps)	
Turbine intermediate stage	
Test gages	
Calibration in instrument shop	Precision gage is accepted in lieu of deadweight tester by some regulatory agencies.
Measurement of static element at online plate differential	
Measurement of static pressure at pressure regulating/reducing valve	
Reid vapor test	
Storage cavern testing	
Accessories	
Hand pumps with test gages (and carrying cases)	Hand pumps with test gages used to check and set relief valves.
Food, Beverage, and Pharmaceutical Industries In addition to typical gage requirements, food, beverage, and pharmaceutical firms also need gages and diaphragm seals that meet sanitary specifications. These requirements are generated by the U.S. Department of Agriculture (USDA) or by industry standards such as "3A" for milk producers. Food and beverage manufacturers usually also use unique stainless piping an fitting systems (such as Ladish Tri-Clamp, cherry Burrell, etc.) that require special diaphragm seals and connections. Stainless steel cases are usually standard in sanitary applications.	
Process gages	
Air compressors	
Centrifuges	
Clarifier discharge	
Continuous mixer and feed pressure	
Evaporator discharge	
Filing and packaging operations	

Table 3–2 Application Notes for Pressure Gages (*Courtesy of 3-D Instruments*) (Continued)

Key Application	Notes
Filter discharge	
Homogenizer feed pressure	Rapid homogenizer cycling usually represents a gage problem. Homogenizers are common in dairy operations, including milk and cheese.
Hydroflex	Hydroflex and retorts are similar to a pressure cooker. Pressure and temperature can be critical.
Ice cream freezer operating pressure	
Process steam lines	
Product mixers	
Receiver gages for pneumatic control	
Refrigeration systems	All dairies require refrigeration suing ammonia or freon compressors. Refrigerant gages are cost effective.
Sanitary pump intake/discharge	
Separator discharge	
"Sparger" pressure	
Table pressing pans	Many processes operate below atmospheric pressure. Vacuum or compound ranges will be used.
Test gages	
Calibration in instrument shop	
Can pressure testing	
Field calibration of pressure transmitters, switches, and relief valves	
Permanently mounted in process area for critical applications	Certain processes require high accuracy monitoring with FDA reliability. Product may be subject to recall if gage fails to holed accurcay.

Table 3–2 Application Notes for Pressure Gages (*Courtesy of 3-D Instruments*) (Continued)

Key Application	Notes
Accessories	
Color-coded cases	Case color coding may help prevent cross-contamination
Extended capillaries for pressure and temperature	
Hand pumps with test gages	
Homogenizer seals	Lower-volume devices mean less movement of the seal. This provides longer life. Seals must match the piping system.
Sanitary seals	Seals may be filled with several fluids, depending on user preference.
Steel Manufacturing Industry Steel mills have a large number of hydraulic operations. They also have their own power plant, instrument shop, water, and wastewater treatment facilities.	
Process gages	
Air compressors	
Cooling water systems	
Coating operations	
Hydraulic equipment	
Power plant	See "Power Generation Industry"
Press tonnage	
Process steam	
Pump discharges	
Sheet rolling operations	
Test gages	
Calibration in instrument shop	
Field calibration	
Accessories	
Hand pumps with test gages	

Table 3–2 Application Notes for Pressure Gages (*Courtesy of 3-D Instruments*) (Continued)

Key Application	Notes
Pulp and Paper Industry Pulp and paper mills handle slurries, have hydraulic operations, perform chemical treatments, and often have their own power plant. These plants also use very large pumps at high velocity, thus offering high-vibration applications for gages.	
Process gages	
Air compressors	
Calendar and drying	
Chemical additive and metering pumps	
Digesters	
Hydraulic shearing equipment	
Mill water systems	
Paper presses	
Power plant	See "Power Generation Industry"
Process steam	
Steam-powered process equipment	
Transfer pumps	
Vacuum boxes	
Water intake pumps	
Waste treatment	Waste treatment and chemical additive pumps are usually reciprocating measurement pumps.
"White water" slurry	
Test gages	
Calibration in instrument shop	
Field calibration	
Accessories	
Hand pump with test gages	

Table 3–2 Application Notes for Pressure Gages (*Courtesy of 3-D Instruments*) (Continued)

Key Application	Notes
Isolators and diaphragm seals	Internal isolators are used in slurry applications. Also offer inexpensive freezing protection. Clay starch, and titanium dioxide slurries (among others) require diaphragm isolators. White water and pulp slurries will require flush-mounted diaphragms.
Water and Wastewater Treatment Industry Municipal water and wastewater treatment plants pump large volumes. Adding treatment chemicals and processing slurries can create gage problems. These plants also often have corrosion problems with hydrogen sulfide and chlorine.	
Process gages	
Booster pumps suction and discharge	Starting and stopping booster pumps generates large surges that destroy gages.
Chemical additive pumps	Additive metering is often done with a reciprocating pump.
Chlorination system	Cannot use liquid fill gages on oxidizing systems such as chlorine.
Cogeneration turbines	
Dewatering operation	
Inlet/outlet of control valves, relief valves, reducing valves	Pressure-reducing valves often generate high frequency and vibration. Pressure-relief valves can cavitate, causing rapid pressure pulsation.
Sludge pumps	Pumping slurries usually create pressure pulsation and "hammer" that rapidly deteriorate C-shape gages.
Test gages	
Calibration in instrument shops	
Accessories	

Table 3–2 Application Notes for Pressure Gages (*Courtesy of 3-D Instruments*) (Continued)

Key Application	Notes
Hand pumps with test gage isolators and diaphragm seals	Isolators are often used in slurry applications and also offer excellent freeze protection.
Tire and Rubber Industry Tire building and rubber manufacturing industries have a large number of pressure applications. They also have a variety of gage problems.	
Process gages	
Air compressors	
Cooling water pumps	
Fire protection system	
Hydraulic presses	
Injection molding presses	
Plant air system	
Power plants	See "Power Generation Industry"
Process steam	
Rubber molding operations	
Slurry pumps	Pumps are subject to pulsation and "hammer" due to the batch nature of the processes.
Tire presses—air and steam	Steam and air presses experience shock and pressure cycling as presses are turned off and on.
Test gages	
Field and bench calibration	
Quality control checks for hydrostatic pressure, safe burst pressure, and accurate inflation	Ask for the Quality Control Manager as well as Maintenance and Process Engineering
Relief valve checks	
Accessories	
Hand pump with test gages	
Vehicle Manufacturing Industry	
Process gages	
Air compressors	

Table 3–2 Application Notes for Pressure Gages (*Courtesy of 3-D Instruments*) (Continued)

Key Application	Notes
General machinery hydraulics	
Motor testing	
Paint and coating systems	
Plant air	
Power plant	See "Power Generation Industry"
Press tonnage	
Process steam	
Robot hydraulic control operations	
Test gages	
Calibration in instrument shop	
Accessories	
Hand pump with test gage	

Temperature

At times it may become necessary to measure process fluids with temperatures beyond the limits specified by the gage manufacturer. For such applications, the gage can be isolated from the process by a short length of pipe or tubing. For steam service a gage siphon or pigtail system can be used to meet the pressure and temperature requirements of the gage.

Curves provided by many manufacturers can be used to specify a protective tubing length for a specific application. The temperature limits at the gage are normally from 0 to 200°F, while process temperature can range from –400 to 1700°F. The following assumptions are normally made when using such curves:

- The pressure vessel is insulated to limit radiant heat transfer to the gage. The major source of thermal energy to the gage is through the connecting tubing.
- The pressure medium has a coefficient of thermal conductivity less than 0.4 BTU/ft/hr/°F. This specification falls within the parameters of most applications.

- The nominal ambient temperature around the pressure gage is below 100°F.
- The convection heat transfer rate from the tubing to still air is 1.44 BTU/ft/hr/°F.

Deviation from these conditions will require slight modifications to the required tubing length. When in doubt, simply select a longer length of tubing.

Some gage manufacturers list specifications that account for and correct error in measurement caused by temperature extremes at the gage body. Tests performed by one manufacturer consisted of selecting several gages of normal production quality that were cooled and heated from –65 to 190°F. At least five upstream and downstream measurements were made with deadweight testers at several test temperatures within the temperature range tested.

The recorded data were reduced by multiple linear regressions to arrive at a correction formula as a function of temperature. An error analysis was performed to determine confidence limits, and a large number of observed data points made it possible to assume normal distribution of data. From the analysis it was determined that the error for data points within one standard deviation from the expected value is 0.31% FS over the temperature range tested. The results are expressed by the following terms:

$$\text{True pressure } (P_T) = \text{Pressure reading } (P_R)(q) \tag{3.1}$$

where $q = 1 - f(\mathrm{T} - 78°\mathrm{F})$, and $f = 0.000158$ between –65 and 120°F and 0.000157 between 121 and 190°F.

From this it can be seen that gages will read lower at lower temperatures and higher at higher temperatures. A random zero shift occurred at each temperature extreme for Inconel X-750, which was the element material. Because of this zero shift, a zero adjustment should be made at the design operating temperature.

The tests also revealed that coil deflection is a linear function of the modulus of elasticity. Over narrow temperature ranges, this is a nearly linear function of temperature, with a slope of approximately 0.000152/°F that corresponds to q.

Exposure of the Pressure Element to Process Fluids

The specific design of a pressure gage and its construction largely depend on the particular manufacturer and may not be an important criterion in the selection of a pressure gage. The materials used in the pressure element, however, may be a consideration. Construction materials depend on the process fluid and will normally be:

- Stainless steel
- Phosphor bronze
- Monel®
- Beryllium copper

When the process fluid is such that degradation of the wetted gage parts will result, more exotic materials will be required for gage construction. Diaphragm separators (also called diaphragm seals) that prohibit physical contact of the process and gage element may be used instead. (Note that the highly viscous fluids used in separators can cause material buildup and plugging in some applications, but may be acceptable in other cases.) The wetted portion of the separator may be constructed of [Ref 1]:

- Tantalum®
- Titanium
- Hastalloy®
- Monel®
- Nickel
- Kynar®
- Teflon®
- Polyvinyl chloride
- Viton®

A diaphragm separator is illustrated in Figure 3–12.

In the construction of a diaphragm separator, the instrument pressure element, connection, and all space above the diaphragm are evacuated and completely filled with a degassed liquid that transmits the process

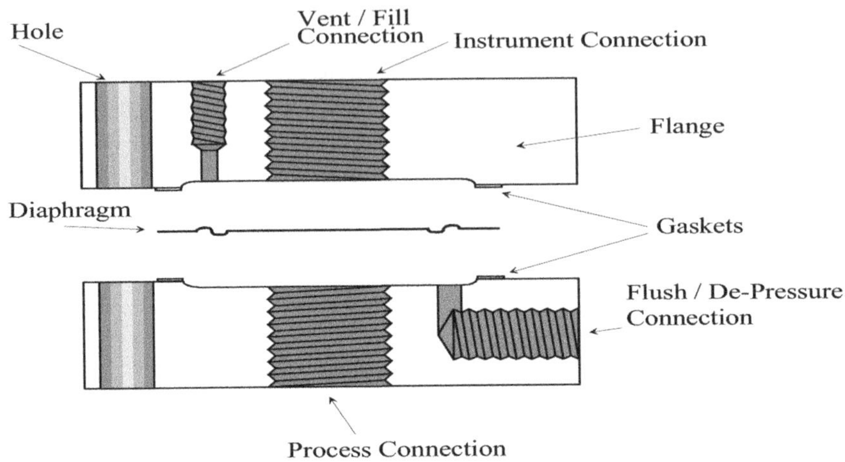

Figure 3–12 Design of Diaphragm Separator (Diaphragm Seal)

pressure to the measuring element by movement of the diaphragm. It is important to realize that any loss of liquid will curtail the transfer of motion from the diaphragm to the measuring element. The diaphragm seal, pressure element, and gage assembly are ordered as a complete unit that has been factory calibrated and tested.

The use of diaphragm seals for vacuum measurement is not recommended.

When properly applied and installed, diaphragm seals can provide the same accuracy as specified for gages without the separator. The chance for error caused by hysteretic and temperature variations is generally more prominent in separator-type gages. Indirect pressure measurement with separators conforms to the hydraulic principles of fluids in a fixed volume (Pascal's law). The pressure chamber of the separator is completely filled with the transmission fluid, and is closed at one end by the pressure element, and at the other end by the elastic diaphragm on the sensor body.

When the process fluid pressure is applied to the process side of the diaphragm, the fluid is displaced from the sensor body to the free end of the pressure element, resulting in motion or force at the free end. The element that separates the process from the fill fluid is normally a thin, concentrically corrugated metal plate (capsule diaphragm) that is permanently attached to the pressure element.

Special precautions should be taken when using filled systems. Changes in ambient temperature will result in volume changes of the fluid, causing the indicated pressure to change. When the element is connected to the gage by a capillary tube, elevation of the element above the gage will result in an error caused by the weight of the fluid acting on the gage element. The gage should be zero adjusted at both the operating temperature and mounting arrangements.

Designs for Specific Applications

Most applications for pressure gages are of a standard nature; in such cases, off the shelf varieties of gages can be used, especially for air, steam, and water. It is important to note, however, that other specific applications require special consideration.

Where mechanical and pressure pulsations would result in extreme gage wear, the gage case may be filled with a liquid to dampen the movement within the gage. The fluid used in liquid-filled gages is clear and inert, with moderate and constant viscosity. Typical fill liquids are glycerin and silicon. These liquids, when combined with strong oxidizing agents such as chlorine, nitric acid, and hydrogen peroxide, can result in a spontaneous chemical reaction, ignition, or explosion, which can cause personal injury and property damage. If gages are to be used in such service, silicon or glycerin should not be used.

Gages are specifically designed for particular applications, including refrigeration service, low-pressure and hot water heating, ammonia, oxygen, boilers, sprinkler systems, and industries such as food, chemical, petrochemical, and others. A primary consideration, then, is the medium to be measured. A corrosion and application chart provided by manufacturers will help in selecting a gage material for a specific application. Some of the criteria for gage selection are:

Characteristics of Process Medium

- Form (gas, liquid, liquid with solids)
- Temperature
- Composition
- Consistency
- pH

- Abrasiveness
- Explosiveness
- Viscosity

Force and Range of Measurement

- Pressure
- Vacuum
- Lowest pressure that must be measured
- Normal measurement range (could be about 50% of the maximum or represent a "12 o'clock" position on the dial)
- Both vacuum and positive gage pressure (compound)
- Maximum measurement that could occur

Rate of Change of Measurement

- Steady
- Varying slowly
- Pulsating violently

The characteristics of the medium will determine the material of the wetted parts, the force will govern the type of element, and the operating range will determine the element strength and will also influence the scale and graduations. The rate of change of measurement will affect the wear on the gage and specify the accessories or type of sensor needed.

The atmosphere and environment on the outside of the gage also have a great deal to do with the gage bezel, case, window, and connection. Most gages are connected to the process by the stem or by a short length of pipe or tubing in the upright position. The above considerations will govern the type of connection, usually 1/4 or 1/2 in. FPT or NPT. The choice of case style and the connection location will be functions of the intended installation.

Figure 3–13 shows various gage connection positions. It should be noted that gages are sensitive to their position. The normal position

(NP) of the gage refers to the position of the dial relative to the horizontal, without taking into account the position and direction of the connection to the movement. Gages should be calibrated in their normal position. When the position is changed from that in which the gage was calibrated, errors can be introduced in addition to those specified under normal operational error limits.

If the gage is to be used as a receiver instrument, it will probably be panel mounted with a colored finish case and matching color of case and dial that conform to other associated instruments.

The specific application will also determine factors such as dial size, type of scale, and overall accuracy. The dial size will be influenced by where and how the gage will be used. This will be a function of the normal distance between the gage and the operator. More important, however, the dial size will be a function of the overall accuracy precision, and resolution required. As these factors increase, the dial size normally will be larger. Test gages will have a narrowed scale to reduce error caused by parallax. Some gages will have a cross line at the zero point whose length indicates the permissible operational error over the entire scale range of 270°, which is determined from the error limits.

Another gage classification is that of standards and accuracy. The two most prominent standard classifications are ANSI B40 1-1985 and DIN, which are referenced to numerous standards. While DIN (German) standards for pressure gages are generally more comprehensive than ANSI (American) standards, both serve much the same purpose. The major differences between DIN and ANSI standards with respect to pressure gages deal with sizing and accuracy.

Gage Size Selection

ANSI standards use dial diameter to determine the gage size, whereas DIN standards use the outer case diameter. Table 3–3 helps to clarify this. Unless considered in panel mounting situations, these differences may have little significance.

Accuracy Standards

ANSI accuracy may frequently be stated as 3-2-3 or 2-1-2. This refers to the full-scale accuracy of the gage and is always expressed as plus or minus (±). In the first example, the gage will have accuracy of ±2% FS

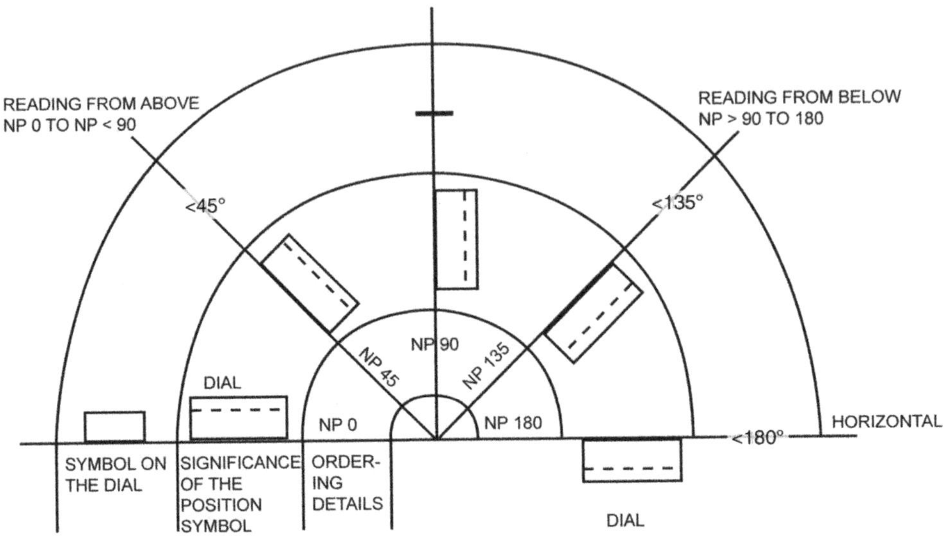

Figure 3–13 Various Gage Positions (*Courtesy of Haenni Instruments*)

Table 3–3 Comparison Between ANSI and DIN Gage Size Standards

ANSI Standard, in.	DIN Standard, mm
1.5	40
2	50
2.5	62
3.5	100
4.5	100
6	160
8.5	200
10	200

at any point in the middle half of the scale. The second example states that the gage will have an accuracy of ±1% FS at any point in the middle of the scale. In both instances, the first number refers to the first quarter of the scale and the last number refers to the last quarter.

DIN accuracy is usually expressed in full-scale statements. Segmented accuracy statements are not frequently used in the definitions of statements regarding accuracy and should not be confused. These terms are defined in Appendix A.

EXAMPLE 3-2

Problem: A pressure gage indicates 260 psig when 250 psig is applied. Find the pressure error and the accuracy expressed in percent of full scale if the range of the gage is 0 to 500 psig and the formula $x/y\,(100) = z$ expresses the relationship of the given terms.

Solution:

$$x = \text{error in pressure indication} = 260 - 250\text{psig} = 10\text{psig}$$
$$y = \text{full-scale indication of the gage} = 500\text{psig}$$
$$= \frac{x}{y}(100) = z$$
$$= \frac{10}{500}(100) = +2\%$$

Likewise, if the indication of the gage is 240 psig when an actual pressure of 250 psig is applied, the error in pressure indication is –10 psig and the accuracy is –2%.

Gage Classification by Accuracy

Pressure gages are often classified by letter designation to specify their accuracy. The following designations are used.

Grade AAAA Precision Test Gages

This gage classification is defined as that wherein the error in pressure indication at any point of scale is not to exceed ±0.1% FS. Such gages are designed for use as pressure transfer standards and for applications requiring exceptional precision and high reliability in the measurement of pressure. These gages are frequently used in research laboratories, metrology laboratories, and as a standard to check less accurate gages.

Grade AAA Test Gages

This gage classification is commonly used for test gages. The error in pressure indication at any point of scale is not to exceed ±l/4% FS. These gages are frequently used in instrument shops, gage repair and calibration shops, and other testing applications where accuracy is important but the greater accuracy of a Grade AAAA gage is not required.

Grade AA Gages

This gage classification is frequently used in monitoring the pressure in an important process. It can also be applied to test gages. Its error in pressure indication at any point of scale is not to exceed ±l/2% FS.

> Note: Classification of a gage by ANSI or DIN standard accuracy may affect other gage specifications. For example, grade AA gages are not generally offered in dial sizes below 4-1/2" diameter, because it is difficult to read this degree of accuracy on a dial any smaller than 4-1/2". In addition, manufacture of gages to this high degree of accuracy is much more difficult as the gage size decreases.

Grade A Gages

In this classification, the error in pressure indication is not to exceed ±l% of the scale range at any point within the middle half of the scale. In the first and last quarters of the scale, an error of ±1-1/2% is permitted. The middle portion of the scale is often referred to as the working portion, because it is in this portion that pressure indications are most frequently made. If you know this will be true, and ±1% accuracy is sufficient for the application (as it often is), you need not specify a more accurate gage. These gages are used in many applications and represent a good compromise between high accuracy and initial cost.

Grade B Gages

In this classification, the error in pressure indication is not to exceed ±2% of scale in the middle half of the scale, with a ±3% error in the first and last quarters of the scale allowed. Most of these gages are commonly referred to as "throw away" gages, as they are not economically repaired and recalibrated. Their accuracy is generally adequate for many applications, and they can be made suitably rugged, as well as compact and economical.

Classification by Case Design and Mounting Arrangement

Drawn Case Gages

This classifies gages with "deep-drawn" cases, which are used in large quantities by equipment manufacturers. The cases are usually steel but also can be made of drawn stainless steel or brass.

Molded Case Gages

The cases of these gages are formed in a mold. The cases are usually made of phenolic (fiber-filled plastic resin), fiberglass-reinforced thermoplastic, or ABS (acrylonitrile-butadiene-styrene) resins. Such cases are frequently used for corrosion resistance.

Cast Case Gages

This type of case, made from cast iron, aluminum, or brass, is generally used for its rugged design features as well as its corrosion resistance and decorative value.

Safe Case Gages

This type of case includes a solid-front/blowout-back construction for increased operator safety should the sensing element rupture.

Plain Case

This is a simple enclosure to protect the sensing element and movement from external damage with a suitable window or lens. Lower mount (LM), center back mount (CBM), or lower back mount (LBM) connections provide the mounting.

Flush Mount

The case design includes a front flange or a narrow bezel with a U-clamp for panel mounting. Back-connected gages are usually used for flush mounting applications.

Surface Mount

Case designs similar to those listed above sometimes are adapted with a rear flange to allow wall or surface mounting. Care must be used to allow for a blowout-back feature on safe case gages when using the surface mounting method.

Classification by Field and Special Use

Commercial Gages

These are generally referred to as equipment gages. Although they may have to be ruggedly constructed, their service conditions are not expected to be severe. Designed for low unit cost without refinements, they are frequently referred to as "throw away" gages, since it is generally not economical to attempt repairs.

Industrial Gages

Industrial gages have heavy-duty sensing elements and case designs, and are typically used in more demanding applications or atmospheres. These gages usually have greater accuracy requirements than commercial gages.

Process Gages

These gages combine the heavy-duty operating and construction requirements of industrial gages with the more exacting accuracy and service-life demands of many process applications. The term "process gage" is frequently used to describe 4 in. phenolic or thermoplastic case gages.

Test Gages

Gages in this classification are generally used to check other gages for accuracy. In addition, they may be used in more exacting situations that demand increased accuracy. They are usually not used in continuous service applications, but rather as a check on another gage.

Ammonia Gages

Such gages are used for agricultural ammonia applications, which are corrosive to brass and aluminum wetted parts. Generally, an AA marking appears on the dial. Ammonia is also used as a refrigerant and it can be corrosive to brass and aluminum. Refrigerant ammonia gages may have additional scales on the dial to indicate ammonia temperature equivalent.

Refrigeration Gages

These are similar to ammonia gages, except that they generally operate with other refrigerant gases (sometimes grouped together and called Freon®). They have special leak requirements, and the dials usually include temperature equivalents in red. An external recalibrator on the dial face is another common feature.

Oxygen Gages

These gages must be manufactured free from grease, oil, or any other substance that reacts explosively with oxygen, and must have the inscription "Use No Oil" on the dial face. A safety device that is commonly referred to as a blowout back is also required on oxygen gages. They are also referred to as welding gages.

Hydraulic Gages

These gages are used in severe service, frequently at higher pressures. Liquid-filled gages are commonly used for this application, as they can better survive cycle requirements that include pulsation and hydraulic shock.

Boiler Gages

These pressure gages for use on home heating boilers include a bimetallic temperature gage.

Fire Truck Gages

Special gages that have been approved or listed by safety agencies for such applications are used for fire apparatus.

Sprinkler Gages

Special gages for sprinkler systems must have safety agency approvals.

Clean Air Gas Gages

These gages are manufactured to very exact cleanliness standards for semiconductor gas applications.

Retard Gages

These gages have a "retarded" or compressed upper scale on the dial.

Duplex Gages

Duplex gages use two independent sensing elements, each connected to a separate pressure source and pointer, and each displayed on the same dial. They are used to show the difference in pressure between two sources [Ref. 2].

Instrument Mounting

In pressure measurement and control systems, the means by which the instrument is interfaced with the process is a prime consideration in achieving reliable operation. The measuring device must come into physical contact with the process but in some instances, special precautions must be taken to prevent damage to the pressure element. The process temperature may be above the design limits of the instrument or the process material may be corrosive. Mechanical vibration or process pressure pulsation can also be detrimental to instrument reliability by causing excessive wear on and mechanical shock to the element and other moving parts. Isolation and bleed valves, seals and purges, as well as temperature insulation and heat tracing are options that must be exercised to maintain instrument reliability:

Zero Suppression and Elevation

When measuring pressures in vapor or liquid processes, the line connecting the process to the instrument generally will be filled with liquid. This is attributable to the likelihood of the vapor in the impulse line condensing into liquid at ambient temperatures in the case of vapor service, and process fluid entering and being trapped in the line for liquid service.

The liquid in the impulse line causes a static hydrostatic head, which results in a pressure applied to the measuring instrument. The total pressure on the instrument is the process pressure and the hydrostatic head. When the process pressure is zero, the head pressure in the impulse lines must be nullified to zero to prevent an error in measurement.

The zero adjustment of most instruments can be offset enough to make minor compensation for such error, but special zero elevation and suppression kits must be added to the instrument for zero shifts of greater than about 1 or 2% of the measuring range. Suppression and elevation kits vary with the design of the instrument for which they are intended, but normally consist of a spring arrangement that is used to add a bias effect to the zero adjustment. It serves simply to allow a greater zero adjustment.

When referring to zero suppression and elevation, it must be understood whether the terms apply to that zero offset caused by the process or to the instrument adjustment made to correct the offset. In this discussion, suppression and elevation describe the corrective instrument adjustment. When pressure caused by a liquid head in the impulse line causes the instrument reading to increase, this is known as elevated zero. The compensatory action taken by an instrument adjustment lowers the instrument's response by suppressing the instrument's output below the elevated zero reference point. Thus, an elevated zero requires zero suppression. Also, when a hydrostatic head in an impulse line is applied to the low side of a differential-pressure instrument, the instrument output signal is suppressed, requiring an elevation of the instrument's output to compensate for the suppressed zero. A suppressed zero, then, requires elevation.

Of the two corrective adjustments required, suppression is more prominent in pressure measurement because it is more common for the static hydrostatic head to increase the instrument reading than to cause a decrease in response. Zero elevation, necessary when the initial zero offset is below the actual zero reference point, would be needed in differential-pressure measurement when a hydrostatic head, called a wet leg, is applied to the low side, decreasing the instrument output. This is a fairly common situation in level measurement with differential-pressure instruments. Such applications will be covered in Chapter 6.

EXAMPLE 3-3

Problem: A pressure transmitter is located 28 ft below the bottom of a steam drum. The steam drum is 12 ft high, with the impulse line connected to the top of the drum. Find the amount of suppression required.

Solution: The impulse line connecting the steam drum to the pressure instrument is elevated 28 ft above the instrument and is filled with steam when the instrument initially is placed in service. Once the steam condenses, the instrument will have a 40-ft hydrostatic head pressure applied to the element. The actual pressure on the element is

$$0.433\frac{\text{psig}}{\text{ft}}(40\,\text{ft}) = 17.32\,\text{psig}$$

The instrument zero must be suppressed 17.32 psig to compensate for the hydrostatic head of the impulse line.

Impulse Lines

The arrangement of the impulse lines connecting the pressure instrument to the process depends on the specific application. For gas service, the line and instrument should be free of liquid. For liquid and vapor service, the lines should be filled with liquid. The reason for this stipulation is to keep a constant static pressure on the instrument. If the amount of liquid in a line changes, the resulting change in hydrostatic head on the pressure element will cause a zero shift. In high-pressure applications, a small amount of zero shift caused by variations in the amount of liquid in impulse lines may be insignificant. For low-range applications, however, the error may be considerable.

Figure 3–14 shows a typical application of a vertical impulse line connecting a pressure instrument to a pressure process. Air or other gases can be vented at point *D* as indicated. Entrapped gases in vertical lines can be vented easily or usually will seek a higher elevation and be returned to the process vessel.

When it is necessary to locate a pressure instrument at a point from the process where horizontal lines are required, care must be taken to prevent entrapped air or gases from accumulating in the lines. For this reason, the horizontal lines should be run at a gradual and continuous slope of at least 1 in. per foot. When initially placing the

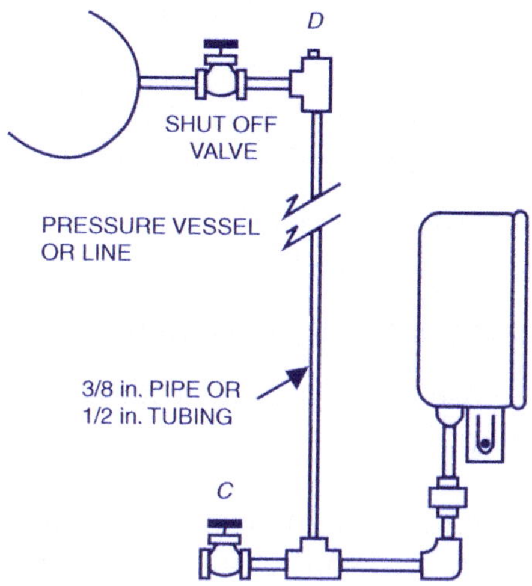

Figure 3–14 Installation of a Pressure Instrument

instrument in service, air and noncondensable gases should be vented through a valve provided for this purpose (e.g., valve *C* in Figure 3–14). There are five possible sources of error resulting from the use of impulse lines. They are: leaks, friction loss (more prominent when purging is used), trapped gases in a liquid line, and changes in density in liquid lines.

The best location for a transmitter with respect to the process being measured is dependent on the particular application, but it is important that the measurement taken should always reflect the actual condition of the process. The following guidelines are recommended:

- Keep the impulse lines as short as possible
- When a horizontal run is necessary, slope the impulse lines at least one inch per foot (8 centimeters per meter) upward from the transmitter to the process connection for liquid service, and slope the impulse lines downward at least 1 inch per foot horizontal run from the transmitter to the process connection.
- Avoid traps and high points for liquid fill.
- For liquid fill applications, the temperature must be constant and equal for differential pressure service.
- The impulse lines should be large enough to avoid friction loss, plugging or bridging by bubbles or liquid droplets.

- Liquid lines must be liquid full and gas lines must be liquid free.
- In steam and other high temperature service, avoid blow-down and purging through the transmitter.
- Prevent sediments from accumulating in the impulse lines. Do not make process connections on the bottom of pipes.
- Prevent freezing of the process fluid.
- Minimize the use of unions or other connections, to minimize potential leak points.

For applications using diaphragm seals to isolate the process from the measuring instrument, the following is recommended for optimum performance:

- The capillary length should be as short as possible.
- For vacuum applications, the transmitter should be mounted below the lower process connection for differential pressure measurement.
- Avoid mounting seals and capillaries in direct sunlight.
- When using two seals, the capillary lengths should be equal.
- Climate changes, season to season, require a re-zero adjustment to the instrument.
- Avoid temperature differences between the two capillaries when using two remote seals. Insulating the capillaries may minimize temperature extremes.

Seal Pots and Instrument Seals

As previously discussed, the instrument and connecting lead lines must be filled with liquid for steam and other vapors that will condense at normal ambient temperatures. Condensate chambers, commonly called seal pots, provide a large area of liquid contact between the process and the instrument.

To minimize the time required for the vapors to form condensate, water or other coolants can be applied to the seal pots to speed the condensing process. This procedure is often used in steam service applications. A seal pot (Figure 3–15) would replace the pipe tee at point *D* in Figure 3–14 for normal applications utilizing such devices.

Process fluid must sometimes be prevented from entering the pressure element of the measuring instrument. Such would be the case in applications where the fluid is highly viscous, causing plugging at lower temperatures. In other applications, the process fluid may be corrosive. In steam and water service, a seal fluid can be used to keep water out of the lines and instrument body to prevent freezing. The shutoff valve is closed for sealing applications; the seal pot, lines, and instrument are filled with a seal fluid by removing the top plug. It is preferred to fill the impulse lines with fluid by injecting the fluid from the bottom connection; this allows the gas in the impulse line to exit through the top vent valve and eliminates the risk of trapping gas in the impulse lines when filling from the top. The instrument should be zeroed with the lines full of the seal fluid, then elevation or suppression performed as required.

Another type of seal used to help isolate the instrument from adverse properties of the process fluid is shown in Figure 3–16. This device, called a pigtail, consists of a complete turn in a section of tubing or pipe, usually 1/4 in. ID, connecting the pressure instrument to the process. A pigtail is normally used with pressure gages to protect the device from high process temperatures. Steam and other vapors are cooled in the device, condense, and form a liquid trap when the liquid settles in the lower portion of the pigtail. When the system is shut down, the liquid forms a seal between the process and the measuring instrument, helping to absorb the pressure and temperature shock when the gage is put back in service. For ease of construction and installation, most pigtails are mounted in the vertical position. When mounted horizontally, as shown in Figure 3–16, mechanical shock and vibration are absorbed to an extent by the pigtail, protecting the instrument from possible damage and excessive wear.

Pulsation Dampers

When sudden changes occur in process pressure (e.g., on the discharge side of reciprocating pumps or compressors), it sometimes becomes necessary to shield the measuring instrument from excessive and repetitive movement that causes undue instrument wear. This is usually accomplished by placing a restriction in the instrument impulse line to reduce the response rate of the pressure measuring

Figure 3–15 Seal Pot

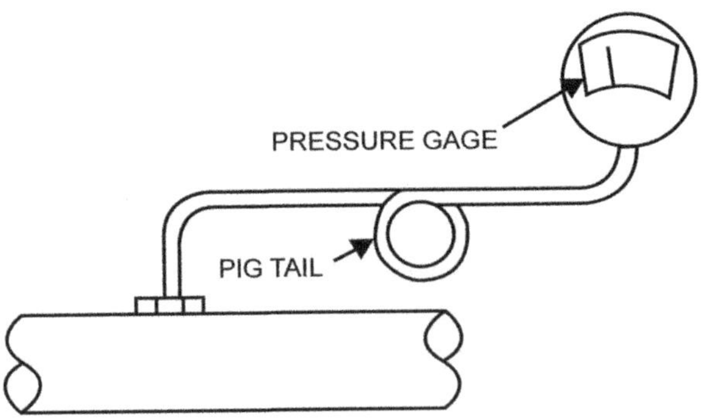

Figure 3–16 Pigtail

system. Any such retardation in the instrument's rate of response must be done with care. Instant measurement of any process response is desirable. In the case of pulsating pressure, an average pressure reading is the usual goal.

Capillary tubing can be used to reduce instrument response time by restricting the rate of material flow between the pressure source and the measuring device. A major disadvantage of capillary tubing is the likelihood of plugging and clogging by foreign matter. In addition, except by changing the tubing length, the amount of restriction is not adjustable.

The devices shown in Figure 3–17 provide a better means of pressure damping. Figure 3–17A shows an adjustable needle valve with a filter to prevent plugging of the needle valve restriction. The desired amount of restriction can also be adjusted to obtain the best damped response. The device in Figure 3–17B offers a somewhat more elaborate approach to damping. A bulb made of rubber or another pliable material is filled with an inert seal fluid, which forms a buffer between the process and the measuring instrument. The transfer of fluid between the sealed chamber and the measuring instrument is retarded by an internally located adjustable restrictor. The amount of restriction is thus more difficult to manipulate. Maintaining the seal fluid constant in the chamber is also burdensome.

Differential-Pressure Instrument Installation

The previous discussion of instrument mounting and installation applies to both pressure and differential-pressure instruments. The latter classification of measuring application requires some special consideration, however, because of the increased sensitivity of the measurement range of differential pressure over pressure-measuring applications. A 10-in. liquid head would be insignificant in a pressure-measuring system with a 100-psi measuring range. The same liquid head could provide 10% of the measurement if the range were only 100 in. Differential-pressure applications will be covered in more detail in Chapter 4.

Special Precautions for Low Temperature Protection

When process fluids freeze at adverse or normal operating temperatures, the connecting lines and instrument must be protected. Liquid seals, described earlier, achieve this purpose. This method cannot be used, however, when traces of the seal fluid enter the process line, a situation that is detrimental to process operation or otherwise undesirable (e.g., in potable water applications). In such instances, other winterizing methods must be used.

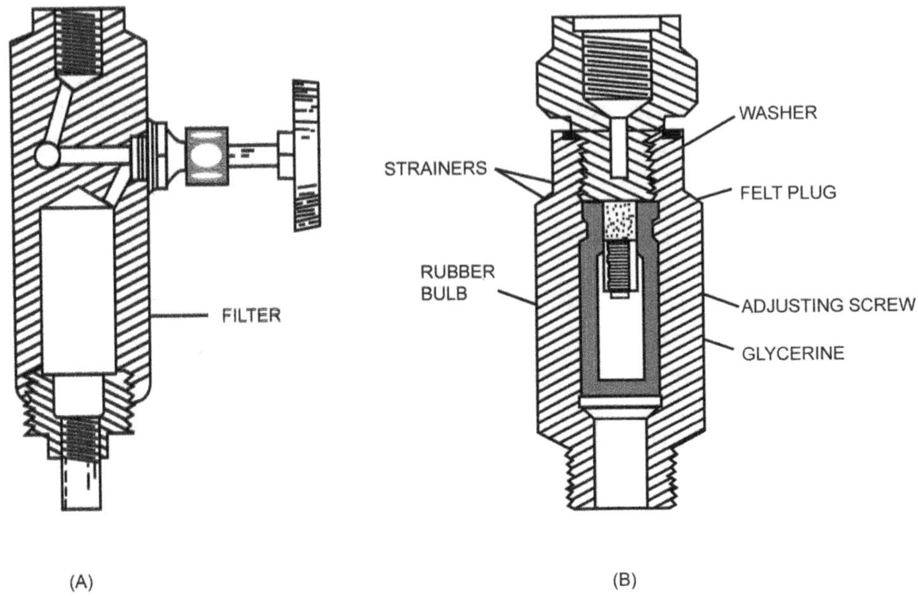

Figure 3–17 Pulsation Dampers

Steam or Heat Trace

Instead of filling the instrument systems with nonfreezing or low-freezing-point liquids, the temperatures of the connecting lines and instrument housing can be maintained above the freezing point. This is accomplished by wrapping the exposed surfaces with tubing, through which an appropriate amount of steam flow is established. The temperature of the steam must be considered and to ensure proper heat transfer, the end of the tracing should be connected to a steam trap to maintain the steam pressure and not waste steam. The entire surface area is then insulated with an appropriate temperature-insulating material. Electrical heat tape can be used instead of the steam tubing when desirable. The service valves normally are left exposed so that field zeroing can be accomplished without disturbing the insulation. The insulated instrument is generally inaccessible for purposes of other adjustment or repair without removing the insulation from the body, unless this can be achieved through the upper portion of the instrument by removing the cover, which usually is not insulated.

In some unusual circumstances where liquid sealing or heat tracing is not a viable option, temperature-controlled housings may be used. A temperature-controlled housing is a container constructed from fiberglass or another suitable material that completely surrounds the instrument. The temperature environment inside the container is then controlled by an appropriate open- or closed-loop temperature-control system.

EXERCISES

3.1 List four pressure elements and give the general operating characteristics and applications of each.

3.2 Explain the term "motion balance" and relate this application to measuring instruments.

3.3 Define the terms "elevation" and "suppression" with respect to instrument calibration, and list an application for each.

3.4 List three metals commonly used for the construction of pressure elements.

3.5 Explain the purpose and operation of separators used with pressure gages.

3.6 List the letter designation for the following gage classifications, along with possible applications:

- Precision test gage
- Test gage
- Error not to exceed ±1/2%
- Error not to exceed ±1 to 1-1/2%

3.7 Give two other classification criteria for pressure gages.

3.8 What precautions should be taken when mounting pressure instruments for gas and liquid service?

3.9 List special precautions that may be required to prevent freezing of material in impulse lines and sensing elements of measuring instruments.

3.10 What are pulsation dampers? List a possible application.

REFERENCES

1. Doro Instruments Company. Catalog No. 1006.

2. Julien, Hermann, 1981. *Handbook of Pressure Measurement,* Alexander Wiegand GmbH & Company.

4

Transmitters and Transmission Systems

4.1 INTRODUCTION

Although the pressure gage is used in many pressure-measurement applications, the fact that it can be used only for local indications sometimes limits its capability in process measurement and especially in control. In most applications, the pressure measurement must be converted into a scaled signal, the signal transmitted some distance, and the signal converted into a pressure reading. Pressure transmitters serve to convert a pressure value to a scaled signal—whether electric, pneumatic, or mechanical. Receiver instruments[1] convert the transmitted signal to a response representing a pressure value. The principle of generating, transmitting and receiving signals for pressure measurement is shown in Figure 4–1.

Regardless of type, all pressure transmitters perform the function of converting pressure applied to the input by the primary transducer into a measurable scaled signal on the output by the secondary transducer. The block diagram in Figure 4–2 illustrates this principle. The primary (input) transducer converts the process measurement to a force or movement which becomes the input of the secondary (output) transducer generating an indication or signal for transmission. The mechanical pressure element can be any of the types mentioned earlier. The transducer in the output portion, which converts the movement of the mechanical pressure element to a scaled output signal, is usually a null-balance or force-balance device. For pneumatic transmission, a flapper-nozzle arrangement is

1. Although the term receiver is not listed in ISA's Process Instrumentation Terminology Standard 51.1, it is generally understood to be an instrument that receives information from a transmitter.

generally used. Electrical transmitters use various electrical quantities or devices as output transducers:

- Inductance
- Reluctance
- Capacitance
- Resistance
- Differential transformers
- Resonant frequency
- Piezoresistive

All of these output transducers are used to convert the force or motion of the input transducer into a scaled output.

Although the terms "transmitter" and "transducer" sometimes are used interchangeably, there is a difference between the two. A transmitter is a device that responds to a measured process value—pressure in the case of pressure transmitters—and produces an input to a controller or other component. A transducer, on the other hand, is in physical contact with the process and produces a response or change in motion or force when the process value changes. When this response is used to generate a scaled signal used for measurement or as an input to a controller, the transducer is a transmitter or sensor. All transmitters thus are transducers, but not all transducers are transmitters. Figure 4–3 is a block diagram of a pressure transducer.

4.2 SECONDARY TRANSDUCERS

Motion detectors produce a response from the primary transducer or pressure element, which is used to generate a scaled signal relative to a measured value. These devices provide an input to the secondary transducer which through signal conditioning circuitry uses this input to generate an output for transmission. Motion detectors can be placed in several different categories, but they are usually categorized as linear motion or rotary motion. The most common is the linear motion detector, which is also referred to as a displacement detector, where displacement is considered to be a change in movement or position.

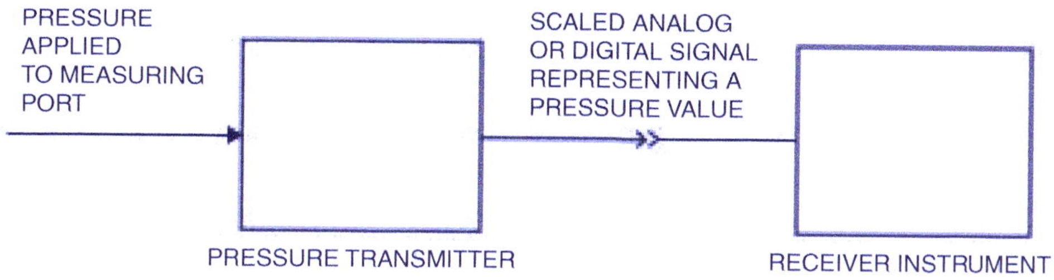

Figure 4–1 Block Diagram of a Transmitter and Receiver

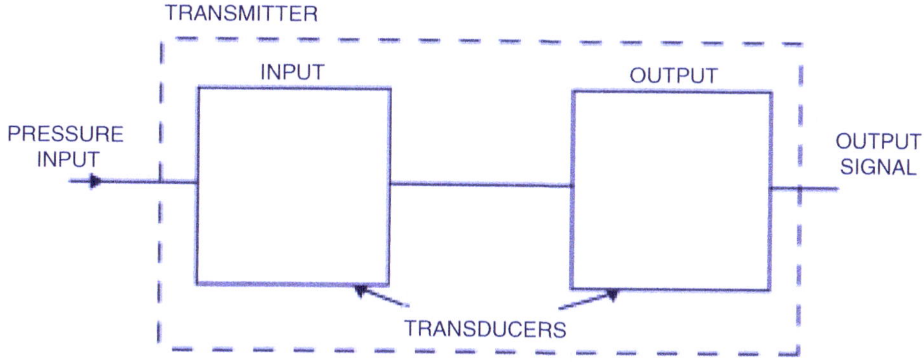

Figure 4–2 Block Diagram of a Pressure Transmitter

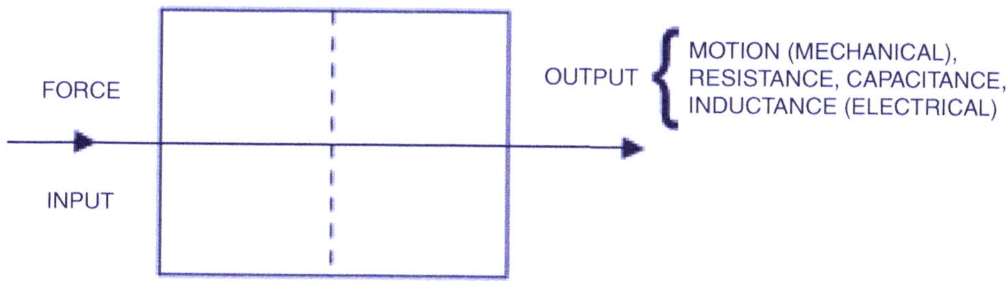

Figure 4–3 Block Diagram of a Pressure Transducer

4.3 POTENTIOMETERS

A potentiometer consists of a slider or movable contact that is attached to a moving point on a pressure measuring element. The movable contact or wiper moves across the resistive element with every variation in pressure on the measuring element. This movement, which causes a change in resistance of the potentiometer, is converted into an electrical response: a change in voltage, or current in a bridge circuit.

This produces an electrical signal that is proportional to the mechanical displacement. By the choice of connection on the potentiometer, the electrical signal can increase or decrease with a given amount of displacement. The resolution, or the amount of resistance change with respect to displacement, is determined by the spacing of the turns of wire on a wire-wound potentiometer. For film-type resistive elements, the resolution is high, with a small amount of contact friction. This reduces the amount of force required for operation. For most measurement and control applications, the electrical signal level is low and the operating force needed to move the wiper is small.

The dynamic response of both linear and angular potentiometers is limited by the inertia of the shaft and wiper assembly. When this inertia is large, the potentiometer may be used for static applications where high-frequency responses are not required.

With wire-wound potentiometers, electronic noise can result as the electrical contact on the wiper moves from one wire turn to the next.

Potentiometers are used to measure relatively large displacements of 10 mm or more for linear-motion devices and 15° for angular-motion devices. They are simple in concept and application, relatively inexpensive, and accurate. Figure 4–4 illustrates a resistance pressure sensor that incorporates a potentiometer.

4.4 SIGNAL-CONDITIONING CIRCUITS FOR RESISTANCE DEVICES

In a resistance-based pressure-measuring instrument, a transducer converts pressure to resistance as discussed, and in a further refinement of the measuring circuit, the resistance value is converted to current or voltage for better manipulation or transmission of the data. A DC Wheatstone bridge circuit is commonly used for this purpose.

A simplified bridge circuit is shown in Figure 4–5 and is in a balanced condition when the voltage between points *c* and *d* is equal to zero ($V_{c-d} = 0$). For a balanced bridge, the current through the detector is also zero.

The voltage from points *a* to *c* (V_{a-c}) is equal to the voltage from *a* to *d* (V_{a-d}) and $V_{b-c} = V_{b-d}$. The following equations relate to and help to explain balanced bridge conditions:

$$I_1 R_1 = I_2 R_2 \tag{4.1}$$

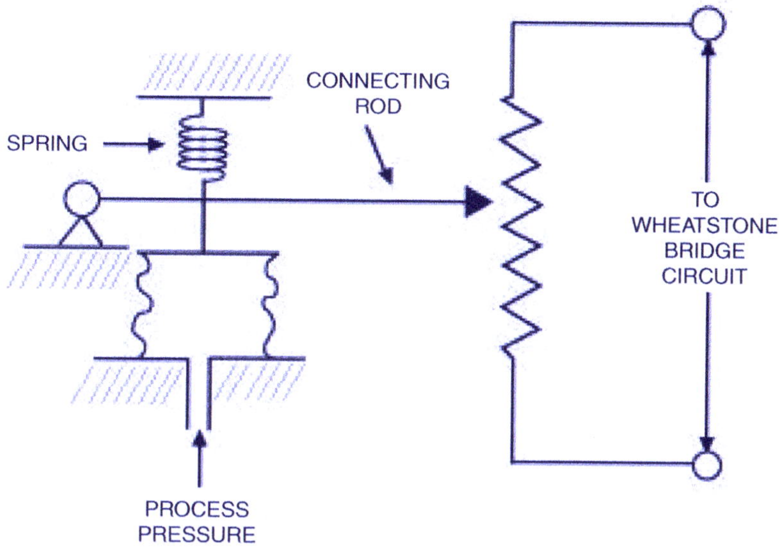

Figure 4–4 Resistance Pressure Sensor (*Instrument Engineer's Handbook: Process Measurement & Analysis, 4th Edition,* edited by Béla G. Lipták, © 2003. Reprinted with permission of the publisher, CRC Press)

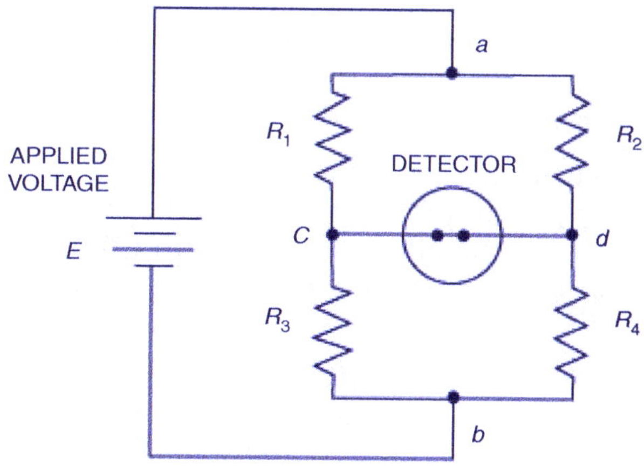

Figure 4–5 Wheatstone Bridge Circuit

$$I_1 = I_3 = \frac{E}{R_1 + R} \tag{4.2}$$

$$I_2 = I_4 = \frac{E}{R_2 + R_4} \tag{4.3}$$

By combining Equations 4.1 to 4.3 and simplifying,

$$\frac{R_1}{R_1 + R_3} = \frac{R_2}{R_2 + R} \tag{4.4}$$

and

$$R_1R_4 = R_2R_3 \tag{4.5}$$

In a practical pressure measurement application, R_4 is the unknown (R_x) variable relating to an unknown pressure. Then,

$$R_x = \frac{R_2}{R_1}R_3 \tag{4.6}$$

R_1 and R_2 are called ratio resistors and are usually sized to limit the current through R_3 and R_x. This is to reduce the effect of resistance changes in the bridge components caused by heating from I^2R losses. They also select the overall bridge sensitivity with respect to resistance. R_3 is normally a variable resistor used to establish initial bridge balance.

Most bridge applications in pressure measuring instruments are constant-voltage applications. By analyzing the bridge with a Thevenin equivalent circuit, the effect of voltage on bridge sensitivity and operation can be seen. Also, the relationship between unbalance voltage and resistance variations can be determined.

By determining the Thevenin equivalent circuit from the detector, the circuit is opened at these points and the power supply is replaced by a short circuit. This will very nearly approximate actual conditions, because the bridge detector, an operational amplifier or measuring instrument, will have a high impedance and the power supply will have a low resistance. The equivalent resistance circuit that results from the procedure is shown in Figure 4–6. The unbalance voltage, E_{cd}, can be determined from Equations 4.7 to 4.9:

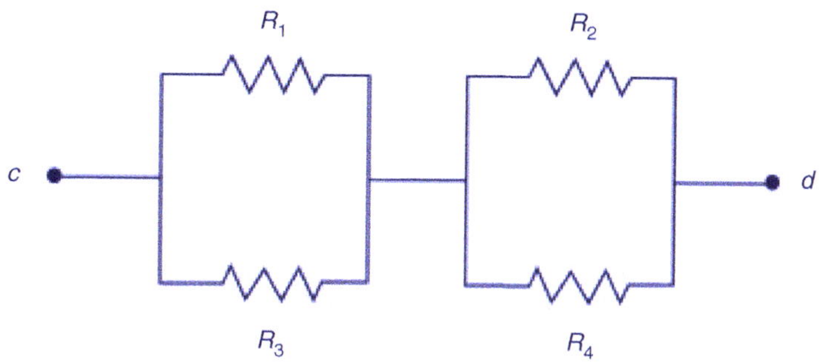

Figure 4–6 Equivalent Bridge Circuit

$$I_1 = \frac{E}{R_1 + R_3} \qquad (4.7)$$

$$I_2 = \frac{E}{R_2 + R_3} \qquad (4.8)$$

$$E_{cd} = \left(\frac{R_1}{R_1 + R_3} - \frac{R_2}{R_2 + R_4}\right)E \qquad (4.9)$$

From Figure 4–6, the total bridge resistance from points *c* to *d* is

$$R_{TH} = \frac{R_1 R_3}{R_1 + R_3} + \frac{R_2 R_4}{R_2 + R_4} \qquad (4.10)$$

The current through the detector circuit is

$$I_D = E_{TH}/R_{TH} \qquad (4.11)$$

As mentioned, the analysis of the bridge circuit assumed zero resistance for the power supply and infinite resistance for the measuring circuit. The resistance of the detector can simply be added to R_{TH}, but if the resistance of the power supply was not neglected, the circuit in Figure 4–6 would be replaced by that shown in Figure 4–7. To find the equivalent resistance of the circuit requires converting it to a more convenient form using the Delta to Wye Transformation Theorem.

Although it may not be necessary to analyze or comprehend the operation of bridge circuits for pressure measurement by variable-resistance

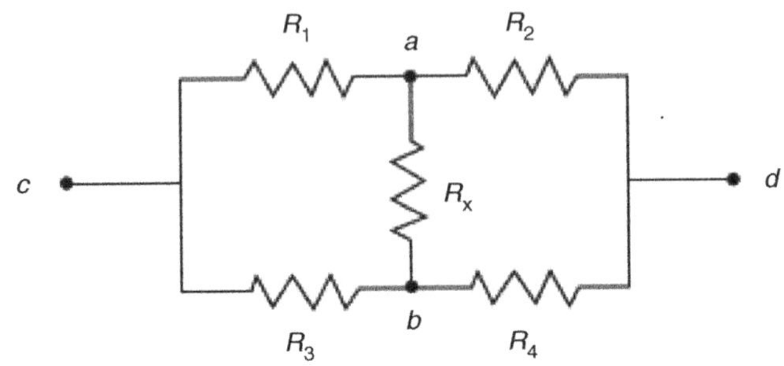

Figure 4–7 Equivalent Bridge Circuit Including the Resistance of the Voltage Source

devices, it is important to realize the variable conversions that take place and to see the bridge components that affect the overall bridge sensitivity. This sensitivity determines the relationship between bridge input (change in resistance) and bridge output (change in voltage or current).

For very low-resistance measurement applications, the resistance of the connecting leads between the measuring resistor and the bridge components can cause errors. For this reason, the distance should be short. In pressure and differential-pressure applications, this generally is the case, because the transmitters with all signal-conditioning circuits are located in the field in close proximity to the point of measurement. If this is not the case, a three-wire bridge configuration should be considered to minimize the effects of wire resistance.

When long leads are necessary, the bridge can be balanced to provide a reference point of measurement. Changes in lead resistance caused by variations in ambient temperature, however, could result in measurement error. The circuit in Figure 4–8 shows a means by which the effect of lead resistance can be compensated. Wire 1 is in the R_X leg of the bridge, wire 2 is in the R_3 leg, and wire 3 is the power supply lead. Resistance changes in wires 1 and 2 will have no effect on bridge unbalance. Because these wires are in opposing arms of the bridge when both change by the same amount, the R_3/R_4 ratio is not changed. If the resistances of the wires change by different amounts, however, an error will result.

It is usually desirable to keep resistance-to-pressure changes small to help eliminate the nonlinearity effect of the bridge circuit compo-

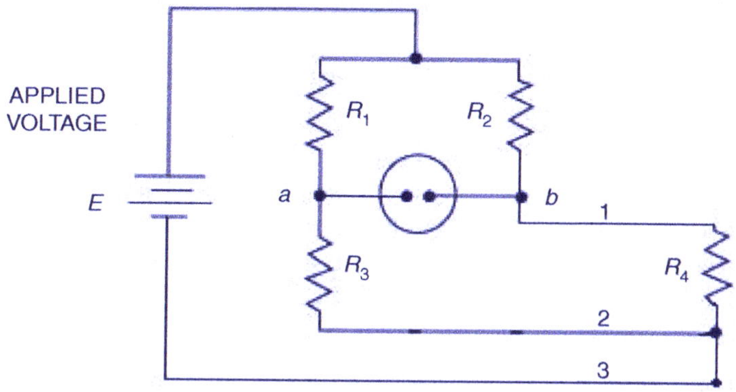

Figure 4–8 Lead Wire Resistance Compensation Using a Three-Wire Bridge Configuration

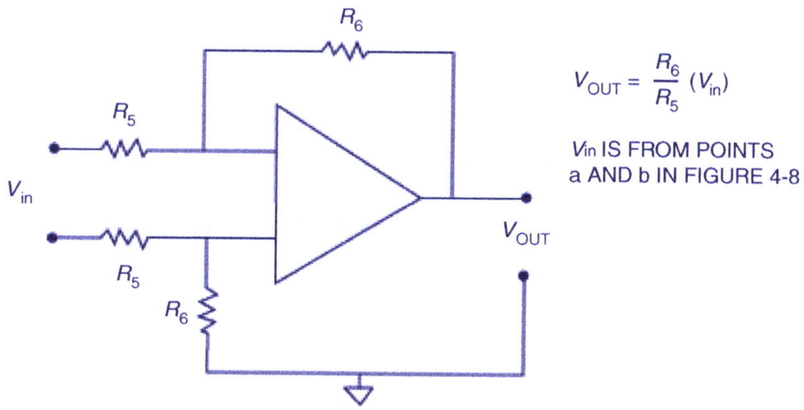

Figure 4–9 Op Amp Used With a Bridge Circuit

nents. As a result, it may be necessary to amplify the bridge unbalance voltage for scaling and stability. Figure 4–9 shows a simple op amp (operational amplifier) circuit for this purpose. The values of R_5 and R_6 are used to set the gain of the amplifier. R_6 may be a variable resistor in order to have a variable gain for scaling purposes.

4.5 VARIABLE INDUCTANCE TRANSDUCERS

A variable inductance transducer consists of a spool or core made of paper or plastic around which is wrapped a coil of wire. Relative motion and change in location of the core with respect to the coil will result in a change in the inductance value of the coil. The inductance change is proportional to the rate and degree of motion of the core in the coil. A change in pressure at the pressure element changes the air gap in the detector assembly, which consists of two pieces of ferrite—one mounted on the force beam and the other on the chassis. A change in the air gap results in a change in inductance in the oscillator circuit, which changes the output current. The output current is fed through the magnetic coil in the feedback motor (not shown in the figure), thus producing an equal and opposite force on the beam to balance the force created by the change in measurement. This negative feedback provides stabilization and is a common characteristic of force-balance systems.

A variable inductance transducer and its associated circuitry are more rugged than a linear variable potentiometer and have less friction, more freedom of movement, and greater life expectancy. The input and output circuits are electrically isolated—a distinct advantage in preventing spurious voltage disturbances caused by electrical interference. Figure 4–10 shows a schematic of a variable inductance transducer. Because of the infrequent use of variable inductance in pressure-measuring instruments, the oscillator and associated tank circuit are not discussed in detail. Many references are available for such circuits.

4.6 LINEAR VARIABLE DIFFERENTIAL TRANSFORMER

The linear variable differential transformer (LVDT) is sometimes used as a secondary transducer in electronic transmitters for signal generation. It has all of the advantages of the variable inductor and is more sensitive to very small displacements—as low as a few ten-thousandths of an inch. Although it is more expensive than most other signal-generating devices, it has several advantages, among which is the fact that a DC output voltage is generated that is polarity-sensitive to core movement. An LVDT is shown in Figure 4–11 (the output diodes are not shown), and a block diagram of a signal-conditioning circuit in Figure 4–12. The displacement of the movable arm and core of the LVDT is generated by the pressure element used for the pressure transducer. An LVDT signal-conditioning circuit is shown in Figure 4–13.

Operation of the circuit can be understood by referring to the circuit in Figure 4–13. The primary winding of the LVDT is connected to an AC voltage source. The two secondary coils are mechanically and electrically similar. They are located symmetrically with reference to the primary coil. When the core is positioned so that it is exactly centered electrically between S_1 and S_2, the two secondary coils, the current through the output resistors R_1 and R_2 will be as shown in Figure 4–13A when the

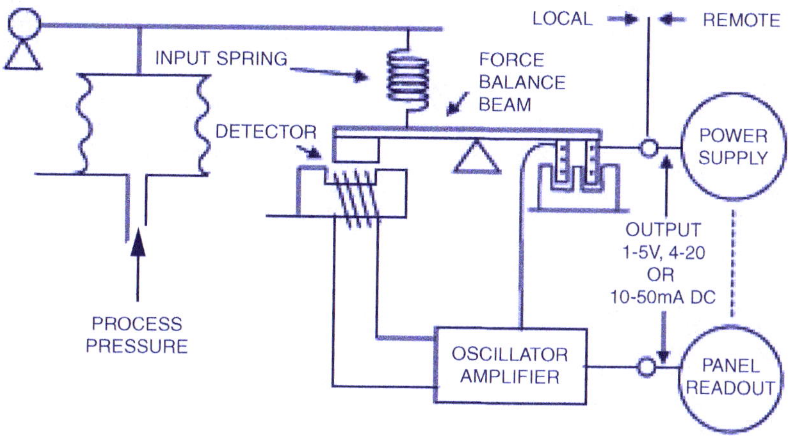

Figure 4–10 Variable Inductance Pressure Measurement (*Instrument Engineer's Handbook: Process Measurement & Analysis, 4th Edition*, edited by Béla G. Lipták, © 2003. Reprinted with permission of the publisher, CRC Press)

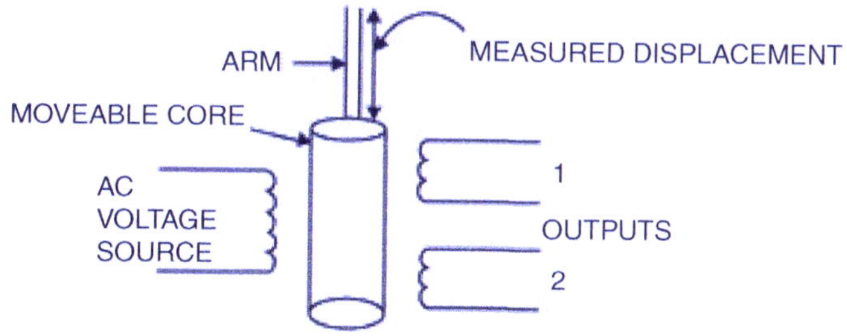

Figure 4–11 LVDT

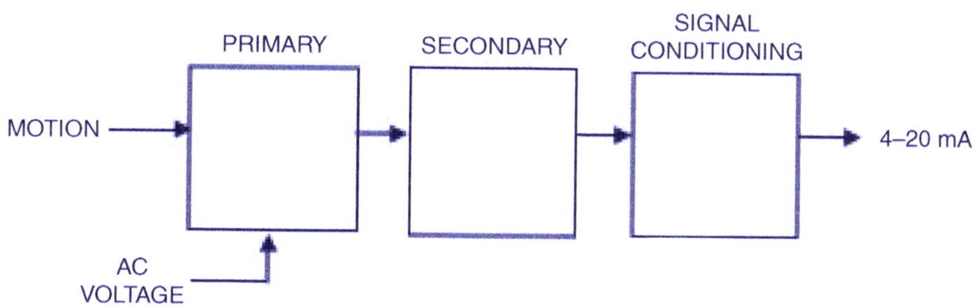

Figure 4–12 Block Diagram of an LVDT Signal-Conditioning Circuit

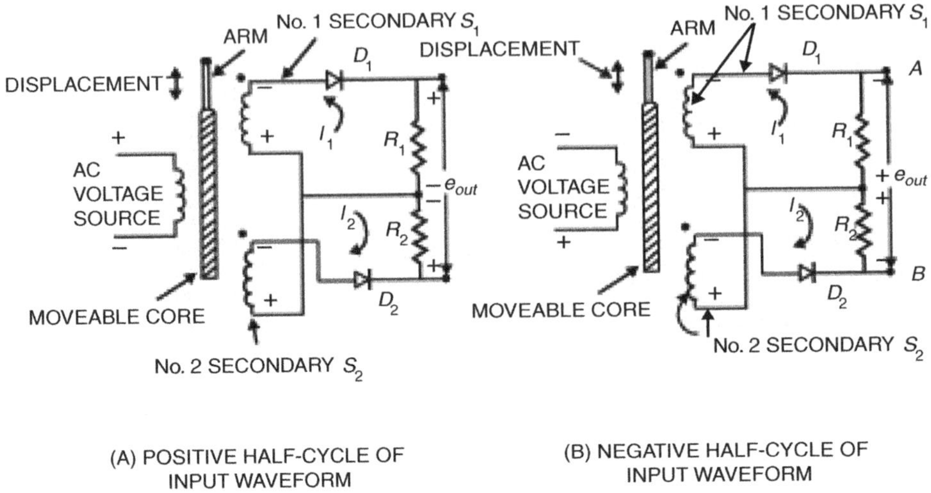

Figure 4–13 LVDT Signal Conditioning Circuit

input waveform is on the positive half-cycle. When the input waveform swings negative on the next half-cycle (Figure 4–13B), the current through R_1 and R_2 is blocked by D_1 and D_2, and the output current is zero. The resistance values of R_1 and R_2 are equal. Because of the symmetrical location of the core with respect to the primary coil, the electrical similarity of S_1 and S_2, and the equal values of R_1 and R_2, the currents through these resistors are equal. The voltage drop IR_1 across R_1 is equal to that across R_2. The polarity of the two voltages is opposite, however, because the current flow is in the opposite direction. For the condition just described, with the core centered between S_1 and S_2, the output voltage (e_{out}) is zero.

When the core is positioned closer to S_1—as would be caused by an increase in pressure in the attached pressure element, for example—the voltage induced in S_1 increases. This increases the current I_1 through R_1. I_1R_1 is greater than I_2R_2, point *A* on the output is more positive than point *B*, and e_{out} will be positive with respect to common. A decrease in pressure in the pressure element will cause the core to be positioned closer to S_2, I_2R_2 will be greater than I_1R_1 and e_{out} will become negative. In most LVDT applications in pressure measurement, the movement of the core is in only one direction from a reference point, and the measured voltage on the output increases from a reference point.

The magnitude of e_{out} is proportional to the position of the core with respect to the center or rest position. The greater the displacement, the greater the output voltage. The polarity of the output voltage indicates the direction of displacement of the core from the center, or rest, position. The core displacement is usually in only one direction as a result of a measurement change. The output voltage, which is a function of the core movement, increases to a maximum value when the applied pressure is a maximum value and goes back to zero when the pressure is reduced to a minimum value.

The output waveform is half-wave rectified by the action of the diodes D_1 and D_2. A filter capacitor across points A and B causes a DC output to be obtained. Although not commonly practiced, an AC waveform can be used as the output voltage by omitting the diodes and output resistors. In such a case, the windings of S_1 and S_2 are in a series opposition arrangement. The AC output is then a phase-sensitive variable peak value whose frequency is the same as the input signal.

The frequency of the voltage applied to the primary winding can range from 25 to 50 kHz. If the LVDT is used to measure periodic displacements resulting from pressure transient response, the frequency should be ten times higher than the highest frequency component in the changing measurement. The highest resolution is attained with excitation frequencies between 1 and 5 kHz, with typical values of 1 kHz. The input voltage can range from 5 to 15 with power requirements of 1 W. Signal conditioning is required for AC excitation voltage from the DC supply of the instrument. LVDTs have been replaced by other transducers for pressure and differential pressure transmitters but still may be used for displacement measurement.

4.7 VARIABLE-CAPACITANCE TRANSMITTERS

A capacitor is formed by two electrically conducting plates separated by an insulating material, or dielectric. The electrical quantity of capacitance and its changes with respect to a measured variable constitute a very useful process-measurement tool.

The basic operating principle involved in capacitance pressure measurement is the measurement of a change in capacitance resulting from the movement of a pressure element that is caused by a corresponding pressure change. Such a measurement system incorporates two capacitor plates (one fixed and one linked to the mechanical output of the pressure element), an oscillator circuit to energize a sensing element, and a capacitance detector circuit.

Among other factors, the value of electrical capacitance depends on the distance between the capacitor plates. When the capacitor is constructed so that the plate separation distance can vary, the resultant change in capacitance can be related to the plate movement.

The output from a variable-capacitance transducer is small; the change in capacitance for changes in plate separation is a slight percentage of the total value. The capacitance-measuring circuit associated with these detectors must be designed very carefully to eliminate the effects of stray capacitance that could swamp the small change of the measuring circuit. When applied to the tank circuit of an oscillator, this variable capacitance results in a frequency variation that is proportional to the measurement.

The capacitance-sensing element requires an AC voltage to generate the capacitance signal. This AC voltage is supplied by an oscillator circuit with a frequency of between about 50 and 100 kHz. The signal from the oscillator circuit is capacitance-coupled to the transmitter case ground through the sensing capacitor. This capacitance between the plate formed by the sensing element diaphragm and the other capacitor plate is converted to a proportional 4–20 mA signal for analog and HART (Highway Addressable Remote Transducer protocol) transmitters and converted to digital for FOUNDATION Fieldbus transmitters.

This pressure measurement principle is based on the following relationships:

$$P = K_1 \frac{C_0 - C_p}{C_p} \tag{4.12}$$

$$I_{DIFF} = fV_{pp}(C_0 - C_p) \tag{4.13}$$

$$fV_{pp} \frac{I_{RFF}}{C_p} \tag{4.14}$$

where P is the process pressure to be measured; K_1 is a constant; C_p is the capacitance of the measuring element at operating pressure; C_0 is the capacitance of the measuring element at zero pressure; I_{DIFF} is the difference in current resulting from the change in capacitance with pressure variations or $C_0 - C_p$; V_{pp} is the peak-to-peak voltage from the oscillator circuit; f is the frequency of the oscillator voltage; and I_{REF} is the current from the constant-current source. The relationship between process pressure, capacitance, and transmitter output is established by the foregoing discussion.

Figure 4–14 shows a block diagram of the electrical circuit of an industrial process pressure transmitter incorporating the principles discussed. A brief functional description of the component modules follows. Figure 4–15 is a schematic of a capacitance pressure transmitter.

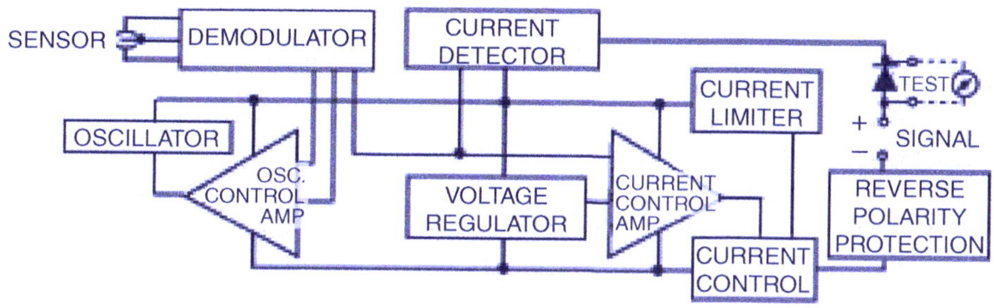

Figure 4–14 Block Diagram of Electrical Circuit of a Capacitance Pressure Transmitter (*Courtesy of Emerson Process Management*)

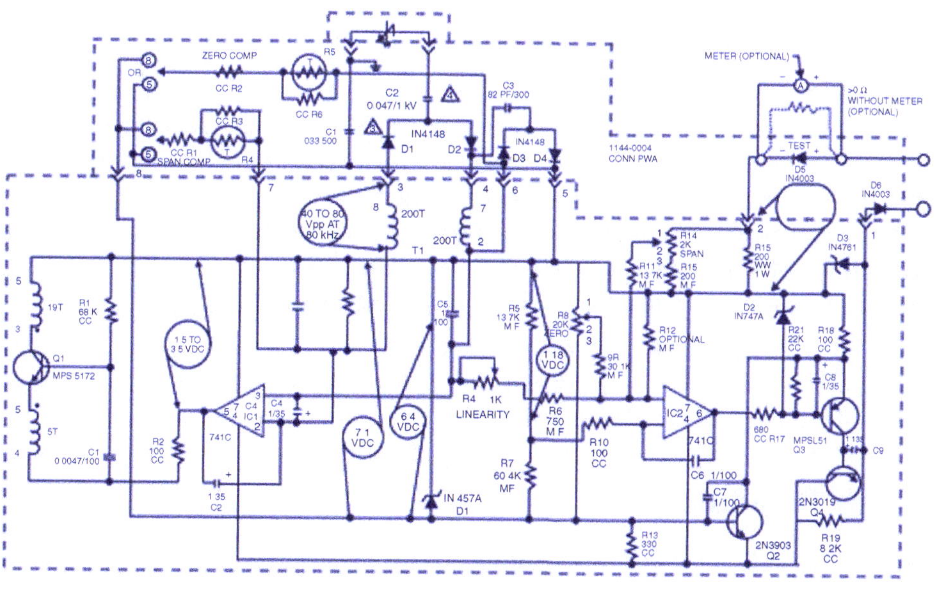

Figure 4–15 Electrical Schematic of a Capacitance Pressure Transmitter (*Courtesy of Emerson Process Management*)

Sensor Module

The sensor module is the variable capacitor to which the process pressure is applied. The pressure, by means of a diaphragm, positions the capacitor plates to change the capacitance of the sensing element. At a measurement value of zero, the capacitance of the sensor is about 80 pF (picofarads). An oscillator circuit at approximately 80 kHz and 70 V pp drives the sensor. The signal from the sensor is rectified by a demodulator circuit. A capacitor cell (sensor module) is shown in Figure 4–16.

Demodulator Circuit

The demodulator circuit consists of a full-wave rectifying bridge circuit that rectifies the AC signal voltage. The direct currents are applied to transformer windings and compared by a 741 op amp. The op amp output is applied to an oscillator control circuit.

The DC signal current detected by a diode in the bridge circuit, D_3 from C_3, is subtracted from the current through another set of transformer windings. The net current is directly proportional to pressure. This current is the I_{DIFF} in Equation 4.13.

Oscillator Circuit

The oscillator circuit consists of components Q_1, T_1, C_1, R_1, and R_2. It has a frequency determined by the capacitance of the sensing element, the reference capacitor C_3, and the inductance of the transformer windings. The only variable component is the capacitance of the sensing element, and the frequency is variable about a normal frequency of 80 kHz. IC_1, a 741 op amp, is used in the feedback capacitor circuit and controls the oscillator voltage such that

$$fV_{pp}\frac{I_{RFF}}{C_p} \tag{4.14}$$

Voltage Regulator

The transmitter uses a Zener diode D_1, a transistor Q_2, and a resistor R_{13} to provide a constant voltage supply of 6.4 V DC for reference and 7 V DC to the oscillators IC_1 and IC_2.

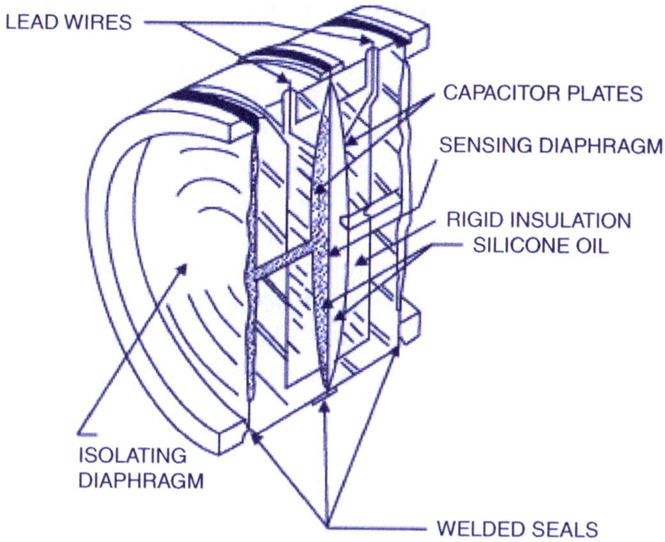

Figure 4–16 Capacitor Sensor Module
(*Courtesy of Emerson Process Management*)

Current Control

The current-control amplifier circuit consists of IC_2, Q_3, and Q_4. The *IC* reference voltage is established between R_5 and R_7. The resulting variable sensor current flows from the junction of R_9 and R_{11} through R_6. The current-control amplifier drives the control to a level such that the detector feeds back a current through R_{11} equal to the sensor current. The current-limiting circuit consisting of D_2, R_{18}, and Q_2 prevents the output from exceeding 35 mA in an overrange condition.

Industrial pressure transmitters (using variable-capacitance sensing elements) have pressure ranges adjustable from zero to approximately 25 in. H_2O, to zero to several thousand psi. An advantage of this type of device is its inherent high overrange capability. Absolute pressure transmitters are also available. Such transmitters have a range of 30 in. Hg vacuum to about 15 psig. This range would convert to 0 to 30 psia. The circuit described is shown in Figure 4–15 and is Emerson Process Management Model 1144 pressure transmitter.

Capacitance and differential-pressure measuring instruments have a proven track record with very good specifications. They are available in a number of ranges and operate over a variety of ambient conditions with a stated accuracy of ±0.5% of calibrated span.

4.8 ELECTRICAL STRAIN GAGE TRANSMITTERS

With respect to pressure-measurement applications, strain is defined as the ratio of the change in length of a stressed component to the initial unstressed reference length. This change is sensed and converted to an electrical quantity by a strain gage, an instrument that causes a change in an electrical resistance as it is stretched or compressed. The stretching or compression is a result of the deformation of a pressure element to which the strain gage is attached. Strain gage electrical transducers include bonded and unbonded types. Most strain gage applications commonly use bonded elements. Transducers which use unbonded strain gages could be sensitive to vibration and could present mounting difficulties.

The operating principle of strain gages as commonly applied to measurement was first discovered by Lord Kelvin, when he reported that a metallic conductor subjected to a mechanical strain exhibited a corresponding change in electrical resistance. This change results from the change in the length-to-diameter ratio of the wire as it is stretched. As the force causes the wire to stretch, its length is increased and its diameter is decreased. Resistance also increases; this quantity is measured and a related value of pressure determined.

Strain gages are either bonded or unbonded. An unbonded strain gage requires the conductive elements to be mounted on a mechanical frame whose parts can move in relation to each other, causing a change in wire tension, length, and resistance with movement of the pressure element.

Most pressure-measurement applications use bonded strain gages, which eliminate the mechanical frame by attaching the sensing element directly to the strained surface. The bonding material insulates the conductor from the strained surface, with limited stresses developing in the bond itself. In bonded gages, the conductor cross-sectional area is small in relation to surface area per unit length. Strain gage wires with diameters smaller than 0.001 in. have a surface area that is much greater than the cross-sectional area. The bond between the strained surface and the wires mounted on paper or plastic carriers is sufficiently strong. Foil gages have been used, with foil thicknesses as small as 0.0001 in.

Semiconductors can also be used as gage elements; they increase the resistance-to-pressure sensitivity above that of metallic conductors. This principle is illustrated by:

$$S = \frac{\frac{\Delta r}{r}}{\frac{\Delta L}{L}} \tag{4.15}$$

where S is the ratio between unit length and extension length and corresponding changes in specific resistance, r is resistance, and L is length.

The term gage factor (GF) has also been used to define sensitivity (S = GF). The change in resistance with strain is also a function of the resistivity of the element, which must be considered in a specific application. Gage factor applies to the strain gage as a whole, to include the dimensional change as well as carrier matrix and element sensitivity.

When strain gage sensing elements are used, compensation for ambient temperature changes may be necessary. Temperature changes can cause both the bare material to which the element is bonded and the element itself to expand or contract. The coefficient of resistance of the element may also vary with temperature. In a near-ambient temperature environment (–100° to 150°F), thermal errors can usually be neglected; for applications where temperatures exceed those limits, compensation is usually provided by using a second (dummy) element in the same temperature environment as the measuring element. The dummy element is connected to the arm of the Wheatstone measuring bridge adjacent to the measuring element and cancels the effect of ambient temperature changes.

The change in resistance with changes in pressure is very small, which is useful for linear, precise operation. The change in resistance is measured with a DC Wheatstone bridge circuit. This circuit was explained earlier and will be discussed here with respect to a typical measuring circuit in a pressure instrument. In Figure 4–17, each arm of the bridge contains a strain-sensitive element (some of which can be dummies). From an initial balance condition, a pressure variation will result in an unbalance of the bridge and a corresponding unbalance voltage (ΔE). The instantaneous value of bridge voltage resulting from a resistance change is

$$\Delta E = \frac{V}{R_0}[(\Delta R_1 + \Delta R_3) - (\Delta R_2 + \Delta R_4)] \tag{4.16}$$

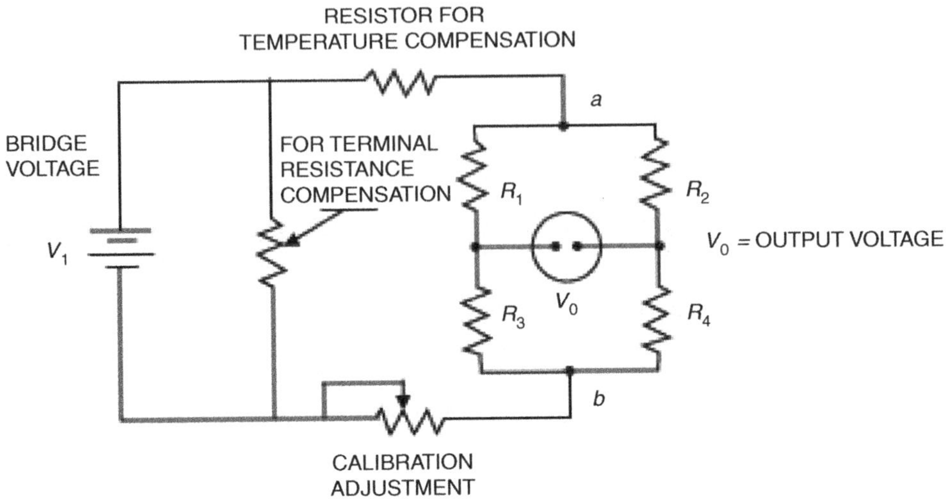

Figure 4–17 Signal Conditioning Circuit for Strain Gage

where R_0 is the initial resistance of each element (which are equal) and V_1 is the bridge voltage.

The total bridge resistance and applied voltage will usually be on the order of 100 to 500 ohms and 8 to 40 volts. The overall sensitivity can be increased by higher voltage values, but excessive values can result in unstable operation caused by small fluctuations or variations in output voltage. Normal values of output voltage may range from 1 to 5 mV per volt of applied voltage.

A simplified diagram of a strain gage transmitter is shown in Figure 4–18, and a functional schematic is shown in Figure 4–19. As has been discussed, an increase in pressure acting on the pressure element develops a force that increases the strain on the sensing element. This causes a change in resistance, which produces a voltage change that is proportional to pressure. The voltage is applied to the input amplifier A_2, which drives the output current regulators Q_2 and Q_3.

One side of the bridge connects to the zero circuitry and the non-inverting input of amplifier A_2. The other side of the bridge connects to the inverting input of A_2 via R_{11} and the span circuitry via R_6. R_{14} is the span adjustment, which sets the closed-loop gain of amplifier A_2. This in turn sets the measurement span of the instrument. The zero adjustment R_{12} establishes the zero reference current to 4 mA at the required lower range limit. A jumper can be positioned between R_{10} and R_{17} to provide increased zero adjustment for elevation or suppression.

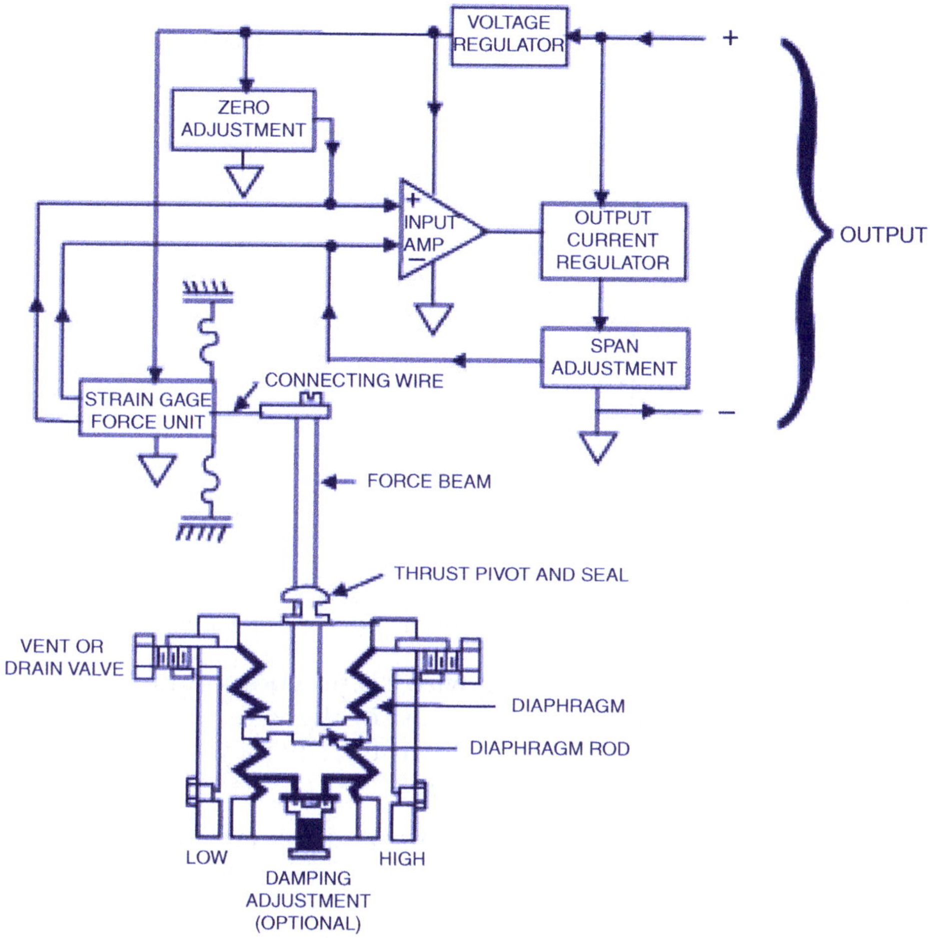

Figure 4–18 Diagram of Strain Gage Transmitter
(*Courtesy of ABB Kent-Taylor Instrument Company*)

4.9 RESONANT FREQUENCY TRANSMITTER

The basic principle of operation of a resonant frequency transmitter is to cause a wire to resonate at its natural frequency and to convert changes in measurement into changes in the resonant frequency.

A wire under tension is located in a permanent magnetic field. This wire is an integral part of an oscillator circuit, which causes it to resonate continuously. For pressure or differential-pressure applications, an increase in pressure causes a pressure element to move, increasing the tension on the wire and thus raising its resonant frequency.

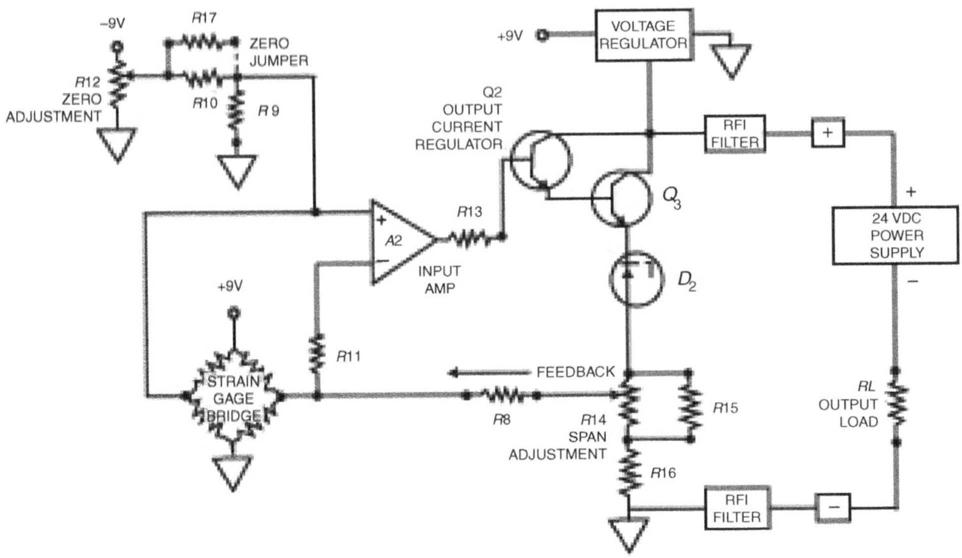

Figure 4–19 Schematic of Strain Gage Transmitter
(*Courtesy of ABB Kent-Taylor Instrument Company*)

The resonant frequency of an ideal wire is a function of the length, the square root of the tension, and the mass of the wire. Since the length and mass are constant, the square of the frequency of oscillation is proportional to the tension on the wire.

The signal-conditioning circuitry performs a squaring operation on the resonant frequency to develop an output signal proportional to pressure. The tension on the wire can be applied only in one direction. For differential-pressure applications, the higher of two measured pressures will normally be connected to the high side.

A block diagram of a resonant frequency transmitter is shown in Figure 4–20. The resonant wire is part of the oscillator circuit, which maintains a continuous movement of the wire at its resonant frequency. The output of the oscillator circuit (its frequency) is applied to a pulse shaper circuit. This produces two complementary frequencies, which are applied in cascade to two frequency converter stages. Each converter produces an output that is proportional to the product of the applied frequency and an input voltage. The output voltage of the second converter is proportional to the square of the frequency and is thus proportional to the tension on the wire. The voltage output is then converted to a proportional 4 to 20 mA or 10 to 50 mA current. All circuit components are encapsulated in a single module. Although most measurement applications require a direct

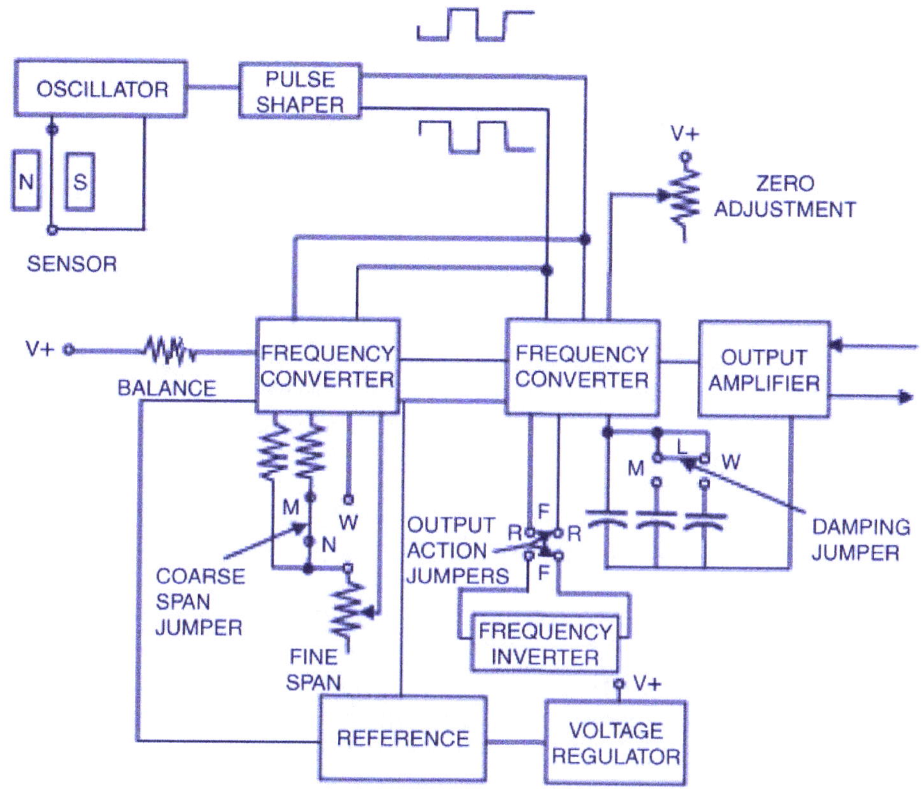

Figure 4–20 Block Diagram of a Resonant Frequency Transmitter (*Courtesy of Invensys Process Systems*)

relationship between measurement and transmitted signal, jumpers on the electronic module can be located such that the action of the transmitter is the reverse of direct operation.

The resonant frequency transmitter just described has also been called a singing wire transmitter. Other transmitters that operate on a similar principle use a quartz crystal for the resonant frequency device. In this type of device, the process fluid compresses the cylindrical quartz pressure sensor, producing internal stress in the resonator. The frequency of the resonator changes in response to the internal stress.

A quartz resonant transducer can be used for very accurate measurement at high pressure; pressure ranges from 0 to 2,000 psi and from 0 to 40,000 psi are available. The quartz resonant crystal is perfectly elastic and responds repeatably without hysteresis or drift through multiple pressure cycles at temperatures up to 175°C. Although the device has a slight, predictable nonlinearity in pressure-to-frequency conversion, the overall residual error at four different temperatures with pressure measurements from 0 to 10,000 psi has been shown to be less than ±0.005%.

The pressure-sensitive quartz resonator vibrates in the thickness/shear mode, which is the same mode used in the time base for computers. A 25-MHz computer speed is specified by the resonant frequency of the time base. This illustrates the long-term stability of thickness/shear resonators.

A block diagram of a signal-conditioning circuit for a quartz resonant pressure instrument is shown in Figure 4–21; the sensor is shown in Figure 4–22. The other two quartz crystals shown are a tuning fork temperature sensor and a thickness/shear mode reference crystal. Both are thermally coupled to the pressure crystal.

The pressure sensor is connected to its circuit by a high-pressure electrical feedthrough. The output of the pressure crystal oscillator circuit is mixed with the 7.20-MHz signal from the reference crystal. The difference between the two frequencies provides a low-frequency pressure signal approximately equal to 3.4 Hz per psi of applied pressure. The temperature sensor frequency is normally 190 kHz, with a sensitivity of 8 Hz/°C. The pressure and temperature signals and the 7.20-MHz reference signal are sent to the frequency counter.

The frequencies of the pressure and temperature signals are counted to determine the temperature-compensated pressure; the three-crystal system enables this to be done with a high degree of precision. The PC-compatible custom plug-in frequency counter uses the low-frequency pressure and temperature signals in a reciprocal counting format to limit counting error to one count of the high frequency ($1/7.2 \times 10^6$ per second gate-time). This method counts the number of pulses of the high-frequency reference during an integral number of low-frequency cycles.

Calibration of the measurement system is provided by empirically determining the pressure/temperature/frequency relationship of the quartz transducer at several different values of temperature and pressure. Each pressure transducer has a unique set of coefficients that are determined from these frequencies and pressures; a 16-term polynomial, third order in both pressure and temperature, is used to convert the measured frequency to a corresponding value of pressure. The pressure values used to determine the calibration factors of the transducer are measured with a deadweight piston gage traceable to NIST standards. The transducer then becomes a secondary standard with sufficient repeatability and accuracy.

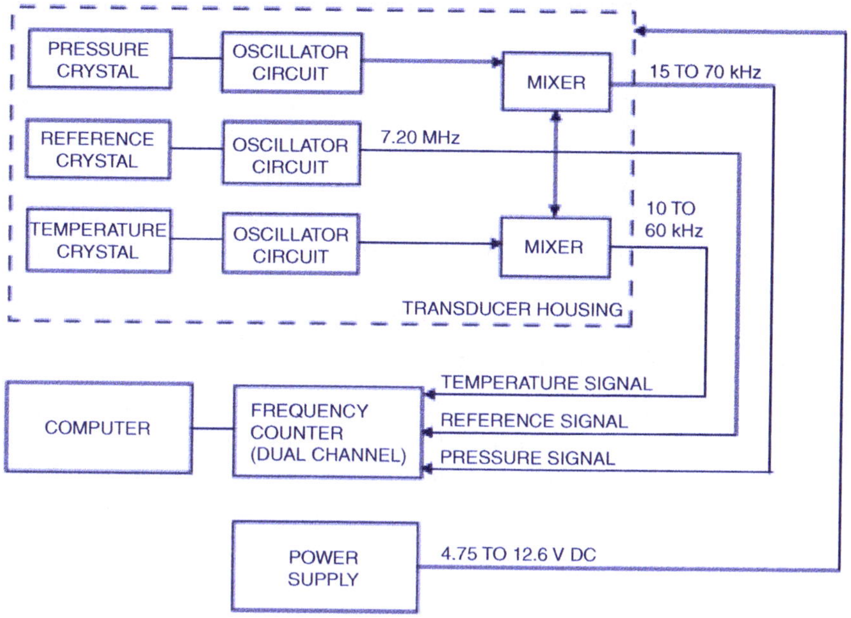

Figure 4–21 Block Diagram of a Quartz Resonant Frequency Pressure Instrument (*Courtesy of Quartzdyne Inc.*)

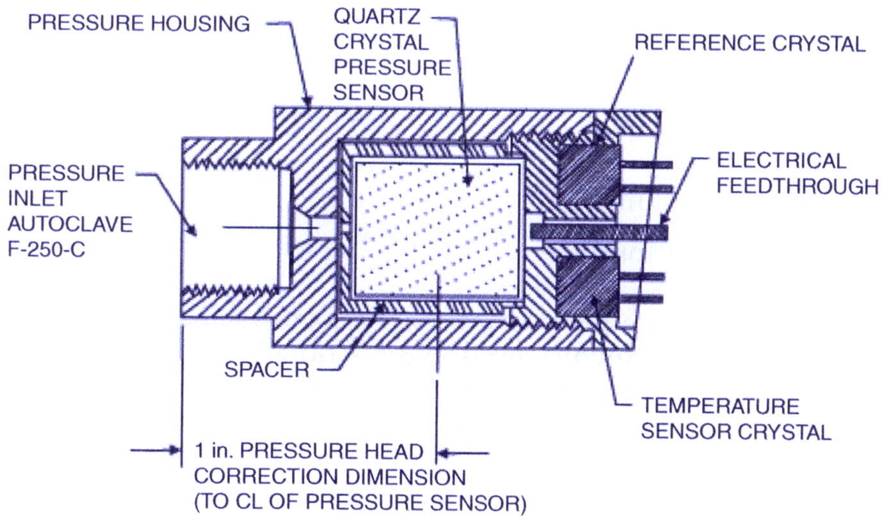

Figure 4–22 Quartz Resonant Pressure Sensor (*Courtesy of Quartzdyne Inc.*)

Traceable temperature measurement is not required for temperature compensation, because only temperature uniformity of the three crystals at each calibration pressure is necessary. Also, traceability of frequency measurement is not required when using the frequency counter, because the internal reference crystal of the transducer ensures that the frequency counted during calibration can be repeated in operations. The limited temperature sensitivity of the reference crystal is automatically accounted for during pressure transducer calibration.

4.10 SILICON RESONANT SENSOR

The output of a silicon resonant sensor is a function of the natural frequency of oscillation. Resonators are constructed from a single silicon crystal in a vacuum cavity using three-dimensional micromasking operations. The frequency of the crystal is determined solely by its physical dimensions, the mechanical constants of Young's modulus and the density of the silicon. The output frequency is very stable and highly accurate.

As pressure or differential pressure is applied to the cell containing the sensor module, the sensor's output frequency is corrected for coefficients and module information. The fully digital signal is applied to signal conditioning circuitry containing a CPU and an electronic module. The CPU provides the functions of:

- Re-ranging
- Damping
- Engineering units
- Communications

The electronic module performs the signal conversion as well as local span and zero adjustments. It should be noted that being a digital sensor, the only signal conversion is 4–20 mA and HART. For digital FOUNDATION Fieldbus and other digital communications, there is no conversion to analog. This improves the overall accuracy of the transmitter since, though small, there is a certain probability for error in each stage of conversion. A silicon resonant sensor is shown in Figure 4–23 and a block diagram of the signal conditioning circuitry is shown in Figure 4–24.

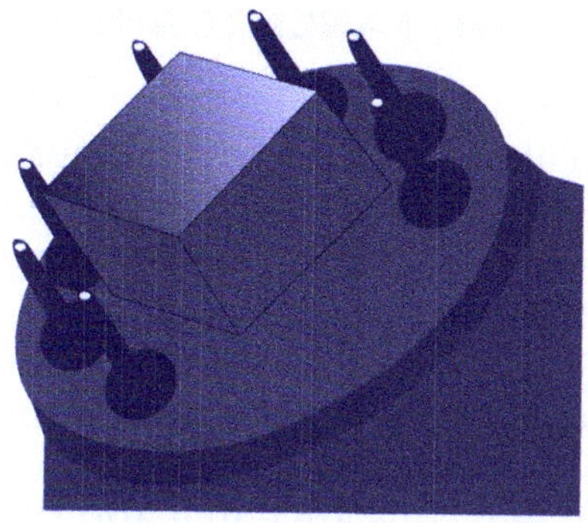

Figure 4–23 Silicon Resonant Sensor (*Courtesy of Yokogawa*)

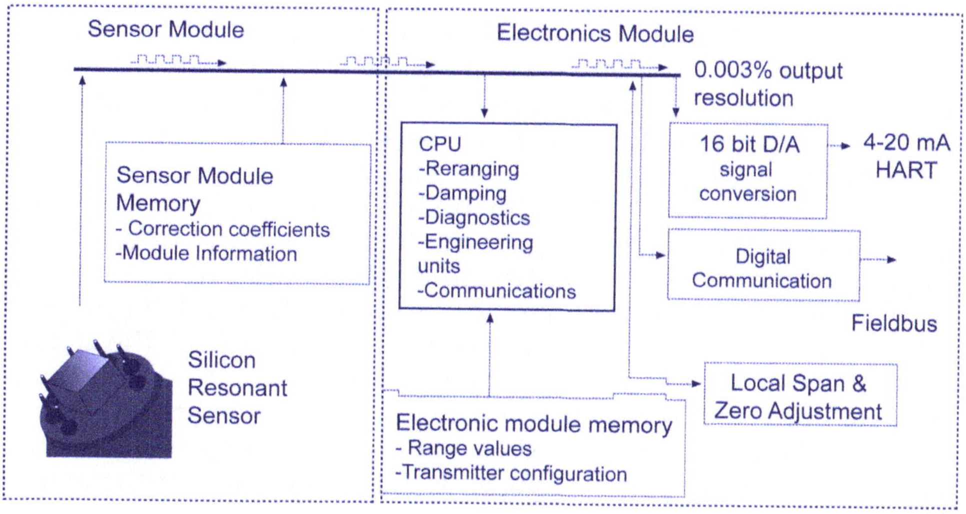

Figure 4–24 Block Diagram of a Signal Conditioning Circuit (*Courtesy of Yokogawa*)

The transmitter shown features three adjustable span/range limits: L, M, and H. The limits are from 0.4 to 40 in. WC (0.1 to 10 kPa) to 10 to 2000 in. WC (2.5 to 500 kPa). Note the significant 100:1 rangeability on the low end calibration to 200:1 rangeability on the high end calibration. The stated accuracy is given as ±0.04% of span. Static pressure measurements as well as differential pressure are provided. A local LCD can be configured to periodically display measured differential pressure, differential pressure in percent of range, scaled differential pressure, and measured static pressure. Diagnostic features will detect CPU failure, hardware failure, configuration error, static pressure and capsule temperature.

4.11 VARIABLE-RELUCTANCE TRANSDUCERS

The operation of a variable-reluctance transducer is depicted in Figure 4–25. Electrical reluctance in a magnetic circuit is equivalent to resistance in an electrical circuit. A change in pressure causes a pressure element to position an armature between two ferrite-core coils. The two impedance-matched coils have a magnetically permeable stainless steel diaphragm mounted between them. The coils are excited with an AC voltage of about 5 V rms at 3 to 5 kHz. At a zero reference pressure, the diaphragm is exactly midway between the two coils, the magnetic flux densities of the two coils are equal and the inductances of the two coils are equal. The coils are wired in series in an AC bridge circuit, and the bridge is balanced.

When pressure is applied to the pressure element, the diaphragm is directed toward one coil and away from the other. The inductance of one coil increases, the inductance of the other decreases, and the bridge is unbalanced. The bridge unbalance voltage is applied to a carrier demodulator (Figure 4–26). The carrier oscillator is a low-distortion Wein bridge type with a differential amplifier detector; a field effect transistor acts as a voltage-controlled resistor and balances the bridge. Both the oscillator frequency and the amplifier are regulated by a stable, temperature-compensated Zener diode. The oscillator supplies 5.00 V rms at 3 or 5 kHz sine wave to the transducer from the center-tapped secondary of a transformer. Grounding the center tap completes the transducer bridge and generates an output that is amplitude proportional to bridge unbalance and sense dependent in the unbalance direction. This AC output is fed to the input of the amplifier/demodulator.

The output of the bridge detector, the zero control, and the auxiliary balance pin are summed into the span potentiometer. The wiper feeds the AC amplifier that drives the demodulator stage. The output uses an active filter to control the carrier ripple on the demodulator output that supplies the final output amplifier, which is stabilized by feedback with a filter. It can also be used to scale the output for different signal ranges.

The full-scale travel of the diaphragm is only 0.001 or 0.002 in. The coils are nominally 20 mH inductance (at zero) and change 1 mH over the full pressure range of the diaphragm. This equates to an output sensitivity for the typical variable reluctance transducer of 20 to 40 mV/V rms at FS (full scale). The accuracy is 0.25%, which includes nonlinearity, hysteresis, and nonrepeatability.

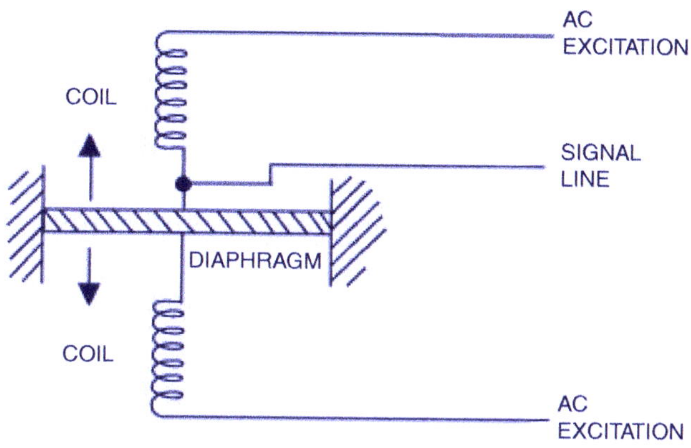

Figure 4–25 Operation of a Variable-Reluctance Detector (*Courtesy of Validyne Engineering Corporation*)

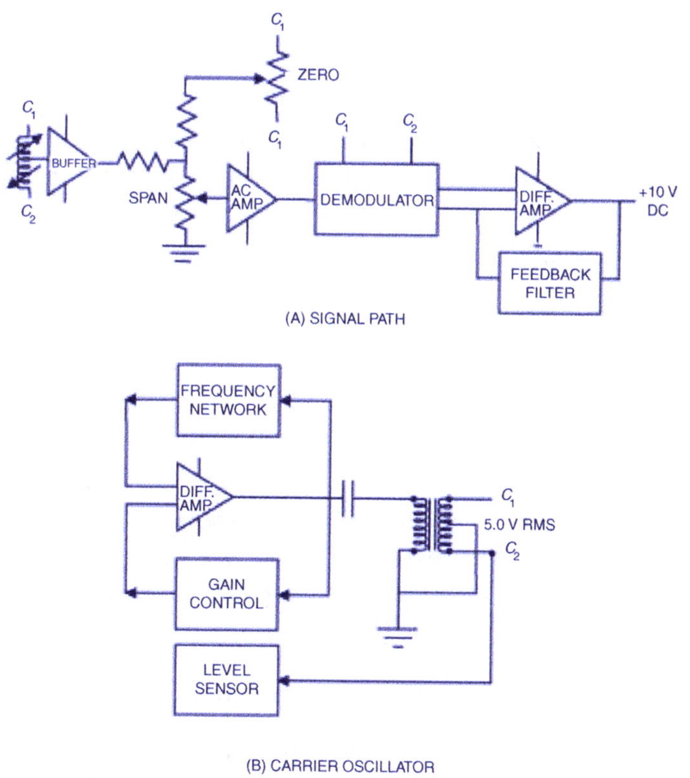

Figure 4–26 Simplified Carrier Demodulator for a Variable-Reluctance Transmitter (*Courtesy of Validyne Engineering Corporation*)

The AC signal from the transducer is sent to a synchronous carrier oscillator/demodulator, which supplies the coil excitation and provides a high-level DC output signal. The carrier demodulator also provides span and zero adjustments for calibration settings, as well as filtering. Optional carrier-phase adjustments allow the transducer to be mounted up to several hundred feet away from the electronics. The many styles of carrier demodulators include high-density rack-mounted versions, two-wire 4–20 mA DC output versions, 0 to 5 V DC and 0 to 10 V DC single-channel versions, alarm relay output types, and many others.

An industrial-style variable-reluctance transmitter set up to control furnace draft on the factory floor can be calibrated to 0 to 0.1 in. H_2O for an output of 4 to 20 mA.

The only moving part in a variable-reluctance sensor is the stainless steel diaphragm, which is accessible by simply removing the housing bolts. The diaphragm is not attached to any linkages, and thus can be changed by the user to suit various measurement ranges. Consequently, a single transducer can cover a wide range of pressures without extreme electronic amplification. Any FS output between 0.125 and 3200 psi is achievable with one transducer, one carrier demodulator, and a sufficient number of diaphragms. About two dozen diaphragms cover this spread of pressures.

A variable-reluctance transducer depends only on the position of the diaphragm to sense pressure. If the diaphragm has a large surface area, a proportionally smaller pressure will produce enough force to deflect the diaphragm to FS. One model has a diaphragm with a diameter of 4 in. and can be calibrated to 0.25 in. H_2O FS. Using special high-gain carrier demodulators, the same sensors have been calibrated to as little as 0.005 in. H_2O FS.

Another advantage of the variable-reluctance transducer is fast dynamic response to low pressures. There are no mechanical linkages or sensor-cavity fill fluids to damp out rapidly changing pressures. A 1-in. H_2O FS transducer, for example, will have a flat response (in air) to about 100 Hz. This makes the variable-reluctance transducer ideal for investigating low-pressure transient phenomena.

The simplicity of the variable-reluctance transducer design has other advantages. Because the sensor consists of only two coils and a diaphragm, the transducer resists shock damage quite well. Another advantage is high-pressure measurement; the sensor housing can be made to handle pressures up to 15,000 psi. Because the sensor uses AC excitation and returns an AC signal, the transmission line can be tuned to work over several hundred feet with negligible signal loss. This allows greater distance between the sensor and its electronics than is typically practical for other types of transducers.

Because variable-reluctance transducers rely on uniform magnetic properties, only a few construction materials are suited to this technology, including 410 stainless steel and 17-7 PH stainless steel. Type 410 stainless steel is susceptible to corrosion in an oxidizing environment and is generally not suited to acid or corrosive chemical service. Type 17-7 PH stainless steel does better in oxidizing environments, but lends itself only to transducers for the higher pressure ranges. Failure of the diaphragm is the most common result of corrosion. In some cases this failure is simply tolerated, and a new diaphragm is installed when required. Nickel or gold plating may also be specified to reduce the rate of diaphragm corrosion.

Variable-reluctance transducer design does not allow for mechanical or hydraulic overpressure protection. Overpressure is usually limited to twice the nominal range of the sensing diaphragm. If this limit is exceeded, excessive zero and span shifts may result. In the worst case, the diaphragm must be replaced and the instrument recalibrated.

Variable-reluctance transducers are often used in applications involving very low pressures. One typical application is clean-room pressure control. In order to keep stray dust particles out of a clean room, the room is kept at a slightly positive pressure with respect to the atmosphere. Variable-reluctance transmitters provide 4 to 20 mA signals to control airduct fans and dampers.

In medical applications, variable-reluctance transducers are used to measure pulmonary pressure, breathing rate, and lung volume. A small restriction is placed in the breathing line, which develops a differential pressure that is linearly proportional to the flow rate. This allows the breathing flow rate to be displayed as a function of time on a chart recorder for clinical diagnosis of pulmonary illnesses. Pulmonary studies of laboratory rats and guinea pigs are also possible using high-gain carrier demodulators and variable-reluctance transducers.

Industrial applications include leak detection on pressure vessels, boiler and furnace draft control, and gas-flow measurements. FM-approved (Factory Mutual Insurance, now FM Global) versions are available for use in hazardous areas. With appropriate electronic displays, variable-reluctance transducers also serve as digital manometers, replacing mercury U-tube types.

Variable-reluctance transducers are available in a variety of configurations. In many models, the carrier demodulator electronics are bundled together with the sensor to provide DC-in/DC-out convenience for the user. One version even looks electrically like a strain gage and will interface easily with existing signal-conditioning systems. For work involving measurement of very low pressures, variable reluctance technology can have advantages.

4.12 PIEZORESISTIVE TRANSMITTERS

Because of new precision manufacturing techniques, major breakthroughs have been achieved in diffused piezoresistive silicon chip performance characteristics and reliability.

In manufacturing, boron is diffused into the crystal lattice structure of the silicon to form a fully active Wheatstone bridge. In this diffusion process, boron and silicon are united at the molecular level, eliminating the need for the mechanical bonding methods normally required with strain gage sensors. This process enables the manufacturer of sensors to achieve exceptional repeatability and long-term stability.

The pressure range of each silicon sensor is determined by the thickness of the silicon directly under the Wheatstone bridge. The thickness of the silicon diaphragm is determined by chemically etching the back of each chip under the bridge to a specific depth. The finished chip is then bonded to a Pyrex® or alumina back plate which supports the chip and isolates it from environmental stresses. For gage and differential pressure applications, a hole is bored through the Pyrex for access to the cavity at the back of the chip. This provides an atmospheric pressure reference for gage pressure sensors and a connecting passage to the low pressure side for the differential pressure (d/p) cell. For absolute pressure applications, the chip cavity is evacuated before the Pyrex backplate is attached, providing a constant absolute pressure reference.

The chip is then mounted onto either a gas-tight ceramic lead frame or a glass-sealed stainless steel header and pin assembly. Gold-wire connections are made from the bridge circuitry on the silicon chip to electrical posts on the lead frame or header. The pressure sensor subassembly is then joined to the sensor assembly packaging.

Isolating diaphragms of various corrosion resistant materials are welded in place over the sensing chip, and the cavities between the diaphragms and the chip are then filled under vacuum with DC-200 silicone oil or Fluorinert® inert fluid. This process totally isolates the silicon sensor from the process fluid without using a mechanical linkage. The isolating diaphragm also provides overrange protection for the silicon sensor in the d/p cell.

The solid-state sensor consists of a Wheatstone bridge circuit (Figure 4–27) which is diffused into a silicon chip, thereby becoming a part of the atomic structure of the silicon. As pressure is applied to the IC chip diaphragm (Figure 4–28), strain is created in the bridge resistors. Piezoresistive effects created by this strain change resistances in the legs of the bridge, producing a voltage proportional to pressure. Output from the bridge is typically in the range of 75 to 150 mV at full scale pressure for a bridge excitation of 1.0 mA.

The micromachined silicon sensor is fabricated into three basic types of pressure sensors. The three types, which are shown in Figure 4–28, are

1. Gage Pressure—sensor referenced to atmosphere.
2. Absolute Pressure—sensor referenced to a full vacuum.
3. Differential Pressure—sensor measures the difference between two pressures (P1 and P2).

Because of the wide rangeability of the sensors, only four different sensor units cover the entire range of d/p spans from 3.5 in. H_2O to 3000 psi (0.87 kPa to 20.7 MPa), thus reducing spare parts requirements.

An internal temperature sensor monitors temperature of the pressure sensor. This temperature is used to compensate the sensor output for the effects of temperature changes. The sensor temperature is also transmitted digitally for monitoring, alarming, and for control of transmitter heat tracing, etc.

A representative piezoresistive pressure transmitter is shown in Figure 4–29.

4.13 FLAPPER-NOZZLE TRANSMITTERS

The heart of most pneumatic transmitters, the signal-generating device, is the flapper-nozzle assembly (Figure 4–30). This device is also

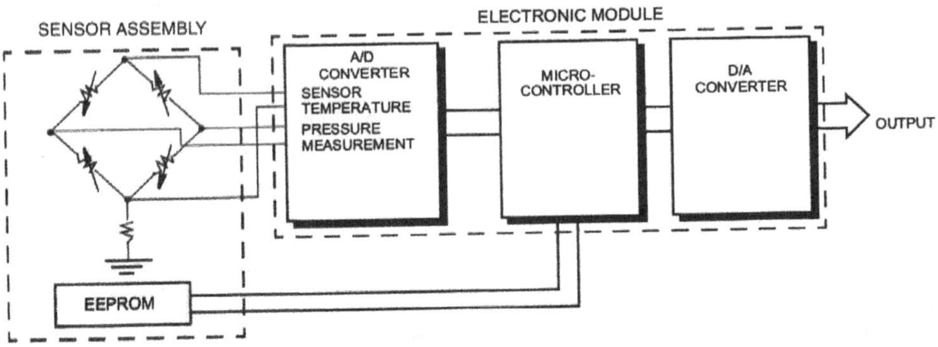

Figure 4–27 Block Diagram of a Piezoresistive Pressure Transmitter
(Courtesy of Invensys Process Systems)

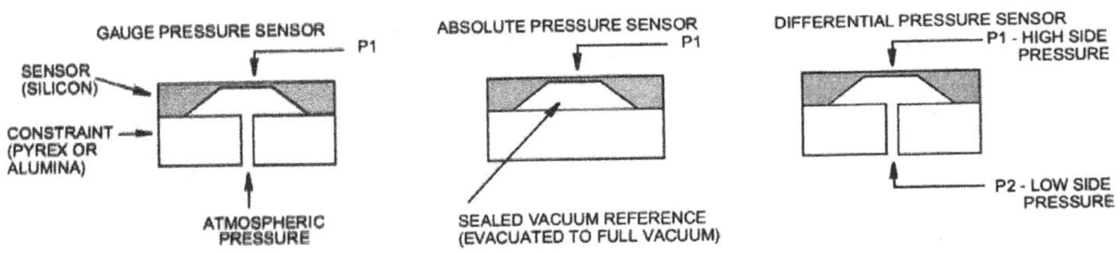

Figure 4–28 Piezoresistive Cell Configurations
(Courtesy of Invensys Process Systems)

called a baffle-nozzle. The initial movement that generates the transmitted signal in pressure-measurement applications is one of the pressure elements previously discussed.

The baffle—or flapper, as it is referred to in this discussion—is normally pivoted on one end to make a class II or class III lever system. In some applications it may pivot at some point between the two ends, resulting in a class I lever system. In all three designs, a change in movement of the pressure element will result in a displacement of the flapper with respect to the nozzle, changing the distance between the two components.

Air is supplied to the nozzle at a fixed pressure of 20 psig, and the airflow through the nozzle is regulated by the fixed-area restrictor. As the flapper is positioned closer to the nozzle, airflow through the nozzle is restricted because of the reduced clearance between the flapper and the nozzle. As the flow through the nozzle decreases, the nozzle backpressure increases. Conversely, as the flapper is positioned farther from the

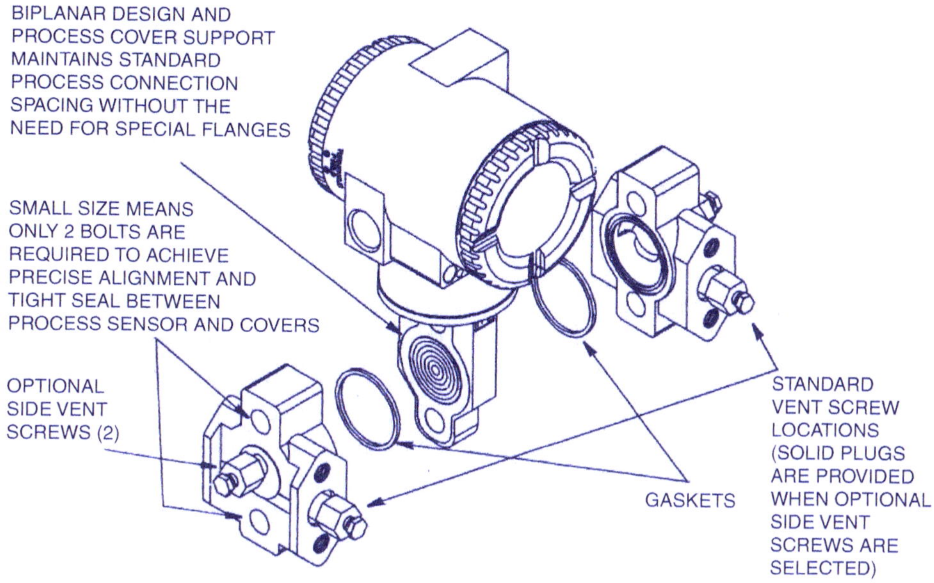

Figure 4–29 A Piezoresistive Pressure Transmitter (*Courtesy of Invensys Process Systems*)

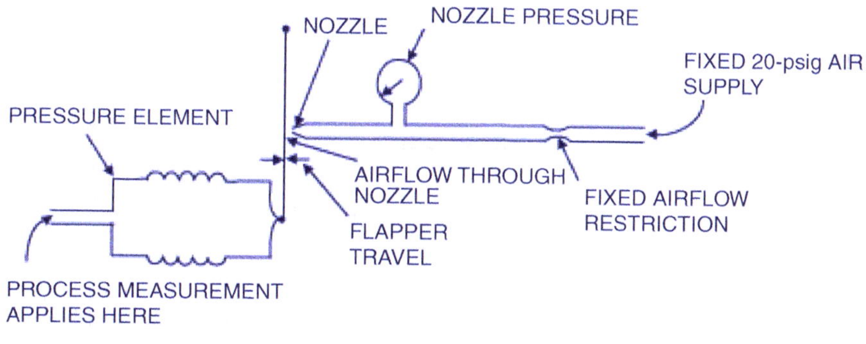

Figure 4–30 Flapper-Nozzle System

nozzle, the flapper-nozzle clearance increases. This causes less flow restriction of air through the nozzle, resulting in an increase in flow and a decrease in nozzle backpressure. From this discussion it can be seen how the nozzle backpressure is a result of flapper-nozzle clearance, which is a function of the mechanical displacement of the pressure element.

The operation of a flapper-nozzle system to generate a scaled pneumatic signal proportional to pressure applied to a measuring element results in a small amount of flapper travel. This movement normally traverses an arc as the flapper is pivoted at one end. The movement is not linear with respect to pressure. However, the amplified nozzle backpressure, the instrument output, must be linear. The problem is solved by limiting the flapper travel to an insignificant few thousandths of an inch.

To detect very small changes in pressure-element displacement and flapper response, the very small resulting change in nozzle backpressure must be amplified. The amplified nozzle pressure is the transmitter output and is proportional to the measured process pressure. The nozzle pressure is amplified by a pneumatic amplifier, or relay. The relay also serves to boost the volume of the air passing through the nozzle. Without this volume boost, the output response of the transmitter would be sluggish because the output air would be required to flow through the nozzle restrictor. The relay not only boosts the volume of the transmitter upon an increase in process pressure measurement, but also quickens the response of the transmitter to a decrease in pressure measurement. Without the relay, all of the air in the output circuit of a pneumatic pressure transmitter would be required to bleed through the nozzle opening.

4.14 PNEUMATIC RELAY

An increase in nozzle pressure, resulting in the flapper moving closer to the nozzle, is applied to one side of the relay diaphragm. This causes the diaphragm to move to the left (Figure 4–31A and B). The movement positions the ball farther from the ball seat and increases the airflow from the air-supply chamber to the output port of the relay. This increases the output pressure at this port, and the same pressure is applied to the valve surface area. The output pressure increases until the pressure on the valve is great enough to create a force equal to that of the nozzle pressure exerted over the much larger area of the diaphragm. The forces on both sides of the diaphragm are equal, resulting in a state of equilibrium about the diaphragm, valve, and valve seat.

It is important to note that nozzle backpressure on the diaphragm will be balanced by the output pressure on the valve. The output pressure will always change in the direction and by the amount required to balance the nozzle pressure on the diaphragm. This is the force-

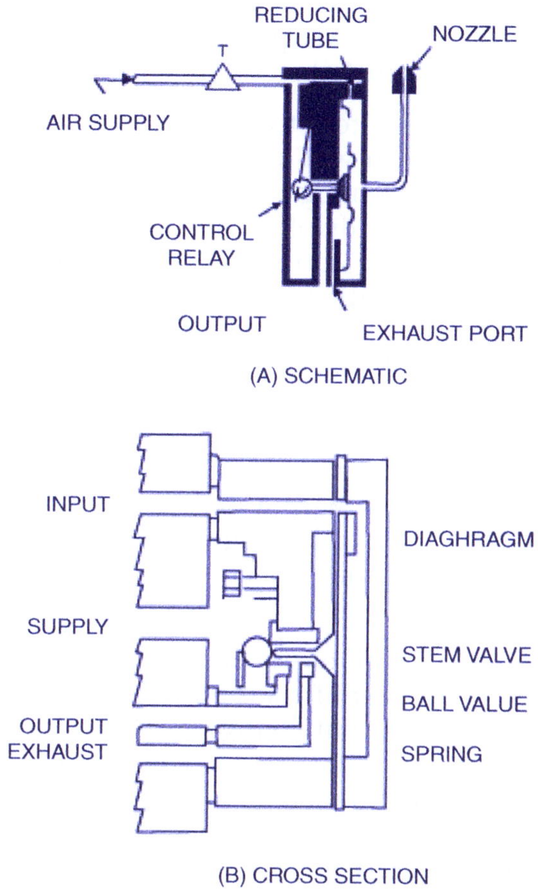

Figure 4–31 Pneumatic Relay (*Courtesy of Invensys Process Systems*)

balance principle, which is the operating principle of many electric and pneumatic pressure transmitters. Relay amplification is a result of the difference between the area of the diaphragm and that of the valve. Remember that force is equal to pressure times area ($F = PA$), and for equal forces, pressure increases as area decreases. A large output pressure acting over the small valve area is required to balance the low nozzle pressure acting over the large surface area of the diaphragm.

When the flapper moves away from the nozzle, decreasing the back-pressure, the cantilever spring forces the ball to the right, causing it to restrict the opening between the supply and output ports. This restricts the airflow between the two ports, reducing the output pressure, which bleeds through the valve seat and exhaust port to atmosphere. The force-balance principle applies to the flapper-nozzle operation for decreases in measurement. The output pressure will decrease until force caused by the output pressure balances that of the nozzle pressure.

The output port is connected to another pressure element in a receiving instrument, and air will flow from the air-supply chamber to the output port only long enough to establish a state of equilibrium between the two sides of the diaphragm after a change in nozzle pressure. The air flowing from the output port to the exhaust port, which is regulated by the valve position with respect to the valve seat, is constant for every condition of equilibrium and will change only when the nozzle pressure changes. The airflow from the output port to the exhaust port will increase with a decrease in nozzle pressure and decrease with an increase in nozzle pressure. Also, the airflow from the supply chamber to the output port will increase with an increase in nozzle pressure and decrease with a decrease in nozzle pressure. The reducing tube or restrictor, which regulates the airflow through the nozzle, limits the flow to that which can flow through the nozzle when the flapper-nozzle clearance is greatest.

The gain of pneumatic amplifiers is about 15 to 20. A common gain used in many pneumatic amplifiers is 16. A change in nozzle pressure of 0.75 psig results in an output change of 12 psi, caused by a change in flapper-nozzle clearance of 0.0006 in. The maximum flapper-nozzle clearance is about 0.003 to 0.005 in.

4.15 NEGATIVE FEEDBACK

As previously noted, the flapper travel required to generate a full change in transmitter output pressure is very small. To eliminate erratic output variations, thereby increasing the operational stability of the transmitter, negative feedback is applied to the flapper-nozzle mechanism. The balance portion in the force-balance principle is the result of this negative feedback. This principle, which was discussed in the operation of the pneumatic relay, is also used in the overall operation of a force-balance transmitter and is illustrated in the force-balance transmitter shown in Figure 4–32.

An increase in pressure on the diaphragm capsule positions the flapper or force bar closer to the nozzle, increasing the nozzle backpressure. This increase in pressure is amplified by the relay and becomes the transmitter output, which is applied to the feedback bellows. The bellows exerts a force on the force bar in opposition to that caused by the

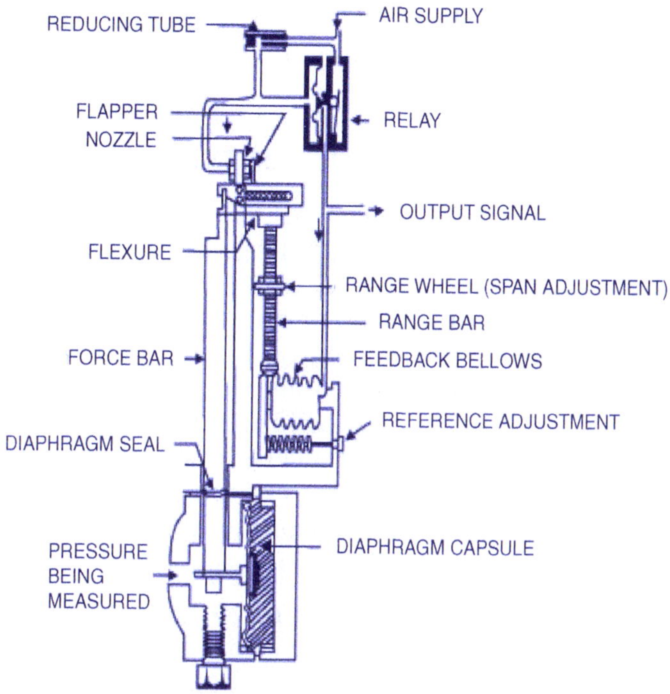

Figure 4–32 Force-Balance Pressure Transmitter Schematic (*Courtesy of Invensys Process Systems*)

diaphragm capsule. During stable operation of the transmitter, when the process pressure is not changing, the two forces are balanced; that is, the force bar moment arm of the feedback bellows is equal to that of the diaphragm capsule. Any change in measurement pressure will upset this balance. First, it will change the flapper-nozzle relationship by unbalancing the forces. Second, the resultant output will change until a great enough force is generated by the feedback bellows to balance the force of the measurement capsule. The feedback force is said to be negative because the force is opposite to or opposes that of the diaphragm capsule which caused the initial disturbance and change in output.

The amount of pressure in the feedback bellows required to balance the force generated by the measuring sensor is determined by the size of the bellows, usually a constant, and the distance of the bellows from the fulcrum. As the distance is increased, the mechanical advantage of the resulting moment arm increases and less pressure is required to balance a given force created by the diaphragm capsule. Conversely, when the distance is decreased, a greater pressure is needed to balance the force of the process pressure. Changing the mechanical advantage of the feedback bellows is a convenient means of changing the gain or span of the transmitter.

4.16 SUMMARY OF TRANSMITTER TYPES

Several types of technology used in pressure measurement and transmission have been explored and discussed in this section. No attempt has been made to mention all methods of detection and signal generation or to promote the application of one over another. No technology is clearly superior or inferior. All methods discussed are available from various manufacturers, with advantages often related to the specific application. While it may not be common practice to choose a transmitter with specific types of transducers, it is nearly always important to consider such factors as measurement range, adjustment capability, accuracy, stability over time, environmental conditions, mounting ease, reliability, maintenance, calibration requirements, and cost.

It would be ideal to find a general-purpose transmitter for all applications. If this were possible, a working knowledge of the available technologies certainly would be most helpful. Conversely, many manufacturers have designed instruments for specific measurement applications, with well-written instruction manuals that document foreseen limitations. Less common applications with very detailed specifications are often difficult to match with a suitable instrument.

Various types of transducers are used in measuring instruments, but few common industry standard characteristics and specifications have been developed for transmitters in the manufacturing and processing industries. Simply as a result of usage and acceptance by various industries, this pseudostandardization has resulted in the common applications of many instruments and to a lesser extent in the setting of standards by a few professional organizations and agencies. A typical transmitter will usually adhere to the adopted industry standard regardless of the measuring transducer and signal conditioning used.

Pressure transmitters are categorized according to the range of pressure expected to be measured and the reference point of measurement. Three types of pressure transducers used in transmitters are generally available (Figure 4–33):

- Absolute pressure

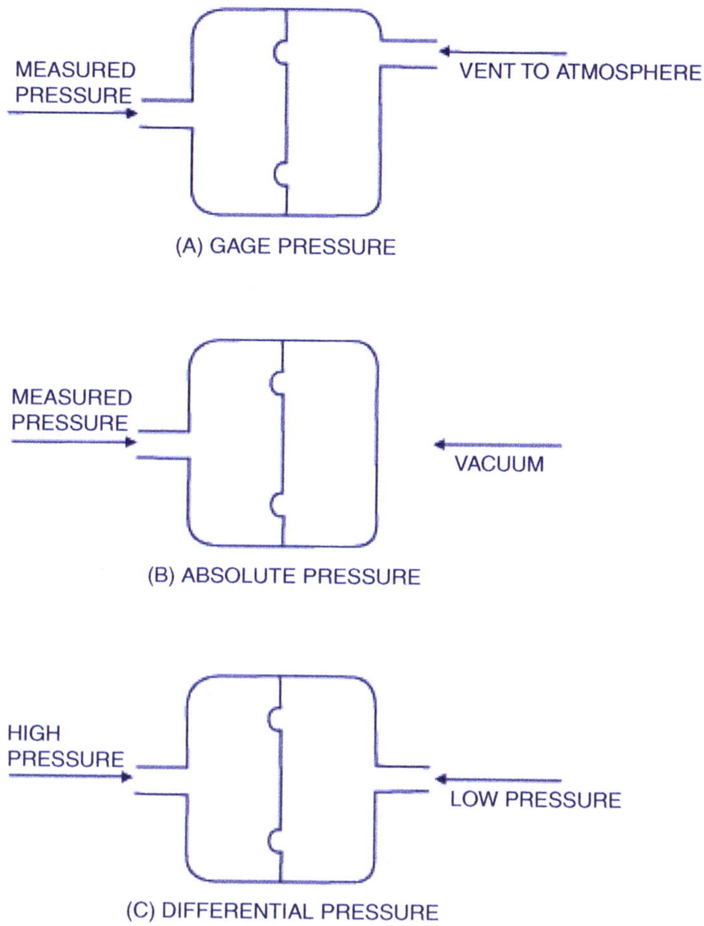

Figure 4–33 Basic Pressure Transducer Configuration

- Relative (gage) pressure
- Differential pressure

Pressure measurement is always made with respect to a reference point. Gage pressure is referenced to atmospheric pressure. A gage pressure measurement represents a pressure level above atmospheric pressure, which is 0 psig. Absolute pressure measurement represents a pressure level above a complete vacuum, which is the absence of pressure, or 0 psia. In either case, a measurement represents the difference in pressure between a value and the reference level. In a strict sense, this is a differential-pressure measurement. If the reference value were to change (a change in atmospheric pressure for gage measurement, for example), the measured value would change by that amount, causing an error in the measured value. When measuring pressure with respect to a reference value, the reference point is assumed to be constant.

Absolute pressure is measured relative to a constant reference pressure, such as a true vacuum. Sometimes instead of a true vacuum a sealed reference pressure is used, and the instrument response is permanently offset to correct for the pressure of the reference. Although such instruments are easier to make, they are temperature sensitive because of the temperature/pressure relationship of a gas in a fixed volume and generally find limited use because of their narrow operating temperature range. Only pressure transmitters with a true vacuum used as a reference pressure can properly be called absolute pressure instruments. The measuring range starts at zero pressure on the absolute scale.

As has been discussed, it is sometimes desirable to measure a pressure value that is independent of a reference point. In such an application, the resulting measurement represents a difference between two pressure points or locations. When either or both pressures change, a new measurement value results. This type of application is called differential-pressure measurement, and the measuring device is known as a differential-pressure instrument.

4.17 SAFETY TRANSMITTERS

Pressure and differential pressure transmitters may be used to detect hazardous conditions. These instruments are often used in Safety-Instrumented Systems (SIS) that monitor processes to bring them to a safe condition if certain thresholds are violated, resulting in potential shutdown situations. For such safety-related applications, and depending on the required Safety Integrity Level (SIL), two or even three transmitters may be used for redundancy. A two of three voting procedure is used by the safety control system to activate an emergency system. Depending on the measuring points and the level of security required, the necessary redundant transmitters can increase system cost by as much as ten-fold.

A new class of pressure and differential pressure transmitters are now available that are specifically designed and independently certified for use in SISs. These transmitters incorporate special redundancy and diagnostic capabilities to increase the level of adaptability to SIL 2 with a single transmitter. Features that enable this are

- A high redundancy level and self-diagnostics.
- A high level of measurement diagnostics.
- The ability to transmit diagnostic information to the system's shutdown controller.

The added redundancy and diagnostics, and the independent certification of transmitters by FM Global or another agency, will increase the cost of the safety transmitter; but, the potential cost savings over a multiple-transmitter system can be substantial.

The sensor in a safety transmitter typically generates two signals representing the process value. A microprocessor in the receiving device validates the information by comparing the two values. And, should the comparison test fail, the output is driven up or down scale for a safe condition. If the two signals match, that is, they are equal to a specified threshold level, a 4–20 mA output is generated that is proportional to the measured value. A supplementary diagnostic circuit enables a shutdown condition when a fault occurs in the analog circuitry of the device or the power supply. The safety transmitter converts the 4–20 mA output back to a digital signal and compares it to the expected value calculated earlier. When they are different, the indication is that the output circuits have failed. Further detection of a failure mode can be provided by the NAMUR NE-43 recommendations (Table 4–1) where the 4–20 mA output serves as a diagnostic indicator of certain types of failure. Many non-SIS transmitters also have the ability to provide fail-safe and diagnostic signals using the 4–20 mA signal as shown in Table 4–1 or by utilizing HART or another digital communication protocol.

In conclusion, the main criterion in using safety transmitters is certification by an independent third party to ensure that they have the required redundancy and diagnostic checks to achieve a stated safety integrity level. Replacing multiple conventional transmitters with fewer safety transmitters provides savings in maintenance and installation costs. However, one should not neglect the potential savings that come from avoiding nuisance trips, a benefit of redundant, voting transmitters.

Table 4–2 shows a checklist for specifying a safety transmitter.

Table 4–1 The NAMUR NE-43 Recommendations

Mode of Operation	Output Signal, mA
Normal	4–20.0
Normal underrange	3.8–4.0
Normal overrange	20.0–20.5
Transmitter failure	3.6–3.8
Transmitter failure	20.5–22.0
Probable open field wire	0–3.6
Probable shorted field wire	22.0 or more

4.18 DIFFERENTIAL PRESSURE MEASUREMENT

Any of the pressure elements previously discussed can be used as transducers for differential pressure measurement. Recall that pressure applied to the inside of a pressure element causes a deformation of the element, resulting in movement at the free end. The only force producing the movement is caused by the pressure exerted over the area on the inside of the pressure element. The force must overcome the elasticity of the element or a spring (sometimes both), which resists or is in opposition to the movement. Imagine, however, what would happen if the element was sealed in a chamber and a pressure other than atmospheric pressure was applied to the chamber. This pressure would create a force on the outside of the element that would oppose the force caused by the pressure on the inside of the element. If both the inside area and the outside area were the same, equal pressure on both sides would create equal opposing forces, which would cancel each other. There would be no movement of the pressure element. A pressure change on either side of the pressure element independent of that on the other side, however, would result in a movement of the element. Such a device is a differential-pressure transducer.

Figure 4–33C illustrates a transducer arrangement used to measure differential pressure. The movement or deflection of the pressure element is a result of the difference in pressure on the two sides of the element. The resultant force on the diaphragm that causes a movement is shown by the following relationships:

Table 4–2 Checklist for Specifying a Safety Transmitter [Ref. 1]

Checklist for Specifying a Safety Transmitter
• Has the transmitter been certified to the desired level of safety by an independent, recognized agency, such as TÜV (Technischer berwachungs-Verein) or FM Global?
• Does the transmitter contain redundant sensor and signal processing circuitry?
• If internal diagnostics detect a failure, can the transmitter signal the controller to take appropriate action?
• Can the transmitter monitor its current output, compare it with the computed value of the output, and send a failure signal if the two values disagree?
• Can the transmitter determine whether its power supply is within specification? A degraded power supply may prevent the transmitter from developing a proper value for the measured variable.
• If the 4–20 mA loop impedance becomes too high for proper operation, can the transmitter detect this condition and send a signal to the controller?
• If self-diagnostics detect a gross transmitter failure, can the transmitter drive its analog signal either below 4mA or above 20mA according to the NAMUR directives?
• Are the hardware and software redundant enough to provide the transmitter specification with a high value for mean time between failure (MTBF)? Safety transmitters with an MTBF value of more than 100 years are on the market.
• Does the transmitter comply with the new Pressure Equipment Directive that ensures mechanical robustness?

$$F_1 = P_1 A_1 \tag{4.17}$$

$$F_2 = P_2 A_2 \tag{4.18}$$

where F_1 is the force on the high-pressure side of the element, P_1 is the pressure on the high-pressure side of the element, A_1 and A_2 are the surface areas of each side of the element, and F_2 and P_2 are similar values for the low-pressure side of the element.

By design of the element, $A_1 = A_2 = K$. The K value is insignificant since it represents the area of the high and low sides of the pressure element which are equal and do not cause a net force acting on the element. This net force is related only to differential pressure. By substituting into Equations 4.17 and 4.18,

$$F_1 = P_1K \tag{4.19}$$

$$F_2 = P_2K \tag{4.20}$$

and by subtracting the results of Equation 4.20 from Equation 4.19,

$$F_1 - F_2 = (P_1 - P_2)K \tag{4.21}$$

it can be seen from Equation 4.21 that the difference in force on the two sides of the element, which causes a resulting movement, is caused by a difference in pressure on the two sides.

In referring to differential-pressure measurement, the pressure on one side of the element, the high side, is the high pressure. That on the low side is the low pressure. The movement of the element is used as an indication or to generate a signal just as in pressure measurement. Most transmitters today are designed such that the two sides are identical, and mechanically, it makes little or no difference which side has the higher pressure. Typically, the transmitter is designed such that increasing the pressure on the high side will produce an increase in the output signal, while increasing the pressure on the low side will produce a decrease in the output signal.

Factors which can cause error in differential pressure measurement are ambient temperature, static line pressure, and drift stability. In most cases, differential-pressure transmitters can tolerate 50°F or more fluctuations. This should be checked by simply blowing hot or cold air over a transmitter on a calibration bench. Static pressure alignments can also be checked by applying a differential pressure to the instrument and changing pressure by equal amounts on both sides and monitoring the output. The signal drift of a transmitter caused by temperature and static pressure should be no more than 0.075% of reference accuracy over a period of several months for a digital transmitter. The reference accuracy of a transmitter is not useful for predicting installed repeatability. Installed repeatability can be improved by calibrating to a higher range when used for flow measurement.

4.19 DIFFERENTIAL-PRESSURE APPLICATIONS

A very common and practical application for a differential-pressure transmitter is in level measurement, where the actual level value is inferred from a pressure measurement. Recall from Chapter 2 the relationship between liquid level and hydrostatic head pressure. It should be remembered that a direct relationship exists between the hydrostatic pressure caused by a column of liquid, the specific gravity of the liquid, and the height of the vertical column. With the specific gravity a known constant, the relationship between the pressure exerted by a column of liquid and its height can be used to infer a value of liquid level from a pressure measurement.

Consider the illustration in Figure 4–34. The pressure applied to the pressure transmitter must be independent of all variables except the liquid level in the tank. Notice that atmospheric pressure is exerted on the liquid surface. Since the pressure instrument measures gage pressure, the measurement is referenced to atmospheric pressure, and a change in this pressure on the liquid surface will be counteracted by the same change in atmospheric pressure on the outside of the pressure element in the transmitter. The measurement is therefore independent of all variables except the hydrostatic pressure caused by the height of the column of liquid (which is the liquid level value).

Now consider an application where the pressure on the surface of the liquid is different from that on the reference side of the pressure element in the transmitter. Such would be the case in the previous example if atmospheric pressure at the liquid surface in the tank was different from that at the measuring transmitter. A more common example, however, would be if the tank were closed. The pressure on the liquid surface would be other than atmospheric. Such an application is called closed tank level measurement (Figure 4–35).

4.20 CLOSED TANK LEVEL MEASUREMENT BY DIFFERENTIAL PRESSURE

If the pressure in the closed tank changes, equal force will be applied to both sides of the differential pressure cell. Since the differential-pressure cell will respond only to changes in a difference in pressure on the two process port connections, a change in static pressure on the liquid surface will not change the response of the transmitter. The output of the transmitter is thus proportional to a difference in pressure caused by the liquid level. If liquid were to become trapped in the low-side

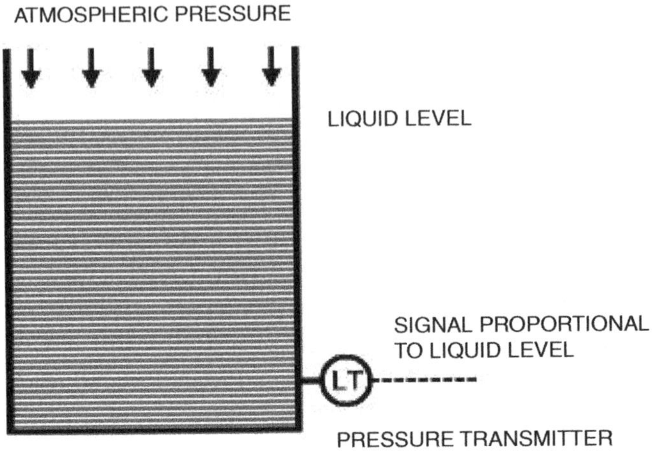

Figure 4–34 Hydrostatic Level Measurement in an Open Tank

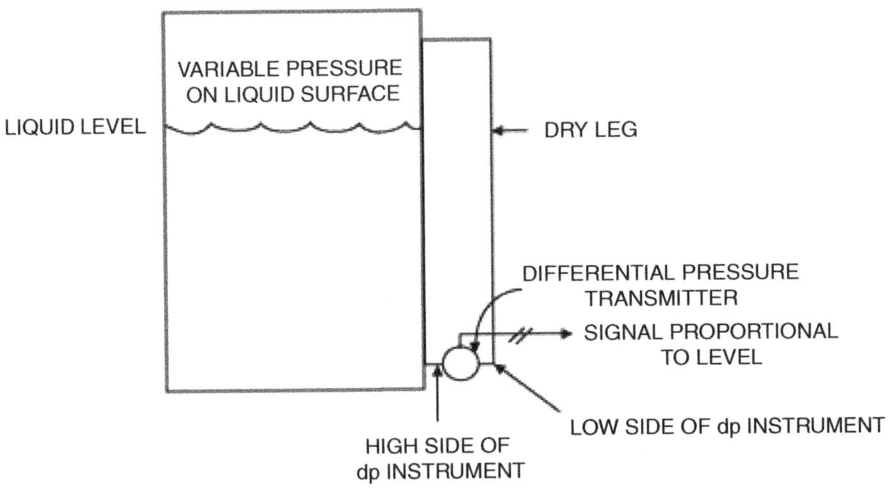

Figure 4–35 Closed Tank Level Measurement by Differential Pressure

connecting pipe, caused by the condensation of vapor or other means, the measurement would be in error by an amount equal to the height of this liquid. Correction for this condition is discussed in Chapter 6.

4.21 FLOW MEASUREMENT BY DIFFERENTIAL PRESSURE

A common application involving differential pressure is flow measurement. As a volume of fluid flowing through a pipe is forced through a restriction of reduced area, the velocity of the flowing fluid increases.

As the velocity increases, resulting in an increase in kinetic energy, certain physical laws enable this change in energy to be expressed as a change in pressure or potential energy. The pressure change resulting from this transformation of energy is used to infer a flow value, and the differential pressure, which is referred to as head, is measured by a differential-pressure instrument.

Figure 4–36 shows the pressure-drop gradient along a section of pipe with a differential producer installed for flow measurement. The pressure taps can be mounted at different locations, but it is apparent that the actual point of measurement will affect the relationship between the differential pressure and the flow rate. The following expressions, based on well-known laws of physics that define the action of the free fall of bodies, show the relationship of flow and differential pressure:

$$V = gt \tag{4.22}$$

where V is velocity in feet per second, g is acceleration due to gravity in feet per second, and t is time in seconds.

$$h = \frac{1}{2}gt^2 \tag{4.23}$$

where h is distance in feet. Solving Equation 4.22 for t,

$$t = \frac{V}{g}$$

Substituting this value of t into Equation 4.22,

$$h = \frac{1}{2}g\left(\frac{V}{g}\right)^2 = \frac{V^2}{2g}$$

and solving for V^2 and V,

$$V^2 = 2gh \text{ and } V = \sqrt{2gh} \tag{4.24}$$

Equation 4.24, sometimes called the velocity of flow equation, is the basis of flow measurement inferred from differential pressure. The

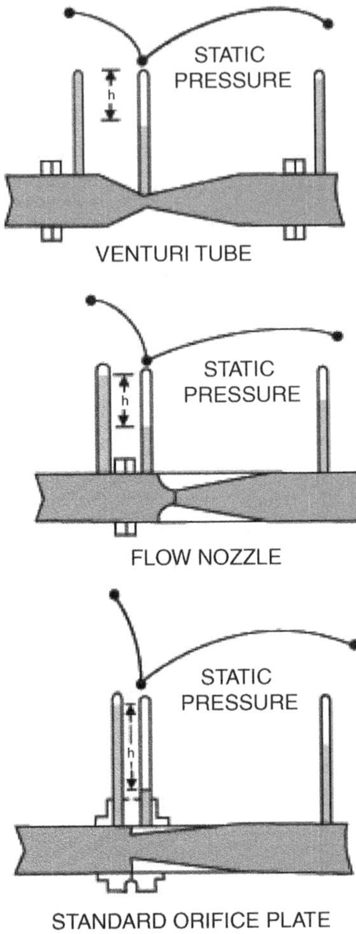

Figure 4–36 Pressure Drop Gradient About a Differential Pressure Producer (*Courtesy of University of TX, Petroleum Extension Service*)

value of $2g$ in Equation 4.24 is a constant, one of many used in flow calculations.

By multiplying both sides of Equation 4.24 by $\pi D^2/4$, the formula for the area of a circle, a volumetric measurement results:

$$V\left(\frac{\pi D^2}{4}\right) = \left(\frac{\pi D^2}{4}\right)\sqrt{2gh} \tag{4.25}$$

Velocity in feet per second times area in square feet gives a volumetric measurement of cubic feet per second, so

$$\frac{\text{ft}^3}{\text{second}} = \frac{\pi D^2}{4}\sqrt{2gh} \tag{4.26}$$

In the treatment of Equation 4.24, the value of *h,* a linear measurement in feet, has been used to represent the vertical height of water in a tank a distance of *h* above a hole (orifice) of area $\pi D^2/4$. The formula is used to determine the flow of water through that orifice by the hydrostatic head *h* of the column of water. In deriving the equation by means of Bernoulli's principle, the factor *h* in Equation 4.24 becomes the differential-pressure head measured in inches of water that exists between two points in a flow line. In this discussion, the pressure is always measured in inches of water because it is a result of the force exerted on a vertical column of water at 60°F caused by the acceleration of gravity. The two points are spaced from a differential producer such as an orifice. The points of pressure measurement are called pressure tap locations. An exact relationship exists for the distance of the tap locations from the differential producer and for the particular application. The amount of differential pressure for a given flow rate decreases as the distance of the tap locations from the differential producer increases. Figure 4–36 shows the relationship between differential pressure and flow. Figure 4–37 shows flow measurement by differential pressure.

Substituting *K* for $2g$ and *A* for $\pi D^2/4$ into Equation 4.26,

$$\text{Flow} = KA\sqrt{h} \tag{4.27}$$

Equation 4.27 is the basic flow equation for head-type flow-measuring devices. It should be pointed out that the *K* value includes many constants based on the physical characteristics of the flow loop and the characteristics of the flowing fluid. The fluid characteristics are both static and dynamic in nature. This discussion on flow measurement is certainly an oversimplification of a complex subject and is included only to show the basis of flow measurement by differential pressure.

It is sometimes desirable to use differential-pressure instruments to indicate a single flow value. High or low points for alarm conditions or control in batch processes and mixing are examples of such applications. Differential-pressure indicating switches are used for this purpose. A signal is transmitted by contact closure at a predetermined differential pressure representing a flow value.

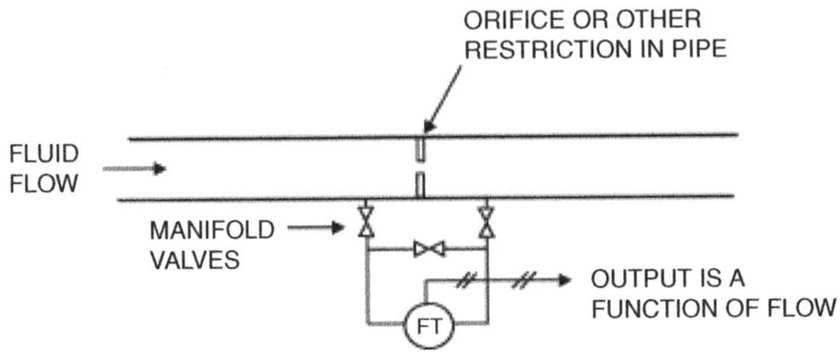

Figure 4–37 Flow Measurement by Differential Pressure

The information given here for flow values inferred from differential pressure is but an instance for a particular application of differential pressure measurement. The material in References 2 and 3 offers an in-depth study of the topic.

4.22 INDUSTRY-STANDARD TRANSMITTERS

The use of pressure and differential-pressure instruments has grown in the last several years. These instruments have become such a commodity that a standard or typical transmitter can be described. While many aspects have been defined by user groups and organizations such as ISA's Standards and Practices Board, others have been set by general acceptance.

Electrical output is a 4–20 mA current loop, and electrical connection is via 1/2-in. NPT for electrical conduit connection. The signal conditioner has controls to adjust the output for calibration over adjustable ranges, including suppression and elevation when required. The unit is constructed of stainless steel for corrosion resistance, with more exotic materials used for the wetted parts when required. Gage pressure transmitters have 1/4- or 1/2-in. NPT male or female process connections. Differential-pressure transmitters have 1/2-in. NPT female connections or a three-valve manifold for isolation and equalization. Many commercial transmitters for pressure and differential-pressure applications will withstand most plant environments with special housings and are rated intrinsically safe or explosion-proof, or have safety barriers inserted in their signal circuit for use in flammable, hazardous, or combustible areas.

Some transmitters have special flanges (such as the ANSI flange) for special mounting requirements. Others are submersible; the transmitter can be lowered into a tank or other vessel and is completely

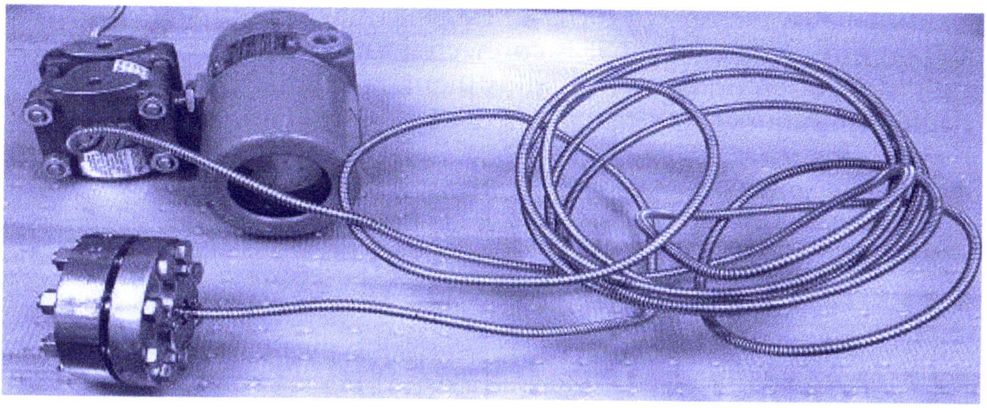

Figure 4–38 A Differential Pressure Transmitter with Remote Seals (*Courtesy of Emerson Process Management*)

liquid-tight at the measured pressure. In some cases, the sensor is remotely located from the signal-conditioning circuitry. Remote seals are available so that both pressure ports of a differential-pressure transmitter or the single port of a pressure transmitter can be separated from the process by a flange connection and a capillary tube evacuated and sealed with the appropriate amount and type of hydraulic fluid. The isolation diaphragm must be very thin and pliable so that all of the pressure on the sensor is passed along to the sensing elements. These diaphragms can be easily punctured, which can result in a loss of fill fluid; thus, the seal fluid is sometimes required to be safe for human ingestion or injection. The standard seal fluid is silicone oil, but vegetable oil, honey, sweet almond oil, or glycerin solutions are also used. A remote seal transmitter is shown in Figure 4–38.

The food and pharmaceutical industries require sanitary ratings. Transmitters must be constructed to inhibit bacterial growth and allow for convenient cleaning of the wetted surfaces. Smooth pressure port connections are necessary for complete cleaning; threads and small crevices can allow food particles to collect and permit bacteria to propagate. Such transmitters must be able to undergo steam cleaning or sterilization, and the exterior parts must be able to withstand repeated washdowns during plant cleaning.

Special signal-conditioning capabilities may be required. As mentioned earlier, explosion-proof and weathertight housings are available, but it may be undesirable to remove their covers for calibration adjustments. One type of signal conditioner incorporates magnetic coupling. In this technique, the technician turns an external magnet with a screwdriver, and an internal magnet that is connected to the controls in the signal conditioner responds to the adjustment. This allows the electronic signal conditioner to be hermetically sealed for underwater or hazardous enclosures. Such devices are giving way to digital transmitters that can be programmed remotely from the control room over signal cables. This technology is covered later in the chapter.

4.23 THE DYNAMICS OF PRESSURE TRANSMITTERS

While transmitters and control systems are typically fast compared to processes, pressure transmitter response times can be an important consideration in some very fast-acting processes. In compressor surge control, for example, the pressure update must be as fast as the compressor's surge rate; otherwise, surges can occur between updates, causing a surge to persist for several cycles before being detected. Different compressors surge at different rates in different applications. Generally, actual surge rates are determined by experimental tests; the surge time is proportional to the ratio of downstream volume to flow rate.

Many use the manufacturer's specification sheets to determine the response time of a transmitter. It is important to realize that this indicated value depends on the instrument's upper range. A transmitter's pressure element is designed for the upper range value. Low pressure range transmitters have flexible diaphragms and tend to react slowly, while high range transmitters with stiff diaphragms respond more quickly.

A transmitter's response time is a function of the mechanical components, a lag time T, and a fixed response time (T_1) that is added by the microprocessor of a digital transmitter; the combination of these times ($T + T_1$) represents the response time of the transmitter. If this were the end of the story, analog transmitters would likely have a longer response time than digital transmitters. This is seldom the case, however, because of the reduced response time for sensors with a stiffer diaphragm, reduced volume of oil fill, and other factors relating to digital transmitters. At pressure ranges where the mechanical response is fast and the micro pressure update rate is dominant (in cases of higher measured values) the analog transmitter is faster. But, when the mechanical response is slow, at lower measured values, the digital transmitter is faster. The importance of this is that when the digital transmitter's response is fast enough, improvements in repeatability are provided.

When a transmitter is fast enough to detect pressure surges, shorter response times do not enhance benefits; repeatability should be improved if possible. For example, when a compressor is operating near its surge condition and the safety margin between flow through the compressor and its pressure drop is less than the reproducibility of the measuring device, the system will enter a surge condition undetected. Also, if the true measurement exceeds the measured value, bypass or recirculated flow will be unnecessary.

When selecting transmitters for control of fast processes like compressors, the response time must be at least as fast as the compressor surge time. The precision of the transmitter should yield measurement demanded by the installed repeatability under the specified conditions of the application.

For safety and efficiency considerations, best practice dictates that a transmitter should respond at least as fast as the process. A process where transmitter response time should be considered is flow and pressure measurement for compressor surge control [Ref. 4].

4.24 DIGITAL TRANSMITTERS AND FIELD COMMUNICATION

The introduction of microprocessors to process transmitters has resulted in a new class of measuring instruments. We have seen the impact of these devices on modern process and manufacturing industries as smart technology has moved to the field and plant floor. This section will address the topic of digital transmitters and discuss various features and applications of communication.

The development of analog measuring instruments resulted in process transmitters that contain analog sensors and signal-conditioning electronic circuits to generate scaled signals for transmission. Such devices have served the industry well, but are limited in terms of versatility, stability, reliability, accuracy, and repeatability. Error caused by component drift, nonlinear physical sensor characteristics that require cumbersome correction factors, maintenance and calibration requirements, limited versatility for application changes and error caused by environmental variations are some of the shortcomings that encouraged the development and application of digital sensor technology.

Most early advances in control technology were based on the development of microprocessor-based controllers. Digital controllers in complex distributed control systems (DCSs), single-loop digital controllers (SLDCs), HART, and FOUNDATION Fieldbus are common in both retrofitting and new installations. The increased performance expected of such control systems cannot be realized in many cases because of limitations in antiquated transmitters. The advantages of digital technology applied to measuring systems have enabled process transmitters to have an expanded role in complex process control systems.

4.25 DIGITAL TRANSMITTERS

Measuring technology once lagged behind the developments in controllers and control systems. The development of digital transmitters has enabled measurement quality to now be commensurate with control quality.

A digital transmitter is a microprocessor-based measuring instrument that provides two major features: improved performance and two-way digital communications between the control station and the field. These features have resulted in increased process profitability by providing improved accuracy, remote communication, configuration for the application, diagnostic capabilities, greater reliability, reduced maintenance, improved control quality, and more efficient communication. The normal accuracy of process transmitters was at one time considered to be ±1% of full-scale value. Measurement precision has always been important because the emphasis generally has been on control of a variable rather than measurement for absolute accuracy. When measurement accuracy can be improved to ±0.1% or even ±0.05%, tighter control around a desired point can be achieved. This improved accuracy can translate to substantial improvements in product quality, reduction in raw materials use, and increased profits. With increasing emphasis on product purity and improved efficiency, the retrofitting of production units is often justified.

4.26 IMPROVED DIGITAL SENSORS

Conventional analog transmitters utilize two transducers. The input transducer converts a quantity relative to the process, such as pressure or differential pressure, to a force or motion. The output transducer then converts this quantity to an electrical or pneumatic signal for transmission. Pressure and differential pressure applications are often affected by ambient or process temperature changes and by variations in process static pressure. The error caused by these disturbances may not be compensated for because of the added power and complexity involved with such analog circuitry. Complementary Metal Oxide Semiconductor (CMOS) electronics technology has changed this and is the key to digital transmitters, offering low power requirements and the ability to function over a wide range of temperature variations.

True digital sensors, primary sensor characterization, and secondary variable characterization have improved the accuracy of digital transmitters. Although it may be an analog world, some sensors operate in the digital domain. This eliminates at least one stage of conversion (and perhaps two), which improves the resolution and accuracy of the measurement. Signal conditioning that is normally performed by the output transducer is performed digitally and requires no conversion. Linearizing, scaling, calibration, range change and other characterizing processes are done in the transmitter by microprocessor circuitry.

A dual-capacitance cell detector that includes reference and sensing capacitors is used in a digital transmitter. A digital circuit converts the process measurement directly into a frequency output corresponding to an applied pressure. A multimode oscillator (an application-specific integrated circuit) converts capacitance values to frequencies which are then converted to serial outputs. This all-digital signal-conditioning circuit eliminates the need for A/D conversion and the errors resulting from parasitic capacitance. Each pressure-detecting capacitance cell is characterized, which also reduces the requirements for pressure sensor calibration. This technique not only improves the overall transmitter accuracy but also can result in turndown ratios as high as 45 to 1 [Ref. 5].

Most analog sensor input transducers exhibit a certain amount of nonlinearity. Often the desired process measurements (e.g., differential pressure, pressure, level, flow, etc.) are distorted by other parameters. Typically, conventional transmitters can have error as great as 2% or more, caused by variations in ambient temperature. Performance improvements are achieved when the computation capability of digital transmitters is used to characterize their output.

By testing a particular instrument at various operating conditions, data can be expressed in an equation that defines the performance of the device. Using inputs from other sensors as variables, the microprocessor circuitry in digital transmitters can compensate for all variables expressed in the equation.

Many applications use the ability of a digital transmitter to characterize the output signal. This capability alone resulted in a revolution in process measurement that is presently under way, and it is a major factor in increased measurement accuracy. Cases have been cited where conventional analog signal conditioning has caused error of 0.5 to 5% and greater. Sensor characterization has resulted in improved accuracy, to 0.1% of span with an 8:1 rangeability. The algorithm that defines the sensor characteristic can be customized for individual transducers and implemented by software. Temperature effects have been reduced from 1% to 0.1%.

In gage-pressure applications, two pressure measurements are made and both are compensated for temperature. The microprocessor uses the barometric-pressure measurement to trim the absolute measurement to gage-pressure specifications. Differential-pressure measurements must be independent of static-pressure variations. A microprocessor receiving a static-pressure measurement from a secondary transducer will compensate for the fluctuations.

4.27 HIGHWAY ADDRESSABLE REMOTE TRANSDUCER (HART)

To enable a means of communication between the field instrument and the control device, a two-way digital/analog hybrid communication network has been developed. In 1993, the HART Communication Foundation (HCF) was established to provide worldwide support for application of the Highway Addressable Remote Transducer technology. Thus, HART, owned and managed by HCF, developed an open standard protocol that is maintained for the benefit of industry. The Automation Research Council (ARC) Advisory Group estimates that a majority of field devices installed worldwide are HART enabled.

HART is a hybrid communication technology in which a modulated, two-way digital signal is imposed on the standard 4–20 mA signal carrying the process information. An AC signal is superimposed on the 4–20 mA signal that signifies the value of the PV (process variable). The frequency of the AC signal is shifted between 2,200 Hz to represent a 0 or low and 1,200 Hz to represent a 1 or high. The digital signal conveys additional information such as device status and diagnostics information that can be routed to asset management for process control systems and safety. This first major two-way simultaneous communication on the 4–20 mA wire is used for fast, reliable and robust control and a digital channel for real-time communication of additional process/device-related information.

HART includes a standardized application layer addressing device status and diagnostics, cyclical process data including floating-point digital value; engineering units, data quality and status. The protocol also enables field devices to continuously publish their process data and standardized operating procedures (e.g., loop test, current loop reranging and transducer calibration).

Every HART device includes thirty-five to forty standard pieces of information which are easily accessible by all HART-enabled systems. These include device identification, basic calibration data, process variables (measured and calculated) and diagnostic alerts. All HART-smart field devices continuously assess and monitor their own performance and return diagnostic status information with every message.

A HART innovation was the creation of a Device Description Language (DDL)—an object-oriented, text-based language for modeling the characteristics and real-time capabilities of intelligent field devices. Instrumentation suppliers use DDL to create a Device Description (DD) file which is similar to an electronic data sheet describing all capabilities of the smart field device so that DD-enabled host systems can communicate with all device features. In early 2004, the International Electromechanical Commission (IEC) approved DDL as an international standard [Ref. 6].

The communication signal of HART networks is low-frequency, so normal instrument-grade cable can be used. This is a definite advantage of the new technology because, as has always been the case with new and emerging systems, users are often not ready to make complete changes such as replacing an existing technology with a new one. End users in the processing industries are conservative and have a wait-and-see attitude, preferring to test small bits of a technology and add to it later as confidence grows. This is an advantage offered by HART. Revisions and upgrades can be added in small bits—a loop at a time if desired.

Another important aspect of HART is that the signal is modulated on the 4–20 mA line at a frequency range that is filtered out by analog devices. The digital signal is symmetrical and does not affect the 4–20 mA signal. This makes it possible to do configuration, diagnostic checks and other functions without disturbing the loop. Previously adopted versions of such systems required the controller to be on manual before performing additional communication functions.

HART transmitters can perform measurements of multivariables. A typical application could be flow inferred from a differential pressure measurement transmitted by the 4–20 mA signal and up to three additional digital signals to represent temperature, pressure and differential pressure. Also, a HART splitter can be used to convert the digital signals to 4–20 mA.

HART can operate in the multidrop mode with the analog signal fixed at 4 mA. Up to fifteen transmitters can be connected to the 4–20 mA line (a bus) and two PV values per second can be represented by the digital signals from the transmitters. Such applications would operate in the master/slave format.

Other features of HART include

- Configuration for
  - Tag I.D.
  - Description
  - Data presentation
  - Setting engineering units
  - Setting and reading URV (upper range value) and LRV (lower range value)
  - Damping constants
- Diagnostics to indicate
  - PV out of limits
  - PV out of range
  - Output current fixed
  - Configuration change
  - Device malfunction
  - Additional device information

- Commissioning functions:
    - Verify tag information
    - Confirm URV, LRV
    - Set engineering units
    - Set damping constants
    - Documentation
    - Set current values for loop test and alarm settings
    - Device start-up procedures

From the previous discussions, it is apparent that HART is both an analog and a digital system, a hybrid means of communication whose use is expanding. When used with a computer as a HART monitor and appropriate asset management technology and applications, HART can result in a more efficient operation. It can operate in analog mode, in digital mode, or in a combined analog and digital mode.

As field devices get smarter, more field intelligence is available for changing parameters and for control flexibility. Most of today's HART devices are multivariable in nature, which enables operations like asset management, compensation techniques, and continuous data validation for improved control loop integrity. Most HART devices are self-calibrating; re-ranging is done by keypad from an HHC (handheld communicator) [Ref. 7].

4.28 DIGITAL TRANSMITTER OPERATION

The operation of digital pressure and differential-pressure transmitters relies on the same basic pressure-detection principles discussed earlier in this chapter and in Chapter 3. In fact, some manufacturers have offered an upgrade kit to convert an analog transmitter to digital by exchanging the electronic signal-conditioning circuitry of an analog transmitter for that of a digital transmitter. Other manufacturers have modified the pressure-detecting cell for digital transmitters.

Figure 4–39 shows a block diagram of a transmitter with a capacitive sensor used in an analog transmitter. An A/D converter translates the analog signal from the sensor module to a digital representation. The resulting digital value is processed by the microprocessor, which also provides sensor linearization, range down, damping and transfer

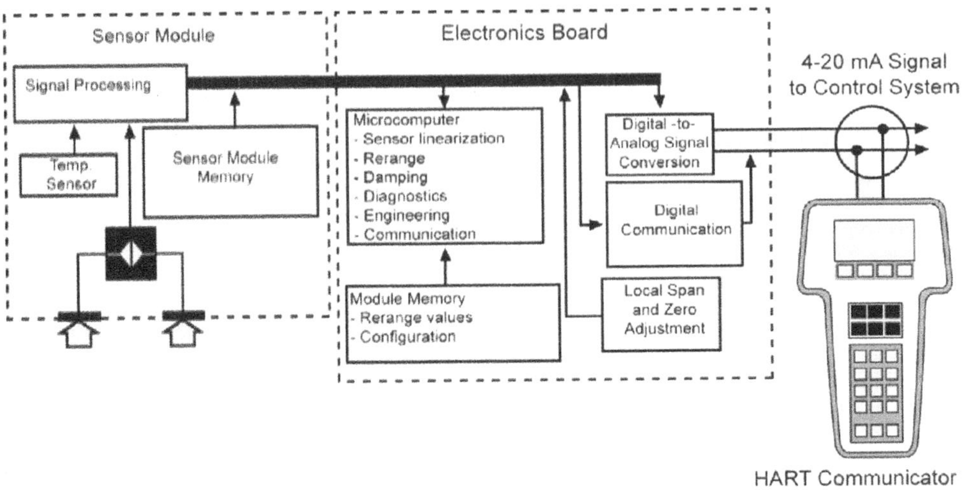

Figure 4–39 A Digital Transmitter and Handheld Communicator with Analog Transmission (*Courtesy of Emerson Process Management*)

functions. The microprocessor also controls the digital-to-analog (D/A) converter for the 4–20 mA output and drives the digital communication which is compatible with the HART protocol. This protocol uses a standard Bell 202 frequency shift keying (FSK) technique. An AC signal is superimposed on the 4–20 mA output communication cable.

The digital pressure transmitters described previously were of the first generation and coverage is included here because many are still in use. A later, more current version such as the one shown in Figure 4–40 is available for the following applications:

- Differential pressure (DP)

  0.1 in. H_2O to 2000 psi (0.02 to 13,800 kPA) with 100:1 rangeability

- Gage Pressure (GP)

  2.5 in. H_2O to 2000 psig (0.62 to 13,800 kPA)

- Absolute Pressure (AP)

  0.167 to 4000 psia (8.6mm Hg to 27,580 kPA)

- Gage and Absolute Pressure (GP and AP)

  0.3 to 10,000 psig/a to 68,900 kPA

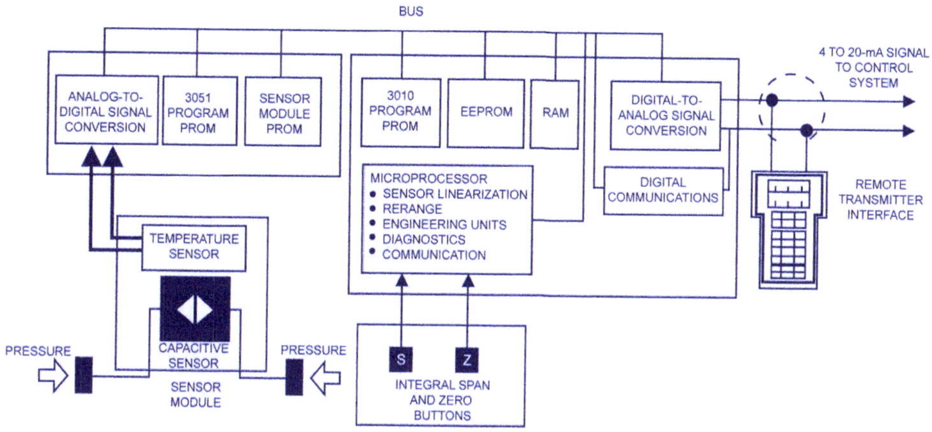

Figure 4–40 Block Diagram of a Digital Transmitter and Handheld Communicator Connected to a Bus
(*Courtesy of Emerson Process Management*)

The sensors in many digital transmitters are a capacitance cell or a piezoresistive silicon sensor. Some will operate at temperatures up to 375°F (191°C).

The major components of later versions of digital transmitters are

- Sensor module—capacitance, piezoresistive, or others
- Sensor electronics
    - Microprocessor
    - Module memory
    - D/A conversion
    - Digital communication network
    - Local span and zero adjustments
- Hardware to contain the process pressures and house the sensor and electronics

The sensor module sends a signal to the signal conditioning circuitry. The signal conditioning circuitry contains an output circuit board that has the microprocessor, memory module and D/A converter for a 4–20 mA output for analog transmission. A HART signal is superimposed on the 4–20 mA line. For digital transmission, the signal stays in the digital domain as is the case for FOUNDATION Fieldbus transmitters. There is no 4–20 mA signal.

All transmitters are calibrated before leaving the factory and when mounting the instrument results in a zero shift caused by the liquid head in the impulse lines, the output is set to zero by a sensor trim procedure.

Many digital pressure and differential pressure transmitters have an RTD (resistance temperature detector) sensor and signal conditioning circuitry to measure the temperature at the pressure cell. Other multivariable transmitters have an RTD probe for installation in the process for a process temperature measurement.

Calibrating a digital transmitter is different from calibrating an analog transmitter. The procedure required for the calibration of an analog transmitter requires only three steps with a digital (smart) transmitter:

- Re-range—sets the 4 and 20 mA points at the desired pressures.
- Sensor Trim—adjusts the position of the factory characterization curve to optimize transmitter performance over a specified pressure range or to adjust for mounting effects.
- Analog Output Trim—adjusts the analog output to match the plant standard or the control loop.

Digital transmitters also operate differently than analog transmitters. A digital transmitter uses a microprocessor that contains information about the sensor's specific characteristics in response to pressure and temperature inputs. A digital transmitter compensates for these sensor variations; the process of generating the sensor performance profile is called factory characterization. Factory characterization also provides the ability to readjust the 4 and 20 mA points without applying pressure to the transmitter.

The re-range and trim functions also differ. Re-ranging sets the transmitter analog output to the selected upper and lower range points and can be done with or without an applied pressure. Re-ranging does not change the factory characterization curve stored in the microprocessor. Sensor trimming requires an accurate pressure input and adds additional compensation that adjusts the position of the factory characterization curve to optimize transmitter performance over a specific pressure range. It is possible, however, to degrade the performance of the transmitter if the sensor trim is done improperly or with inaccurate equipment.

The "recall factory trim" feature enables the restoration of the as-shipped factory settings of the sensor trim and analog output trim. This procedure can be useful for recovering from an inadvertent zero trim adjustment to an absolute zero pressure transmitter. Trim settings can also be made to differential pressure transmitters to compensate for output changes that result from static line pressure changes. All trim and calibration adjustments should be made in accordance with procedures established by the manufacturer.

Many digital pressure and differential pressure transmitters have diagnostic features that will help in finding faults during times of equipment failures. Such features may include the following diagnostic messages on the local LCD meter display:

- ERROR

  Error messages appear on the LCD meter display to inform the operator of serious problems affecting the operation of the transmitter. The meter displays an error message until the error condition is corrected and the analog output is driven to the specified alarm level. No other transmitter information is displayed during an alarm condition.

- FAIL

  The transmitter CPU board and the sensor module are incompatible. Factory support may be needed to correct this problem.

- FAIL MODULE

  The sensor module is disconnected or is malfunctioning and the user must verify that the sensor module ribbon cable is connected to the back of the electronics board. If the ribbon cable is properly connected, there is a problem within the sensor module. Possible sources of problems include

  - Pressure or temperature updates are not being received in the sensor module.
  - A non-volatile memory fault that will affect transmitter operation has been detected in the module by the memory verification routine.

Some non-volatile memory faults are user-repairable. A HART Communicator can be used to diagnose the error and determine if it is repairable; any error message that ends in "FACTORY" is not repairable. In cases of non user-repairable errors, the sensor module must be replaced.

- FAIL ELECT

  The transmitter electronics board is malfunctioning due to an internal fault.

  Some Fail Elect errors are user-repairable. A HART Communicator can be used to diagnose the error and determine if it is repairable. Any error message that ends in "FACTORY" is not repairable. In cases of non user-repairable errors, the electronics board must be replaced.

- FAIL CONFIG

  A memory fault has been detected in a location that could affect transmitter operation, and is user-accessible. To correct this problem, a HART Communicator can be used to interrogate and reconfigure the appropriate portion of the transmitter memory.

Warnings appear on the LCD display to alert the operator of user-repairable problems with the transmitter or current transmitter operations. Warnings appear alternately with other transmitter information until the warning condition is corrected or the transmitter completes the operation that warrants the warning message.

The warning messages on previous versions of the LCD display may vary slightly from those listed here, but they represent the same warning:

- PRESS LIMIT

  The process variable read by the transmitter is outside of the transmitter's range.

- TEMP LIMIT

  The secondary temperature variable read by the transmitter is outside of the transmitter's range.

- CURR FIXED

  The transmitter is in multidrop mode. The analog output is not tracking pressure changes.

- CURR SATURD

  The pressure read by the module is outside of the specified range, and the analog output has been driven to saturation levels.

- LOOP TEST

  A loop test is in progress. During a loop test or 4–20 mA trim, the analog output is set to a fixed value. The meter display alternates between the current selected in milliamps and "LOOP TEST."

- XMTR INFO

  A non-volatile memory fault has been detected in the transmitter memory by the memory verification routine. The memory fault is in a location containing transmitter information. A HART Communicator can be used to interrogate and reconfigure the appropriate portion of the transmitter memory. This warning does not affect transmitter operation.

- OPERATION

  Normal operation messages appear on the LCD meter to confirm actions or inform the operator of transmitter status. Operation messages are displayed with other transmitter information and warrant no action to correct or alter the transmitter settings.

- ZERO PASS

  The zero value, set with the local zero adjustment button, has been accepted by the transmitter and the output should change to 4 mA.

- ZERO FAIL

  The zero value, set with the local zero adjustment button, exceeds the maximum rangedown allowed for a particular range or the pressure sensed by the transmitter exceeds the sensor limits.

- SPAN PASS

  The span value, set with the local span adjustment button, has been accepted by the transmitter and the output should change to 20 mA.

- SPAN FAIL

  The span value, set with the local span adjustment button, exceeds the maximum rangedown allowed for a particular range or the pressure sensed by the transmitter exceeds the sensor limits.

- LOCAL DSBLD

  This message appears during re-ranging with the integral zero and span buttons and indicates that the transmitter local zero and span adjustments have been disabled. The adjustments may have been disabled by the transmitter security jumper on the transmitter circuit board or through software commands from the HHC.

- WRITE PROTECT

  This message appears if an attempt has been made to change the transmitter configuration data while the security jumper is in the **ON** position [Ref. 8].

4.29 HHC TRANSMITTER SMART FAMILY INTERFACE

The field communicator shown in Figure 4–41 is both HART and FOUNDATION Fieldbus compatible; it will also support a field device programmed for earlier revision. Operator interface is available through a touch screen or keypad entry. The arrow keys enable the navigation through the menu structure of applications. Both letters and digits can be selected and entered via the alphanumeric keypad. On start-up, either the HART or FOUNDATION Fieldbus application is selected. The HHC will then poll for a device. The system properties and license information are available.

Figure 4–42 shows an HHC connected for a HART application. When the on-line menu is selected, PV (Process Variable), AO (Analog Output), URV (Upper Range Value) and LRV (Lower Range Value) can be read and changed. Basic set-up provides quick access to a number of configurable parameters including tag name, unit, range values, and damping. The detailed menu provides access to all editable device parameters and all device functions such as characterization, configuration, and sensor and output trim. The review menu lists all of the static parameters read from the connected device, which includes information about sensor setup and limitations. Information such as tag name, materials of construction and the device software revision

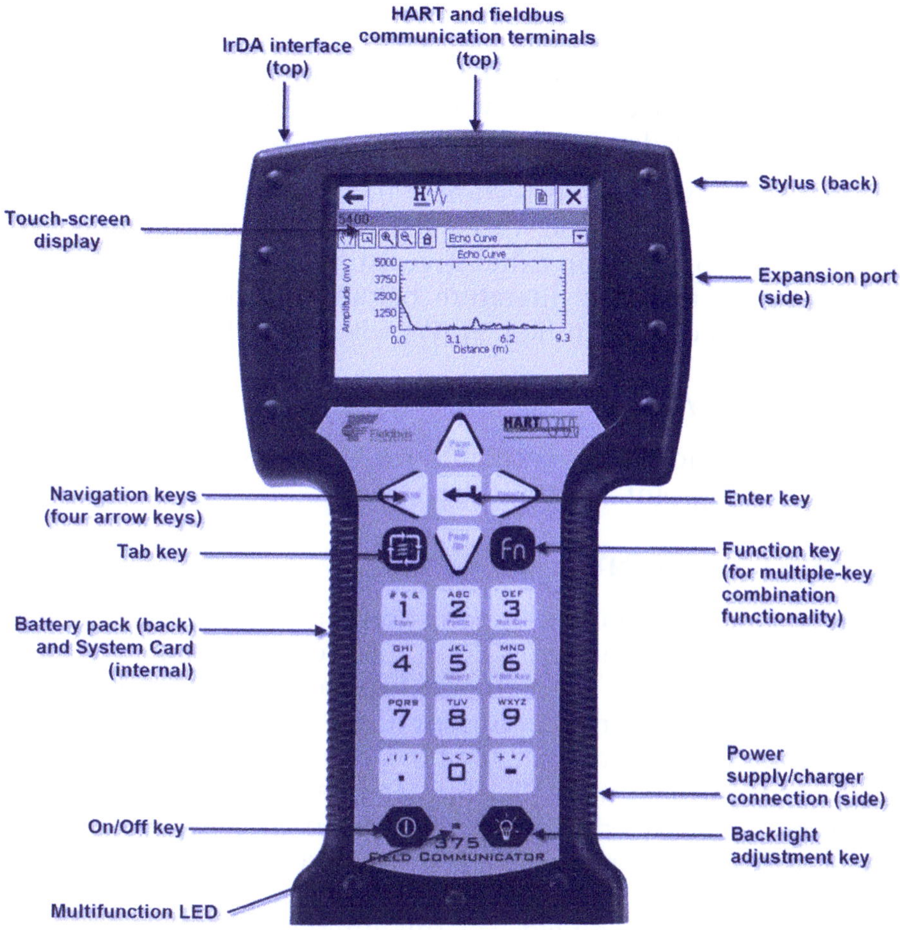

Figure 4–41 A Field Communicator for HART and FOUNDATION Fieldbus (*Courtesy of Emerson Process Management*)

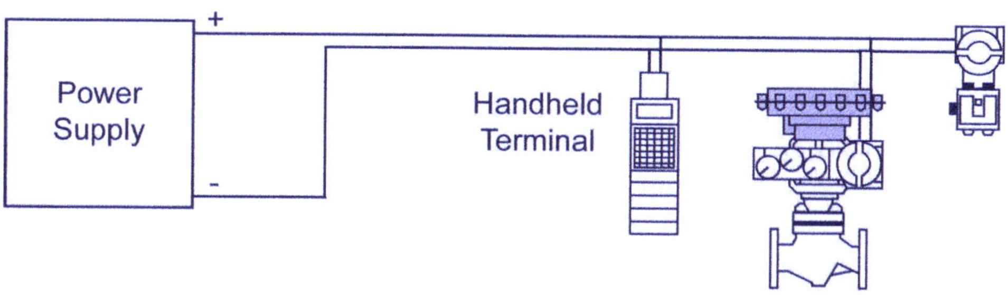

Figure 4–42 Handheld Communicator Connected for HART Application (*Courtesy of ISA*)

may also be included. Other functions include changing polling options, changing ignore status message, deleting certain unused items, viewing the available device description, simulating an on-line connection to a HART device, viewing HART diagnostics, and disconnecting from a HART device.

The information given in this section is to show basic operations and functions that can be provided with an HHC for HART and FOUNDATION Fieldbus. It is by no means an exhaustive list, and the user is required to consult the manufacturers' literature before performing checks and working with the equipment [Ref. 9]. We will start with HART.

4.30 TOPOLOGY

Two possible topologies are available for the HART communication protocol: point-to-point and multidrop.

4.31 POINT-TO-POINT

The most common topology is point-to-point wiring that in many cases is a traditional two-wire 4–20 mA DC current loop. The ripple limit can be met by any modern stabilized and regulated DC power supply. The resistance in the current loop is usually the standard 250 ohms but can be as low as 230 ohms. As many as two masters can be used in the HART network. The primary master sometimes becomes a permanently connected interface such as an RTU (remote terminal unit) or a multiplexer for the host. The secondary master is generally a portable handheld terminal that is connected temporarily when performing checks in the field. For this interface to work it must be connected between the field device and loop resistor. The signal is shorted if the HHC is connected next to the power supply.

In the point-to-point topology only a single field device is connected on each pair of wires. A transmitter may be loop powered or separately powered. The connection, however, is the same as that of a conventional analog transmitter. Since recorders, single-loop controllers, indicators, and other auxiliary loop devices may be used, adding resistance to the loop, an additional resistor may not be needed. As in conventional analog loops, it is good practice to check and adjust the total loop resistance as required. In this point-to-point topology, the analog devices in the current loop are wired in series, which results in a voltage drop. Care must be taken to ensure that the power supply is adequate to maintain the device voltage above the minimum requirement.

4.32 MULTIDROP

In the multidrop topology, several field devices are connected in parallel on each pair of wires. The field devices may be loop powered or powered separately. Loop powered transmitters have a fixed current of 4 mA and the current does not reflect a process variable value. Communications in a multidrop situation are entirely digital. Standard HART commands are used to communicate with field instruments to determine process variables or deviation parameter information. The typical cycle time required to read information from a single transmitter is about 500ms. This limits the number of devices on a network to fifteen for a 7.5 second response time from all transmitters. Connecting HART field devices in a multidrop network can provide significant savings in installation costs. Figure 4–43 shows a HART multidrop connection.

The total cable length for multidrop networks is generally less than the maximum cable length for point-to-point connections because the capacitance of the additional devices reduces the distance the HART signal can be carried.

To save on installation costs, HART multidrop networks can be used for remote monitoring stations, tank farms, pipeline systems and other monitoring applications where fast updates are not required. Using the polling address structure of the HART protocol, up to fifteen devices can be connected in a multidrop network. The analog current of a HART device can be fixed by setting its polling address to a number other than 4mA. With the multidrop protocol, each field device must be configured with different polling addresses or tag numbers before

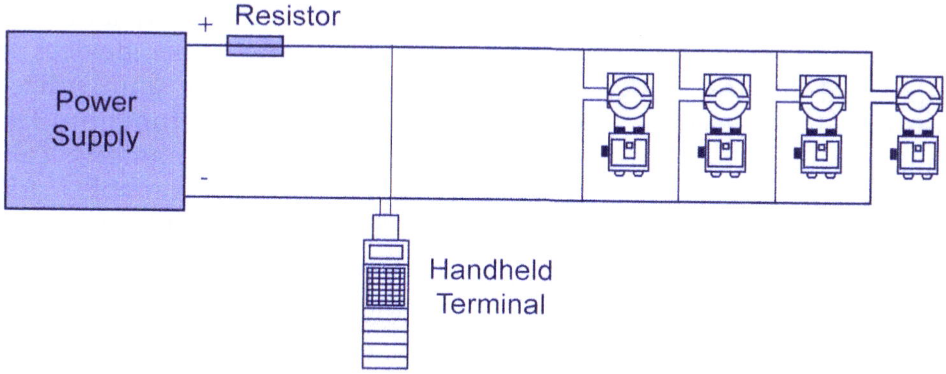

Figure 4–43 A Multidrop HART Application

being attached to a multidrop network; otherwise, the host will not be able to establish communication with the slave devices.

HART-compatible networks can be used to interface with a large number of HART devices. Such multiplexers can be modular and capable of supporting both point-to-point and all digital (multidrop) HART communications.

When installing HART multiplexers, the following capabilities should be considered:

- Number of HART channels supported.
- Number of HART channels that share a HART modem.
- Burst mode support.
- Multidrop support.
- Method of communication with the host computer or control.

Many HART enabled products are able to perform more than one measurement, for example, pressure and temperature. All such information can be accessed digitally through existing controllers. For interface equipment that does not have the ability to read HART data, products are available that can read the HART digital information and convert it to analog data. This enables any traditional analog/digital I/O to realize the advantages of the benefits of HART communicating devices.

4.33 CALIBRATING HART FIELD DEVICES

All field devices require periodic calibration due to the physics of the sensor, the measurement techniques employed and the dynamics of the process variables. The process of calibrating these devices is supported by the HART 6 Protocol which provides standardized commands and related commands to enable this critical function. The calibration process consists of three major steps or functions: calibrating the process variable, scaling the process variable, and producing/developing the process signal. This is shown in Figure 4–44.

The first step is commonly referred to as the transducer block and is responsible for generating an accurate digital value representing the connected process. In this step, properties like the transducer upper and lower limits, trim points, and transducer characterization data employed to generate this digital value are checked. In general, the field device is

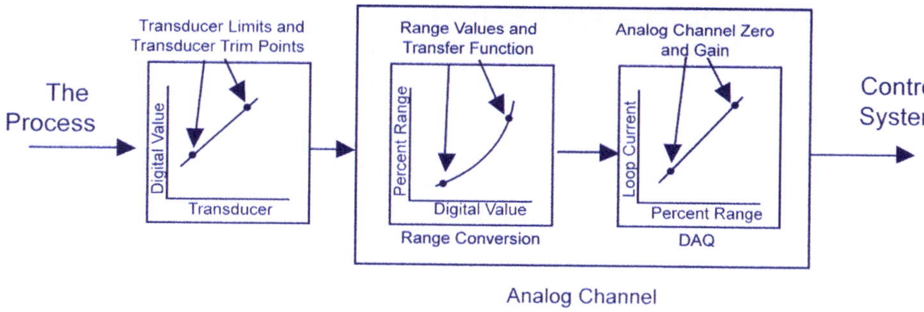

Figure 4–44 Calibrating HART Devices
(*Courtesy of HART Communication Foundation*)

calibrated by providing a value from a traceable reference near the upper and lower limits and trimming the internal mathematical calculations to compensate for any error. Process calibrators are commercially available to perform this function using the HART protocol.

The second step is referred to as the range block or "zero and spanning the device." In this step, the upper and lower range values are used to produce the transducer value previously set to correspond to a 4 mA signal for the lower range and a 20 mA signal for the upper range. Percent of range is then developed in this step. In addition, an appropriate transfer function (e.g., linear, square root, quadratic, cubic measurement, etc.) may be applied. Square root functions, for example, are used to approximate a flow measurement from a differential measurement. It is important to note the difference between calibrating and re-ranging a field device.

The third step, performed by the data acquisition (DAQ) block, is to actually produce the 4–20 mA signal, ensuring that 0% equals exactly 4 mA and 100% equals exactly 20 mA. The most critical function of the field device, from a system perspective, is to produce a signal that is agreed upon by the control system as corresponding to the process value [Ref. 10].

4.34 A HART CALIBRATOR

Handheld Terminals or Communicators (HHT or HHC) were developed to communicate with field instruments using the HART protocol. Multifunction calibrators are available to enhance HART calibration procedures by providing database software to retain information for many device tags and device description numbers. This can help maintain a history file and calibration record for field instruments. These multifunction calibrators can be used to check and calibrate/trim both analog and digital information from HART devices.

Calibration procedures for the multifunction calibrator depend on the method by which the HART field instrument is connected to the calibrator.

- The calibrator can supply power for the field instrument and measure the output current. The HART terminal is used for digital communication. A 270 ohm resistor is automatically connected in the loop for the calibrations.
- For external supply applications, the output current of the field device is measured. The HART terminal is used for digital communication. An external resistor must be included to give a loop resistance of 230 ohms.
- When the HART calibrator is used as a HART communicator, the output current is not measured and a 250 ohm resistor must be connected between the power supply and the communicator.
- By using a diode connected from the plus to minus terminal of the field device, the transmitter current is measured by connecting the test instrument parallel to the diode.

When a HART instrument has been added to the database, the following data is available to be read:

- Input quantity
- Input range
- Output quantity
- Output range
- Transfer functions
- Set point delay
- Input unit
- Input method
- Output unit

- Output method
- Device ID

Once the database has been created, the calibrator opens the instrument windows, from where the instrument's data can be checked and the instrument can be calibrated. The HART instrument's tag is assigned to the Device ID in the calibrator for record and future reference. Anytime after a database has been created for an instrument, the tag will appear in the database where it can be selected and added to the calibration window for editing or calibration. A multifunction HART calibrator is shown in Figures 4–45A and B [Ref. 11].

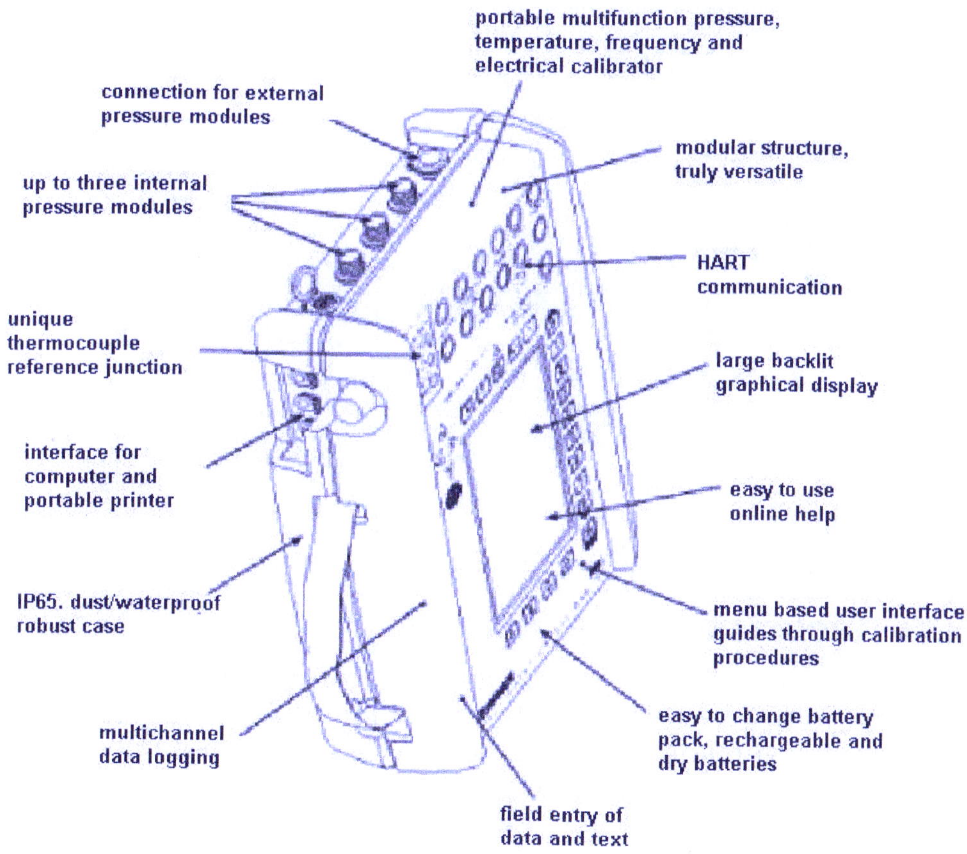

Figure 4–45A Features of a Multifunction HART Calibrator
(*Courtesy of Beamex*)

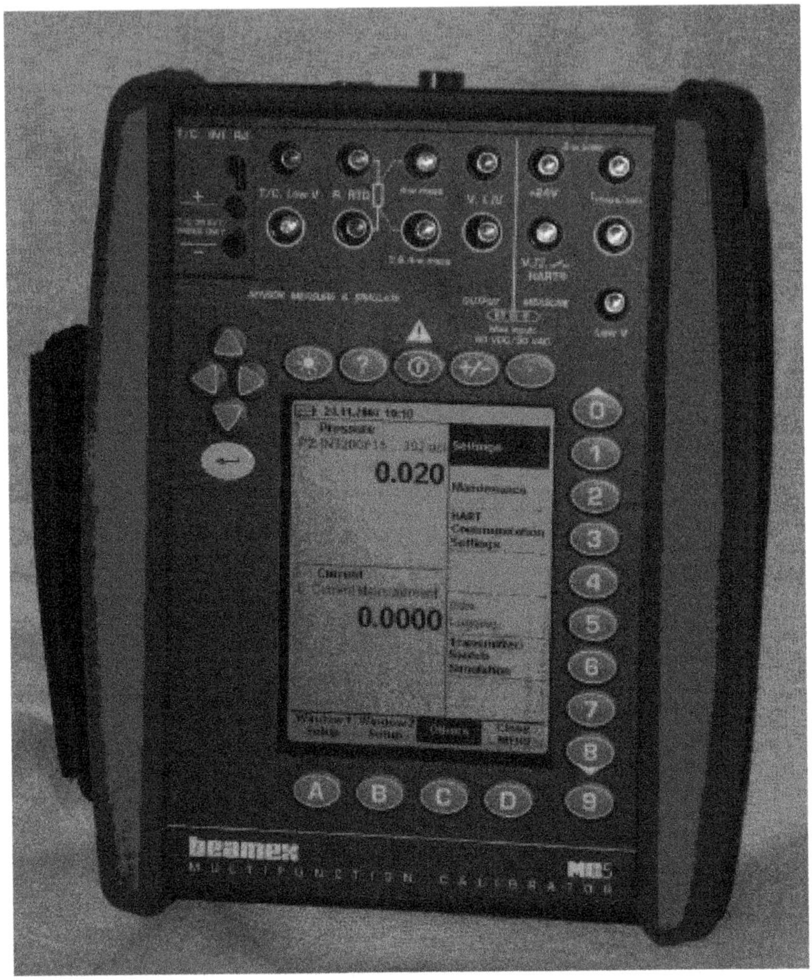

Figure 4–45B A Multifunction HART Calibrator (*Courtesy of Beamex*)

4.35 TROUBLESHOOTING HART

As a HART field device communicates with a host or HHC, it is continuously reporting back its status. In applications that have permanently connected communication interfaces, checks are continuous; any problem is reported immediately. Most presently used HART devices are communicating with the 4–20 mA signal and the HHC is only connected for short periods of time for configuration and planned checks. In such cases, the controller or readout device will monitor the transmitted 4–20 mA signal for alarm conditions. The technician will use the HHC to detect the faults. A transmitted signal out of the 3.8–20.5 mA range (Table 4–2) is a general indication of a fault.

There are a number of standard diagnostic messages and the device manual should be used to acquire more specific details as to the nature of the problem associated with the diagnostic message. Some of the common diagnostic messages are listed below.

Field Device Malfunction

This message indicates that the device is unable to perform its function. For a transmitter, this is generally an indication that the primary sensor, the secondary sensor, or the sensor circuitry has failed.

Configuration Changed

This is an indication that the device configuration has been changed; some HHCs will not display this. It is primarily intended for use by the host (primary master) to detect that a HHC (secondary master) has been used to change the configuration of the device. This prompt is to enable the host to update its database to the device configuration so the configuration file is synchronized with the actual device.

Cold Start

This indicates that the instrument's power has been removed and has then again been applied. The device may have been removed or separately reconfigured and reconnected. The power supply voltage could be too low for the transmitter to operate properly and it is switching off and on. The more status available flag provides an overview of the condition of the transmitter. Detailed diagnostics must be retrieved separately and the flag indicates that the diagnostics are available and should be checked.

Analog Output Current Fixed

This indicates that the 4–20 mA output is in a constant current mode such as loop testing or simulation. The output does not respond to or correspond to the sensor value. It could also indicate that the 4–20 mA output is deactivated in a multidrop mode.

Analog Output Saturated

This indicates that the 4–20 mA signal is saturated at either extreme. It may not represent an actual error, but that the range of the transmitter has been set too low for the specific application. Such a condition could occur in elevation or suppression applications and the transmitter is undergoing a bench test, or perhaps a low range absolute pressure transmitter could be vented to atmospheric pressure.

Nonprimary Variable Out of Limits

Many digital transmitters have secondary and even tertiary values that can be used for compensation of the primary sensor. In such cases, a pressure transmitter will have a temperature sensor to measure the temperature of the pressure sensing module. This message indicates that the sensor is located in a temperature environment out of the measurement range or the temperature sensor is not ranged properly.

Primary Variables Out of Limits

All sensors have a limit to their range of operation. This flag indicates that the applied input is beyond the range of the sensor and a greater sensor range should be selected.

HART transmitters have a feature to simulate a 4–20 mA signal independent of the sensor output (Figure 4–46). This enables a technician to use the loop test feature to scale receiving instruments by remotely setting the output current of the transmitter and verifying the readings. This is useful to set and check alarms, shut down devices, and perform troubleshooting features.

Following are several significant features of HART.

- HART technology can be used as a part of a plant network to improve plant operations. Some of these improvements include signal verification, device status, secondary variables and other operational data on a continuous basis.
- Communication with HART devices can be provided by I/O components, handheld communicators, such as Palm Pilots™ or other personal portable devices. PC-based configurators can also be used to communicate with HART devices via a HART-compatible modem.

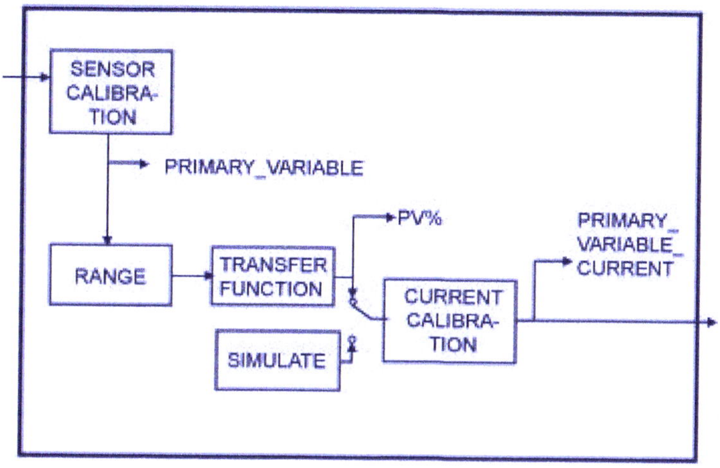

Figure 4–46 A HART Transmitter Block Diagram

- There is a wide range of HART-compatible devices in process measurement and control product categories including valve positioners; digital valve controllers; actuators (pneumatic and electric); and some 4-wire devices that include magnetic flowmeters, level devices, analyzers (liquid and gas), control devices (control systems, PLCs, multiplexers) and others.
- HART Protocol allows the simultaneous communication of the continuous 4–20 mA signal as well as a second digital communication path resting on top of the analog signal but not interfering. The digital information is communicated via the field proven Frequency Shift Key (FSK) technology that superimposes a digital signal on top of the analog signal.
- A Device Description (DD) is an optional element of the HART communication technology and is not required to communicate with a HART device. DDs are mostly used for device set-up and are not required for routine device communication.
- With HART, the analog 4–20 mA signals are there full time to address the needs of fast changing or critical process measurements. Data to support multivariable measurements, remote diagnostics, and device status can be accessed digitally while the analog signal is being used for control.
- Accessing the data in HART devices on a full-time routine basis increases the performance, integrity, and reliability of plant control and enterprise level systems. Getting digital data direct

from the device avoids the filtering that may occur in acquiring this information through intermediate systems. Process variables are provided in engineering units. Status indicators provide information on the quality of data and health of the device [Ref. 12].

4.36 HART SUMMARY

HART communication was developed as a hybrid system to bridge the link between the analog domain (4–20 mA) and a digital communication protocol that was yet to be developed. Although several standardization committees worked for many years to adopt a standard protocol for field communications, progress was made but such a standard was never established. The need for an all digital control network led to the formation of the Fieldbus FOUNDATION Organization, which developed the presently used standard for communication between digitally operated field instruments. With the adoption and use of FOUNDATION Fieldbus (FF), many predicted that the long awaited all digital system had at last arrived and HART applications would give way to FF. A survey in 2004 by the Automation Research Council (ARC), however, confirmed that HART adoption by users was actually expanding instead of declining as some predicted. Automation applications include multiple communication protocols to meet the diverse needs of both manufacturers and users. These protocols for the processing industries are largely HART and FF with Ethernet.

FOUNDATION Fieldbus

Networking was introduced into the nearly all analog world in the 1970s and was first utilized in direct digital control (DDC) systems between computer and I/O. Later, it was used in distributed control systems (DCS) and programmable logic controllers (PLC) to connect the controllers and operator consoles. In the 1980s, digital communications were used in field devices (digital transmitters). HART was developed as a hybrid communication system between the analog control instruments and field instruments over the 4–20 mA transmission system. True digital communication bus networking of field instruments began to gain wide acceptance in the 1990s. Many networks, including television, radio, telephones and industrial control systems, are now analog but the trend is definitely toward all digital communication systems.

Digital communications enable data to be transmitted serially between field devices and controllers, workstations and servers. A major advantage of such communications is that a great deal of information can be transmitted over a single cable instead of the two-wire twisted pairs required for each control loop with 4–20 mA. Also, digital communications carry not only the process and manipulated variables but operational information such as set point, mode, alarms, and tuning information. This has enabled distributed processing, diagnostics, configuration, transmitter calibration, valve configuration, and other functions of both field and control stations to be manipulated and coordinated from one single point or location.

FOUNDATION Fieldbus technology is an all-digital, serial, two-way communication system which interconnects field equipment such as sensors, actuators and controllers. It is a local area network (LAN) for instruments used in manufacturing and process automation with built-in capability to distribute the control applications across the network. FOUNDATION Fieldbus incorporates a field control system that can enable the control strategies to be distributed among the field devices. The use of function blocks enables control system strategies to be configured according to the needs of the users. Function blocks within the fieldbus devices perform the various functions required for process control. Because each function is different, the mix-and-match of the configurations of functions are different. A range of function blocks has been designed, each to address a different need, and all function blocks for process control applications have been developed. For example, the function blocks needed for a single loop control system are AI, AO, and PID.

With digital communications, there are many different methods of representing, encoding, and transmitting data. Each method is called a protocol. Manufacturers have devised many different protocols and products designed for one protocol are not compatible with those designed for another.

ISA's SP50 committee, which developed the 4–20 mA standard and others, has worked to define a standard protocol that all devices can follow. Such a standard protocol would make it possible for equipment from different manufacturers to interoperate and thus be able to communicate with each other. HART, FOUNDATION Fieldbus (FF), and Profibus are all based on the Open System Interconnection (OSI) reference model as defined in the ISO 7498 standard. While HART and FF are prominent in the process industries, Profibus is used for the faster applications found in automated manufacturing. The OSI model as applied to fieldbus technologies is shown in Figure 4–47. Notice that layers 3, 4, 5, and 6 are not used in FOUNDATION Fieldbus. The Fieldbus Foundation is an independent organization comprised of more than 100 instrument and control suppliers and which provides the specifications for the fieldbus communication stack and user. DD technology provides the interoperability between devices from different suppliers. To ensure this interoperability, conformance and interoperability testing software kits have been made available to the product developer. Conformance testing by FF ensures that fieldbus products conform to the communication stack and technology standards established by FF. All devices submitted to FF for interoperability testing must pass FF conformance testing. Only products that pass this testing can be approved by FF. A lack of conformance and interoperability testing does not prevent manufacturers from selling products with FF technology; some suppliers have sold FF compatible products that have not undergone the full range of testing from FF. Users will no doubt want to make certain that any equipment purchased for FF application has been certified to be in compliance.

Process-related networks include FOUNDATION Fieldbus, Profibus PA (application profile) and HART. As a category, all of these buses are referred to as "fieldbus" without the capital F. FF specifically represents FOUNDATION Fieldbus. All of these networks were designed for bus powered field instruments with predefined parameters and commands for asset management information such as identification, diagnostics, materials of construction and functions for calibration and commissioning. Generally, the networks used in automation and control are considered local area networks, usually a mile or so in diameter.

In these small networks however, there is still a need for different network characteristics, and FF is comprised of a two-tier control system hierarchy. In the field there are instruments such as transmitters and valve positioners with specific characteristics and needs. At the host level, workstations, linking devices, and controllers have other requirements.

For this discussion, the two-tier network consists of the field network, connecting the field instruments, and Ethernet, connecting the host devices. Figure 4–48 shows the linking device (H1 card) for the two networks. The fieldbus has two wires that are used for both communication and power to supply the bus. Bus power supply voltage is from 9–32 VDC, which is the voltage range required to supply the fieldbus instru-

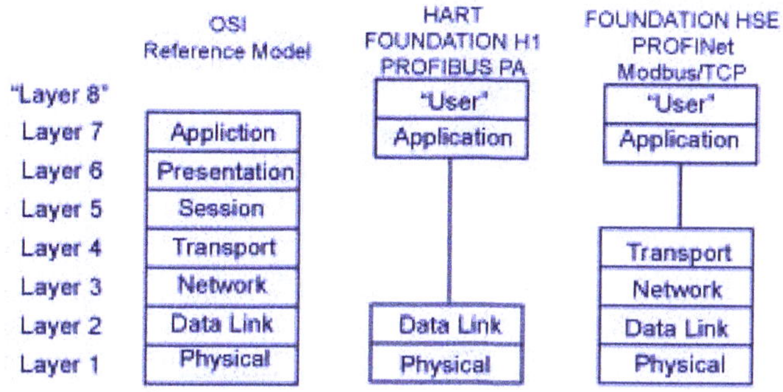

Figure 4–47 Open System Interconnection Reference Model

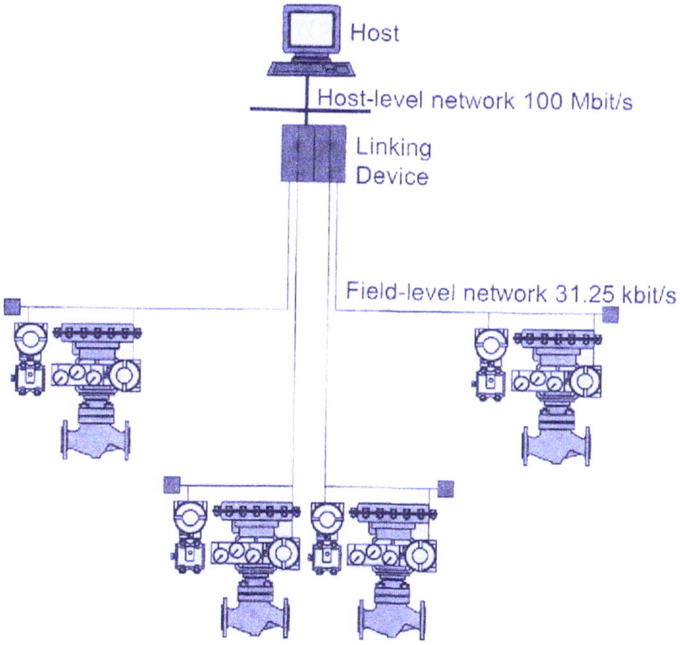

Figure 4–48 A Two-Tiered Automation Network Architecture Showing the Linking Device (*Courtesy of ISA*)

ments. The instruments are connected to the two-wire bus in parallel and each field device connected forms a spur. The spurs can be connected to the trunk by any appropriate connecting method but it is helpful to use specially designed connection terminals for ease in connecting and disconnecting field instruments. Individual instruments

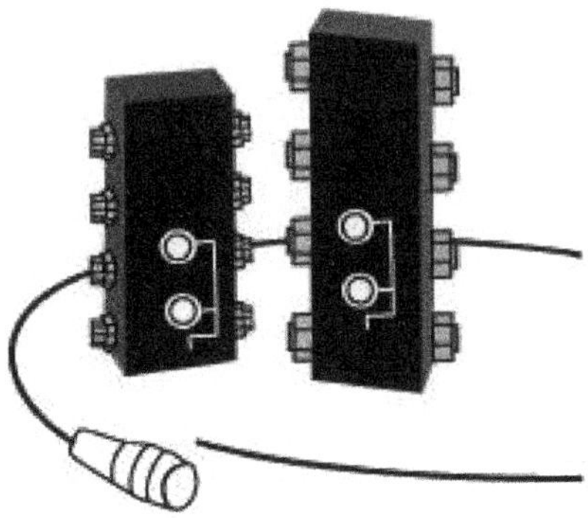

Figure 4–49 Junction Box (*Courtesy of ISA*)

can be disconnected from the trunk without disrupting the operation of other instruments on the bus. Figure 4–49 shows a junction box for connecting instruments to the bus.

The components, trunk, and spurs comprise a segment. A maximum of 32 devices can be on one segment if the instruments are separately powered. Most devices are bus powered and because of the voltage drop along the bus caused by the impedance of the devices creating a load, the maximum number of devices for bus-powered applications is 16.

Each segment must have two terminators, usually one at each end. The terminator is made of a 100 ohm resistor and a 1 microfarad capacitor. It does not communicate, but prevents signal communication errors caused by reflection and converts the current on the network to voltage, which is the receiving signal. Terminators, like other bus devices, are not polarity-sensitive.

The bus power supply must have an output impedance in accordance with the Digital Data Communication for Measurement and Control standard series IEC 61158. A power conditioner, usually in the power supply, matches impedance to the bus to prevent the signal from being shorted by the normally near-zero ohms impedance of the power supply. Figure 4–50 shows the application of a conventional power supply with an internal power conditioner and an added terminator. The interface unit is the H1 card.

Devices can be connected in two basic topologies, bus and tree. (Figures 4–51 and 4–52). For bus topology, the main cable, called the trunk, runs from the marshalling panel (usually near the control

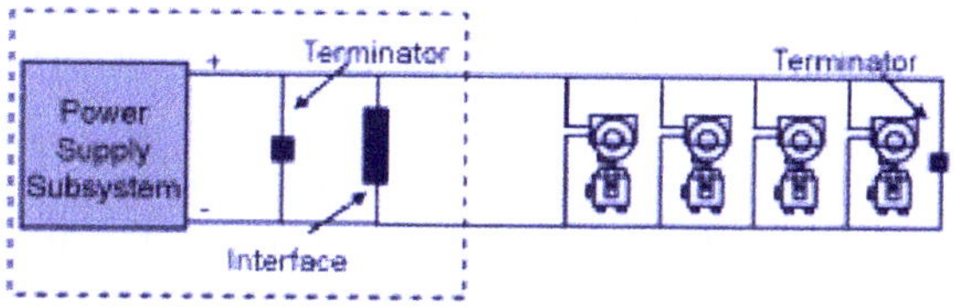

Figure 4–50 Application of a Power Supply with an Internal Power Conditioner and an Added Terminator (*Courtesy of ISA*)

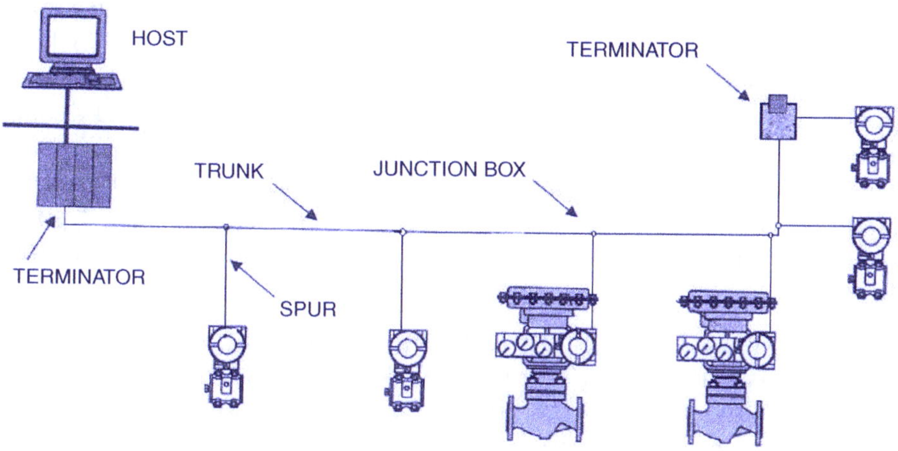

Figure 4–51 Bus Topology (*Courtesy of ISA*)

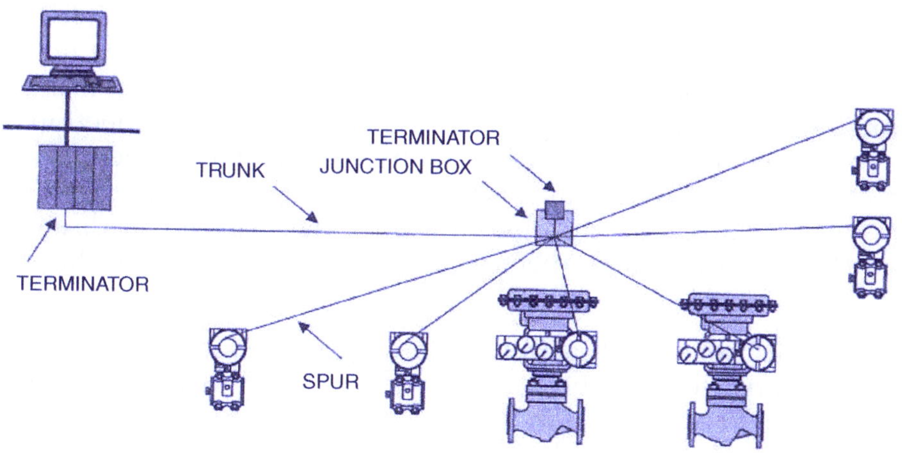

Figure 4–52 Tree Topology (*Courtesy of ISA*)

room) into the field to the last device, servicing each device along the way. Field devices are connected to the trunk by shorter wires called spurs. For tree topology, the trunk runs into the field, where it branches out into the individual devices from a single junction box. This topology minimizes the number of connections but the cable length of the trunk is longer, especially when the field instruments are not close together. Tree topology is more common in upgrades in older plants.

The IEC 61158-2 standard does not specify the type of cable to be used for the trunk and spur. Because of cable resistance, the fieldbus signal is attenuated as it travels down the trunk until it becomes too low to power the instruments. Normal current on the bus is 12 mA. The expected total length for four types of cable is listed in Table 4–3. Types A and B are recommended for new installations and type B is particularly suited for home runs from the field junction box to the marshalling panel. FF cable requirements were specifically designed so that wire in existing 4–20 mA applications can be used when upgrading to fieldbus.

For FOUNDATION Fieldbus devices, the mode address assignment is automatically handled by the linking device or another interface unit so the field device can be connected directly to the network. This is done after the device support files have been loaded or installed in a dedicated folder for support files, with subfolders for manufacturer and device type. The support files are normally shipped with the device or otherwise obtained from the manufacturer. Once the field device is installed and connected to the network, it will be detected and displayed on a live list with all the other devices on the same network (Figure 4–53). Should the device not appear on the live list, it may be connected to the wrong network, have a faulty wire or connection, not be connected properly, or be a faulty device. Once the device is detected, and to ensure that it is the correct device in the correct location, it is good practice to disconnect the device, see if it disappears from the live list, and connect it again to see if the correct tag reappears on the live list.

Once the device is identified, it will be assigned a mode address noted for temporary use until the configuration is continued for assignment in a permanent address location.

Each FOUNDATION Fieldbus device has a unique 32-character device identifier (Device ID) that is like a hardware address that is one-of-a-

Table 4–3 Typical Cable Types and Range

Pair	Shield	Twisted	Size	Length	Type
Single	Yes	Yes	0.8 mm^2 (AWG 18)	1,900 m (6,200 ft.)	A
Multi	Yes	Yes	0.32 mm^2 (AWG 22)	1,200 m (3,900 ft.)	B
Multi	No	Yes	0.13 mm^2 (AWG 26)	400 m (1,300 ft.)	C
Multi	Yes	No	1.25 mm_2 (AWG 16)	200 m (650 ft.)	D

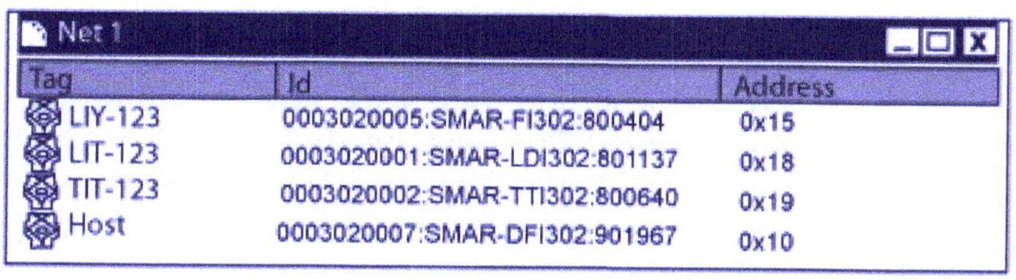

Figure 4–53 Live List

kind and is used to distinguish a particular device from all others. This address is set in the circuit board by the manufacturer and is permanent. During configuration, a particular device is set in a control or measurement strategy associated with its configuration tag and device ID. This concept is known as OLE—Object, Linking, and Embedding—for process control. Once all devices are associated with their respective device tags, the configuration can be downloaded for the entire network or individually. Some manufacturers preconfigure device tags so they will automatically be matched to the configuration.

Parameters are set in the resource and transducer blocks. These blocks have several characteristics in common with the function blocks that are used to build the control strategy. Each device must be configured with one resource block. All input and output devices can be configured with one resource block and they can be configured with at least one transducer block, one for each measurement or actuation. Transducer blocks act as an interface between physical I/O hardware and function blocks. The resource block contains information about the device as a whole. Transducer and resource blocks have no input or output parameters and cannot be linked. Each block should be identified by a block tag (FB_TAG); these tags should be the same name as the device tags on the P & ID (Process and Instrumentation Diagram).

During the configuration of FOUNDATION Fieldbus devices, the range is normally set. There is a difference between range setting and calibration. Calibration principles were given in Chapter 2. Recall that calibration required a known value of the process variable, or a simulated value, to be applied to the input of the instrument under test. The value and the corresponding output are then measured with a test instrument. Corresponding values of input and output, at least three and usually five, are established at various points along the measurement range of the instrument. From this, a data table and calibration plot are made to confirm the instrument's accuracy. This procedure formerly required the instrument under test to be removed from service and transported to a calibration laboratory. This calibration procedure also required accurate test instruments, three to five times as accurate as the IUT (instrument under test), and a calibration apparatus for the simulated inputs. This was needed to correct the sensor reading to match the applied quantity and to check linearity. As in HART calibration, setting the range with FF devices means establishing the values for zero percent output and one hundred percent output or the lower range value (LRV) and the upper range value (URV). Test instruments and a calibration apparatus are not needed.

For both HART and FOUNDATION Fieldbus, a technician can set the range remotely without applying any inputs, but by definition calibration means that a known input from a standard must be applied; a corresponding output is compared and matched to the input.

Damping can be used to filter out measurement signal noise by configuring a filter time constant in seconds. It should be noted, however, that when signal noise is a result of an equipment malfunction, filtering simply masks the problem and can hide actual process responses.

Digital pressure transmitters contain pressure transducer blocks that indicate the primary value type and parameter. This specifies whether the sensor is for differential pressure, gage pressure, or absolute pressure. The sensor type parameter also indicates which technology is used in the pressure sensor: strain gage (resistive or resonant mode), capacitance, or another technology. The primary value parameter is always the measured pressure that is passed on to the analog input (AI) function block by way of the hardware channel. The AI block can convert the pressure measurement into flow, level or some other inferred measurement [Ref. 13].

Summary for FOUNDATION Fieldbus

The FOUNDATION Fieldbus standards effort began in 1985, and the standards were mostly set in 1987. The FOUNDATION Fieldbus chip was developed in 1992 and after a beta test, the first plant was controlled by FOUNDATION Fieldbus in 1994. In 2000 the first high-speed Ethernet (HSE) was registered and the fieldbus age was here in 2002. More than 100 manufacturers are currently supplying FOUNDATION Fieldbus-compatible equipment [Ref.14].

FOUNDATION Fieldbus is becoming the digital technology of choice in a majority of oil and gas companies, pharmaceutical manufacturers, chemical producers, pulp and paper companies and food and beverage companies. These companies are not only in North America but worldwide. The greatest cost savings are realized in installation by reduced wiring and reduced hardware, and in operation over a projection of a few years. This technology is used in projects of all sizes, including retrofit and new installations. Multidrop wiring schemes are generally advantageous over 4–20 mA technology, reducing commissioning cost. As the technology evolves to a mature one, system complexity will be simplified to a common engineering approach across the entire plant site.

Other features of FOUNDATION Fieldbus include single loop integrity with short-circuit protection in wiring blocks. In addition, batch control, logic control, single loop and cascade PID (proportional-integral-derivative) control, feedforward and lead/lag control are included in the field control system (FCS) whereby the control strategy can reside in the transmitter, smart control valve positioners or fieldbus to control transducers.

Enhanced Device Description Language (EDDL) technology eliminates the need for proprietary device support files and proprietary testing and certification. All FOUNDATION Fieldbus-compliant devices are interoperable, as certified by Host Interoperability Support Tests (HIST). System interrogation risks are greatly reduced.

H1 technology is now accepted but high-speed Ethernet (HSE) is yet to be fully recognized and used. Field control systems (FCS) are often preferred because of a decreased response time realized by reducing the network operational response time. Full implementation of H1 and HSE technology will continue to be used for communication and control in the processing industries [Ref. 15].

4.37 APPLICATIONS

Some applications and functions of digital transmitters include configuration and re-ranging by remote communications, complex calculations, versatile configurations, diagnostics, new maintenance structure, built-in control capability and added control efficiency by true distributed control. The fact that true smart digital transmitters do not have physical adjustments—knobs to twist or switches to throw for calibration purposes—makes them ideal for operation in remote and inaccessible locations. Calibration adjustments are made, digitally stored in nonvolatile memory and performed by an HHC usually connected at the receiving instrument, making smart transmitters very attractive for remote re-ranging and self-diagnostic checks.

Previous discussions have given a glimpse at the features of the remote adjustment capabilities of digital field devices. A more detailed investigation is given below to stress the importance of these instruments and to review the distinction between calibration and re-ranging.

The terms lower range value (LRV) and upper range value (URV) are often used in calibration applications and procedures to avoid the confusion and misunderstanding that sometimes result from using the terms zero suppression, suppressed zero, zero elevation and elevated zero. LRV is the net pressure applied to the high and low ports of a transmitter for a zero output. URV is the net pressure applied to the measuring ports for an output of 100%. When considering single port pressure transmitters, the terms relate to the single net pressure applied. As used in this context, span is the difference between URV and LRV.

Some handheld communicators can be used in FOUNDATION Fieldbus installation and troubleshooting. An HHC connected to a fieldbus device is shown in Figure 4–54 for bench checking and in Figure 4–55 for field checking. When connecting for field wiring, care should be taken to ensure that the maximum permissible current on the segment that would result in a loss of communication has not been surpassed. Functions that can be performed with the HHC are:

- Displaying devices on the live list
- Displaying the block list
- Configuring the device blocks
- Modifying parameters

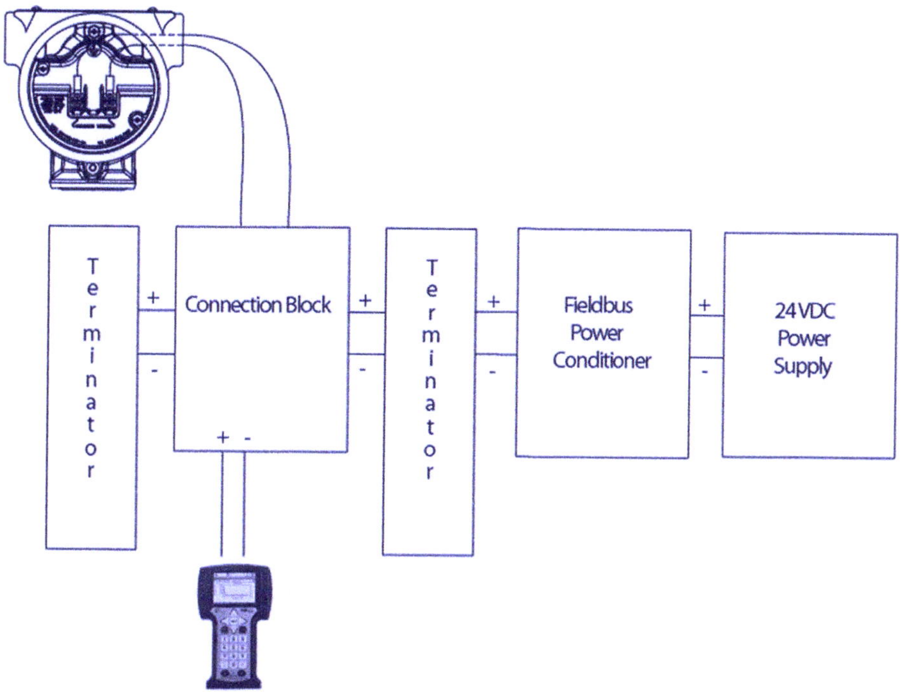

Figure 4–54 Bench Wiring Diagram
(*Courtesy of Emerson Process Management*)

- Displaying device status
- Calibration and sensor trim adjustments
- Displaying device status and physical tag
- Advanced features such as diagnostics
- Aid in various troubleshooting procedures

Digital transmitters have a microprocessor that contains information about the sensor's specific characteristics in response to pressure or differential inputs. The transmitter compensates for sensor variations and the process of generating the sensor profile called characterization which also has the ability to readjust the 4–20 mA points without applying pressure to the measuring ports.

Trim and re-ranging differ in the function performed. Re-ranging sets analog outputs to the LRV and URV values and can be done with or

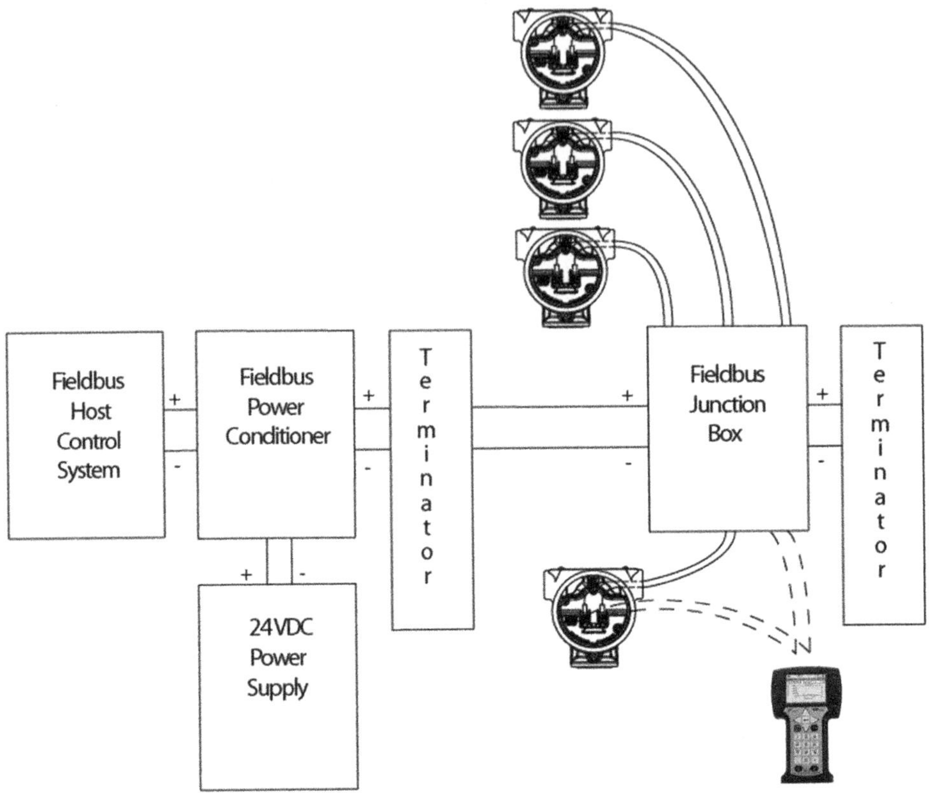

Figure 4–55 Field Wiring Diagram (*Courtesy of Emerson Process Management*)

without an applied pressure. Re-ranging does not change the factory characterization curve stored in the microprocessor.

Sensor trim requires an accurate pressure input and adds additional compensation that adjusts the position of the factory characterization curve to optimize performance over a given pressure or differential pressure range. It should be pointed out that since trim adjusts the position of the factory characterization curve, it is possible to degrade the performance of the transmitter. If the trim adjustment is not performed properly, inaccurate instrument performance can result.

As was stated in Chapter 2, one of the overriding considerations in trim adjustment is the required accuracy of the test instruments used. In earlier times calibration procedures dictated that the test instruments should be at least ten times as accurate as the IUT. As field instruments became more accurate, it became increasingly difficult to acquire tests instruments many times more accurate than the IUTs.

The accuracy ratio dropped to about three or five, with transmitter composite accuracy in the order of 0.1% of rate. Many times the test instruments are not a great deal more accurate. ISO 17025 Best Practices states that specification limits should be set three times higher than calibration uncertainty. For this reason, care and caution should be exercised in making sensor trim adjustments. However, when and if it is desired, a NIST-traceable standard can be applied to the input to override the manufacturer's sensor characteristic. Failure to adhere to this practice can result in the degradation of the instrument's performance [Ref. 16].

The advent of digital transmitters has improved measurement quality beyond that provided by conventional analog instruments. Features that have contributed to this improved performance include greater accuracy, precision, reliability, turndown and, probably most advantageous, convenient re-ranging capabilities. One of the most common applications where this enhanced performance of digital transformers has yielded a considerable improvement in measurement quality is in differential pressure flow measurement.

The turndown of orifice plates is potentially very high without significant error, assuming that applications and parameters are such that the Reynolds number is sufficiently high. The quality of the transmitter used to measure the differential pressure is usually the limiting factor affecting the accuracy of the measurement. When compensation can be made in the transmitter for ambient temperature and static pressure, zero and range drift are minimized and the signal is maintained in the upper 50% of the range, head flow measurement accuracy is greatly improved, and characterization of the sensor and transmitter permits a much wider rangeability without loss of measurement accuracy.

Automatic re-ranging or on-line rangeability depends on the amount of acceptable uncertainty, but generally the flow should, as stated, be 70% of maximum value. The error increases as the flow is reduced. The 3:1 turndown limitation for orifice meters is usually based on the fact that the measurement error becomes excessive for flow rates below 30% of maximum volume. On-line rangeability is limited to 3:1 for conventional transmitters. By using the HHC to re-range the transmitter, the uncertainty error at low flow rates is decreased. The turndown can increase to 5:1 for 1% uncertainty and 7:1 for 2% uncertainty. Equivalent rangeability using conventional transmitters for continuous operation would require a minimum of three stacked transmitters on a single orifice run or three parallel orifice runs with automatic valving.

The on-line re-ranging of digital transmitters has placed orifice meters in a more favorable position, considering the disadvantages of limited turndown capability and degradation of accuracy at decreasing flow rates. A word of caution is in order, however. Some engineers still do not feel comfortable about setting the range on a transmitter without using a deadweight tester or other standard to verify the accuracy. This is especially true for critical area measurement, custody transfer and other such applications, but this precisely defines situations where digital technology is advantageous. As digital transmitters continue to prove reliable with repeatable calibration ability, however, true remote ranging has become state of the art [Ref. 17].

The computing power of digital transmitters adds to the versatility of these measuring devices. Applications utilizing this capability include linearizing and characterizing nonlinear level applications, linearizing signals for head flow measurements, averaging and totalizing flow values, configurable initializing and setting the failure mode of transmitters to fail at 0 or 100% of span or at last reading, as desired.

Diagnostics is another feature of digital transmitters that adds to the advantages of smart devices. Neglecting the failure of the microprocessor, this feature enables the detection of a faulty input circuit, a failed sensing element, failed ROM or RAM or extreme process or environmental conditions, and determines the likelihood of an unreasonable output. By setting an output signal, operators can set and verify alarm and shutdown values from the control room or from any point in the loop using the HHC.

Maintenance procedures are simplified by these diagnostic features, and frequent trips to the field are eliminated. Transmitters can be set by standards in the shop and changed remotely for various applications.

When few applications exist for the added versatility of digital and networking technology, the added expense of a communication network cannot be justified. Such facilities often use digital transmitters with analog transmission.

The discriminating user should evaluate each application and choose the best product based on value and application. As always, the choice should not be based on purchase price alone, but should also consider cost of ownership, transmitter design and options available for future integration into more advanced systems.

Advances in digital measurement and control have resulted in the decentralization of many control systems. Total digital technology on a single-loop basis with many of the advantages once reserved for DCS is now a reality. CMOS (Complementary Metal Oxide Semiconductor) technology and low power consumption have resulted in the common use of solar-powered transmitters or those that utilize the ambient heat in process units. This is presently realized in remote applications in the pipeline industry.

In addition, field level networks have made it possible to retrieve much more data from the field instruments, resulting in an information explosion. This expanded data retrieval is possible because the bus wars have ceased and proprietary control limitations no longer exist. Instruments and systems are interoperable and the development of all-digital applications will expand to realize the advantage of two-way communication and information exchange between control centers and field instruments.

4.38 INTRODUCTION TO WIRELESS COMMUNICATION

Wireless technology and communication are certainly not new to the applications that surround us. Most of us take advantage of and are affected each day by wireless technology through radio, television, cellular telephones and remote control for the many electronic gadgets that we use. Also, wireless is not new to the processing industries; remote monitoring has been used for many years in the oil field and pipeline industries. Not until recently, however, has wireless technology been used to a significant degree for in-plant communication.

The greatest impediments to the development of wireless in-plant communications have been reliability, security, device battery life and economic expedience. As improvements in wireless have made these obstacles easier to overcome, traditional wired applications are giving way to wireless. As with every new technology, users must evaluate their own application to determine if they have a practical use for it. With the many wireless technologies offering solutions and cost savings, most process industries can find applications where wireless offers an attractive return on investment.

When data needs to be gathered over a distance of several thousands of feet through areas where cables cannot be buried or wires installed, wireless can prove to be very successful and economically feasible. Monitoring inventory in remote storage tanks, remote pump monitoring, data retrieval for asset management and alarming are but a few examples where wireless technology can be advantageous. Other situations, such as physical structures and barriers like pipelines, railways, highways and open waterways, characterize applications where communication by hardwire can be difficult. For both near-plant and in-plant applications that are unstaffed or in hard to reach areas of peripheral locations, wireless can prove beneficial.

As with other technological expansions, the reluctance to abandon standard methods and procedures has curtailed development. Digital transmitters have been available since the late '80s but even today, many control systems rely on 4–20 mA, a standard which itself took decades to develop. The bus wars of the '90s are now being settled by FOUNDATION Fieldbus; ISA's SP50 committee, which set the 4–20 mA standard, has worked diligently to develop a fieldbus standard.

The IEEE 802.11b specification establishes a 2.4-GHz band used to transmit approximately 11 megabits per second of wireless data from several dozen to several hundred feet of indoor distance and tens of miles of outdoor distance. This specification is an extension of wired Ethernet and it brings the same principles to wireless communication. It is primarily used for transmission control protocol/internet protocol but can also handle AppleTalk or PC file sharing standards. The standard is backward compatible to the earlier 802.11 standard and allows speeds of 1, 2, 5.5 and 11 megabits per second on the same transmitters. Multiple 802.11b access points can operate in the same overlapping area over different channels which are subdivisions of the 2.4-GHz band available in each country. Fourteen channels are staggered at a few MHz intervals from 2.4000 to 2.4835 GHz. Different channels are legal in different countries and only channels 1, 6, and 11 have no overlap among them [Ref. 18].

4.39 SPECIFICATIONS AND STANDARDS FOR WIRELESS TECHNOLOGY

Bluetooth is an industrial specification for wireless personal area networks (PANS). It provides ways to connect and exchange information between devices like mobile phones, laptops, PCs, printers and other components in the commercial arena and has recently been adopted in applications in the manufacturing and processing industries. The Bluetooth specifications are developed and licensed by the Bluetooth Special Interest Group. Bluetooth is a standard and a communications protocol primarily designed for low power consumption and a short range. Range is power-class-dependent and enables devices to communicate with each other when they are in the designated range of 1 meter with 1mW, 10 meters with 2.5mW and 100 meters with 100mW. To enable the use of Bluetooth, a device must be compatible with certain Bluetooth profiles that define the possible applications and uses of the technology.

Wi-Fi (Wireless Fidelity) Alliance® is a trade group; Wi-Fi refers to a family of related specifications (IEEE 802.11) that specify methods and techniques of wireless local area networking. This group owns and controls the Wi-Fi Certified® logo, a registered trademark which is permitted only on compliant equipment.

Bluetooth and Wi-Fi have slightly different applications in today's commercial and industrial communication and use versions of unlicensed spread-spectrum technology. They use the same frequency range and employ different multiplexing methods. The Bluetooth specification is based on frequency-hopping spread-spectrum technology and adheres to IEEE protocol 802.15.1. Wi-Fi falls under IEEE 802.11 specification; therefore, they are not interoperable. They differ in the fact that Wi-Fi provides higher throughput, covers greater distances and requires more expensive equipment with higher power consumption.

Bluetooth is generally considered to be suitable for cable replacement for several applications and is thought of as wireless USB. This analogy may not be entirely correct as Bluetooth devices can usually host other Bluetooth devices; that is not universal to USB devices. Wi-Fi is usually understood to be cable replacement for local area networks only, and is thought of as wireless Ethernet. Both Bluetooth and Wi-Fi operate at lower bandwidths than the cables they replace [Ref. 19].

The ZigBee Alliance is an association of companies working together to enable reliable, cost-effective, low-power, wireless networked monitoring and control products based on an open global standard. This is to provide the user with ultimate flexibility, mobility, and ease of use by building wireless intelligence and capabilities into every device. Companies will have a standards-based wireless platform optimized for the unique needs of remote monitoring and control operations to include simplicity, reliability, low cost and low power.

The ZigBee network layer supports star, tree and mesh topologies. A star network is controlled by a single device called the ZigBee coordinator that is responsible for initiating and maintaining the devices on the network, and all devices or end devices directly communicate with the ZigBee coordinator. For mesh and tree topologies, the ZigBee coordinator starts the network and chooses the key parameters; the network may be extended through the use of ZigBee routers. Mesh networks will allow peer-to-peer communications [Ref. 20].

Bluetooth wireless technology has adapted a U.S. military battlefield communications system: frequency-hopping spread-spectrum. The radios in Bluetooth hop randomly among 79 different frequencies at approximately 1,000 times per second—that is an incredible amount of security. Messages are divided into small packets, sending one packet per hop. If the receiver is unable to correctly interpret the packet, it sends a message to the transmitter, which resends the message to the transmitter, which resends the message and reinserts it in the proper sequence. On top of the entire message structure, a 128-bit encryption provides more security.

Frequency hopping also aids in robust operation. The chances of interference trespassing on all frequencies are much smaller than on any individual frequency. Some of the frequencies will get through at all times. Because the packets retransmit as needed, all packets will get through. Bluetooth provides three radio options. One can use Class 3 operation in a PC space and get reliable signaling in a one-meter radius. Class 2 operation will give a mobile networking and office environment with a ten-meter radius. Class 1 Bluetooth operation is for enterprise operations such as the factory floor and provides 100-meter signaling.

Bluetooth operates in the frequency range of the 2.4-GHz industrial, scientific and medical (ISM) band, which is available world-wide for user license-free equipment operation. Manufacturers must still certify equipment, making sure it meets technical requirements for operation in each jurisdiction. The universal applicability of this ISM band means lower costs for Bluetooth devices because manufacturers can build equipment in volume with no special variations for specific political jurisdictions.

The general trend is for technology to develop from simple, straightforward applications to more complex ones. Manufacturers have begun to provide devices to communicate with Bluetooth instrumentation without proprietary or custom test equipment. This interoperability enables a user to calibrate, test and monitor without plugging in and opening loops or mechanically connecting to instruments. All of these features are brought to plant units and factory floor models with 4–20 mA circuits from point to point for indoor and outdoor environments. For manufacturing, devices have been made that monitor machinery performance, diagnostics, temperature monitoring, vibration monitoring and other optional signals.

Users are comparing Bluetooth to IEEE 802.11b wireless fidelity (Wi-Fi). Bluetooth consumes less power, can be battery operated, is smaller and is easy to connect to a host for measurement and control strategies. Also available are serial connections for interface and Wi-Fi for parallel bus connection. 802.11b is used more for small office and home wireless applications and does have security issues. Both technologies are needed to enable the device market to develop. The basic difference between the two wireless standards are power consumption and communication range. They both connect devices by a radio signal [Ref. 21].

The time for wireless communication may have arrived with built-in encryption, authentication, verification, anti-jamming, security measure, low power consuming devices and simplicity of installation and maintenance. Familiar instruments, little pre-engineering and design, available plant webs for digital plant architecture and familiar networks are considerations that make wireless communication attractive.

4.40 TOPOLOGIES

The topology of a wireless network is the way that the network components are arranged. It includes the physical layout of devices, routers and gateways, as well as the layout of data flow paths for connection between these components. The three most common wireless topologies for applications at a plant site are star, mesh and cluster-tree. A good definition and understanding of the strengths and weaknesses of each will enable one to determine which topology is best for a given application.

Before the design and installation of a wireless network, it is important to consider a site survey to help in the planning of the network. This will help identify the location of devices so line-of-sight transmission or another appropriate transmission configuration, as well as the required range of the gateway, can be ensured. Even with site surveys that may result in additional initial expense, many conditions are subject to change and cannot be predicted. For example, new construction and changing environmental conditions, which may result in limitations in a star network, can be accommodated with other topologies that enable more than a single transmission path between gateways and devices.

Star

Star topology is a point-to-point or line-of-sight architecture where individual wireless devices or nodes communicate directly with a gateway or central hub. The data is transmitted to a central collection point such as a control room or control station, either directly or by connecting to another network. Star topologies can use the least amount of power because of the simple wireless connections. It should be noted, however, that the transmission distance from the wireless device to the gateway is usually limited to between 30 and 100 meters. Another limitation of star topology is the possible loss of data with path disruption between the device and gateway caused by RF interference, physical structures, environmental factors or various types of temporary obstructions such as mobile vehicles and structures. A star topology model is shown in Figure 4–56.

Mesh

Mesh topology devices are configured such that the devices (nodes) can communicate with other nodes in the network in a point-to-point multipoint fashion using multi-hopping. A message can hop from node to node until it reaches the assigned gateway. This technique enables longer communication range and a decrease in the loss of data. The multi-hop capability enables data to find alternate paths should blockage between nodes, or even failed node, occur. Mesh topology is shown in Figure 4–57.

Cluster-Tree

The cluster-tree topology is a hybrid where wireless devices in a star topology are clustered around routers or repeaters that communicate with each other and the gateway in a mesh topology. This topology combines the advantage of the star topology (lower power consumption) with the greater range and fault tolerance of the mesh portion of the cluster-tree. The limitations of path blockage still exist in the star pattern, so a site survey is still recommended to make certain that end

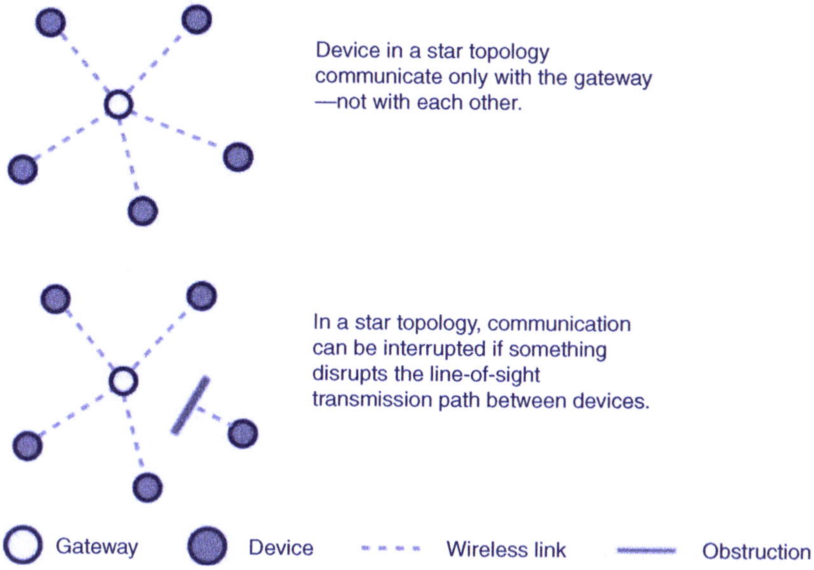

Figure 4–56 Star Topology Model (*Courtesy of Emerson Process Management*)

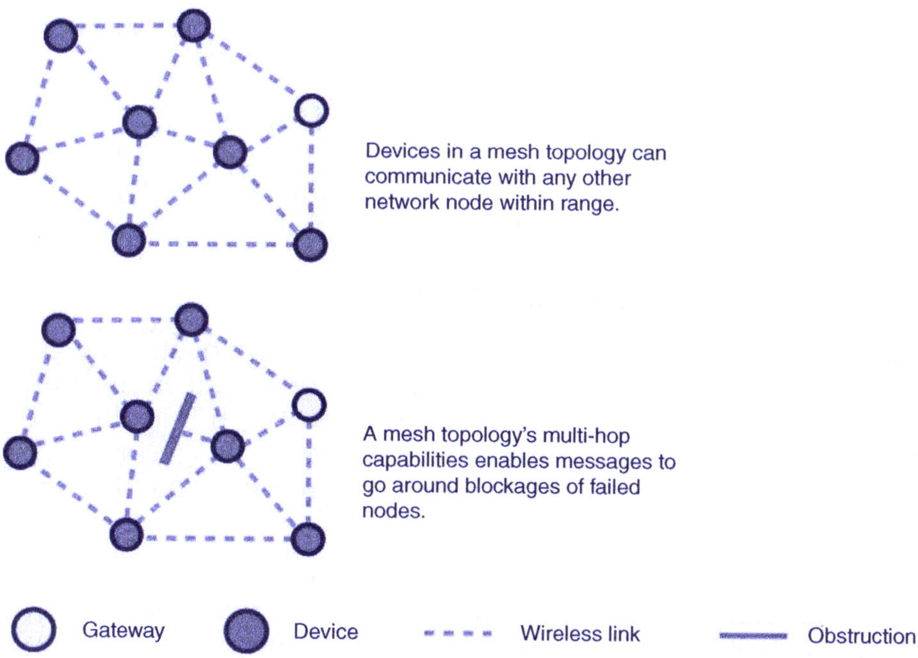

Figure 4–57 Mesh Topology Model (*Courtesy of Emerson Process Management*)

devices in each star have an unobstructed line-of-sight to the assigned router and that an alternate path exists to cope with changes or interference. A cluster-tree topology network is shown in Figure 4–58.

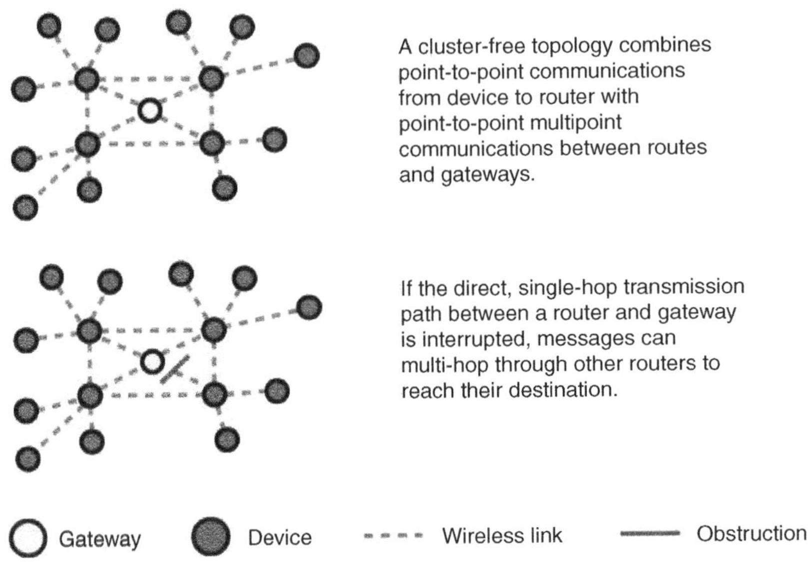

Figure 4–58 Cluster-Tree Topology Model
(Courtesy of Emerson Process Management)

Figures 4–59, 4–60 and 4–61 will help in the understanding of the application and configuration of instruments using wireless technologies. Figure 4–62 provides a comparison of the installation costs of systems using wired and wireless technologies.

4.41 SELF-ORGANIZING NETWORKS

No single topology is likely to be the best or even adequate for a given application because of the dynamic situations of a plant environment. Self-organizing networks with built-in intelligence and flexibility can automatically choose a star, mesh or cluster-tree topology, or a combination of all three, to provide maximum communication efficiency in a continuous operation.

Self-organizing networks can let every wireless device act as a router for a nearby device. Devices and networks can work together to identify and select the most appropriate communication path for each message. Communication is not dependent on a dedicated device,

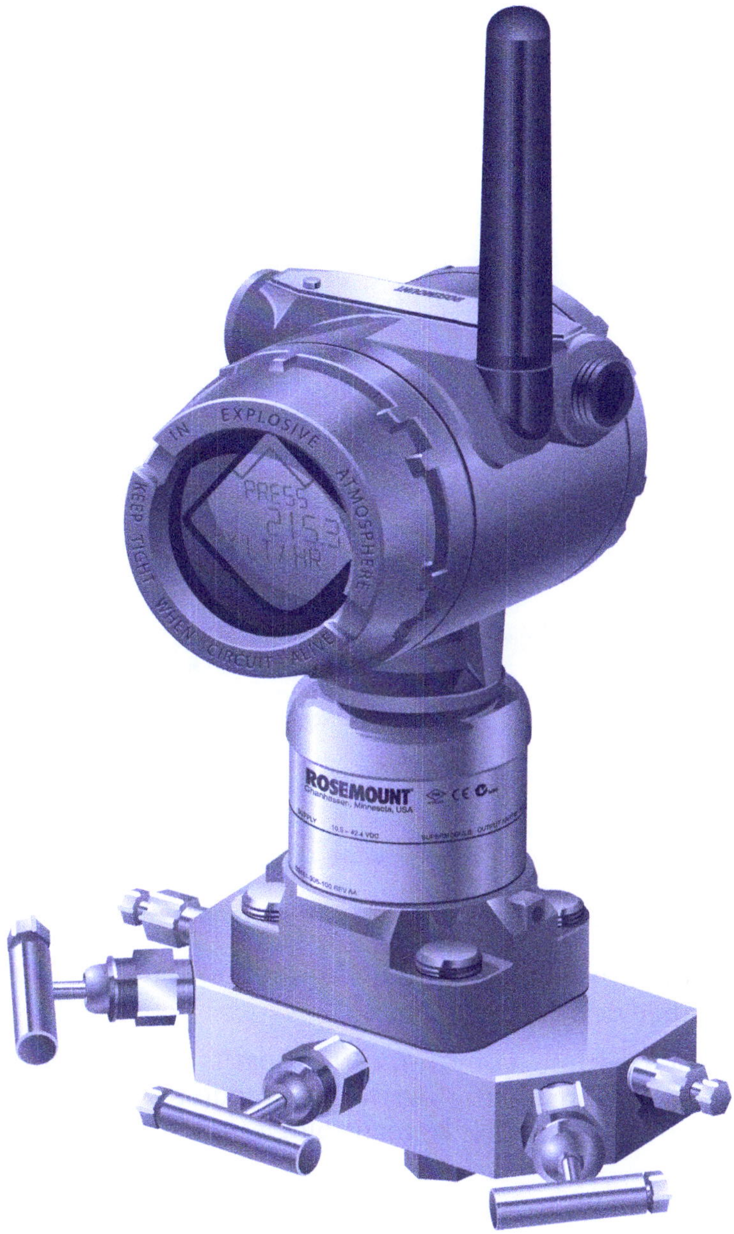

Figure 4–59 A Wireless Differential Pressure Transmitter (*Emerson Process Management*)

Figure 4–60 A Wireless Gateway (*Emerson Process Management*)

component or path, but is distributed throughout the network via the best path and device.

By providing multiple communication paths and automatic path configuration, self-organizing networks provide several advantages in data transfer over other wireless technologies including the following:

- Better than 99% communication reliability compared to 40% in networks that cannot provide redundant communication paths.
- Flexibility in network expansion is enhanced by the network's ability to recognize new devices and to provide a message structure that knows how to get messages to the proper destinations.

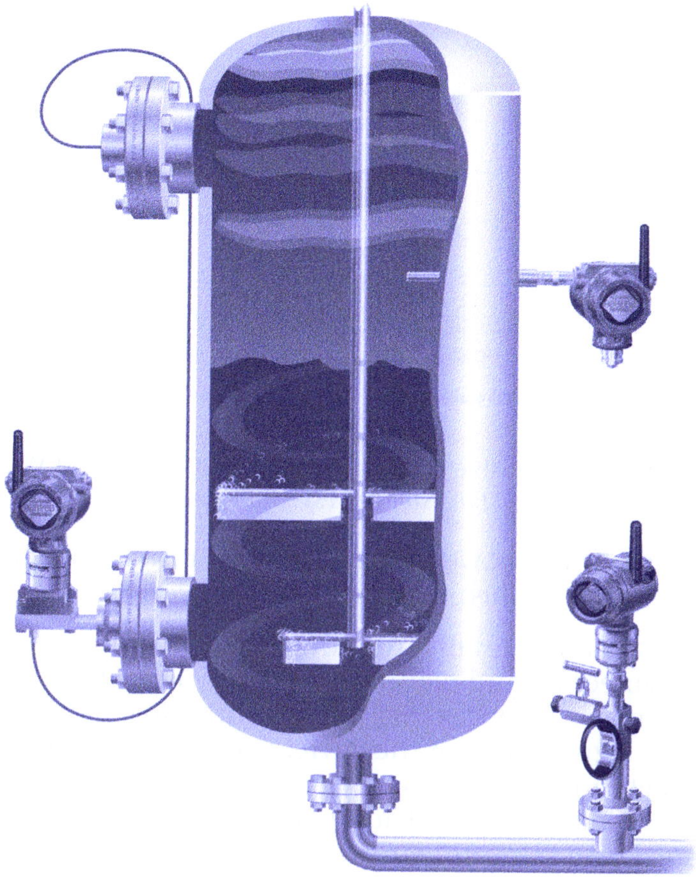

Figure 4–61 A Level Application with Wireless Transmitters (*Emerson Process Management*)

- When a path is blocked or a device fails, an alternate is found. When a new device is recognized and authorized, another or better path is detected and used. Such constant searching for better paths of communication and bypassing failed devices has been described as self-healing.
- Limited or no site surveys are required since the network automatically seeks the best path of communication and adjusts for temporary or permanent modifications.
- Energy efficiency is improved over point-to-point wireless networks. For example, a one-watt transmitter in a point-to-point network can have a coverage of 1500 meters, while ten 1-mW devices can cover the same area with 150 hops. Also, a sleep

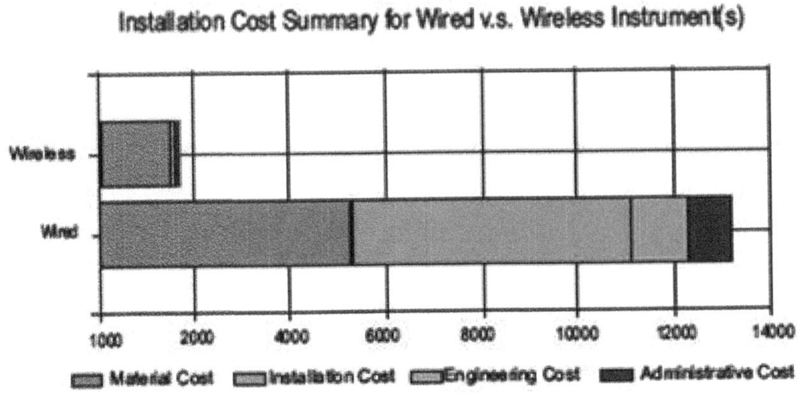

Figure 4–62 The Comparative Installation Costs of Wired and Wireless Technologies (*Emerson Process Management*)

mode minimizes power consumption by turning off communication electronics when they are inactive.

- Easy installation results from the lack of site surveys and less pre-engineering and planning. The only knowledge needed is the number of devices that the network can support and the wireless signal range of the devices and gateway. Once it has been determined that the device or gateway is within range, the installed devices will automatically join the network, self-organize and transmit data.
- Diagnostic data can be communicated directly to other devices such as those that are HART-compatible.

The update rate for the first generation of wireless technology is no faster than several times per minute. This compares to a second or less for conventional wired technology. This means that secure, well designed, self-organizing networks are well suited for process monitoring only, and should not be applied to high-speed factory automation, but offer much improvement over the hundreds or even thousands of manual monitoring applications in present use. These manual monitoring techniques can be on data collection schedules ranging from almost never to several times per shift. By comparison, once per minute with wireless may seem like real-time.

Aside from the fact that wireless can greatly reduce sampling time, many problems and inconveniences connected with manual data logging can be eliminated with automated wireless techniques. For example, notation errors, poor repeatability and accuracy in local readout devices such as pressure gauges and handheld measuring devices, and the cumbersome methods of lead lines and dip sticks are but a few of the characteristics of manual monitoring that can be overcome with automated wireless technology. Improvements can also be made in open loop control applications with wireless devices.

Of course, it can be argued that present day technology is available to apply transmitters to any measurement application. This is true, but often the expense of running transmission lines prohibits the use of such devices for continuous and unattended monitoring. Case studies have shown that opening vessels for manual gaging and monitoring does pose safety and environmental risks that are eliminated with wireless monitoring, without the cost of hard-wire transmission lines. Wireless technology can reduce the cost of a system by as much as 90% over hard wire.

An example of the argument for wireless technology is monitoring pressure safety valves (PSVs) to track the release of volatile organic compounds (VOCs). The EPA requires very strict reporting of VOC entry into the atmosphere, and any violation can result in penalties. Because of the very wide area monitoring required in large tank farms, plant operating personnel previously resorted to less costly, but far less reliable methods to monitor PSV activation. An example of this is a sock attached to the discharge of a PSV to indicate a VOC release.

Case histories have shown that wireless pressure transmitters can be installed on the high side of PSVs and fed through wireless gateways connected to an existing network for event monitoring and for calculating VOC release rates. A system like this can provide high resolution data for EPA compliance. As an added bonus for actual process monitoring, instead of a "sock off" criterion, the pressure measurement can be used for proactive procedures to prevent a VOC release.

Realizing the reluctance of some facilities to try a new technology reaching to the next level of technical development, some manufacturers offer a pre-engineered, checked-out and configured network. This starter network will usually contain a variety of topologies, a choice of from five to one hundred wireless transmitters and a several-tag system with appropriate software and a gateway package to study management techniques. When the capability of wireless technology is realized, many opportunities become apparent to engineers, who quickly envision applications that are out of reach with conventional communication techniques.

With the development of wireless digital transmitters providing two-way communication between the process and control centers, operating personnel have access to data that has previously been unavailable, enhancing process performance. Self-organizing wireless technology and gateways with wireless smart measurement systems, AMS™, and supporting software have given new dimensions to integrated automation systems, statistical process control and predictive maintenance.

Plant operators can now have a much greater understanding of their process, see more of what is actually happening in process units and equipment, and achieve automation improvements that are not possible with wired instruments.

As with wired systems, security issues are a concern with regard to the adoption of wireless technology in the processing industries. Following are several topics to be considered to address security matters:

- Security methods can be applied when overall measurement and control security objectives have been defined.
- A systematic and comprehensive approach to secure a system requires an in-depth defense. Installing a firewall or virus scanner may not provide adequate protection.
- Complete and adequate data must be available. Unauthorized access must be prohibited when permitting outside web availability.
- State-of-the-art authentication and encryption must be ensured.
- Networks should be segregated for shared HMI (human-machine interface) and control resources.
- All nonessential and inappropriate applications such as e-mail, games, media players, etc., should be removed from the system.
- It is required that a firewall be configured by competent individuals with proven capabilities in process control and IT.

The difficulty of providing absolute security is well understood by network and process control engineers, and absolute security can seldom be completely achieved. It is imperative, however, that all measures to seek absolute security be taken.

EXERCISES

4.1 Explain the function of a transmitter and the difference between a transmitter and a transducer.

4.2 List and explain the operation of six types of electrical transducers used in pressure transmitters.

4.3 Explain the function of a DC Wheatstone bridge circuit and the means by which the circuit can be used in a pressure transmitter.

4.4 Explain the function and general operating principle of a differential-pressure transmitter.

4.5 List two common applications of differential-pressure measurement.

4.6 State the general specifications of industry-standard transmitters.

4.7 Explain the principle of the following with respect to transmitters:

- Force-balance
- Negative feedback
- Motion balance

4.8 What advantage do digital transmitters have over conventional transmitters?

4.9 What is sensor trim?

4.10 What is an HHC or HHT? List the functions it performs.

4.11 Explain the concept of configuration with respect to smart transmitters.

4.12 Since digital transmitters can communicate by digital transmission, why is 4–20 mA analog transmission still used?

4.13 Why is the HART protocol said to be a hybrid between digital and analog?

4.14 It was once thought that HART would serve as a bridge between an all analog system and one that is all digital and that HART applications would decline as digital systems became more prominent. Has that occurred?

4.15 What is FOUNDATION Fieldbus?

4.16 How is security provided in Bluetooth wireless technology?

4.17 Explain the concept of the field control system (FCS) available with FOUNDATION Fieldbus.

4.18 List the common topologies used for wireless communication.

REFERENCES

1. "Redundancy and Self-diagnostics Enhance Process Safety Systems," Mike Cushing, *Sensors*, October 2004.
2. Miller, Richard, *Flow Measurement Engineering Handbook*, McGraw-Hill, 1996.
3. Technical Paper 410, Crane.
4. Menezes, Mark, "Improving Compressor Safety and Efficiency with the Right Pressure Transmitter," *Control Solutions*, November 2001.
5. Moore Bulletin 34-1, "Mycro XTC™ Transmitter Controller."
6. Helson, Ron, "What is HART?" *Control for the Process Industries*, September 2004.
7. Fussell-Policastro, Ellen, "HART Beats Stronger," *InTech*, ISA, October 2004.
8. Model 3051 Series Pressure Transmitter Technical Literature, Emerson Process Management.

9. Berge, Jonas, "Fieldbus for Process Control," Chapter 3, ISA, 2002.

10. Holiday, Kenneth L. "Calibrating HART Transmitters" Southwest Research, Inc., published in HART Technical Applications, March 2005.

11. Beamex MC5 Multifunction Calibrator Manual

12. Helson, Ron, "Top Ten Misconceptions About HART Technology", HART Technology Applications, March 2005.

13. FOUNDATION Fieldbus Technical literature.

14. FOUNDATION Fieldbus Report Volume 5, Issue 1.

15. Caro, Dick, "Delivering on the Promises of Fieldbus," *InTech*, ISA, June 2006.

16. "Guidelines for Evaluating and Expressing the Uncertainty of NIST Measurement Results" NIST Technical notes 1297, 1994.

17. Whitman, Stanley L., "Microprocessor-Based Transmitters Increase Orifice-Meter Rangeability. *Measurement and Control*, April 1988.

18. "Giving in on Wireless," Protocol/*InTech*, ISA, April 2003.

19. Wikipedia, the free encyclopedia, en.wikipedia.org/wiki/wi-fi, December 2007.

20. Zigbee.org, December 2007.

21. Drake, Bill, "Bluetooth Gets a Filling," *InTech*, ISA, May 2003.

5

Level Measurement Theory and Visual Measurement Techniques

5.1 INTRODUCTION

Few measurements are as common and widespread in terms of application, operation and variety as is level. This measurement is defined as "the determination of the position of an existing interface between two media." These media are usually fluids, but they may be solids or a combination of a solid and a fluid. The interface can exist between a liquid and a gas, a liquid and its vapor, two liquids, or a granular or fluidized solid and a gas.

Many techniques are available for the measurement of these interfaces, each with its own advantages and limitations. The best selection depends on the nature of the specific application, including the process to be measured, the degree of accuracy and dependability desired, and economic considerations and constraints. The design engineer must have a working knowledge of the various types of measuring devices available as a guide for the selection and implementation of a system best suited for a particular application.

It should be emphasized from the start that there is probably no one best method or specific rule to follow for the selection of a measuring device. Material presented here can be helpful as a guide, although it must be pointed out that instruments presently in use may become outdated because of the constant introduction of newer techniques. Generally, however, these newer devices simply represent more modern and efficient means to accomplish the same measurement based on established principles.

Level applications have changed somewhat in recent years because of the dynamic economic constraints and stipulations brought about by material cost, international market strategies, and product changes. In the past, the measurement of level did not usually demand highly accurate and sensitive devices. Many applications simply required maintaining a process level within widely varying limits. Such level systems consisted of surge tanks between process units or vessels—merely a wide place in a process line to take up the slack in the change of production rates in the various units of a complex facility.

Because of the increased costs of raw material, power, and other resources, along with stricter quality standards, more precise level measurement is now required. Custody transfer may be based on the quantity of material in a process vessel, which often is based on level measurement. This also requires more accurate level measurement systems. Hydrostatic tank gaging is an example of a technology that was born of necessity. Beyond such advances, however, the advent of modern electrical components, other hardware, and software has resulted in process and manufacturing measurement accuracies moving from ±1% FS (full scale) to ±0.1% actual value. While this improvement may not represent the norm in process applications, level measurement accuracy to this high degree can be achieved.

Like many variables, level can be measured by both direct and inferred methods. Direct methods employ physical principles, such as fluid motion, floats and buoyancy, and optical, thermal, and electrical properties. Inferential methods involve the use of hydrostatic head, weight quantities, radioactive properties, density and sonic detectors. The selection of a particular level device, however, is generally made with regard to characteristics more closely related to requirements of a specific application.

The direct measurement of level is possible because of the straightforward and unique simplicity of this dimension. Pressure is force per unit area; temperature is the measure of the statistical activity of molecules and almost always requires inferential measurement schemes; flow is a measure of volume per unit time. Level is merely a measure of length. This length measurement is often and easily converted to a volumetric measurement, an important factor when one considers that the determination of liquid level can have a direct bearing on the amount of money involved in the exchange of material or product.

5.2 VISUAL MEASUREMENT METHODS

Perhaps the most direct approach to measuring level is by visual means. Level is one of the few variables that can be detected by sight observation. While simple, accurate and reliable, visual measurement is normally limited to backup measures or to applications where constant monitoring is either unnecessary or is undesirable.

5.3 DIPSTICKS, LEAD LINES, STEEL TAPES WITH BOBWEIGHTS

The earliest instruments used for level measurement were perhaps sticks or poles with calibrated scales to test the depths of streams. This simple method is still in common use. Consider the dipstick used to measure the oil level in internal combustion engines and the long calibrated poles used at service stations to measure the fuel level in underground storage tanks (Figure 5–1 is an example of a dipstick). Flexible lines fitted with end weights, called chains or lead lines, and steel tapes with special weights are used to gage fuel levels and quantities in large petroleum storage tanks. Crude as this method seems, it is accurate to about 0.1% with ranges up to about 20 ft.

Application of a tape and bobweight assembly (Figure 5–2) for level measurement is simple and straightforward. The weight is lowered through the fluid in a vessel until it reaches the bottom and then reeled back up on a spool. The point at which the liquid in the vessel marked the tape is noted and the level determined. An obvious problem that may be encountered is the difficulty of knowing when the weight hits the tank bottom, especially if the fluid is highly viscous or the depth is great enough for the liquid to create a buoyant force that offsets the force of gravity on the weight. Inaccuracies caused by thermal expansion of the tape are probably negligible, offset by the thermal expansion of the liquid being measured. This assumes that the tape and liquid are at the same temperature and have about the same coefficient of thermal expansion.

Although the dipstick, lead line and steel tape methods of level measurement are unrivaled in terms of accuracy, reliability and dependability (at least in skilled hands), there are drawbacks. First, these techniques require an action to be performed, which means that an operator must interrupt other duties. Also, there cannot be a continuous representation of the measurement. Another limitation is the difficulty of successfully and conveniently measuring level values in pressurized vessels. Such disadvantages limit the effectiveness of these types of visual level measurement.

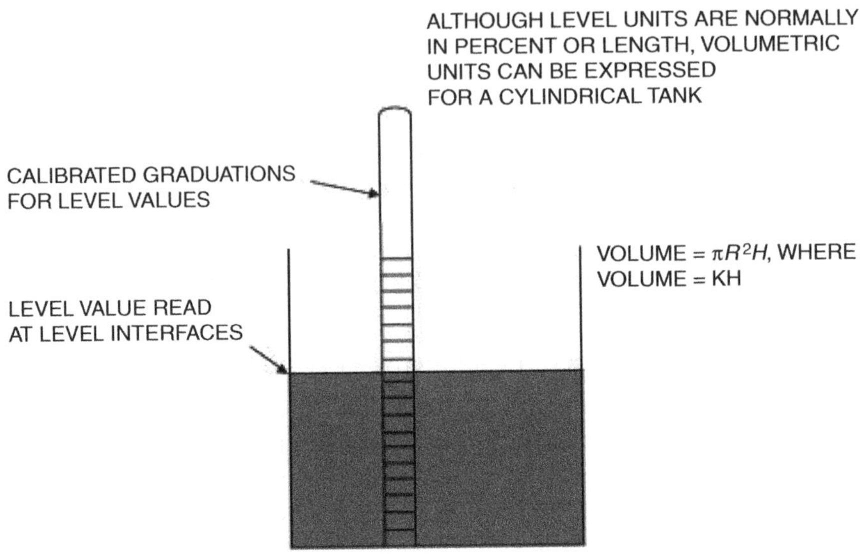

Figure 5–1 Dip Stick Level Measurement

Figure 5–2 A Steel Tape and Bob

5.4 SIGHT GLASSES

A sight glass, also called a gage glass, is an important instrument for visually determining level quantities. To understand its simple principle of operation, consider a U-tube manometer. With equal pressure on both legs of the manometer, the levels in the two legs will have the same amount of vertical displacement. One of the legs is represented by the process, and the other is represented by a transparent tube, located on the outside of the process vessel, that is available for visual inspection. As the process level fluctuates, the level in the transparent tube changes accordingly and is a true representation of the process level.

Figure 5–3A and B depicts the operation of sight glass level measurement; as shown, this technique can be adapted to either open- or closed-tank applications, although open-tank measurement by this means is less common in many industrial facilities. The closed-tank sight glass, on the other hand, is used in many pressurized and atmospheric processes. Its greatest use is in pressurized vessels, such as boiler drums, evaporators, condensers, stills, tanks, distillation columns, liquid accumulators, bootlegs, and traps.

Gage (sight) glasses can be generally classified in terms of low- and high-pressure processes. A low-pressure gage glass consists of a clear, round tube fitted between two valves of a special type. The valves serve to isolate the tube from the process so that the tube can be removed for cleaning or replacement. As shown in Figure 5–4A, balls are inserted within the valve chambers to prevent flow that would result from a sudden pressure drop across the ball ports if the tube were ruptured. The balls permit the free passage of fluid when the process level is changing, allowing the level in the tube to change.

The tubular sight glass is usually protected by a shield constructed from metal or plastic rods. Care should be exercised when working around this device, however, to avoid mechanical shock and breakage of the tubes, as the shield provides only minimum protection. The tube is normally limited to about 70 in. in length and a pressure of a few hundred pounds per square inch. As the length decreases, the pressure rating can increase. Extended range can be provided by mounting several shorter tubes in series, end to end. The problem with this arrangement is the inability to see the level when it is between glasses. The solution to that problem is to "gang" the sight glasses so that the visible sections overlap, as shown in Figure 5–3C.

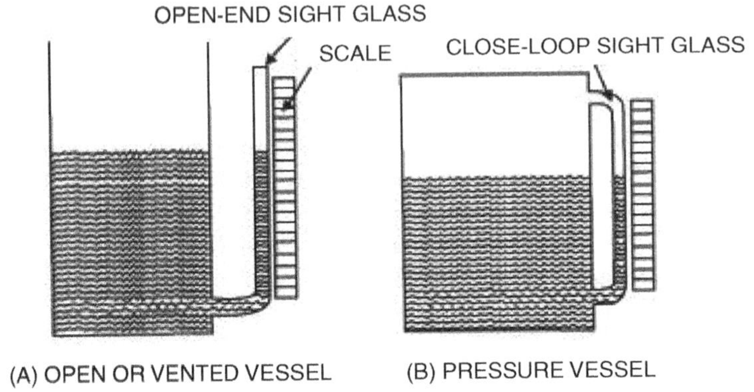

Figure 5–3A and B Basic Sight Glasses

The gage glass shown in Figure 5–4B, sometimes called a reflex glass, is used for higher-pressure applications and in areas that require a more rugged device. The armored appearance of the reflex type indicates that it can withstand a much higher pressure and a greater amount of mechanical shock and vibration. The operation of this gage is based on the optical law of total reflection of light when it passes from a medium of greater reflective power into a medium of lesser reflective power. The groove facets, cut in the inner surface of the glass at appropriate angles, make it possible to eliminate all light from the vacant space (back portion) of the glass. At the same time, light is permitted to pass through the portion of the glass that is covered with the process fluid. Light is reflected by the grooves in the glass above the liquid but not below the liquid surface. The liquid thus appears to have the same color as the background of the chamber (usually black) to give the greatest contrast. A sharp, clear line marks the height of the liquid, above which the air or gaseous space has a bright, mirror-like appearance. When natural light is not sufficient to see the level, a lighted plastic strip is placed in the backside opposite the viewer. This allows the gage glass to be used in low-light areas and at night. The length of reflex glass gages ranges from a few inches to 6 or 8 ft, but like the tube-type gages, they can be ganged together to provide nearly any length of level measurement.

Special precautions should be observed before installing a sight glass. To avoid imposing piping stresses on the gage chamber, especially for the low-pressure non-armored type, the mounting should be such that the gage does not support the piping. Also, differential rates of thermal expansion between the vessel and gage can cause severe mechanical

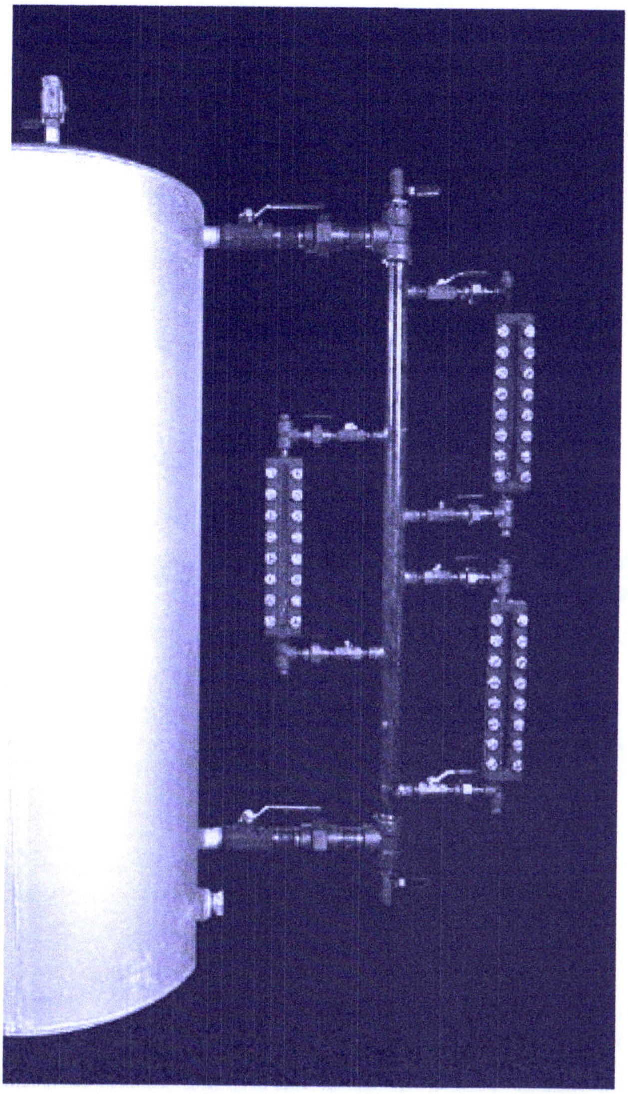

Figure 5–3C Overcoming Length Limitations

stress, especially for extremely high or low process temperatures. An expansion loop should be installed between the gage and the vessel, or a reasonably long run of piping should be used. Support brackets should be installed for long non-armored gages greater than 4 ft in length or weighing more than 100 lb.

Shutoff valves should be provided to isolate the gage from the vessel, and drain valves should be provided to permit depressurizing the gage

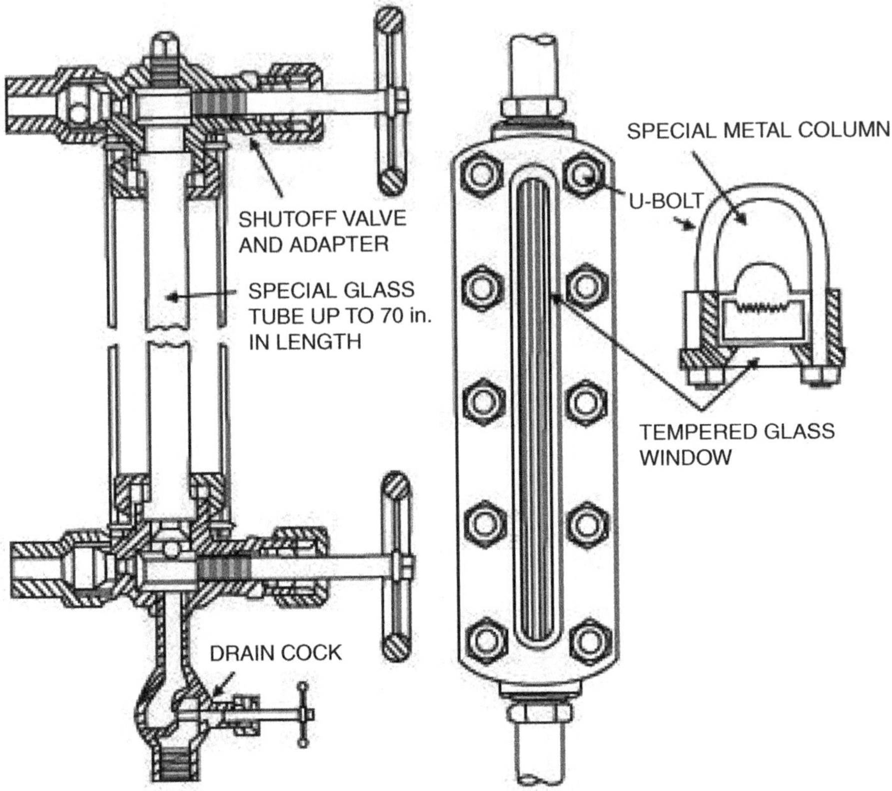

Figure 5–4 Closed-Tank Sight Glasses

for maintenance. When placing the gage in service, the shutoff valves should be opened slowly, especially in hot fluid applications, to avoid damage caused by thermal shock. Most industrial gage material is tempered glass that can withstand sudden thermal transition without breaking, but because of additional loads and stress on the gage, the ability to withstand thermal shock is reduced. While the gage is in operation, the isolation valves (labeled shutoff valves in Figure 5–4A) should be fully open; partially restricted valves can prevent the ball checks from seating, which could result in personnel injury or loss of product should the glass break. After putting a reflex sight glass in service, it is usually recommended to re-torque the bolts to reduce the risk of leaks, due to the difference in thermal expansion of the glass, the bolts and the body.

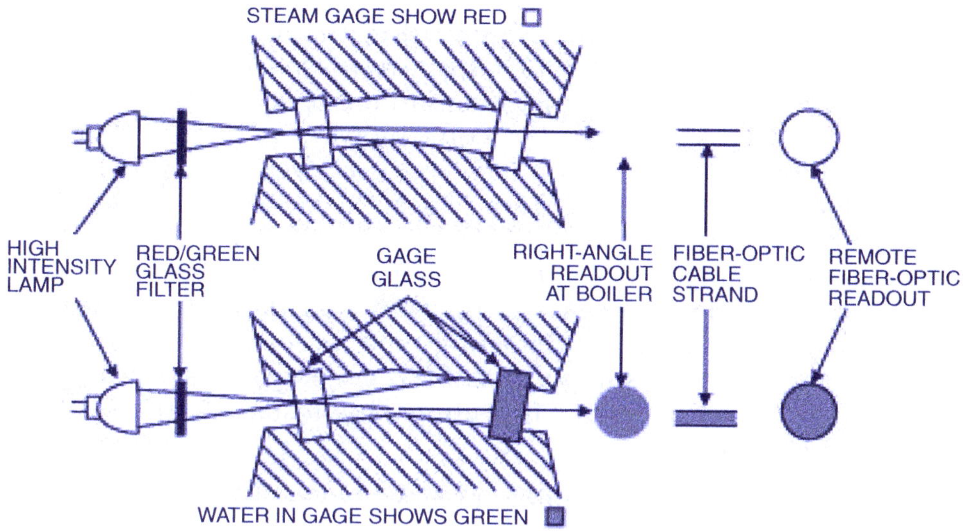

Figure 5–5 Light Color Passage through a Filter Level System (*Courtesy of Clark-Reliance*)

A recent innovation in sight glass technology involves the use of fiber-optic cables to transport light generated by a bicolor boiler-water-level gage system. The sensing portion of the fiber-optic system is a gage that operates on the principle of light refraction. Light is refracted by different amounts when it travels through different media. When light from a high-intensity lamp is strategically positioned to pass through steam and water, the amount of light refraction will determine the path direction through red- and green-colored filters (Figure 5–5). Light shows red when traveling through steam and green when traveling through the water. A fiber-optic cable is used to transmit the light to a fiber-optic readout for remote induction. Fiber-optic cables can be used for transmission lengths of 3 to 500 ft, depending on the configuration.

The simplicity and reliability of gage-type level measurement have resulted in the widespread use of such devices for local indication. When level transmitters fail or are out of service for maintenance or due to power failure, gage glasses allow the process to be measured and controlled manually. Except for occasional cleaning or maintenance of the lighting system (when one exists), gage glasses provide relatively carefree operation [Ref. 1].

5.5 AUTOMATIC TANK GAGES

An automatic tank gage (ATG) is defined in the American Petroleum Institute (API) *Manual of Petroleum Measurement Standards* as "an instrument which automatically measures and displays liquid level or ullages in one or more tanks, continuously, periodically or on demand." From this description, an ATG is a level-measuring system that produces a measurement from which the volume and/or weight of liquid in a vessel can be calculated. API Standard 2545 (1965), "Methods of Gaging Petroleum and Petroleum Products," describes float-actuated or float-tape ATGs, power-operated or servo-operated ATGs, and electronic surface-detecting-type level instruments. These definitions and standards imply that ATGs encompass nearly all level-measurement technologies, including hydrostatic tank gaging (covered in Chapter 9).

The ATG devices discussed are float-tape types, which rival visual level-measurement techniques in their simplicity and dependability. These devices operate by float movement with a change in level. The movement is then used to convey a level measurement. Many control devices are float-actuated and will be presented in a later section. This section deals only with the measurement aspect.

Many methods have been used to indicate level from a float position, the most common being a float and cable arrangement (Figure 5–6). The float is connected to a pulley by a chain or a flexible cable, and the rotating member of the pulley is in turn connected to an indicating device with measurement graduations. As shown, when the float moves upward, the counterweight keeps the cable tight and the indicator moves along the circular scale. When chains are used to connect the float to the pulley, a sprocket on the pulley mates with the chain links. When a flat metallic tape is used, holes in the tape mate with metal studs on a rotating drum.

Floats for these systems can be installed in a standpipe in the tank or in a long section of pipe running up the side of the tank and connected to the top and bottom by an appropriate valve arrangement to isolate the gage chamber for accessibility. This chamber not only serves isolation purposes but also acts to dampen spurious changes in level that would cause erratic float movement.

Two major types of readout devices are commonly used with float systems. Perhaps the simplest type is a weight connected to the float with a cable. As a float moves, the weight also moves by means of a

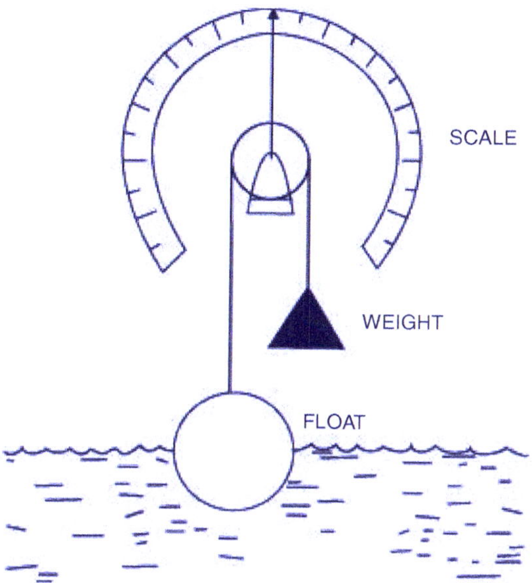

Figure 5–6 Float and Cable Level Arrangement

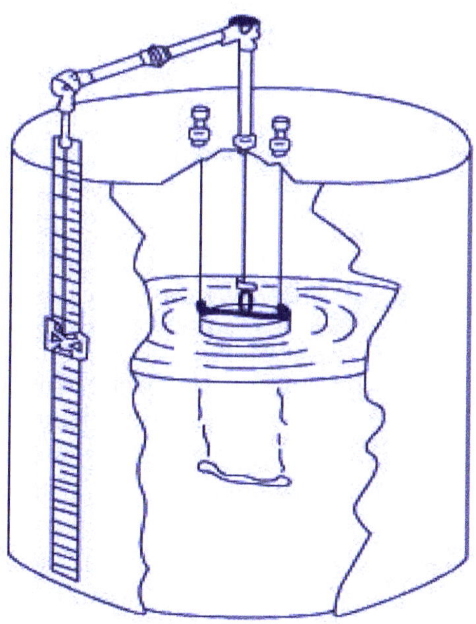

Figure 5–7 Closed-Tank Operation of Float, Cable, and Weight

pulley arrangement (Figure 5–7). The weight, which moves along a board with calibrated graduations, will be at the extreme bottom position when the tank is full and at the top when the tank is empty. This type is generally used for closed tanks at atmospheric pressure.

To add a bit of sophistication to the measurement system, for ease of reading the level from a location near the bottom of the tank, and to gain a bit more accuracy, the system shown in Figure 5–8 can be used. To understand the operation of such a device, assume that the installation example is a straightforward application: using a side-mounted gagehead on a standard vertical cylinder with a fixed roof, 40 ft high. The stored product is gasoline. Working pressures of the vessel range from 2 oz (.125 lbs/in^2) pressure to ½ oz (.031 lbs/in^2) vacuum. A 14 to 16 in. diameter type 316 stainless steel disc-shaped float rests on the surface of the gasoline. Immersion depth of the float in gasoline is approximately 50%. The float tracks the rising and falling liquid level with an accuracy of ±1/16 in. over the full range of 0 to 38 ft (maximum safe fill level of a 40-ft-high tank). As liquid level rises (thermal expansion or product filling/pumping in), the float rises. Constant tension is maintained on the perforated type 316 stainless steel tape by a negator spring motor. The negator motor has a pullback of 35.20 oz, or 2.2 lb. As liquid level falls due to thermal contraction or drawing down of product, the weight of the float exceeds negator motor pullback and the float continues to seek the immersion depth.

The perforated tape is played out from or retracted into the gagehead itself. It travels openly in the ullage space of the tank until it enters the l-1/2-in. tank top piping, where it is carried by two 90° sheave elbows with freely rotating wheels, through the vertical piping run on the side of the vessel, into the gagehead. Once in the gagehead, it is stored on an aluminum tape storage sheave, which is powered by the negator spring motor.

As it passes into and out of the gagehead, the perforated tape engages a smaller sheave called a sprocket wheel. The sprocket wheel contains stainless steel pins positioned 1 in. apart that engage the precision punched holes on the stainless steel tape. This engagement permits mechanical shaft rotation to position the mechanical counter assembly to display level units in English fractional units, decimal feet, or metric. It also enables a pin on the back of the sprocket wheel to drive a mechanical coupling attached to a transmitter or limit switch assembly.

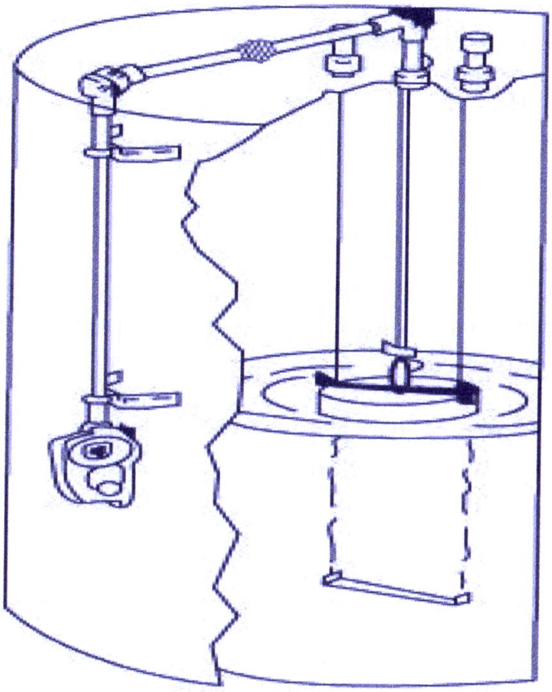

Figure 5–8 Float-Actuated Device

The float travel is guided vertically by two guide wires, spaced 16 to 18 in. apart, which run parallel to each other and plumb from the floor to the roof of the tank. The guide wires are generally type 316 stainless steel and are secured by a welded bottom anchor and a spring-loaded top anchor on the roof (adjacent to the manhole and float access cover).

Key ATG Components

Gagehead. A gagehead is typically manufactured with an aluminum housing and internal parts, including a tape-storage sheave, sprocket sheave negator spring motor, negator motor assembly, tape keeper components, dial and counter assembly, and support shafts. Gageheads are also available from most suppliers with steel or stainless steel housings.

Float. Typically of a circular, hollow shell design, a float consists of two halves welded together at the equator. Also included are guide eyes of various types and an eyelet for attaching the perforated tape. Floats usually are manufactured of type 316 stainless steel, but may also be available in aluminum, type 304 stainless steel, Alloy 20, Monel®, Hastelloy®, or plastics such as polyvinyl chloride/chlorinated polyvinyl chloride (PVC/CPVC).

Tape. Tape is generally manufactured of type 316 stainless steel material, usually about 1/2 in. wide, with holes precision punched 1 in. apart over the entire length of the tape. The tape is attached physically to the float via a tape clamp at one end and to the gagehead tape-storage sheave at the other.

Float and tape systems are also applied to large spherical tanks used to store volatile chemicals with relatively high vapor pressures. The tape runs over the side of the sphere, with the readout at ground level. When used in vapor service, the head and connecting piping are pressurized. A soft-seat gate valve is usually installed at the top of the sphere in a section of the pipe close to the point where the tape enters the tank.

Should the tape break at a point between the valve and the readout device, and if the tape has not fallen back into the tank, the valve can be closed on the tape, thus isolating the tank from the readout and connecting piping.

Once the tank has been isolated, the covers can be removed from the pulleys at the turn of the pipe, and each end of the broken tape can be spliced together. When this procedure is performed, the 6 or so inches of the tape used in the splice must be recovered to sustain the accuracy of measurement. To do this, tape is recovered from the take-up drum at the readout. During initial installation, excess tape will be coiled on the take-up drum for such emergency maintenance procedures. The readout device must be zeroed to compensate for the tape used in the splicing procedure. Provisions will usually be made at the take-up drum for this purpose. When splicing the tape, the amount of tape used for the splice must be noted so that compensation can be made during the splicing procedure. Failure to do this can result in the need to empty the tank to establish the zero reference point.

If the tape breaks at a point inside the tank, of course, the tank will have to be emptied and entered to perform the splice. This is a major undertaking, especially in the case of large tanks. If the tank contains an explosive or volatile substance, great care must be exercised to empty the tank and purge the contents. Before entry into any tank, extreme caution must be taken to ensure the presence of a life-sustaining atmosphere.

In cases where the tank is not pressurized and contains no toxic or harmful vapors, a manhole-type cover can be positioned in close proximity to the float guide cables and tape to accommodate tape maintenance and repair from the top of the tank without entering it. Regardless of the situation, safety precautions should always be observed.

Transmitters for Automatic Tank Gages

In earlier times some manufacturers made non-absolute encoders that were thought to solve some problems encountered with previously used types of brush encoders. Some such devices that have been used were a series of optical encoders; however, some applications suffered with less than satisfactory results. Some optical encoders have the following drawbacks and disadvantages:

- Calibration in direct sunlight is difficult, requiring level offset in nonvolatile RAM.
- Many optical encoders presently used require factory servicing.
- Optical encoders are by nature sensitive to dust, moisture, condensation and airborne pollutants.
- Alignment of optical pick-ups is often subject to critical alignment procedures.
- Light diffusion within the coded disk can result in errors.
- Optical disks are subject to electrical interferences, resulting in failure or errors.

Non-absolute encoder devices count up or down in increments, holding the counts in volatile memory which is subject to power glitches and surges that result in zeroing of the transmitter or loss of power. Back-up power systems are used and can help in these situations, but not all have good reliability. Secondary level alarms and more-reliable level measurement sources are sometimes used for auxiliary service.

The advent of digital and microprocessor-based technology, as now used in most field instruments, has enabled the development of reliable transmitters for mechanical gage devices.

The basic components of the transmitter are the encoder, transmitter card, and housing. Three types of encoders are absolute brush, optical (linear and reflective) and non-absolute up-down counters. The transmitter shown in Figure 5–9 is an all digital transmitter of the absolute brush design. This is accurate, dependable and easy to maintain in the field.

The absolute encoder shown in Figure 5–10 is designed to convert the shaft rotation, produced by the up and down movement of the float, to Gray code for integration by the transmitter card. The encoder consists of two identical copper/nickel/gold-plated disks. One disk measures feet and the other inches in 1/16 inch segments. A removable brush block is inserted into the encoder to make contact with etchings on the disks as the float moves in the tank. The brushes are always in contact with the disks so the transmitter is always able to keep track of the current level. Battery back-up maintains continuous operation even during times of power failure.

The transmitter cards in the transmitter can accommodate communication protocol for telemetry applications by hard wire, fiber optics, radio, cellular, modem or satellite to receiving devices such as transponders in satellites, fiber multiplexers, programmable logic controllers (PLCs), remote terminal units (RTUs) and distributed control systems (DCSs) , and various supervisory control and data acquisition (SCADA) systems. Transmitter cards can be selected for many different electrical standards and are adaptable to retrofits and new installations. Two transmitter cards can be provided for users that have two independent communications requirements. Dependability is enhanced by using MIL spec and industry specification for components including gold pin sockets. A protective coating on the card protects from moisture and corrosion. Temperature sensing and transmissions are also available. Communication via 10BASE-FL and Bluetooth wireless are available options.

An ATG using a digital transmitter with an absolute brush encoder can measure level in 1/16 in. increments, with a repeatability of 1/16 in. over a 96 foot range. Specific applications are shown in Figure 5–11 [Ref. 2].

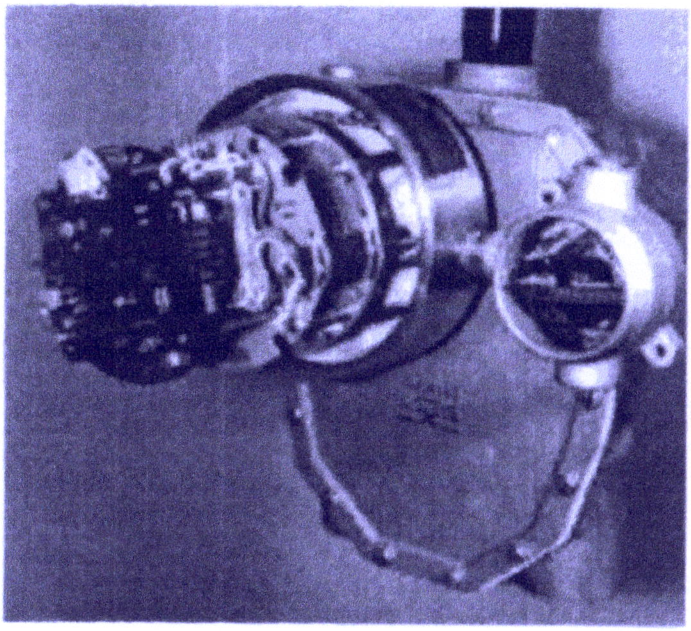

Figure 5–9 Digital Transmitter for ATG (*Courtesy of Gauging Systems Inc.*)

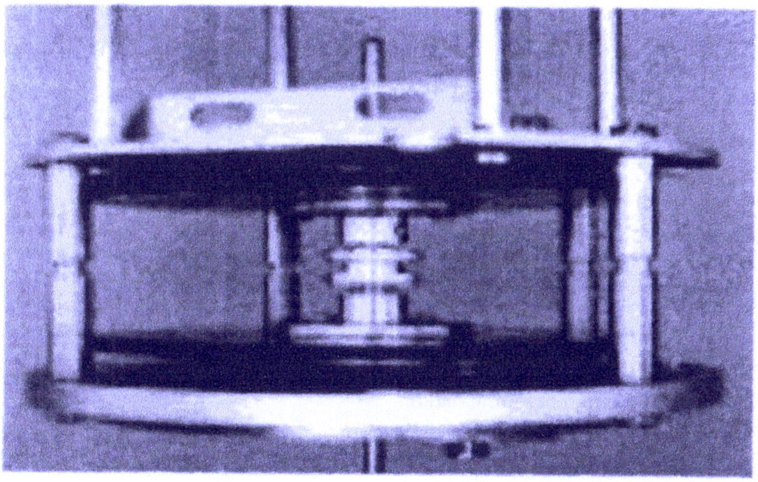

Figure 5–10 Absolute Encoder (*Courtesy of Gauging Systems Inc.*)

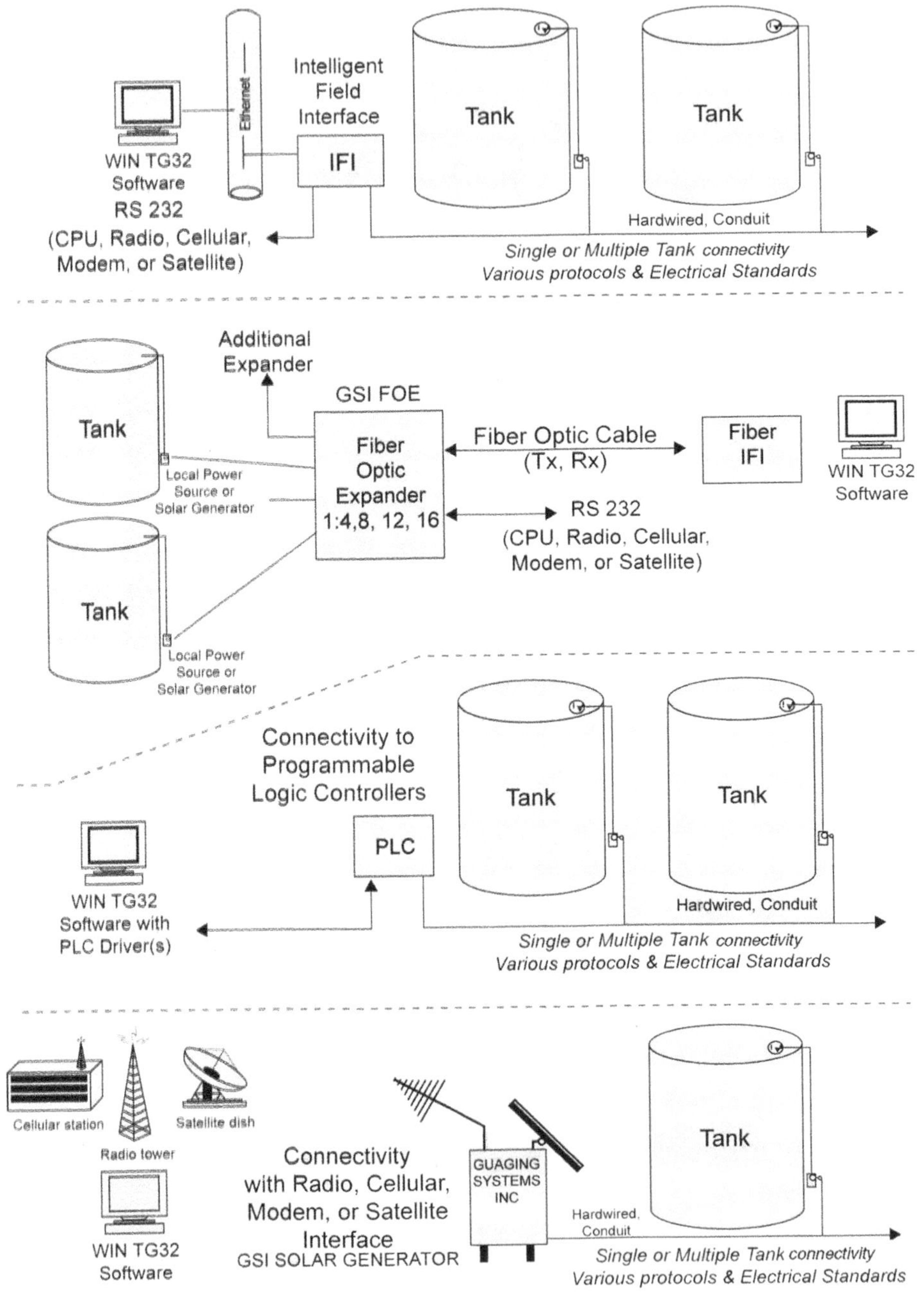

Figure 5–11 ATG Transmission and Telemetry Applications
(Courtesy of Gauging Systems Inc.)

Misapplication

Float and tape ATGs are suitable for a wide range of services on a wide variety of tanks; however, certain applications can cause problems for float-type gages. These include use on products such as highly viscous or waxy liquids that will accumulate on the float, guide wires, or both. In service such as the measurement of asphalt, bitumen, and bunker C oils, care should be exercised to ensure that the gage will operate reliably over time. Since the float and guide wires are in constant contact with the product, materials should be selected that will not be affected metallurgically by such contact. Extreme cases such as hydrofluoric acids are best served by noncontacting technologies.

Another concern involves the presence of excessive turbulence. This may be the result of agitation due to mixing operations or the result of pumping product into or out of the vessel, especially if the float gage is located close to the inlet or outlet piping system.

As most float and tape ATGs are designed for relatively low-pressure operation (i.e., 2.5 to 5 psi), care should be exercised to ensure that the pressure will not damage the float or gagehead, or cause the sheave elbows to leak vapors into the atmosphere. High-pressure versions of the standard float and tape gage are available that can operate in the range of 150 to 300 psi. Examples of these types of applications include pressurized cylinders or spheres that store products such as butane, butadiene, or propane. Float-type ATGs are designed primarily to measure liquid level, not the level of grains, pellets, or other bulk solids.

Application Guidelines

Automatic tank gages are most often applied within the petroleum and chemical industries, but they are also used in water and wastewater, food and beverage, pulp and paper, and paint and ink applications. Generally, they are used to measure liquid level in large API field-erected tanks storing crude oils or refined products. They are also frequently used on smaller vessels, including chemical storage tanks in petroleum refineries, chemical process plants and water-treatment facilities. Products typically measured include crude oils, gasolines, diesel fuels, jet fuels, kerosene, fuel oils, acids and caustics.

The gagehead can be mounted at grade level for ease of reading the mechanical level display, checking gage operation and performing maintenance. The head can also be mounted on top of fixed-roof vessels to eliminate the two 90° sheave elbows and long run of vertical pipe that carry the perforated gage tape. Fixed-roof tanks, open floating roof tanks, closed floating roof tanks, underground tanks and certain specialized tank shapes are all suitable candidates for a float and tape ATG.

Installation

Float and tape ATGs can be installed in out-of-service and many in-service vessels. The general steps for installation are as follows:

- Locate an appropriate site for the gage.
- Weld the top anchors to the vessel roof or to a split section of the manway or inspection cover.
- Drop a plumb line and mark locations to weld the bottom anchors to the vessel floor.
- Weld bottom anchors to the floor.
- Attach guide wires to top and bottom anchors, ensuring tight connections and true parallelism.
- Weld a half coupling for 1-1/2-in. pipe to the roof of the split section of the inspection cover for the tape piping assembly.
- Weld mounting brackets to the side of the vessel to support the 1-1/2 in. tape piping.
- Install, mount, and leak test the tape piping system.
- Attach the gagehead to the tape piping at ground level and position.
- Wind the perforated tape on the tape-storage sheave and pull it through the piping system.
- Attach the perforated tape to the tape clamp assembly.
- Place the float on the guide wires by rotating and enabling the wires to pass through the eyelets.
- Slowly lower the float to the base of the vessel or to the product surface.
- Bolt shut all mating surfaces, including the inspection cover, elbow covers and gagehead back cover.

- Calibrate the mechanical gagehead dial counter to match the actual level.
- Fill the counter housing, body housing or both with the appropriate grade of oil, if desired.

Maintenance and Preventive Maintenance

When properly applied and installed, a float and tape ATG will offer many years of reliable service at low cost. The maintenance suggestions offered by manufacturers of float and tape ATGs are basically these:

- Regularly inspect the gage for signs of unusual wear. Check for scale and rust from the inside diameter of the tape piping system that may have fallen into the head. An alternative is to use stainless steel piping.
- Calibrate the gage on an as-needed basis. Take several hand-dip measurements when the tank is static and average them for a gage calibration value.
- Using the check knob, ensure that the tape, float, sheave elbow wheels and internal sheaves operate freely.
- Drain accumulated water from the gagehead at regular intervals, even if a condensate reservoir is used.
- Check the perforated tape for any signs of stress due to excessive turbulence within the tank, friction and so on. Replace if necessary [Ref. 3].

Conclusion

Factors that affect ATG accuracy include the physical properties of the process liquid; the overall accuracy of the level-measurement system to include calibration shifts, electrical interference and so on; and, in some applications, the temperature in the vapor space of the vessel. ATG systems are subject to errors resulting from inadequately designed and installed components. Careful consideration of the selection, design and installation of these systems can result in very reliable and accurate measurements.

API Standard 2545 originally stated, "If mutually agreeable, properly designed, installed and maintained automatic tank gaging devices of known accuracy which meet the following standards may be used to determine oil levels in atmospheric or low-pressure, field storage and working tanks for the purpose of custody transfer." Based on this, many users think that ATGs can be used for custody transfer.

The rewritten standard defines the method to measure the installed accuracy of the ATG, which must be calibrated against strict manual gaging at multiple levels. When ATG measurements are within the tolerance established by the standard, the requirements for custody transfer have been met [Ref. 1].

5.6 MAGNETIC-TYPE FLOAT DEVICES

The measurement of concentrated acid, sodium hypochlorite or caustic in a vessel is a difficult application. A system with components constructed of exotic materials such as tantalum or Monel®, which could withstand the corrosive nature of the tank fluid in many circumstances, would be cost prohibitive. To overcome this shortcoming, a magnetic float system is often used (Figure 5–12). In such a device, the only portions of the system in contact with the corrosive material in the tank are the float, which is a magnetic material coated with Teflon®, a suitable plastic, Hastelloy®, Kynar®, CPVC, zirconium, titanium, or some other type of protective coating, and a nonmagnetic dip tube. Coupling between the float and a movable indicator is supplied by a magnetic field produced by the magnet in the float. In Figure 5–12A, the doughnut-shaped float moves up and down with the level, guided by the dip tube, which is constructed from a suitable nonferrous material. The resulting movement of the magnet or magnetic material is conveyed to an indicator on a scale or used to generate a signal for transmission. A stop must be provided on the dip tube to prevent the magnet from falling from it under low-level conditions.

Other applications of magnetic float gages include auxiliary measurement for boiler drum level and condensate return tanks. Most boilers have a means to measure drum level for control purposes, but a visual backup is required. A magnetic float gage can provide a local visual indication for operators; float gages have replaced sight glasses in many instances because a sight glass can be difficult to read and can present safety problems for high-pressure and high-temperature systems. Compressors generally have dipsticks to enable operators to check the oil level, but a visual indication is sometimes desirable. Magnetic float gages can serve this purpose. These systems can also be equipped with electrical contacts so that pump switching can be

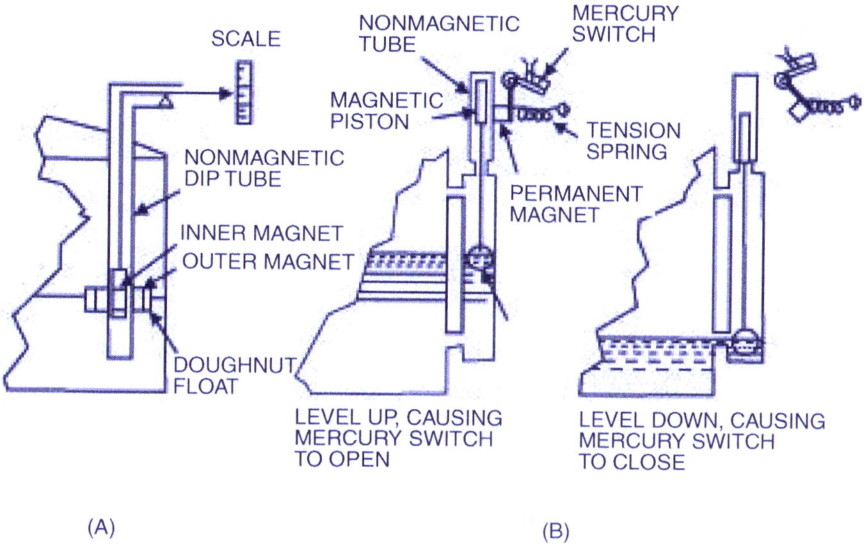

Figure 5–12 Magnetic Float Switches

provided; pump up and pump down with alarm arrangements are possible configurations.

Another arrangement utilizing a magnetic float system (Figure 5–12B) operates a position-sensitive switch (such as a mercury switch) to start or stop a pump, energize an alarm or provide some other function. The magnetic follower can also trip flags to a position that shows a particular color when the direction of movement is upward and another color when the movement is downward. The point at which the two colors meet represents the position of level or interface in the tank.

In a switch application, when the magnet moves up, positioned by the float riding on the liquid surface, the magnetic field attracts a segment of magnetically-permeable iron fixed to the mercury switch carrier, and the switch moves to a new position (which may be "on" or "off"). This movement also distorts a spring; in the case shown in Figure 5–12B, a tension spring is used. As the level moves downward, causing the magnetic field to move away from the magnetic segment on the carrier, the attraction of the magnetic field decreases and the spring tension returns the switch carrier to its original position. Note that in magnetic float devices, the actual float design is not limited to the doughnut configuration illustrated, but can be spherical, disc-like or of some other design.

A magnet can be used on both the float arm and the switch carrier. When the float arm movement causes like poles of the magnets to be positioned together, a repelling force results that causes the switch carrier to move away from the magnetic tube. In this case, a compression spring must be used to return the carrier to its original position when the float arm moves downward. For liquid/liquid interface applications, such as oil and water separators or with any other liquids with a specific gravity difference of at least 0.1, specially weighted floats that sink in the lighter fluid but float in the bottom layer can be used to indicate or control the interface level.

5.7 MAGNETIC TANK GAGE

A visual indication method of measuring level is the magnetic level gage, which was originally designed to replace sight glasses. These gauges are relatively easy to install and do not have the maintenance problems associated with sight glass etching, cleaning, lighting, leaks, etc. They are suited for oilfield service, high pressure in power plants, corrosive conditions, and the like. They can also be equipped with switches and clamp-on transmitters.

The float chamber is a column of two- or three-inch pipe with process connections to mate with existing storage tanks, reactors, distillation towers, or other process equipment where level measurement is required. Various connection configurations are available.

The float design is determined by the process fluid's specific gravity, pressure, and temperature. The float houses a cylindrical magnet assembly and the indicator on the gage is horizontally scaled in a glass or polycarbonate tube. Two types of indicators are used: a red follower type and a flipper type. As the float moves up or down with the process fluid, it positions the follower or flipper to indicate the level of the fluid. The float can be designed to operate with a fluid specific gravity as low as 0.25. A magnetic float gage is shown in Figure 5–13.

Transmitters designed for use with magnetic gauges mount externally to the gage chamber and consist of a stainless steel sensor tube with a string of resistors and magnetically operated reed switches spaced each ¼ or ½ inch along the tube for the desired resolution. The length of the circuit board with resistors and switches determines the range of the transmitter and can be any portion of the measured range, up to the full length. The transmitter is mounted at the top of the tube with a threaded connection and is rated for explosive atmospheres. Intrinsic safety can also be provided with approved safety barriers. For more accurate and repeatable measurement applications, the magnetic gage can be equipped with a magnetostrictive transmitter like the one shown in Figure 5–14. The operation of magnetostrictive devices is explained in Chapter 7.

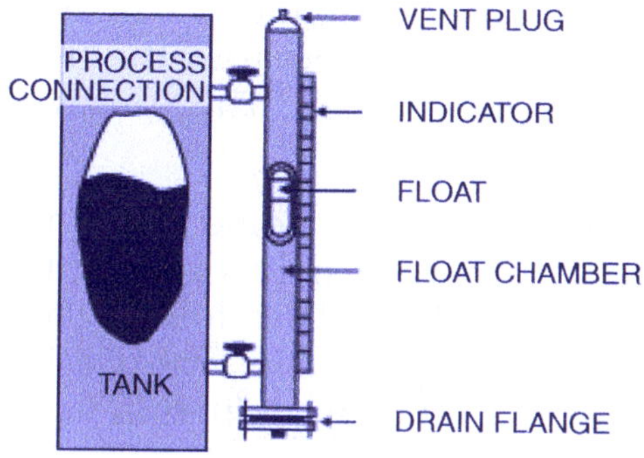

Figure 5–13 A Magnetic Float Gage (*Courtesy of ISE-Magtech*)

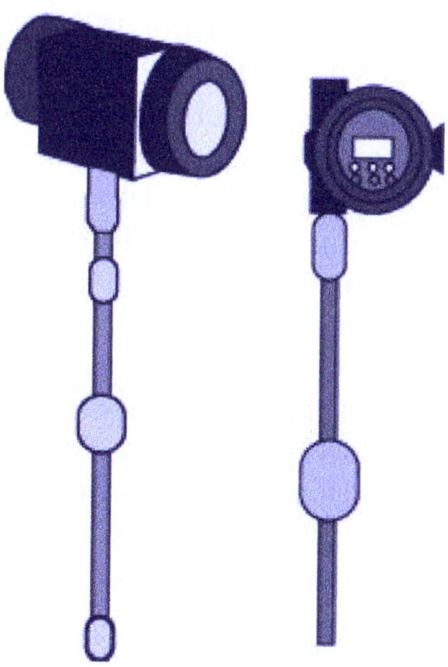

Figure 5–14 Magnetostrictive Transmitter (*Courtesy ISE-Magtech*)

5.8 DISPLACEMENT PRINCIPLES FOR LEVEL MEASUREMENT

The type of measurement devices previously discussed operate on the displacement principle; that is, the float used for measurement displaces its own weight in the liquid in which it floats. The float will sink into the fluid until a volume of liquid is displaced that is equal in weight to that of the float. The weight of the float is generally adjusted by a weight, either internal or external to the float, so that the float maintains a half-submerged position. This causes the float to attain a maximum operating force.

When the specific gravity of the liquid remains constant and the weight of the displacer or float remains constant, the float rises and falls the same amount as the level. Therefore, regardless of the position of the level, the float will assume a constant relative position with the level. This displacement principle defines a constant-displacement device.

5.9 VARIABLE-DISPLACEMENT MEASURING DEVICES

When the weight of an object is greater than an equal volume of the fluid into which it is placed, full immersion results and the object never floats. As the level moves up and down along the length of the displacer, the displacer undergoes a change in weight caused by the buoyancy of the liquid. Buoyancy is explained by Archimedes' principle: any body partially or completely submerged in a fluid is buoyed up by a force equal to the weight of the fluid displaced by the body. The upward pressure acting on the area of the displacer creates the force called buoyancy. The force is of sufficient magnitude to cause the float (displacer) to be supported on the surface of a liquid in float-actuated devices. However, in variable-displacement level systems, the immersed body or displacer is supported by arms or springs that allow some small amount of vertical movement or displacement with changes of resulting buoyancy forces caused by level changes. This buoyancy force can be measured to reflect level variations.

A displacer arrangement is shown in Figure 5–15. The vessels shown are open to atmosphere, but the principle described also applies to closed-tank measurement.

In the vessel depicted in Figure 5–15A, the displacer is suspended by a spring scale that shows the weight of the displacer in air. This would represent the 0% level in a measurement application. The full weight of the displacer is entirely supported by the spring and is shown to be 3 lb. In the center vessel (Figure 5–15B), the water is at a level that, in this case, represents 50% of the full measurement span. Note that the scale indicates a weight of 2 lb. The loss in weight of the displacer (1 lb) is equal to the weight of the volume of water displaced.

When the water level is increased another 7 in. to a full-scale volume of the 14 in. displacer, the net weight of the displacer is 1 lb, which represents a change of 2 lb; that is, when the water level changes from 0 to 100% (0 to 14 in.), the weight of the displacer changes from 3 lb to 1 lb (Figure 5–15C). As the net weight of the displacer decreases, the load on the spring scale decreases by an amount directly proportional to the increase in water level. For the displacer in question, a 14 in. increase in level is equal to about 55 in^3 of water displaced. This is the volume of the immersed portion of the displacer, which is determined by multiplying the cross-sectional area by the submerged length of the displacer.

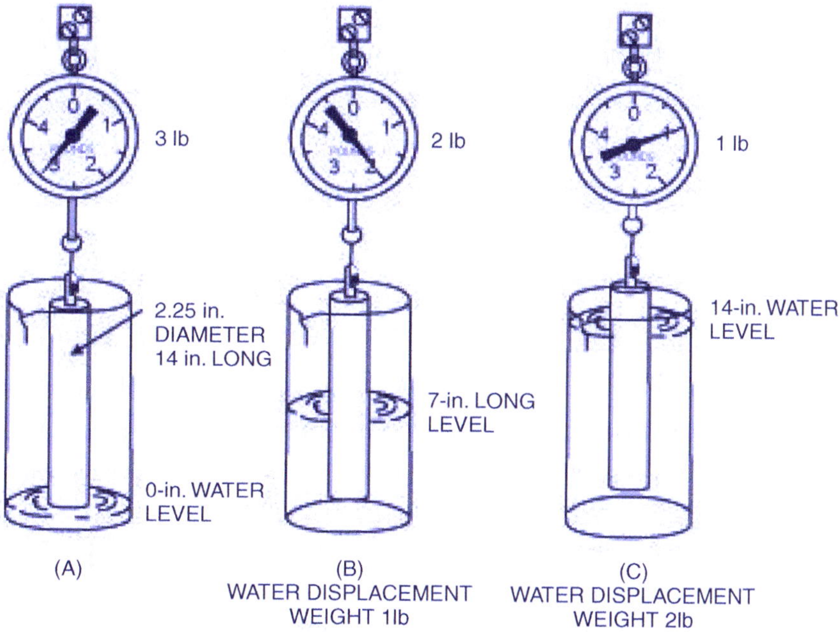

Figure 5–15 Operating Principle of Displacers
(*Courtesy of Dresser-Masoneilan Industries*)

Also, it can be seen that 55 in^3 represents 23.8% of 1 gal:

$$\frac{55 \text{in}^3}{231 \frac{\text{in}^3}{\text{gal.}}} = 0.238 \text{gal.} = 23.8\%$$

The weight of water is 8.33 lb/gal, and 23.8% of 8.33 is 2 lb. From this, it is evident that the volume and density of the displacer and the specific gravity of the displaced fluid are important considerations in displacer application. The following examples will help to clarify this point.

EXAMPLE 5-1

Problem: A displacer has a submerged length of 14.0 in., a diameter of 2 in., and is immersed in water. The weight of the displacer in air is 5.00 lb. One gallon of water weighs 8.33 lb, and the volume of 1 gal is equal to 231 in^3. Find the weight of the displacer when it is completely submerged in water.

Solution: The volume of the displacer is found by the formula

$$V = \frac{\pi D^2}{4}(L)$$

where D is diameter and L is length.

$$V = \frac{(3.14)(2 \text{in.})^2}{4}(14 \text{in.}) = 44 \text{in}^3$$

To find the weight of water displaced by 44 in^3

$$\frac{44}{231} = 0.19 = 19\% \text{ of 1 gal.}$$

and

$$0.19 \text{ gal } (8.33 \text{ lb/gal}) = 1.6 \text{ lb}$$

The weight of the submerged displacer is 5.0 – 1.6 lb = 3.4 lb

EXAMPLE 5-2

Problem: The displacer in Example 5-1 is submerged in a liquid with a specific gravity of 0.5. Calculate the weight of the displacer when it is completely submerged.

Solution: From Example 5-1, the volume of the displacer is 44 in^3. This represents the amount of liquid displaced when it is 100% submerged. The weight of liquid displaced is

$$(0.19 \text{ gal})(8.33 \text{ lb/gal})(SG) =$$

$$(0.19 \text{ gal})(8.3 \text{ lb/gal})(0.5) = 0.80$$

The weight of the submerged displacer is 5.0 – 0.80 lb = 4.2 lb

From the previous examples, it is clear that the difference in weight of the displacer is directly proportional to the specific gravity of the displaced fluid. A decrease in specific gravity of 50% resulted in a 50% reduction in weight change of the displacer for the same increase in level. Another example illustrates the relationship between displacer volume and weight change.

EXAMPLE 5-3

Problem: A displacer with a submerged length of 14 in. and a diameter of 1 in. is completely submerged in water. The weight of the displacer suspended in air is 5.00 lb. Calculate the weight of the displacer when it is completely submerged in water.

Solution: The volume of the displacer is

$$V = \frac{\pi D^2}{4}(L)$$

$$= \frac{3.141 \text{in}^2}{4}(14 \text{in.}) = 11 \text{in}^3$$

The weight of water displaced is

$$\frac{11\,\text{in}^3}{231\,\frac{\text{in.}}{\text{gal.}}} = 0.05\,\text{gal.}$$

$$0.05 \text{ gal } (8.33 \text{ lb/gal}) = 0.42 \text{ lb}$$

The weight of the submerged displacer is 5.0 – 0.42 lb = 4.6 lb

Comparing the results of Examples 5-3 and 5-1 reveals that the amount of buoyancy caused by equal amounts of level change varies with the volume of the displacer. The displacer weight change is less when its volume is decreased, with all other factors being equal. Of the same importance as volume, however, is the ratio of diameter to length of the displacer. The length is determined by the desired measurement span; the diameter is governed by the amount of fluid that must be displaced for a desired amount of buoyancy in a fluid with a certain specific gravity.

Although the discussion of displacer operation up to this point has focused on the relationship between variations in weight and changes in level, more is involved to increase accuracy. Using accurate spring scales, the scale dials could be calibrated in terms of level. This arrangement would provide an accurate means of level measurement for liquids with a constant specific gravity for level ranges of the total length of the displacer which is normally up to 10 ft. The limitation to such a measurement scheme is that it is suitable for open-tank applications only.

Another disadvantage is that use is limited to local indication; however, displacers can be used as transducers to produce proportional, repeatable, and accurate representations of level variations. Also significant is the fact that by appropriate and careful selection of displacers, along with consideration of fluid density, wide ranges of level applications are feasible.

The system in Figure 5–16 illustrates the concept of variable-displacement level measurement in an open tank using a displacer and torque tube (torsion bar). Note that the scale graduations are in pounds to signify the principle of weight variations with the vertical change of level. In an actual level application, the pointer movement would be slight because of typically small displacer movement. The pointer

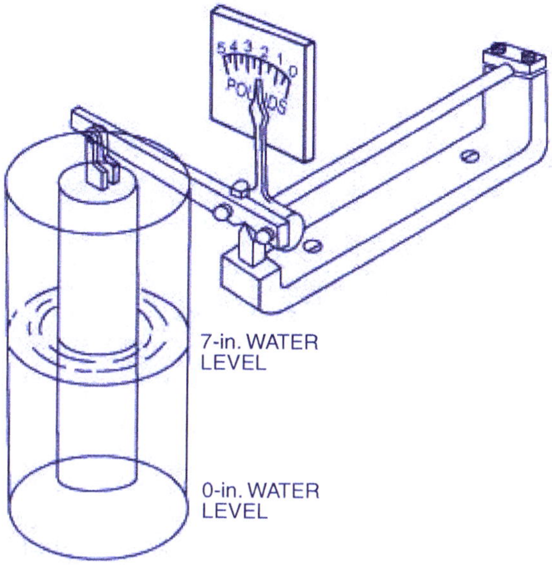

Figure 5–16 Displacer and Torque Tube for Open-Tank Level Measurement (*Courtesy of Dresser-Masoneilan Industries*)

movement must be amplified for scale resolution. Generally, however, the displacer movement is converted to a scaled electrical or pneumatic signal for transmission.

For closed-tank level measurement, the contents of the process vessel must be sealed from the instrument. Providing such a sealing device for frictionless operation that can be used over a wide range of pressure and temperature variations and under a variety of corrosive conditions has been a basic problem for manufacturers of displacement level instruments. Most companies use a torque-tube type of seal for this purpose. A torque-tube displacer level instrument is shown in Figure 5–17. A torsion spring (torque tube) and a torque rod have replaced the spring scale. The diameter and length of the torsion spring must be selected so that the full weight of the displacer can be supported in the absence of buoyancy, as is the case for zero-level conditions.

The hollow tube provides the required frictionless pressure seal. This type is generally used because it is not only able to support the displacer weight, but the displacer movement also can be transferred from the inside of a pressurized vessel to the instrument readout mechanism, which is at atmospheric pressure. The device's typical 4° to 6° of

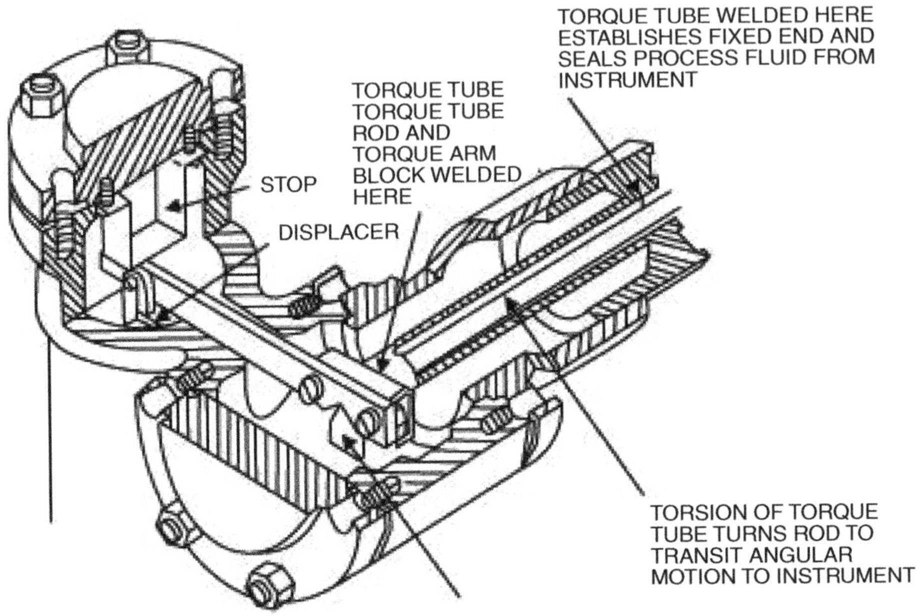

Figure 5–17 A Torque Tube Level Instrument
(*Courtesy of Dresser-Masoneilan Industries*)

angular rotation is used to establish a flapper-nozzle relationship for pneumatic transmission or to drive an electrical transducer for electrical transmission. The knife-edge bearing at the free end of the torque tube provides a nearly frictionless support for the displacer. Because of the small movement of the displacer and torque tube assembly, wear at this point is minimal, and trouble-free service of many years is common.

5.10 DISPLACERS USED FOR INTERFACE MEASUREMENT

It was mentioned earlier that level measurement is the determination of the position of an interface between two fluids or between a fluid and a solid. It can be clearly seen that displacers for level measurement operate in accordance with this principle. The previous discussion concerning displacer operation considered a displacer suspended in two fluids, with the displacer weight being a function of the interface position. The magnitude of displacer travel is described as being dependent on the interface level change and on the difference in

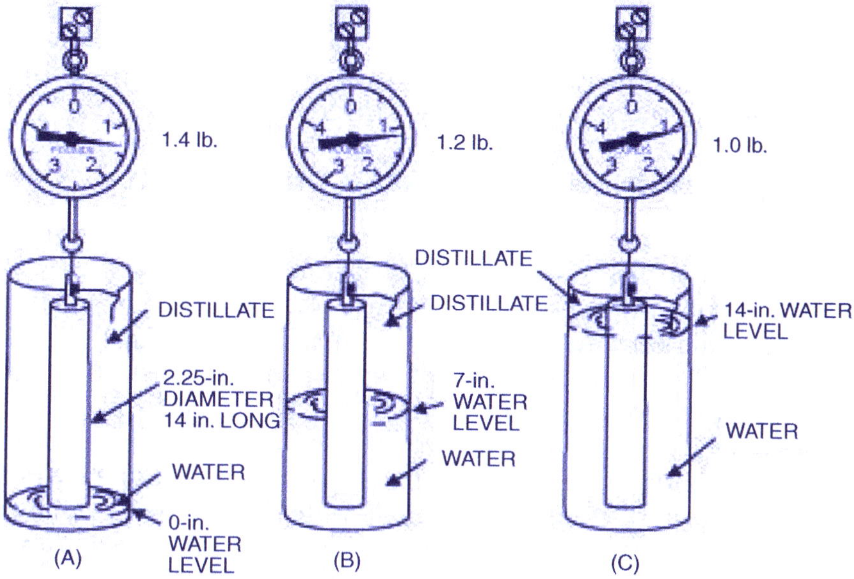

Figure 5–18 Liquid Interface Measurement
(*Courtesy of Dresser-Masoneilan Industries*)

specific gravity between the upper and lower fluids. When both fluids are liquids, the displacer is always immersed in liquid. The principle of liquid interface measurement is illustrated in Figure 5–18.

A 14-in. range displacer is shown immersed in liquid and is used to measure the position of the interface between water and a liquid hydrocarbon or distillate. The specific gravity of the distillate is 0.8 and that of water is 1.0. With the displacer completely immersed in the distillate, the weight of the displacer is 1.4 lb, as shown in Figure 5–18A. The weight of the displacer in air is 3.0 lb. The net weight of the displacer as shown on the scale, 1.4 lb, is the weight of the displacer in air minus the weight of the volume of distillate displaced. As shown in Figure 5–18B, the water level is increased to 7 in. and the displacer weight is decreased to 1.2 lb. This decrease in weight is a result of the increase in buoyancy caused by the difference in specific gravity between the distillate and water. When the water level is increased further to the 14-in. level shown in Figure 5–18C, the displacer weight is reduced to 1.0 lb. This further reduction in weight is the result of a greater portion of the displacer being immersed in the heavier fluid—in this case, water. From this discussion, it should be clear that, just as when the lighter fluid is air or vapor, the weight of the displacer is a function of the interface position. Also, recall that as the displacer weight changes, the twisting force on the torque tube changes, resulting in a rotary motion of the tube and causing the torque rod to transmit the motion to the instrument. This motion is directly proportional to the change in interface position.

5.11 FIELD-MOUNTED INTERFACE CONTROLLERS

In many applications, especially in the chemical and petroleum industries, interface control is more important than measurement. However, for feedback control, measurement must be provided. For example, if the water distillate interface level in a separator is controlled at some point around the midlevel value, the water can be drawn off through a valve whose position is a function of the interface level. The main concern in such an application is that the valve be closed when the distillate level is low to prevent the loss of product, and open when the level is high to permit the removal of water from the distillate. The field-mounted liquid-level controller shown in Figure 5–19 can be used for the liquid interface applications shown in Figure 5–20.

A level controller used for interface service should be able to respond over a wide range of adjustments of specific gravity differences. The operation of a displacer system is a function of the difference of the specific gravity in the two liquids forming the interface. A standard controller and displacer may be used in applications where there are very small differences in specific gravity values that would otherwise be beyond the adjustable limits of the devices. For example, a controller with a displacer suited for a limit of 0.5 for a specific gravity difference can be used for a difference as low as 0.1. In such applications, the specific gravity index would be set on 0.5, and the proportional-band setting of the controller would be set on 20% if the actual desired proportional band were 100%. The gain of the level controller must be 5 times as great for a specific gravity of 0.1 as for a specific gravity of 0.5 because the transmitter gain varies directly with the specific gravity of the fluid. When the desired gain of the controller (K_C) is 1 or proportional band (PB) is 100%, the controller gain can be multiplied by 5 to account for the related decrease in the level instrument response. The plot in Figure 5–21 illustrates this principle. A plot such as this is usually located in the door of the controller to help facilitate the procedure.

When it is desired to measure as well as control a liquid level, a duplex-type controller-transmitter is available (Figure 5–22). This type of device operates from a single displacer and torque-tube assembly to generate two completely independent signals. One is used by a control

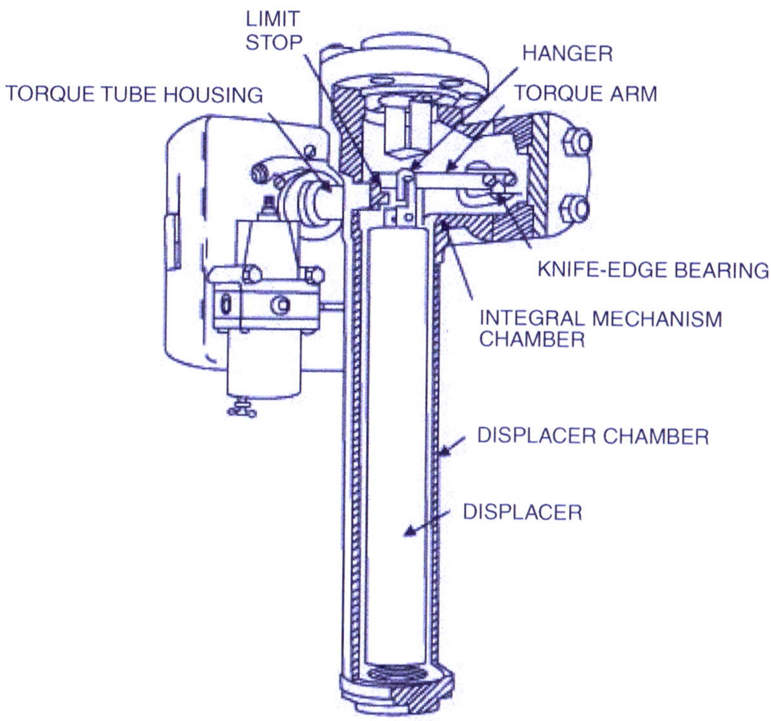

Figure 5–19 Field-Mounted Liquid Level Controller
(*Courtesy of Dresser-Masoneilan Industries*)

segment that positions a final control element for control purposes, and the other is used for the remote indication or recording of the level value. The duplex controller shown, a pneumatic type, incorporates a four-arm reversing arc to transmit angular motion of the torque tube to the pneumatic mechanism that generates the outputs. By adjusting a linkage mechanism along these arcs, specific gravity values can be set.

When the linkage is moved from one side of the arc to the other, the controller action is changed from direct to reverse or vice versa.

The pneumatic displacer level transmitter-controller shown in Figure 5–22 has been used in many applications, predominantly in the oil- and gas-related industries, and pneumatic displacer instruments are still being manufactured. As is true with other modern field instruments, microprocessor-based displacer transmitters and controllers are available.

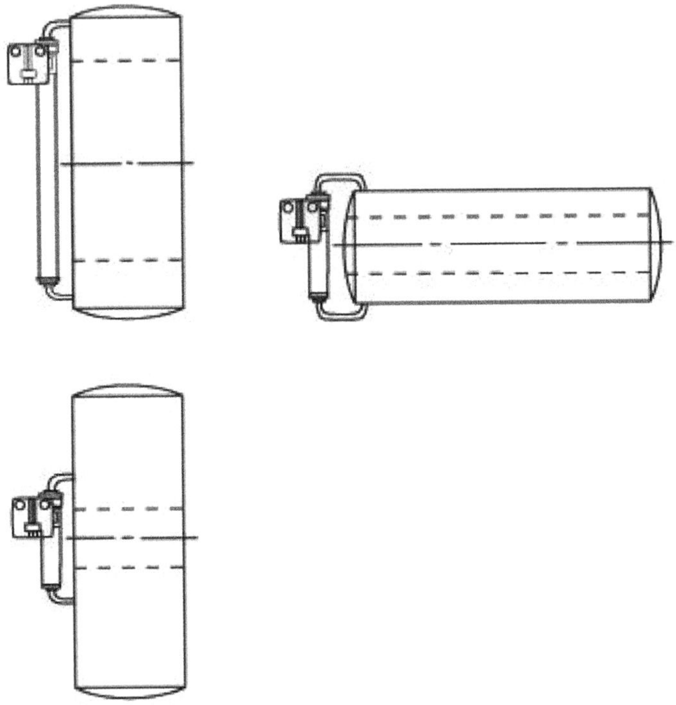

Figure 5–20 Liquid Interface Control
(*Courtesy of Dresser-Masoneilan Industries*)

Figure 5–23 shows a digital displacer level controller and displacer that can be used with caged or uncaged displacer sensors. It can be used in any displacer applications that include liquid level, interface level and density measurement. This type of controller and compatible displacer can be connected to industry standard or custom process vessel connections.

The operation of the digital transmitter is such that the linear motion of the displacer is converted to rotary motion by the torque tube arrangement. The rotary motion moves a magnet attached to the lever assembly, changing the magnetic field; this change is sensed by a Hall-effect sensor. The sensor converts the magnetic field signal to an electronic signal that is conditioned and converted to a 4–20 mA output signal.

A HART- (Highway Addressable Remote Transducer protocol) or Asset Management Solutions (AMS) compatible calibrator can be used to provide setup and calibration of the transmitter/controller. Level and

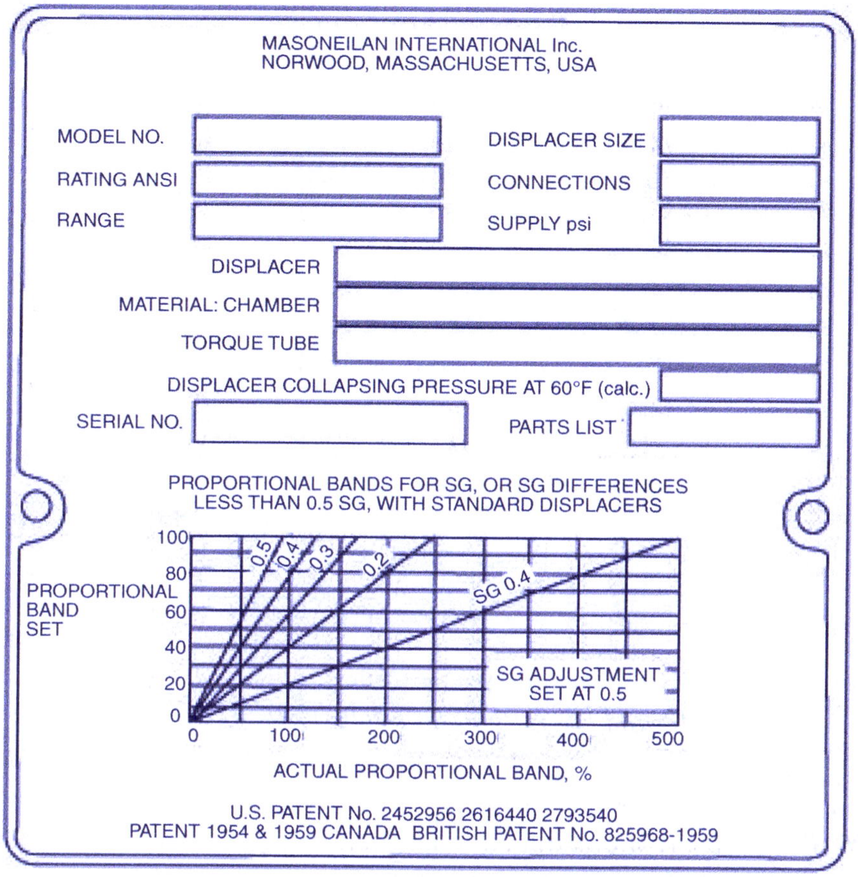

Figure 5–21 Proportional Band and Specific Gravity Relationship (*Courtesy of Dresser-Masoneilan Industries*)

temperature alarms, specific gravity tables, calibration trim, trending, and re-ranging can be configured with an HHC.

The digital level controller is loop powered, has high resolution, and provides input filtering and output damping for noise attenuation for mechanical disturbances or liquid turbulence at the displacer.

Field wiring connections are in a compartment separate from the electronics. This makes the electronic section immune to moisture introduced in the field wiring. Should the digital level controller ever need to be removed for in-shop maintenance, the field wiring does not need to be disconnected.

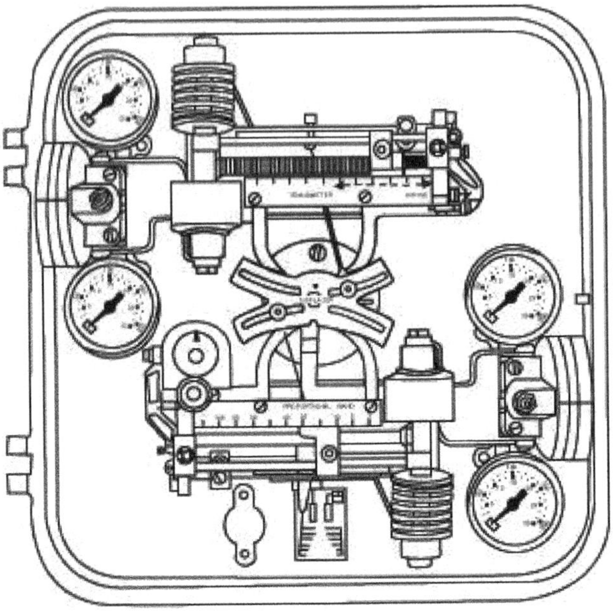

Figure 5–22 A Duplex Transmitter-Controller
(*Courtesy of Dresser-Masoneilan Industries*)

A guided wave radar transmitter is available to replace displacer-operated transmitters and often replaces the displacer and transmitter when mounted on the displacer chamber. In other cases, the displacer chamber may be replaced with the guided wave radar transmitter and cage assembly, completely replacing the displacer assembly. Such a system is shown in Figure 5–24. Many replacement guided wave radar assemblies are HART- and FOUNDATION Fieldbus-compatible.

5.12 APPLICATION OF DISPLACER ACTUATED LEVEL CONTROLLERS

The following examples will help to illustrate several common applications that incorporate field-mounted displacer-operated level controllers and transmitters. Figures 5–25 through 5–27 are included to illustrate the many applications of displacer actuated level instruments. The transmitter can be any type and will most commonly be digital (Figure 5–23). Perhaps the greatest limitation of displacer application is the stipulation that the displacer length must equal the level span. Displacers usually vary in length from 14 in. to 10 ft or more.

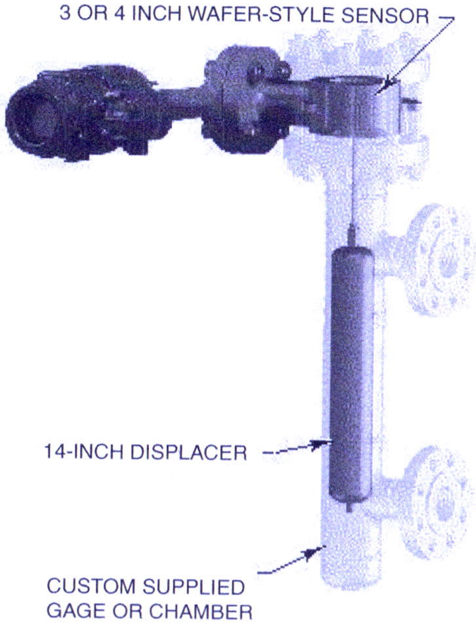

Figure 5–23 A Digital Displacer Level Controller (*Courtesy of Emerson Process Management*)

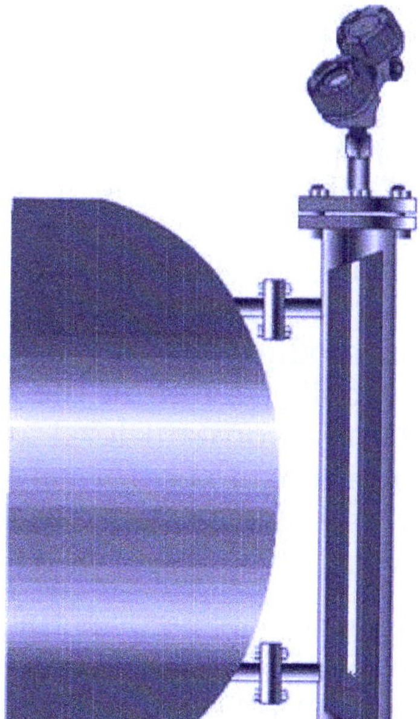

Figure 5–24 Guided Wave Radar Replacement for Displacer Assembly (*Courtesy of Magnetrol*)

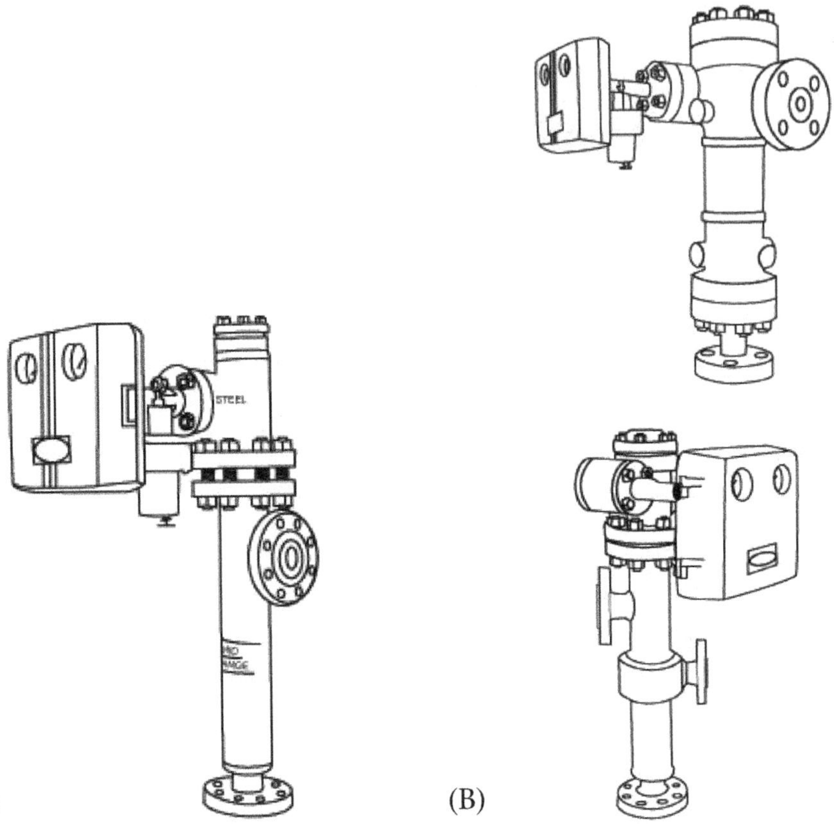

(A) (B)

Figure 5–25 Flange-Mounted Displacers
(*Courtesy of Dresser-Masoneilan Industries*)

In fractionating tower (distillation column) operation, heat energy must be added at different levels along the tower to vaporize the distillate from the feed and bottom product. This is normally accomplished by a heat exchanger called a reboiler. A weir in the kettle-type reboiler shown in Figure 5–26 maintains a constant level around the heating tubes. The level controller positions a control valve to maintain the rate of product leaving the vessel in accordance with level changes in the downstream side of the weir.

A duplex level controller-transmitter is used in the crude-tower level arrangement shown in Figure 5–27. The control valve is manipulated to regulate the flow of the bottom product of the crude fractionating column in accordance with the level variations in the column. An independent signal is generated and transmitted to a receiver instrument at a central location for level monitoring.

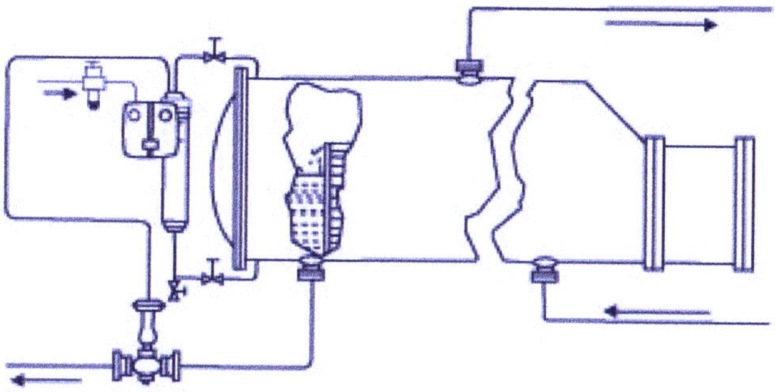

Figure 5–26 Control of Product Level in a Reboiler
(*Courtesy of Dresser-Masoneilan Industries*)

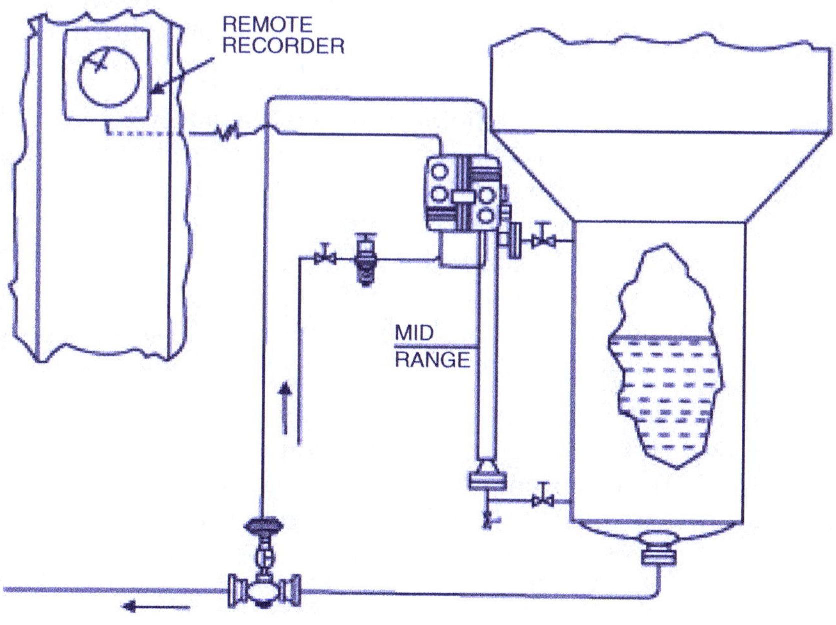

Figure 5–27 Crude-Tower Level Control
(*Courtesy of Dresser-Masoneilan Industries*)

It is sometimes desirable to implement a cascade arrangement to control level. In Figure 5–28, a field-mounted level controller regulates the set point of a flow controller to maintain the level. This cascade system adds stability to the system.

In applications where three fluid phases exist, as is common in gas/oil and oil/water separators, two level controllers can be used. In Figure 5–29, the lower level controller maintains the interface level of the two liquids, and the upper controller controls the level of the lighter liquid.

In applications where the flow from a level process must be maintained constant (within the storage capabilities of a feed accumulator, for example), a transmitter is used to transmit level changes to a centrally located panel. A flow controller and level receiver instrument are used to permit an operator to adjust the flow rate in accordance with wide variations of level. The transmitted signal from the level transmitter can also be used for local indication if desired. This operation is similar to the cascade arrangement of Figure 5–28, except that the operator manipulates the set point of the flow controller. The flow is maintained at a more constant value, and the level is allowed to fluctuate. Such a system is shown in Figure 5–30.

A common application of a field-mounted displacer-actuated level controller is the level transmission and control of a reflux accumulator on a fractionating column. In Figure 5–31, exchangers are used to maintain the level in an overhead accumulator to produce a reflux supply but no liquid overhead product. The level in the accumulator is controlled by regulating the flow of cooling water to the overhead condenser. In this application, a duplex controller-transmitter is used to provide remote monitoring of the accumulator level.

In large process refrigeration units, it is necessary to measure and control the level of the refrigerant in the evaporator portion or chiller of the refrigeration system. In the chiller chamber shown in Figure 5–32, the displacer chamber is built into the vessel and connected to the vessel by the external piping as shown. The liquid in the displacer is maintained at the same temperatures as the liquid in the vessel. This arrangement prevents false or erratic level measurement, which can result when the measurement is made at externally mounted chambers where flashing of the liquid could occur as the liquid is warmed by the ambient atmospheric temperatures. Hand valves are provided to isolate the chamber from the vessel.

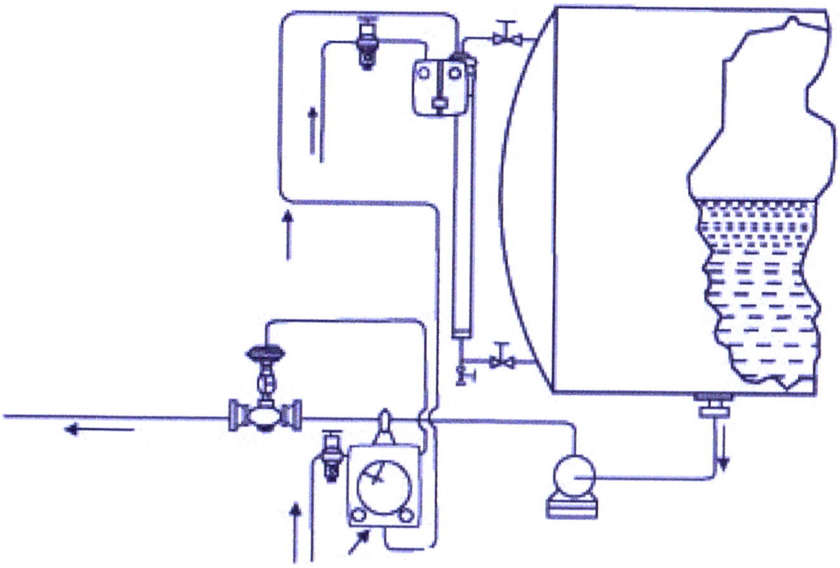

Figure 5–28 Level-Flow Cascade Control
(*Courtesy of Dresser-Masoneilan Industries*)

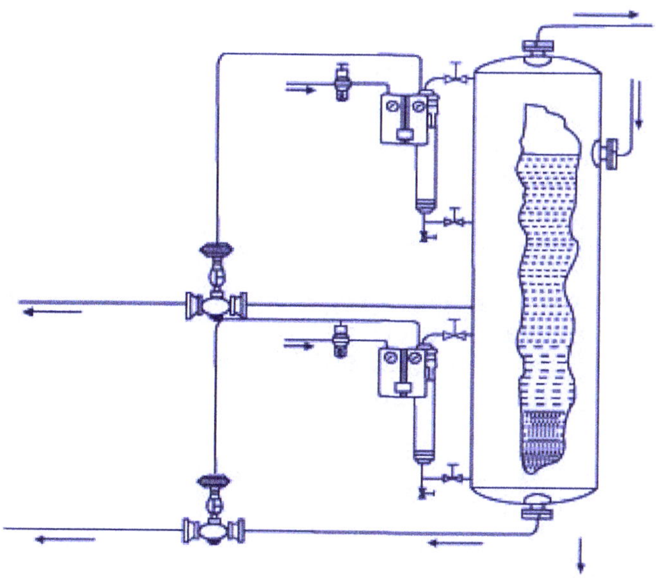

Figure 5–29 Dual Level Control in a Gas-Water Separator
(*Courtesy of Dresser-Masoneilan Industries*)

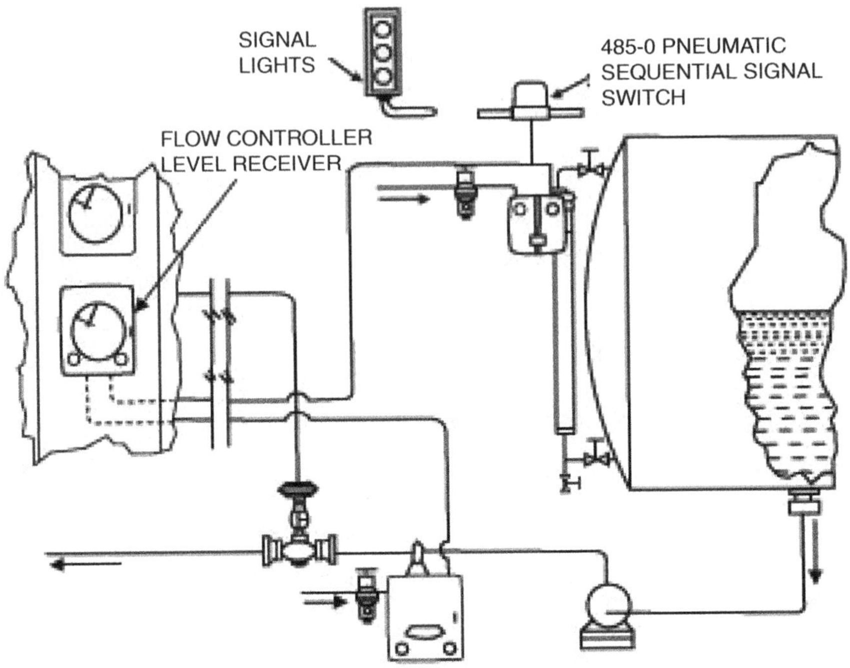

Figure 5–30 Level Transmission to a Remote Recorder (*Courtesy of Dresser-Masoneilan Industries*)

A counterflow de-aerating feedwater heater is shown in Figure 5–33. In this application, steam enters the vessel at a side port and is directed by baffles to the upper section, where it contacts the water entering from the top. The water condenses the steam. The condensate and cooling water are collected in the lower portion of the vessel. The cooling water flow is regulated by a displacer level controller to maintain the level in the lower portion.

In Figure 5–34, the interface level in a settling tank is controlled by a field-mounted displacer level controller. In this example, a solvent enters the tank at the top and is mixed with untreated oil, which is introduced into the tank at the bottom. The displacement level controller on the side of the tank measures and controls the interface level. In solvent extraction, a backpressure regulator on the raffinate (remaining liquid) discharge line controls the backpressure in the settling tank.

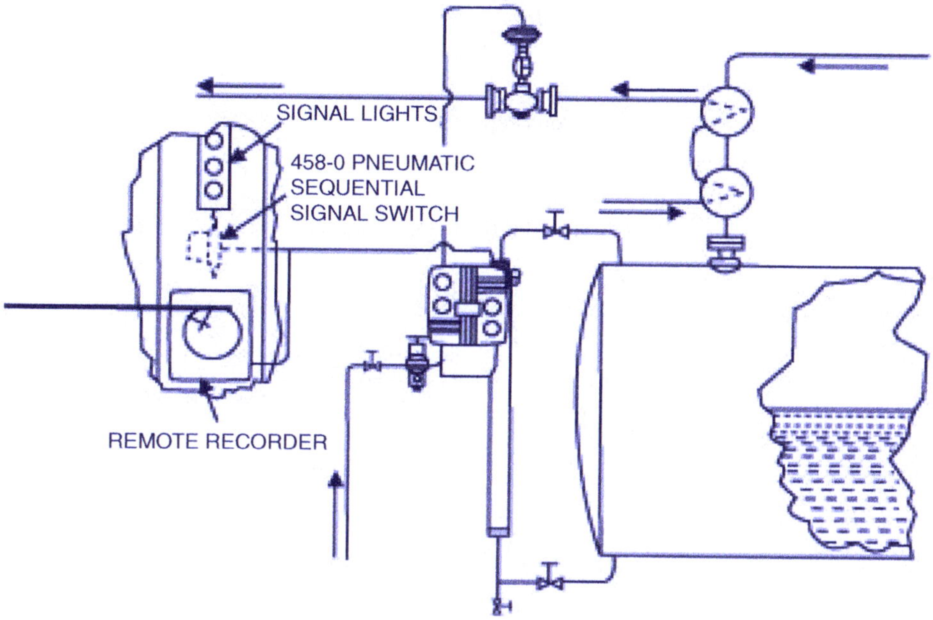

Figure 5–31 Transmission and Control of Level in a Reflux Accumulator (*Courtesy of Dresser-Masoneilan Industries*)

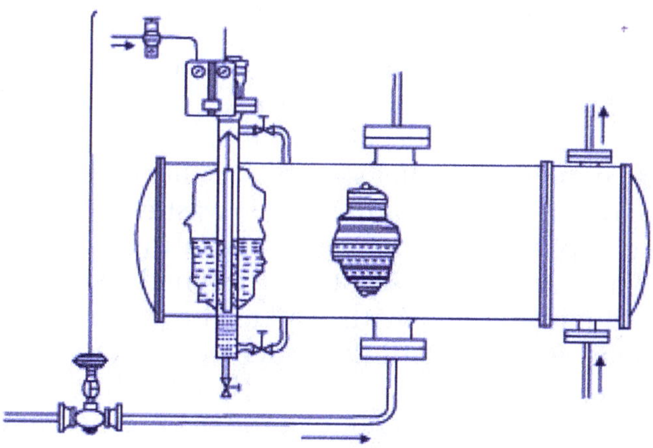

Figure 5–32 Refrigerant Level Control in a Chiller (*Courtesy of Dresser-Masoneilan Industries*)

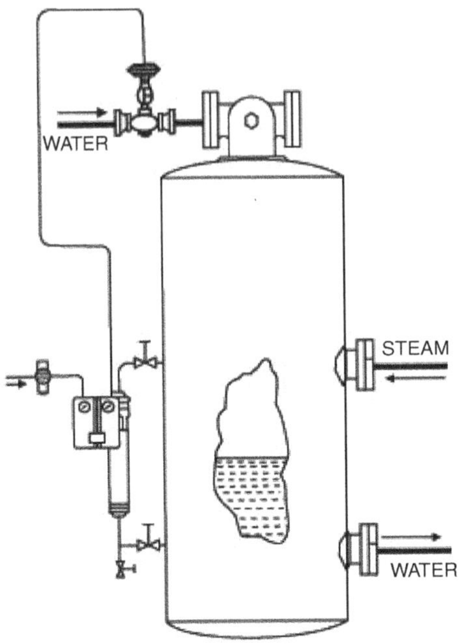

Figure 5–33 Level Control in a Counterflow Deaerating Feedwater Heater (*Courtesy of Dresser-Masoneilan Industries*)

5.13 MAINTENANCE AND CALIBRATION

The most prevalent maintenance procedure required for displacer level instruments consists of flushing the displacer chamber with a solvent for cleaning purposes and checking calibration. Cleaning procedures will naturally depend on the specific application, but they often simply call for the displacer chamber to be washed with water. This can usually be accomplished by closing the isolating valves and opening the bottom drain. A water hose can be fastened to the drain connection and the chamber filled with water. The water is then emptied. This procedure is repeated until the water leaving the chamber is clear, indicating that the contaminants in the chamber have been dissolved. It may be necessary at times to remove the displacer from the chamber for cleaning with another type of solvent.

Once the chamber has been cleaned and emptied, the instrument is checked for proper zero adjustment. This procedure calls for the displacer to be immersed in the lighter of the two fluids. If the application is a liquid/air interface, the displacer is hanging in air, with no buoyancy force created by a liquid. If the application is a liquid/liquid

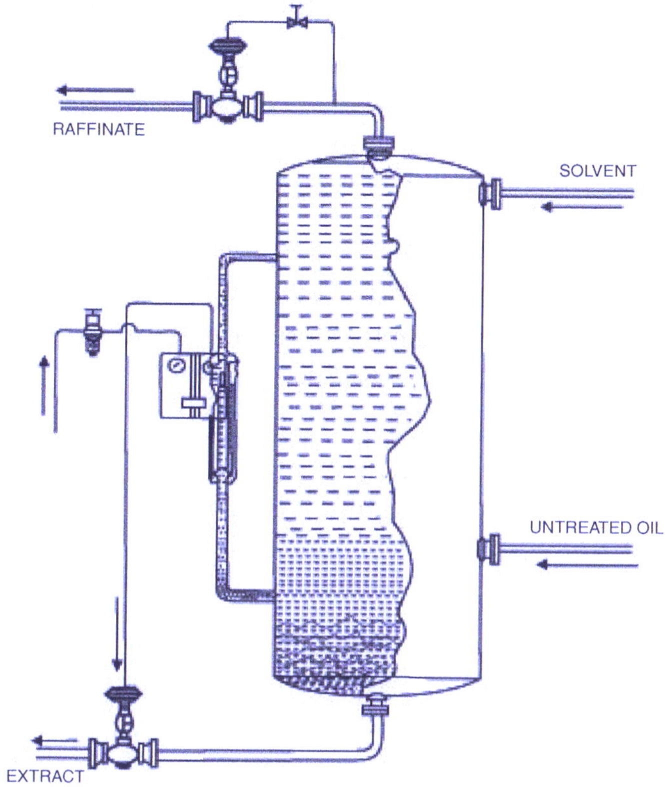

Figure 5–34 Control of Interface Level in a Settling Tank
(*Courtesy of Dresser-Masoneilan Industries*)

interface, the displacer is immersed in the lighter liquid. With the displacer fully immersed in the lighter liquid, the output is adjusted for 0% with the zero adjustment provided.

The span of the instrument can be checked by immersing the displacer completely in the heavier of the two fluids, in which case the output should be 100%. If the reading is not correct at this point, the span should be adjusted to achieve a 100% output. Older pneumatic or analog electronic displacers may require adjusting the zero and span several times to achieve the desired accuracy. Newer microprocessor-based instruments can usually be calibrated using water as a level standard, then reconfiguring the device for interface measurement, and entering the two fluids' SGs. It is necessary to verify that the configuration changes are correct and the transmitter is accurately measuring interface level. The span is determined by the displacer weight and the specific gravity settings. The specification chart for the instrument should be checked for these values.

Once the proper specification and selection have been made for a specific application, and the instrument span is known to be correct, the calibration procedure is usually reduced to setting the zero adjustment for 50% output, with the level at the midrange mark on the displacer chamber. This mark is provided by the manufacturer. It is always good practice to follow the manufacturer's recommendations and procedures when maintaining or calibrating equipment.

5.14 MULTI-DISPLACER APPLICATIONS

In addition to the various applications involving proportional level measurement and control discussed, displacers are widely used for point level indication and alarm. Spring-supported displacer-type level switches are used in sumps, tanks, and vessels where long insertions and switching over wide variations in liquid level are required. For wide-ranging or multiple-function level applications, two or more displacers can be mounted on a single cable or rod connected to a spring, as shown in Figure 5–35. By positioning the displacers on the cable or rod, the level points relative to switch operation can be determined. The spring tension, displacer density, and rod or cable length are established in accordance with the process liquid's specific gravity.

Some units are equipped with three or four displacers. These devices can produce multiple independent outputs as the liquid level crosses each displacer. Common applications for multiple displacer units are multistage pump control with alarm functions, high-low shutdown control with alarm points and other such functions.

In the multi-displacer device shown in Figure 5–35, widely adjustable differential action is accomplished by setting the stroke rod stops so that an amount of slack equal to the movement caused by one displacer exists between the attractor and the stop. The attractor is against the lower stop and is moved into the field of a switch station magnet when two displacers are covered. The station actuates and the attractor remains in that position until the lower displacer is uncovered and total spring movement is enough to pull the attractor away from the magnet with the upper rod stop.

While supported displacer-type level switches are less sensitive to wave action than float-activated units, flow or currents in the liquid can affect displacement action. For such conditions, splash guards or stilling wells should be used to protect the gage from fluid disturbances.

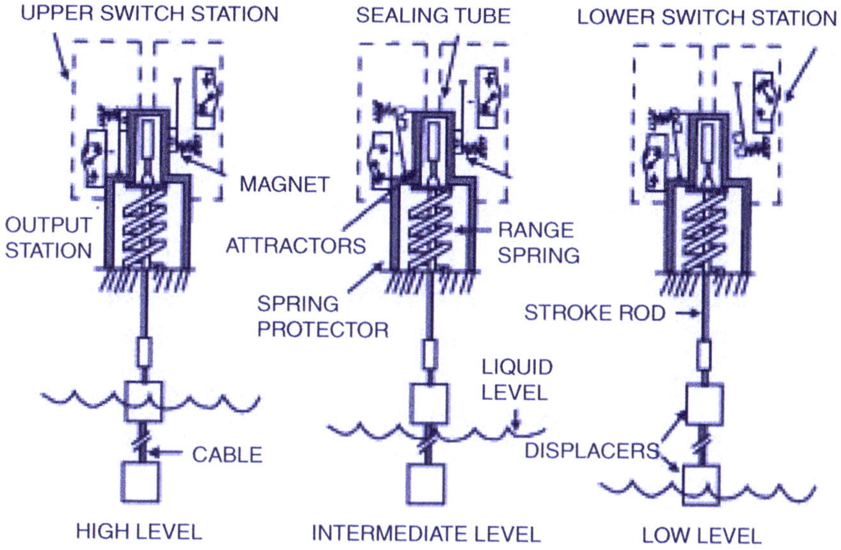

Figure 5–35 Level Instrument with Multiple Displacers (*Courtesy of Delta Controls*)

5.15 INSTRUMENT MOUNTING AND SPECIAL APPLICATIONS

The usual procedure for mounting a displacer on the process vessel involves connecting the displacer chamber to the process vessel by connecting piping (Figure 5–36); other applications may call for the displacers to be mounted directly in the process vessel without a displacer chamber (Figure 5–37B). A stilling well may be used (shown in cutaway in Figure 5–37A) to provide stability in turbulent conditions.

EXERCISES

5.1 List two visual methods for measurement level and two applications of each.

5.2 Explain the general principle of level measurement in terms of float-type ATG systems.

5.3 List the key components of an ATG system.

5.4 Explain the operation of a variable-displacement level device.

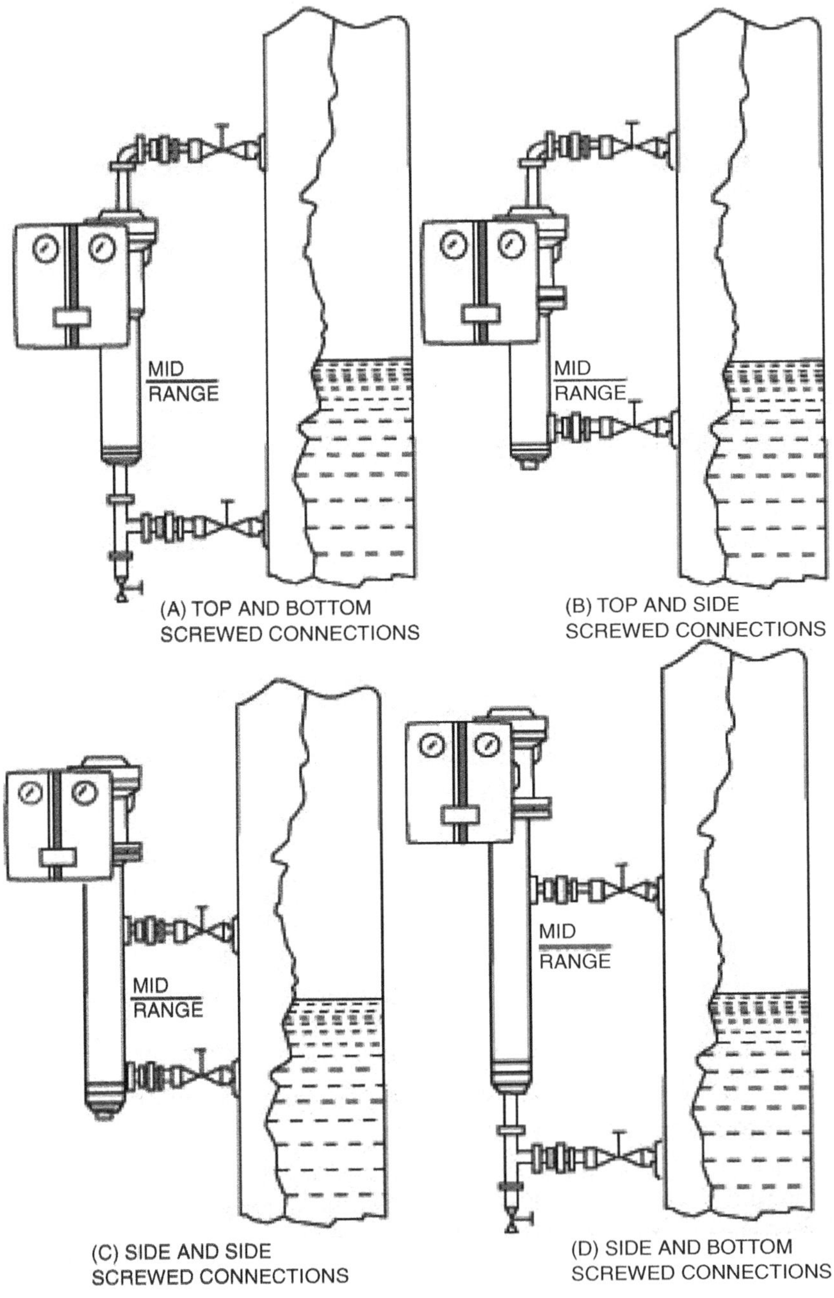

Figure 5–36 Chamber Mounted Applications
(Courtesy of Dresser-Masoneilan Industries)

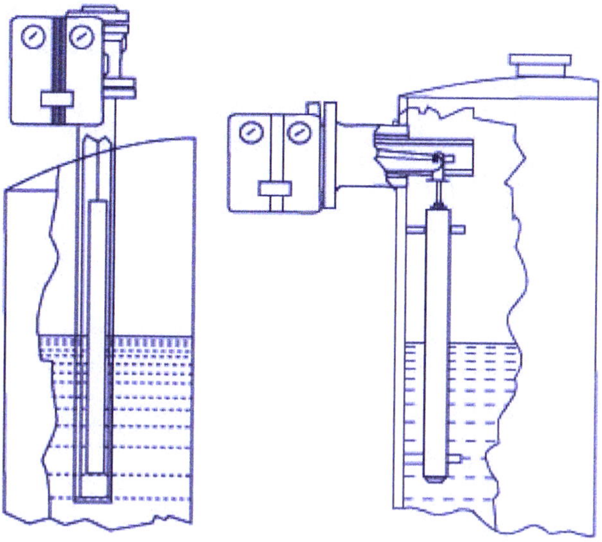

Figure 5–37 Displacers Mounted in a Process Vessel
(*Courtesy of Dresser-Masoneilan Industries*)

5.5 What factors determine the buoyancy force on a displacer?

5.6 A displacer 7.0 in. long and 2.0 in. in diameter is used to measure the interface position of oil and water. The specific gravity of the oil is 0.60. What is the weight difference of the displacer for a 100% level change?

5.7 Why is the length/weight ratio of displacers a more important consideration than weight alone?

5.8 What function is served by the torque tube in a displacer application?

5.9 Give a calibration procedure for a displacer level device.

5.10 Distinguish between direct and inferential methods of level measurement.

REFERENCES

1. Mei, Kenneth W. "Automatic Tank Gages Can Be Used for Custody Transfer," 1989. *Technology, Oil and Gas Journal.*

2. Gauging Systems, Incorporated Technical Literature, 2007.

3. Miller, Dave. *Float-Type Continuous Measurement,* Research Triangle Park, NC: ISA, 1992.

6

Hydrostatic Head Level Measurement

6.1 INTRODUCTION

Many level measurement techniques are based on the principle of hydrostatic head measurement. From this measurement, a level value can be inferred. Such level-measuring devices are used primarily in the water and wastewater industries; in the oil, chemical and petrochemical industries; and to a lesser extent in the refining, pulp and paper, and power industries.

6.2 PRINCIPLE OF OPERATION

As shown in Figure 6–1, the weight of a 1 ft^3 container of water is 62.342 lb, and this force is exerted over the surface of the bottom of the container. The area of this surface is 144 in^2; the pressure exerted is

$$P = F/A \tag{6.1}$$

where P is pressure in pounds per square inch, F = 62.342 lb, and A = 1 ft^2 = 144 in^2

$$P = \frac{62.342\text{lb}}{144\text{in}^2} = 0.433\text{psi/ft} \tag{6.2}$$

This is the pressure caused by the weight of a 12 in. column of water pushing downward on 1 in^2 of the surface of the container. The pressure exerted on 1 in^2 of area by a 1 in. column of water is then

$$P = \frac{0.433\text{psi}}{12\text{in.}} = 0.0360\text{psi/ft} \tag{6.3}$$

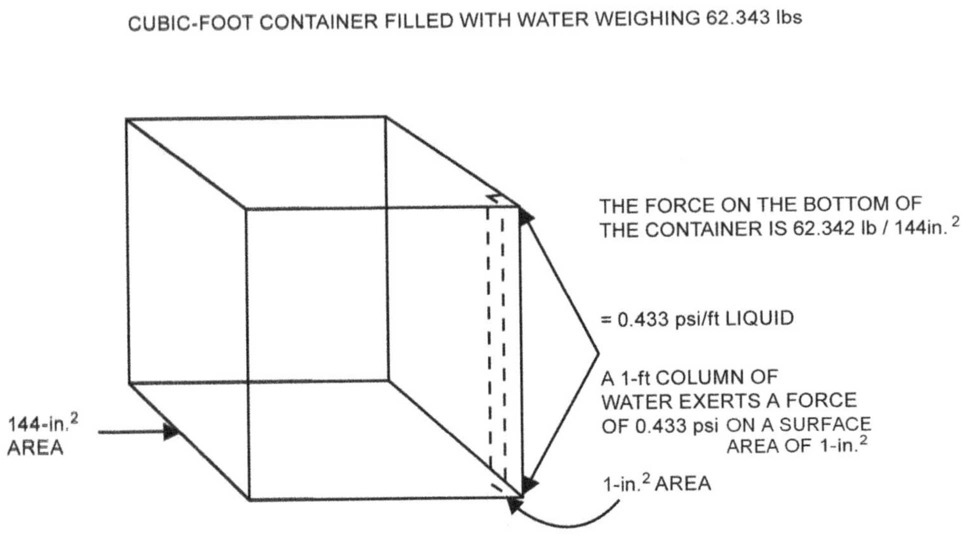

Figure 6–1 Relationship between Pressure and a Column of Liquid

By the reasoning expressed in Equations 6.1 through 6.3, the relationship between the height of a column of water and the pressure exerted on the supporting surface is established. This relationship is important, not only in the measurement of pressure, but also in the measurement of liquid level.

Extending the discussion one step further, the relationship between level and pressure can be expressed in feet of length and pounds per square inch of pressure:

$$1 \text{ psi} = 2.31 \text{ ft wc} = 27.7 \text{ in. wc} \tag{6.4}$$

(The term *wc*, which stands for water column, is usually omitted as it is understood in the discussion of hydrostatic pressure measurement.) It is apparent that the height of a column of liquid can be determined by measuring the pressure exerted by that liquid.

6.3 OPEN-TANK HEAD LEVEL MEASUREMENT

Pressure measurements are made with respect to atmospheric pressure, a vacuum or to another pressure. The terms "pounds per square inch gage" (psig) and "pounds per square inch absolute" (psia) are used to express the reference point of measurement. In gage-pressure applica-

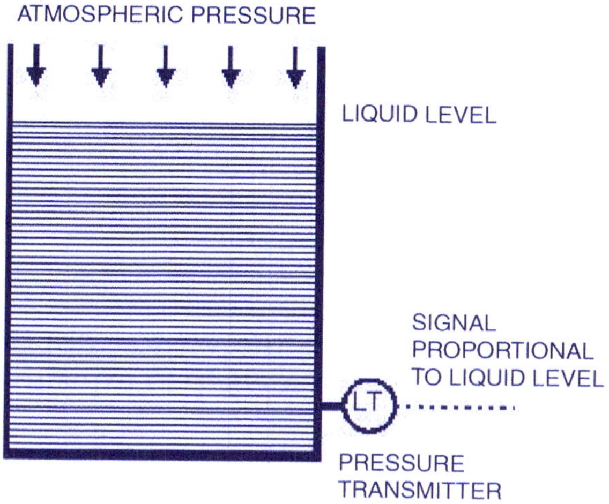

Figure 6–2 Hydrostatic Level Measurement in an Open Tank

tions, the measured pressure is applied to one side of a pressure transducer, usually a pressure element, while the other side is exposed to atmospheric pressure. When atmospheric pressure is applied to both sides of the transducer, a zero point of measurement is established because atmospheric pressure changes will create opposing forces on the transducer, and no movement or measurement will be recorded. A measurement will result only when the applied pressure is elevated above atmospheric value. This measurement principle will now be applied to level measurement.

Figure 6–2 illustrates an application where the level value is inferred from a pressure measurement. When the level is at the same elevation point as the measuring instrument, atmospheric pressure is applied to both sides of the transducer in the pressure transmitter, and the measurement is at the zero reference level. When the level is elevated in the tank, the force created by the hydrostatic head of the liquid is applied to the measurement side of the transducer, resulting in an increase in instrument output. The instrument response caused by the head pressure is used to infer a level value. The relationship between pressure and level is expressed by Equation 6.5. If the measured pressure is 1 psi, and the liquid is water, the level will be 2.31 ft, or 27.7 in. Changes in atmospheric pressure will not affect the measurement because these changes are applied to both sides of the pressure transducer.

When the specific gravity of a fluid is not 1, Equation 6.1 must be corrected for specific gravity because this equation is based on the weight of 1 ft^3 of water. If the fluid is lighter, the pressure exerted by a specific column of liquid is less. The pressure will be greater for heavier liquids. Correction for specific gravity is expressed by Equation 6.5:

$$P = \frac{0.433\text{psi}}{\text{ft}}(G)(H) \tag{6.5}$$

where G is specific gravity and H is the vertical displacement of a column in feet. The following examples help to explain the principle of head-type level measurement.

EXAMPLE 6-1

Problem: A pressure gage is mounted on a water tank 30.0 ft below the liquid level. Calculate the pressure value on the gage in pounds per square inch.

Solution: From Equation 6.5,

$$P = \frac{0.433\text{psi}}{\text{ft}}(1.0)(30.0\text{ft}) = 13.0\text{psi}$$

EXAMPLE 6-2

Problem: A pressure gage is mounted on an oil tank 30.0 ft below the liquid level. If the specific gravity of the oil is 0.65, find the pressure on the gage in pounds per square inch.

Solution: From Equation 6.5,

$$P = \frac{0.433\text{psi}}{\text{ft}}(0.650)(30.0\text{ft}) = 8.44\text{psi}$$

EXAMPLE 6-3

Problem: A pressure gage is mounted on a salt brine tank 30.0 ft below the level. If the specific gravity of the brine solution is 1.30, find the pressure on the gage in pounds per square inch.

Solution: From Equation 6.5,

$$P = \frac{0.433\text{psi}}{\text{ft}}(1.3)(30.0\text{ft}) = 16.9\text{psi}$$

EXAMPLE 6-4

Problem: The pressure on a gage in a level application similar to that in Figure 6–2 is 4.62 psi. If the specific gravity of the liquid is 0.89, find the level in the tank in feet above the point of measurement.

Solution: Using Equation 6.5 and solving for *h,*

$$h = \frac{P}{0.433\frac{\text{lb}}{\text{in}^2}(SG)} = \frac{4.62\frac{\text{lb}}{\text{in}^2}}{\frac{0.433\frac{\text{lb}}{\text{in}^2}(0.89)}{\text{ft}}} = 12.0\text{ft}$$

In these examples, a pressure gage calibrated to read in pounds per square inch was used for level measurement. In such applications, however, the measuring instrument is usually calibrated in a linear measurement such as feet or inches and the level is read directly, eliminating the need for conversions. The relationship expressed in Equation 6.4 or Equation 6.5 may be used to construct the scale graduations on the gage. For example, instead of reading in pounds per square inch, the markings on the gage would express graduations of feet and tenths or inches. Many head level transmitters are calibrated in inches of water. The receiver instrument is also calibrated in inches of water or linear divisions of percent.

For ease in measuring levels in tanks, wells, and reservoirs, a submersible or flange-mounted pressure transducer or transmitter can be used to hydrostatically measure level. Equations 6.4 and 6.5 show that hydrostatic pressure measurement can easily be converted to a corresponding level.

A strain-gage pressure transducer can be used to hydrostatically measure liquid level in tanks, wells, reservoirs, and similar vessels where other head measurement techniques would be difficult. A compatible microprocessor-based controller is used to supply excitation voltage, display level and transmit a signal for remote indication. The submersible transducer is lowered into the vessel, supported by a connecting cable. The unit can be moved to desirable locations and, by fixing the length of the supporting cable, a zero reference point is easily established. Available ranges are from 0–14 ft. to 0–690 ft. of water.

Calibration of the transmitter can be accomplished by applying a known pressure with a deadweight tester or another calibration instrument before installation, or by empirical means after installation. The transducer can be raised above the liquid level or set to some reference point for zero level. The level can be raised to an established 100% value, or the transducer can be lowered a distance equal to the measurement span of the device for setting the 100% value. The empirical method just described compensates for the specific gravity of the process fluid. When applying a known pressure, specific gravity compensation must be done by calculating the pressure/distance relationship (Equation 6.5) and following Examples 6-1 to 6-4. The instrument can also be scaled to a level value as a percentage of an established range.

When desired, a 4 to 20 mA transmitted signal is available to read a level value at a remote location. The analog output is derived from the digital display. Once the scale factor has been set, the analog signal is calibrated in accordance with the range set for the digital display. For example, a meter calibrated for a pressure of 0 to 21.65 psi creates a meter display of 0 to 100%, which corresponds to 0 to 50 ft of level and an analog output of 4 to 20 mA.

Flush and pipe-mounted pressure transducers are also available for level measurement applications. They can be directly mounted on the side or bottom of a vessel for various measurement schemes. Several head-type level-measuring applications are described in the following paragraphs.

Facilities such as power plants and waste-treatment plants bring in large volumes of water through channels, either for use or for treatment. Large bars or screens, called traveling screens, are placed in the channels to collect debris from the water. This prevents the debris from being drawn into the process, where it could damage pumps and other equipment. Eventually, however, the screen becomes clogged and reduces flow.

A differential-measurement traveling screen can be used to activate alarms or automatic cleaning equipment when clogging occurs (Figure 6–3). Two submersible transducers (4-wire millivolt only) are connected to a meter to provide differential measurement. One transducer is placed upstream of the screen and one downstream, each the

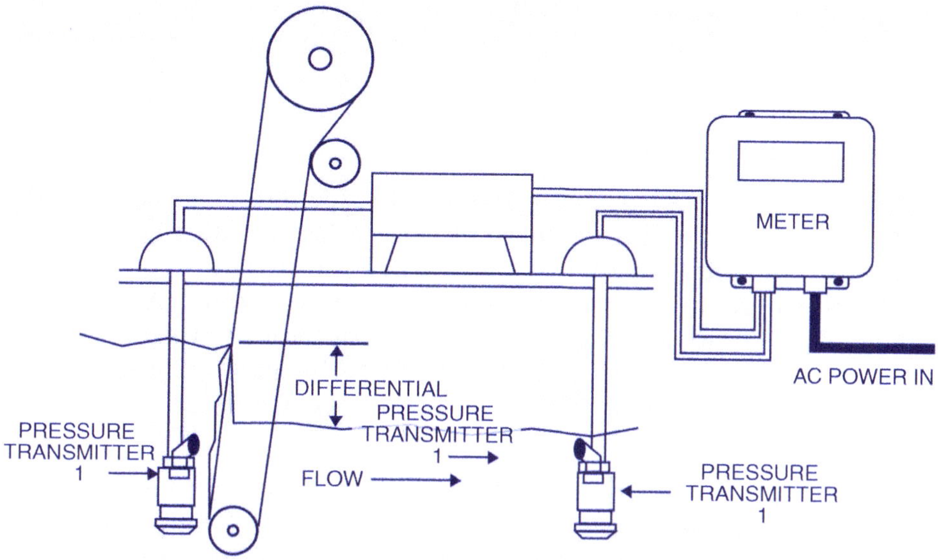

Figure 6–3 Differential Measurement Traveling Screen (*Courtesy of Ametek, Inc.*)

same distance from the floor of the channel. The meter is programmed to see and display only the difference between the two transducers. As the screen becomes clogged, the water level downstream of the screen becomes lower, resulting in a lower pressure-measurement signal from the downstream transducer. When the difference between the two transducers reaches the programmed set points, the meter relays activate an alarm or an automatic cleaning system to remove the debris and restore full flow. This method can be used for other differential pressure measurement applications within certain accuracy limits.

Many landfills are made up of a series of non-permeable cells used to hold the fill. A separate semi-permeable barrier is positioned several feet from the inner walls of each cell, allowing natural runoff and rainwater to seep through the barrier and accumulate in the bottom of the cell. The liquid seepage must then be eliminated.

A submersible transmitter is used to measure the liquid seepage and is positioned in the bottom of the cell by means of a PVC pipe. Both the transmitter and a local pump are connected to a controller. The transmitter sends a signal representing the level of liquid to the controller and, at the desired level, the meter/controller activates the pump. The pump then expels the liquid to a holding tank for future treatment and disposal (Figure 6–4).

A similar system is used to measure and control the level of a storm-water runoff lift station (Figure 6–5) by sequentially turning the primary and secondary pumps on at the high level and off at the low level. A 4 to 20 mA signal linear to the level is sent to the telemetering equipment in the control room.

A transmitter and controller are used to regulate the lift station (Figure 6–5). The transmitter provides both a local indication of level and a 4–20 mA output signal to the control room. The signal is regulated by the high and low set points in the controller. At high set point 1, relay 1 sends the signal to the control panel to turn on pump 1. Similarly, at high set point 2, relay 2 sends the signal to the control panel to turn on pump 2. The reverse process occurs at the low set points.

To measure and control the water level in city wells and to prevent the pump and system from operating if the water level falls below the pump, the system illustrated in Figure 6–6 has proven useful. This system has a meter with relay control and a submersible transducer. The submersible transducer is lowered in a 1-1/2 in. PVC tube. The transducer sensor monitors the well level, and the readout indicates level in feet. The low set point and hysteresis values are adjusted to switch the well pump on and off at desired levels. The low set point is set so that the well cannot run dry. Optional 4–20 mA output from the meter is sent to the remote control room for a remote readout. As a precaution, a relay should be wired such that a 115 VAC failure will disable the pump, preventing it from pumping dry.

6.4 HYDROSTATIC HEAD LEVEL MEASUREMENT

The system in Figure 6–7 is used to monitor the levels in acetic acid storage tanks and supply a 4–20 mA signal to the control room computer for the control of the fill-and-empty process. Transmitters with mounting adapters provide a direct 4–20 mA signal over a full pressure range. The computer interprets the signal to determine the level of liquid relative to the tank height in order to control the fill-and-empty process.

6.5 DIAPHRAGM BOX

Another type of open-tank level measurement using the hydrostatic head principle is the diaphragm box (Figure 6–8). Although this technique is not as common as some other applications, it is worthy of

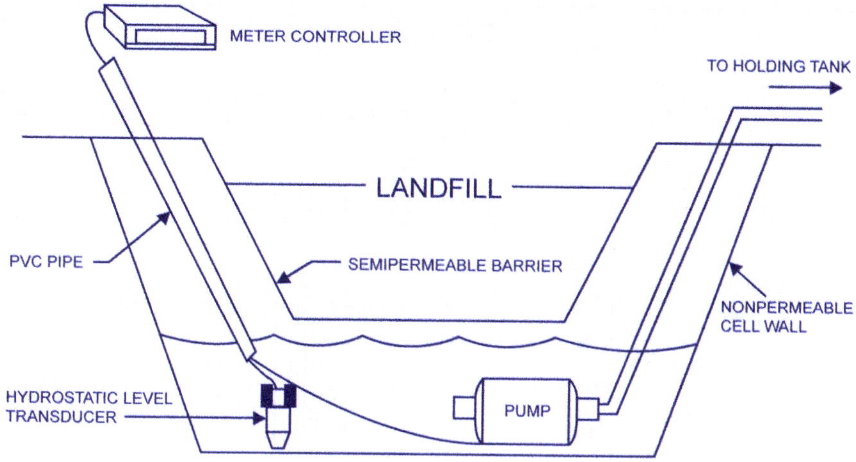

Figure 6–4 Landfill Level Measurement
(*Courtesy of Ametek, Inc.*)

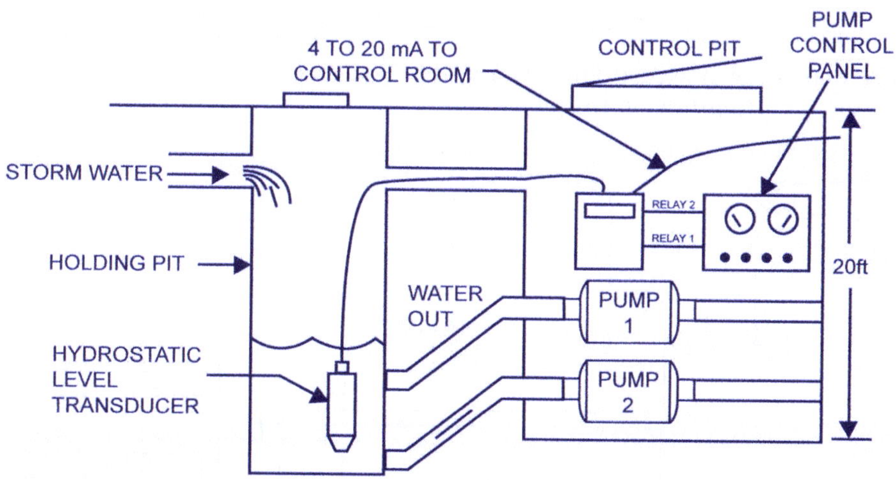

Figure 6–5 Lift Station Level Measurement
(*Courtesy of Ametek, Inc.*)

mention. Hydrostatic pressure is applied to a soft neoprene diaphragm, which causes the diaphragm to flex and deform. When this occurs, the upper side of the diaphragm compresses the fluid in the sealed system, which causes an increase in pressure. The pressure is applied through a connecting tube; a pressure element is part of a recorder or indicator which is calibrated in liquid level units.

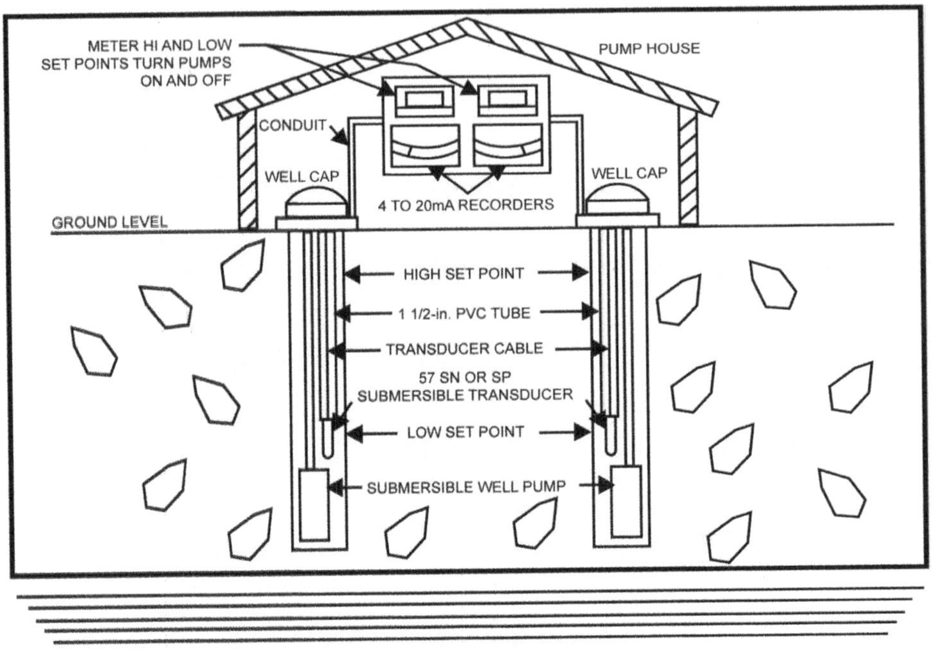

Figure 6–6 Level Measurement in Water Wells
(*Courtesy of Ametek, Inc.*)

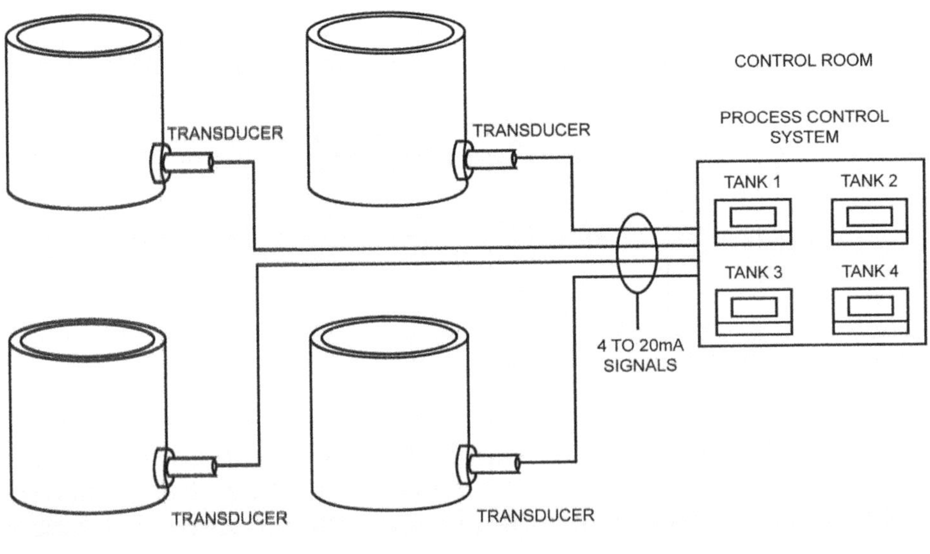

Figure 6–7 Multiple Tank Measurement System
(*Courtesy of Ametek, Inc.*)

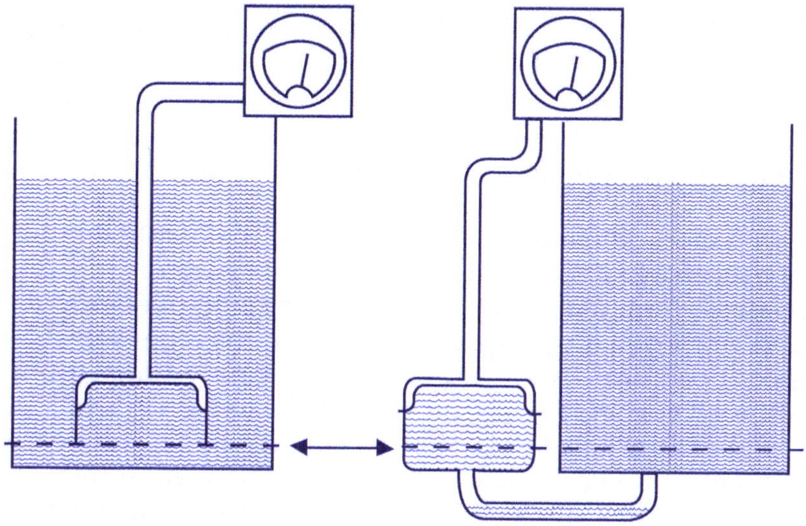

Figure 6–8 Diaphragm Box
(*Courtesy of The University of Texas, Petroleum Extension Service*)

A leak-free box and connecting tubing are essential to the successful operation of the diaphragm box system. If the air inside the tube should leak to the atmosphere, there would be no increase in pressure with an increase in liquid level, and the system would be inoperative. This is a very important consideration. The diaphragm box system can be placed in operation by lowering the liquid to the zero level, which is the same elevation as the diaphragm. The air trapped in the upper chamber of the box, tubing, and the pressure element is then at atmospheric pressure. This traps a constant quantity of air in the system.

As the level of the liquid in the vessel rises above the zero reference line, the resulting hydrostatic head will exert a pressure on the diaphragm in direct proportion to the rise in level. The pressure will be related to level by Equation 6.5. The diaphragm must be flexible to the extent that negligible resistance is offered to movement caused by the hydrostatic pressure. When this is the case, the diaphragm will be deflected until the air pressure in the upper chamber is equal to the pressure on the lower side of the diaphragm.

A very important advantage of the diaphragm box method of level measurement is the fact that the actuating element position with respect to the level being monitored is not critical. The only consideration concerning this relationship is the pressure on the box caused by the column of air above the box. This is generally insignificant and causes no appreciable error. The system is, however, sensitive to wide variations in temperature because of the changes in volume and pressure with temperature as defined by the ideal gas laws. Also, the diaphragm may be affected by extreme temperature.

6.6 AIR-TRAP METHOD

The air-trap method (Figure 6–9) is similar to the diaphragm box. This method uses an open box with no diaphragm; therefore, the disadvantages related to the diaphragm—namely, extreme operating temperatures and corrosive fluid applications that may be detrimental to diaphragm operation—are overcome. Air is simply trapped under the box; as the level rises, the air is compressed and generates a pressure that is proportional to level. The system requires a nearly perfectly sealed system to prevent the escape of air, as this will result in system measurement error. A distinct disadvantage of this system is the loss of air due to air absorption by the liquid. This is more prevalent at elevated temperatures, and in many situations will preclude the use of this type of measurement system.

6.7 AIR PURGE OR BUBBLE SYSTEM

The system shown in Figure 6–10 is known by various names: air purge, air bubble, or dip tube. It is a variation of the air-trap method discussed above, but there is no box and the air is replenished at a constant rate. With the supply air blocked, the water level in the tube will be equal to that in the tank. When the air pressure from the regulator is increased until the water in the tube is displaced by air, the air pressure on the tube is equal to that required to displace the liquid and is also equal to the hydrostatic head of the liquid in the tube.

The pressure set on the regulator must be great enough to displace the air from the tube for maximum pressure, which will coincide with maximum level. This will be indicated by a continuous flow, which is evidenced by the formation of bubbles rising to the level of the liquid in the tank. As it may not be convenient to visually inspect the tank for the presence of bubbles, an airflow indicator will usually be installed in the air line running into the tank. A visual flow indicator such as a rotameter is generally used for this purpose. The importance of maintaining a flow through the tube lies in the fact that the liquid in the tube must be displaced by air, and the backpressure on the air line provides the measurement, which is related to level.

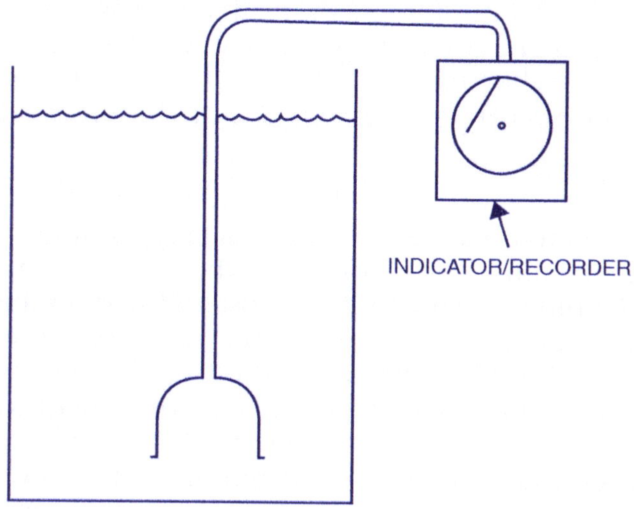

Figure 6–9 Air-Trap Method
(*Courtesy of The University of Texas, Petroleum Extension Service*)

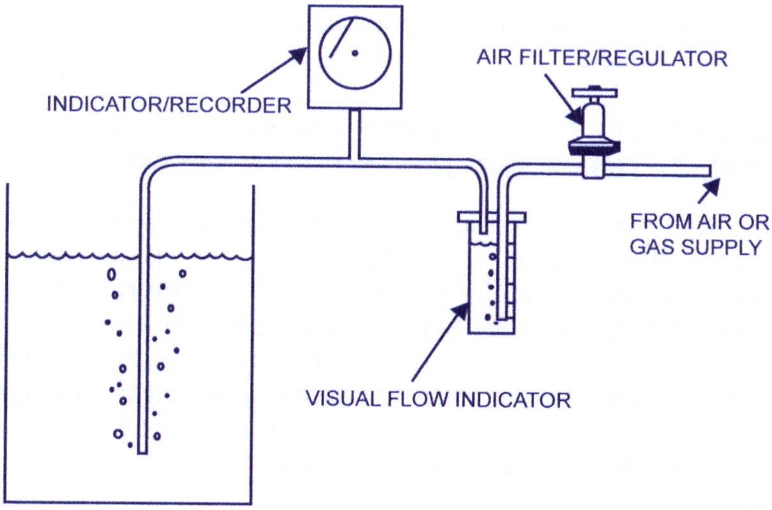

Figure 6–10 Air Bubble System
(*Courtesy of The University of Texas, Petroleum Extension Service*)

The amount of airflow through the dip tube is not critical, but should be fairly constant and not too great. In situations where the airflow is great enough to create a backpressure in the tube caused by airflow restriction, this backpressure would signify a level, resulting in a measurement error. For this reason, 3/8-in. tubing or 1/4-in. pipe should be used.

An important advantage of the bubble system is the fact that the measuring instrument can be mounted at any location or elevation with respect to the tank. The connection where the transmitter or indicator senses the pressure should be as close to the dip tube as possible to minimize measurement errors due to pressure loss of the bubbler gas as it flows to the dip tube. This application is advantageous for level-measuring applications where it would be inconvenient to mount the measuring instrument at the zero reference level. An example of this situation is level measurement in underground tanks and water wells. The zero reference level is established by the relative position of the open end of the tube with respect to the tank. This is conveniently fixed by the length of the tube, which can be adjusted for the desired application. It must be emphasized that variations in backpressure on the tube or static pressure in the tank cannot be tolerated. This method of level measurement is generally limited to open-tank applications but can be used in closed-tank applications with the special precautions described below.

6.8 HEAD LEVEL MEASUREMENT IN CLOSED-TANK APPLICATIONS AND PRESSURIZED VESSELS

The open-tank measurement applications that have been discussed are referenced to atmospheric pressure; that is, the pressures (usually atmospheric) on the surface of the liquid and on the reference side of the pressure element in the measuring instrument are equal. When atmospheric pressure changes, the change is by equal amounts on both the measuring and the reference sides of the measuring element. The resulting forces created are canceled, one opposing the other, and no change occurs in the measurement value.

Suppose, however, that the static pressure in the level vessel is different from atmospheric pressure. Such would be the case if the level were measured in a closed tank or vessel. Pressure variations within the vessel would be applied to the level surface and produce an accumulated effect on the pressure instrument, thus affecting level measurement. For this reason, pressure variations must be compensated for in closed-tank applications. Instead of the pressure-sensing instrument shown in Figure 6–2, a differential-pressure instrument is used for head-type level measurements in closed tanks. This application is illustrated in Figure 6–11.

Since a differential-pressure instrument responds only to a difference in pressure applied to the measuring ports, the static tank pressure on the liquid surface in a closed tank or vessel has no effect on the measuring signal. Variations in static pressure, therefore, do not cause an error in level measurement as would be the case if a pressure-sensing instrument were used.

Certain precautions should be taken when using dip tubes in closed-tank applications. Variations in backpressure on the dip tube caused by changes in static pressure in the vessel could interrupt or stop the flow of air through the tube. To prevent this, it is sometime necessary to use a system to maintain tube flow independent of backpressure. Such a device is a constant-differential flow controller. Also, check values should be installed in the dip tube supply to prevent process fluid from entering the instrument when the air supply pressure is interrupted.

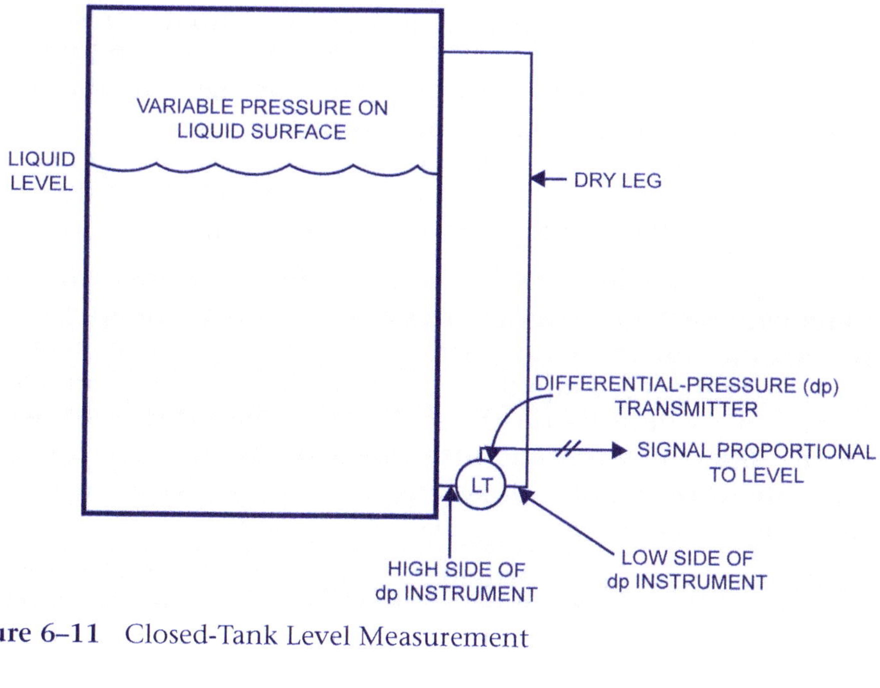

Figure 6–11 Closed-Tank Level Measurement

6.9 MOUNTING CONSIDERATIONS: ZERO ELEVATION AND SUPPRESSION

Unless dip tubes are used, the measuring instrument is generally mounted at the zero reference point on the tank. When another location point for the instrument is desired or necessary for accessibility or when process conditions are too harsh, the head pressure caused by liquid above the zero reference point must be discounted. Such an application is illustrated in Figure 6–12.

As shown, the high-pressure connection of the differential-pressure instrument is below the zero reference point on the tank. The head pressure caused by the elevation of the fluid from the zero point to the pressure tap will cause a measured response or instrument signal. This signal must be suppressed to make the output represent a zero-level value. The term "zero suppression" defines this operation. This refers to the correction taken or the instrument adjustment required to compensate for an error caused by the mounting of the instrument to the process. The zero suppression adjustment can be made in the field by turning an adjustment screw on the instrument or using an HHC (handheld communicator) to configure a smart transmitter's LRV (lower range value) with the fluid in the impulse line. With the level at the desired zero reference level, the instrument output or response is made to represent the zero-level value. In the case of transmitters, this would be 3 psi, 4 mA, 10 mA, or the appropriate signal level to represent the minimum process value. More commonly, however, the zero suppression adjustment is a calibration procedure carried out in a calibration laboratory or shop, sometimes requiring a kit for the transmitter that consists of an additional zero bias spring.

When using differential-pressure instruments in closed-tank level measurement for vapor service, quite often the vapor in the line connecting the low side of the instrument to the tank will condense to a liquid. This condensed liquid, sometimes called a wet leg, produces a hydrostatic head pressure on the low side of the instrument, which causes the differential-pressure instrument reading to be below zero. Compensation is required to eliminate the resulting error. This compensation or adjustment is called zero elevation.

To better illustrate the principle of suppression and elevation, a number of examples follow. Figure 6–12A shows an open-tank application requiring suppression; Figure 6–12B illustrates a closed-tank application requiring elevation. Dry legs in this type of application should

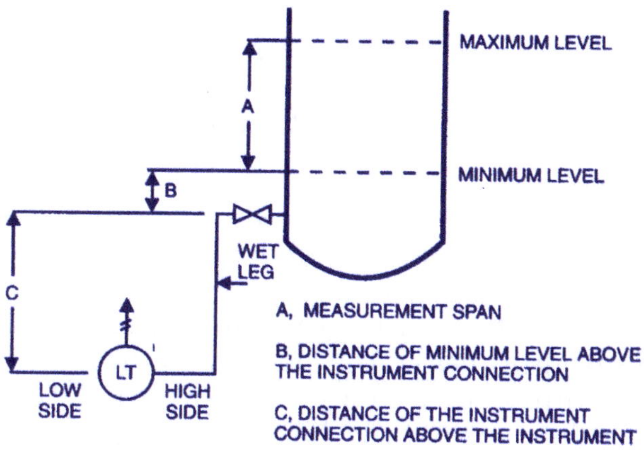

(A) ZERO SUPPRESSION IN AN OPEN TANK

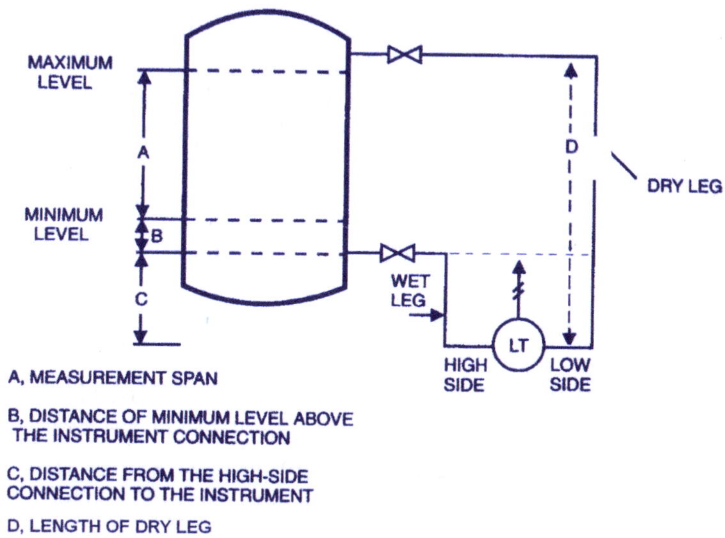

(B) ZERO ELEVATION IN A CLOSED TANK

Figure 6–12 Applications Requiring Zero Suppression and Elevation

be approached cautiously, as any condensation will produce a measurement error. Impulse piping can be configured with block and drain valves on the low leg to ensure that the dry leg remains dry.

It is important to remember that the process fluid in the impulse lines is frequently at ambient temperature, which can have a different density than the same fluid in the process vessel. This difference in density must be considered when calculating zero elevation or suppression.

As shown in Figure 6–12A, the LRV may be at an elevation higher than the process connection. This is sometimes done to eliminate problems created if the transmitter zero should drift upward. If the process connection elevation is at the LRV, and the transmitter zero drifts high, the vessel level could drop below the process connection while the transmitter would continue to read a level above the LRV elevation. One needs to remember that if the level drops below the process connection, the transmitter output will remain fixed and the true level cannot be measured.

EXAMPLE 6-5

Problem: Referring to Figure 6–12A, A = 80 in., B = 5 in., and C = 15 in. The specific gravity of the fluid in the tank is 0.8. Calculate the calibrated span of the instrument and the amount of suppression, both in inches of water. Also, list the LRV and URV (upper range value).

Solution: The span is equal to

$$(A)(G_L) = (80)(0.8) = 64 \text{ in. } H_2O$$

Suppression is equal to

$$(B)(G_L) + (C)(G_L) = (5)(0.8) + (15)(0.8) = 16 \text{ in. } H_2O$$

$$\text{LRV} = 16 \text{ in. } H_2O$$

$$\text{URV} = 80 \text{ in. } H_2O$$

$$\begin{aligned} \text{Span} &= \text{URV} - \text{LRV} \\ &= 80 \text{ in. } H_2O - 16 \text{ in. } H_2O \\ &= 64 \text{ in. } H_2O \end{aligned}$$

EXAMPLE 6-6

Problem: Using the same values given in Example 6-5, but referring to Figure 6–12B, calculate the calibrated span of the instrument and the amount of suppression, both in inches of water.

Solution: From Figure 6–12B, it can be seen that the outside dry leg is filled with gas above the liquid in the tank. When the density of this gas is disregarded, the force on the low side of the measurement, caused by the column of gas in the low-side connecting pipe, is the same as in Example 6-5, as is the solution.

EXAMPLE 6-7

Problem: Refer to Figure 6–12B, but consider D a wet leg on the low side with a fill fluid of SG 0.9 = G_S. A = 80 inches, B = 5 in., C = 15 in., D = 100 in. and G_L= 0.7.

Find LRV, URV and the calibrated span.

Solution:

$$\text{LRV} = (\text{D})(G_s) - (\text{B} + \text{C})(G_L)$$

$$= -90 - 14$$

$$= -76 \text{ in. } H_2O$$

The minus sign denotes a pressure on the low side.

$$\text{URV} = [(G_L)(\text{A} + \text{B} + \text{C})] - (\text{D})(G_s)$$

$$= [(.7)(80 + 5 + 15)] - (100)(.9)$$

$$= -20 \text{ in. } H_2O$$

$$\text{Calibrated span} = \text{URV} - \text{LRV}$$

$$= -20 - (-76)$$

$$= 56 \text{ in. } H_2O$$

The instrument in the above example can be calibrated using two pressure sources, one for the low side and one for the high side. In this procedure, 90 inches of water [(D)(G_S)], is applied to the low side and 14 inches [(B + C)(G_L)] is applied to the high side; the instrument's output is adjusted to 0%. While maintaining the pressure on the low side, increase the high side pressure to 70 in. H_2O [(G_L)(A + B + C)] and adjust the span for 100% output. Another procedure is to apply 76 in. H_2O to the low side with the high side vented, and adjust the output to 0%. Apply 20 in. H_2O to the low side and adjust the span for 100% output.

EXAMPLE 6-8

Problem: Referring to Figure 6–13, A = 100 in., B = 10 in., C = 120 in., D = 20 in., and G_L = 0.73. Calculate the calibrated span of the instrument and the amount of elevation, both in inches of water.

Solution: The span is equal to

$$A(G_L) = (100)(0.73) = 73 \text{ in. } H_2O$$

The outside wet leg is filled with condensed vapor, and the resulting liquid head creates a pressure on the low side of the measuring instrument. The measurement is driven below the normal zero value, and compensation must be made to elevate zero to the normal value. Elevation is equal to

$$(C)(G_L) - (B)(G_L) = (120)(0.73) - (10)(0.73)$$
$$= 87.6 - 7.3 = 80.3 \text{ in. } H_2O$$

EXAMPLE 6-9

Problem: The values are the same as in Example 6-8, except that a seal liquid (seal fluid) is used in the outside wet leg. The specific gravity of the seal fluid (G_S) is 0.6. Calculate the calibrated span of the instrument and the amount of elevation, both in inches of water.

Solution: The calibration span is independent of the wet leg seal fluid, and the span value is the same as in Example 6-8. Elevation is equal to

$$(C)(G_S) - (B)(G_S) = (120)(0.73) - (10)(0.6)$$
$$= 87.6 - 6 = 81.6 \text{ in. } H_2O$$

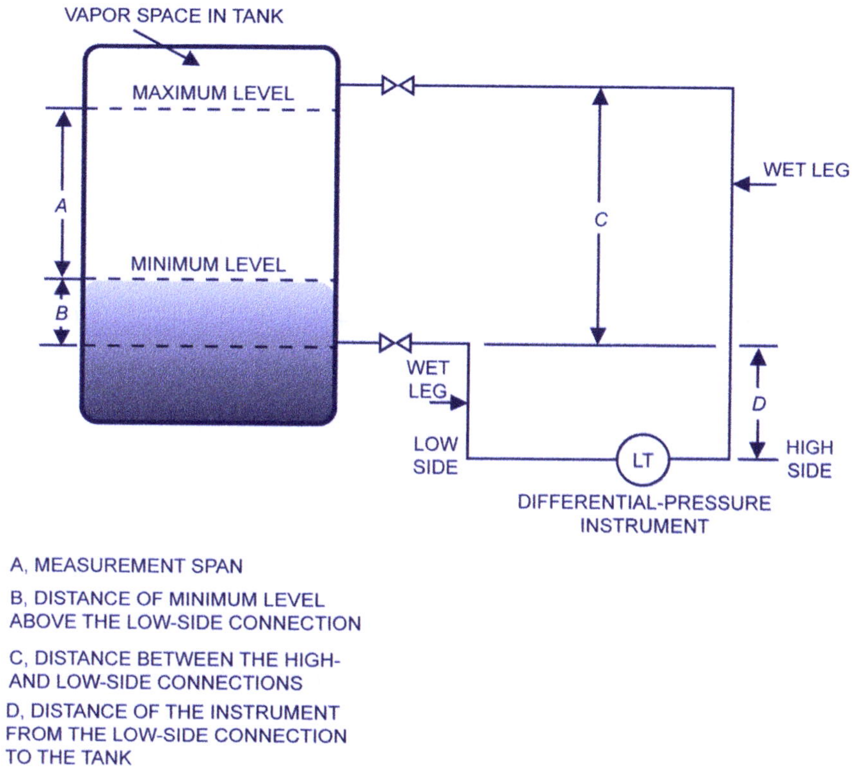

Figure 6–13 Closed Tank Application with Process Connection to the Low Side with Two Wet Legs

The operation of some differential-pressure transmitters requires that the net pressure applied to the sensing element always be positive; that is, the pressure on the high side should always be greater than that on the low side, or the pressure on the low side should always be less than that on the high side. Under these circumstances, the arrangement in Figure 6–13 would be used. Range values would then be determined as shown in Example 6-10.

For this application, the transmitter output is maximum at minimum level and minimum at maximum level. This is because the wet leg on the high side keeps the high-side pressure always higher than the low-side pressure. The output of the transmitter is usually inverted to show an increase in signal as the level increases.

EXAMPLE 6-10

Problem: Referring to Figure 6–13, A = 4.0 ft, B = 0.5 ft, C = 4.5 ft, D = 2.0 ft, G_L = 0.8, and G_S = 0.9. Calculate the span and the calibrated range, both in inches of water. List the LRV and URV.

Solution:

$$H_W \text{ at maximum level} = [(C)(G_S)] - [(A)(G_L) + (B)(G_L) + (D)(G_L)]$$

$$= [(4.5\,\text{ft})(0.9) + (2.0\,\text{ft}((0.9)] - [(4.0\,\text{ft})(0.8) + (0.5\,\text{ft})(0.8) + (2.0\,\text{ft})(0.8)]$$

$$= 5.85\text{ ft} - 5.2\text{ ft}$$

$$= 0.65\text{ ft }(12\text{ in./ft})$$

$$= 7.8\text{ in. }H_2O$$

$$H_W \text{ at maximum level} = [(C)(G_S)] - [(D)(G_S) + (B)(G_L) + (D)(G_L)]$$

$$= [(4.5\text{ ft})(0.9) + (2.0\text{ ft}((0.9)] - [(0.5\text{ ft})(0.8) + (2.0\text{ ft})(0.8)]$$

$$= 5.85\text{ ft} - 2\text{ ft}$$

$$= 3.85\text{ ft }(12\text{ in./ft})$$

$$= 46.2\text{ in. }H_2O$$

The calibrated range is 7.8 in. H_2O to 46.2 in. H_2O.

The calibrated span is 46.2 in. H_2O – 7.8 in. H_2O = 38.4 in. H_2O.

$$\text{LRV} = 7.8\text{ in. }H_2O$$

$$\text{URV} = 46.2\text{ in. }H_2O$$

At times it is desirable to prevent a corrosive process fluid from entering the instrument and connecting lines, or to prevent water or other condensed vapors from the process from entering. In such cases, the instrument and connecting lines are filled with a seal fluid (as in Example 6-9). This fluid must be one that is heavier than the process fluid, and will form an interface with the process fluid so as not to become diluted. Even though the seal fluid is static when the instrument is in service, during times of installation and process upsets, the seal fluid may enter the process. The effects of contamination of the process with a seal fluid must be managed by selecting a seal fluid that will not harm or react with the process.

Information about seal fluids was presented in Chapter 3, and procedures for sealing the instrument will be summarized here. It should be noted that any method used must ensure that all lines and the instrument are completely filled with the seal liquid. If the instrument has a vent plug, it can be opened to vent the air while filling from a T at the process connection. The system can also be filled by pumping the fluid in at the instrument and venting at the process connection. For sealed systems in liquid and vapor service, the instrument is usually located below the process. Refer to Figure 3-14 for clarification of the above procedure.

6.10 A MULTIVARIABLE LEVEL CONTROLLER

This device is a multivariable microprocessor-based analog and digital output transmitter for use in single loop level process measurement. It can be used for open- or closed-tank applications to infer a level value from a differential pressure measurement. This level measurement then becomes the input to the PID (proportional-integral-derivative) algorithm which computes an output to position a control valve to maintain a level in the process. Because it is a multivariable device, temperature measurement is available with an RTD (resistance temperature detector) sensor and a HART (Highway Addressable Remote Transducer protocol) output. The transmitter can be bench calibrated or ranged using a HART HHT (handheld terminal).

Recall from Equation 6.5 (here, *H* in Equation 6.5 is shown as L):

$$L = \frac{\frac{\#}{\text{in}^2}(\text{measured})}{\frac{0.433\frac{\#}{\text{in}^2}(\text{SG})}{\text{ft}}}$$

$$L = \frac{\text{feet of level}}{\text{specific gravity}}$$

The pressure, specific gravity, and feet of level values are derived by a technician, based on specific process conditions at the point of the high pressure process connection for open-tank applications and both high and low pressure connections for closed-tank applications. The technician uses an HHT to set the following parameters:

- Unit of measurement
- Density
- Specific gravity
- URV
- LRV
- Set point (for control purposes)

The level transmitter assumes a constant fluid density and does not correct for density changes caused by process variations or temperature changes. Of course, density corrections can be made based on temperature measurements provided by the transmitter after reconfiguring, using the HHT.

The PID function of the transmitter has an auto tuning feature to compensate for changes in the tank's geometric characteristics and other nonlinear situations in the control loop. For the auto tuning operation, the following parameters can be set:

- Level limits around set point to initiate the tuning process
- Allowable overshoot
- Limits in the controller output
- The number of cycles for settling time

An optional local operator interface (LOI) is available to give a local indication of:

- Process Variable (PV)
- Differential pressure
- Controller output
- Process temperature

- Current controller mode
    - Auto
    - Manual
    - Out of service (OOS)
    - Failure alarm—predetermined high and low controller output values can be set for failed conditions

Mode and set point values can be selected locally.

Transmitter/controller configuration consists of the following

- Set-up level calculation consists of setting values for the specific application that include:
    - Process density
    - Measuring units
    - Vessel configuration and transmitter mounting
    - Damping if required
- Controller configuration
    - Range settings
    - Control modes
    - Controller action
    - Output limits
    - Fail features
    - Set point values
- Level trim
    - Automatically adjusts the density and controller height so the PV value matches the actual level value
- Controller tuning
    - Proportional
    - Integral
    - Derivative (if used)

This configuration can be done by bench configuration or by using the HHT. The controller must be in the OOS or manual mode [Ref. 3].

6.11 DIAPHRAGM SEALS

Diaphragm seals function as an extension of the transmitter and are used to eliminate wet legs. They do not alter the theory of operation of head level measurement previously discussed. Transmitters with remote diaphragm seals prohibit the process fluid from contacting an otherwise wetted portion of the instrument. The use of such seals is advantageous when

- There is a need to eliminate wet legs to reduce maintenance problems encountered with zero shift because of loss of seal fluid.
- Sanitary conditions are required.
- Impulse line plugging may occur due to the presence of solids or viscous fluids.
- The process is corrosive.
- Process temperatures are beyond the temperature limits of the instrument.
- There is a need for easy cleaning of the instrument to prevent contamination between process batches.

A transmitter with remote diaphragm seals is shown in Figure 6–14. The same calibration, elevation, and suppression principles apply for sealed systems as explained in previous discussions and examples. However, when compensating for elevation and suppression, the instrument is normally adjusted for zero with the high and low pressure mounting flanges located the same distance from the process connection as they would be in actual operation. This is because the specific gravity of the seal fluid used in the system is generally not known.

Diaphragm seal systems respond to changes in the process pressure at both measuring taps, as the static pressure and process level change. These changes in pressure are transmitted to the measuring element in the transmitter through liquid-filled capillary tubes. The tubes and seals are filled with a noncompressible liquid compatible with the process temperature and pressure. The wetted portion of a diaphragm seal system is the surface area of the diaphragm that contacts the

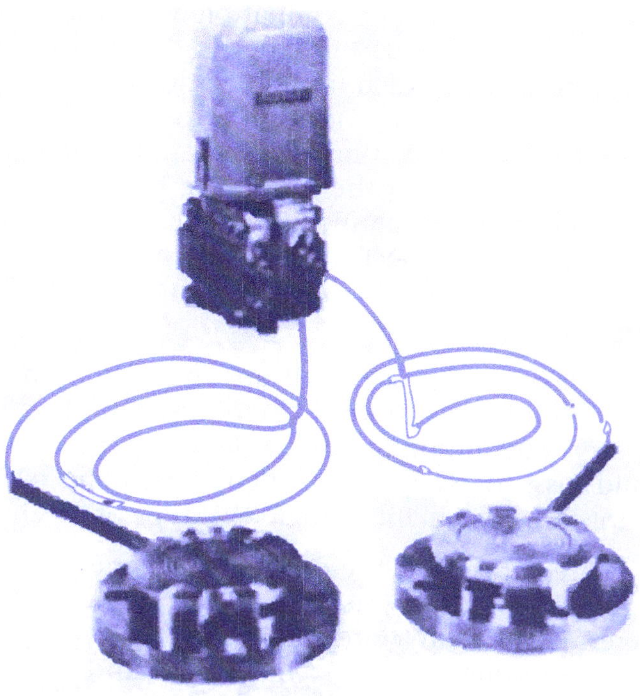

Figure 6–14A Transmitter with Seals—Not Connected to Process
(*Courtesy of Invensys Process Systems*)

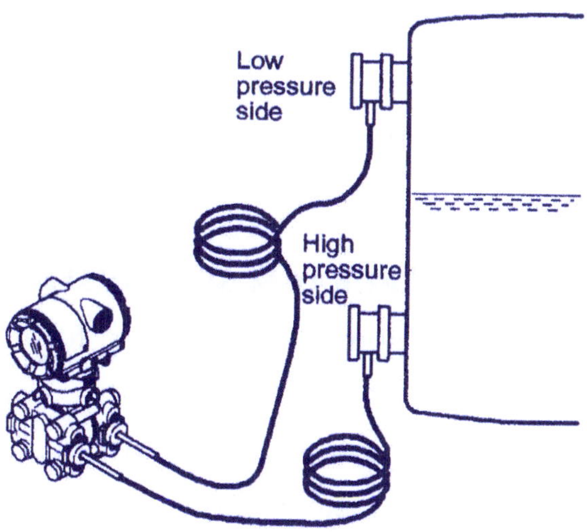

Figure 6–14B Transmitter with Seals Connected to Process
(*Courtesy of Invensys Process Systems*)

process. This portion must also be compatible with the process fluid. The transmitter can be mounted at ground level or in close proximity to the high pressure tap. For vacuum applications, the transmitter is mounted below the high pressure connection to reduce vacuum effects on the transmitter fill fluid. The minimum capillary length is determined by the distance between the mounting position of the transmitter and the low pressure connection. Manufacturers select the appropriate fill fluid with respect to viscosity, coefficient of thermal expansion, specific gravity and the physical dimensions of the capillary tube.

During the manufacturing process, all air and other gases must be removed from the entire instrument cavity and capillary to prevent erratic transmitter operation. Should a leak in the system occur, resulting in fluid loss, the system must be repaired and refilled. Users normally will not have the facilities to do this type of repair.

Although great care is taken in the design and manufacture of sealed diaphragm systems, some inherent temperature-induced error can exist. For differential pressure transmitters, with both capillary tubes identical and at the same ambient temperature and elevation, error caused by temperature in each tube will cancel. For level application, however, the vertical offset of the high and low pressure process connections may result in a head temperature effect error caused by density changes of the fill fluid with temperature. When the high pressure diaphragm is directly flange-mounted to a mating flange at the process connection, the effect of temperature error is reduced. Also, the absence of the capillary tube improves the instrument response time which is proportional to tube length. Since the temperature effect is a function of tube volume, reducing tube length also reduces head temperature effect error.

6.12 SUMMARY OF DIAPHRAGM SEAL SYSTEMS

Following are important points to consider for diaphragm seal systems:

- Temperature-induced error is reduced by
  - Seals with large diameter surface area
  - Short capillaries
  - Low coefficient of expansion seal fluids

- Use of asymmetrical configuration (direct mounted high pressure flanges)
    - Reduces high pressure fill volume
    - Increases low pressure fill volume
    - Minimizes the stiffness of high pressure diaphragm
    - Increases the stiffness of the low pressure diaphragm

Advantages of diaphragm seal transmitters are:

- Extended the capabilities of pressure and differential pressure transmitters
- Mounting flexibility
- Reduced maintenance on
    - Wet leg problems, such as
        - Plugging
        - Freezing

Limitations of diaphragm seal transmitters are:

- Temperature-induced errors
- Small spans that may be difficult to measure
- Increased response time caused by fill fluid damping [Ref. 4]

6.13 REPEATERS USED IN CLOSED-TANK LEVEL MEASUREMENT

Some level applications require special instrument-mounting considerations. For example, sometimes the process liquid must be prevented from forming a wet leg. Some liquids are nonviscous at process temperature but become very thick or may even solidify at the ambient temperature of the wet leg. For such applications, a pressure repeater can be mounted in the vapor space above the liquid in the process vessel, and a liquid-level transmitter (a differential-pressure instrument, for example) can be mounted below in the liquid section at the zero reference level. The pressure in the vapor section is duplicated by the repeater and transmitted to the low side of the level instrument. The complications of the outside wet leg are avoided, and a static pressure in the tank will not affect the operation of the level transmitter. Using a pressure repeater is limited to relatively low-pressure applications. The vessel pressure must be less than the instrument supply pressure and the limits of the 1:1 pneumatic relay. This is shown in Figure 6–15.

If electronic instrumentation is used, two pressure transmitters can be mounted directly on to the vessel, and the difference in pressures can be calculated by the control system and used to infer the vessel level. This application is illustrated in Figure 6–16.

6.14 SUMMARY OF HYDROSTATIC HEAD LEVEL MEASUREMENT

Hydrostatic head or differential pressure provides a simple approach to level measurement. This is because pressure measuring devices are readily available and well understood; the relationship between the pressure exerted by the height of a vertical column of fluid and the corresponding level is well known.

The accuracy of the actual measuring devices, pressure or differential pressure transmitters, can be as good as ±0.1% of rate. It should be noted, however, that the inferred value of a corresponding level may not be as good. This is because of the process conditions and the means by which the measuring instrument is interfaced to the process.

Application problems can occur when the process fluid is highly viscous, likely to freeze or contains solids that can settle in the instrument and impulse lines, forming a gel or polymer in the dead-end cavities. Precautions can be taken to alleviate the resulting circumstances but because head level measurement is dependent on process and wet leg fill fluid densities, changes in process fluid and temperature can cause considerable errors. Such errors that result in changes in LRV and URV values with density changes can be compensated for with multivariable transmitters, but failure to consider this in system design and application can cause a degradation of the measurement.

Notwithstanding these disadvantages, however, head level measurement is a viable technology that has been successful in the water, petroleum, power, and, to a lesser extent, in the food, pulp and paper industries.

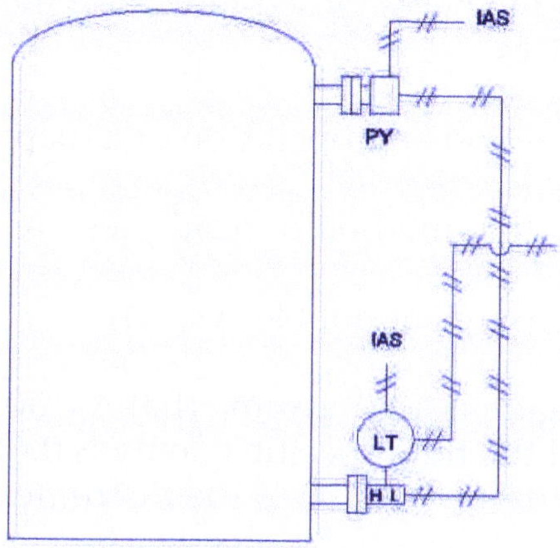

Figure 6–15 Differential Pressure Level Measurement Using a Repeater

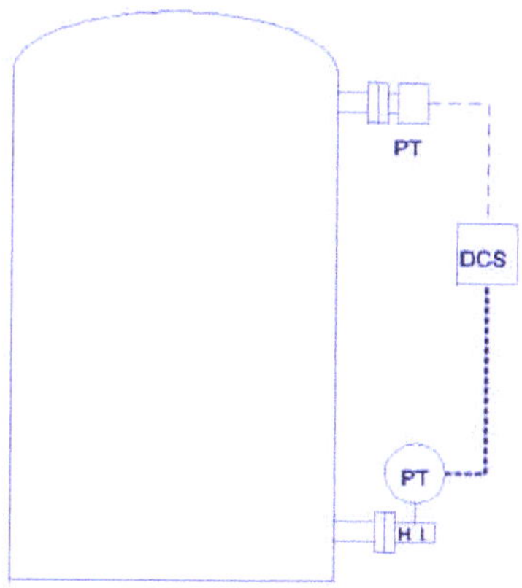

Figure 6–16 Level Measurement Using Two Differential Pressure Transmitters

EXERCISES

6.1 State the principle of hydrostatic head level measurement.

6.2 Define open-tank head-type level measurement.

6.3 A pressure transmitter similar to the application in Figure 6–2 is used to measure level. The specific gravity of the process liquid is 0.80. What is the level when the pressure on the transmitter is 3.60 psi? What is the pressure when the level is 30.0 ft?

6.4 Describe the principle of closed-tank level measurement with head-type devices.

6.5 Explain the function and operation of a diaphragm box.

6.6 What is the air trap method of head pressure measurement?

6.7 What is the air purge or bubble system?

6.8 Referring to Figure 6–12A, assume that A = 50 in., B = 2 in., and C = 10 in., and the fluid specific gravity is 0.6. What amount of suppression is needed? What is the instrument span in inches of water?

6.9 Referring to the process in Figure 6–13, assume that A = 80 in., B = 5 in., C = 85 in., D = 12 in., and G_L = 0.79. What is the instrument span? What is the elevation required?

6.10 Explain the operating principle of diaphragm sealed instruments and list an application.

REFERENCES

1. Anderson, Norman A., *Instrumentation for Process Measurement and Control,* 3rd edition, Radnor, PA: Chilton Book Company, 1980.

2. Lipták, Béla G., Ed., *Instrument Engineer's Handbook: Process Measurement & Analysis,* 4th edition: CRC Press, 2003.

3. Fisher-Rosemount Model 3095 Multivariable Level Controller technical literature

4. Improving Differential Pressure Diaphragm Seal Performance and Installation Cost, A Rosemount White Paper by Michelle Hursh.

7

Electrical Level Measurement

7.1 INTRODUCTION

The level-measurement techniques discussed thus far were those first commonly used in level-measurement applications. Certain process characteristics and considerations can preclude the use of such devices, however. It is then necessary to adapt other level-measuring techniques to these more difficult applications.

7.2 RESISTANCE LEVEL MEASUREMENT

Because of the simplicity of electrical resistance transducers and the ease of measuring a resistance quantity and converting that measurement to an inferential value representing a process condition, resistance devices may be used to measure level. One such device utilizes a resistance-tape sensing element (Figure 7–1).

The level sensor is in the form of a tape that is about 1.5 in. wide and 0.3 in. thick. Its length can vary from about 4 to 100 ft. The sensor has an internal helical resistance winding, similar to a slide wire, with one turn per 1/4 in. and a resistance value of about 6 Ω per turn, equivalent to 1 Ω/mm. This wound helix is spaced away from a gold-plated contact or shorting strip, and the two electrical elements are held within a sealed flexible plastic sheath.

The sensor is suspended within the process vessel; when the vessel is empty or at a 0% level condition, the full resistance value of the sensor helix appears across the two lead wires coming from the sensor top. For example, if the sensor is 10 m in length with a resistance/length (or

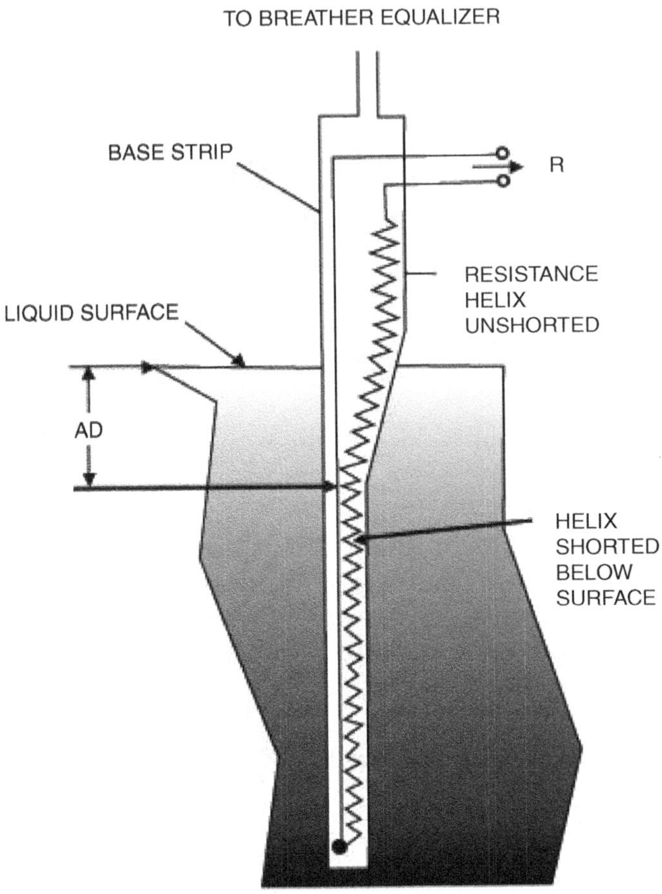

Figure 7–1 Schematic of Resistance Tape Level Sensor

"gradient") relationship of 1000 Ω/m, the total resistance of the sensor is 10,000 Ω.

As the level increases, immersing the sensor in liquid or slurry, the hydrostatic pressure exerted by the liquid on the flexible plastic sheath causes the helical winding to be displaced, pushing against the contact strip and shorting the windings near the liquid surface and below. If the level increase is 25% of the sensor length, the total resistance value of the sensor will be reduced by 25% of the empty tank value. If the sensor is half-submerged in liquid, the resultant decrease in resistance value of the sensor, as measured between the lead wires, will be 50% of the resistance value at 0% tank level. When the sensor is fully submerged to a 100% level value, the full length of the wound helix is shorted and the resistance value of the sensor is nearly 0 Ω. Compensation for the resistance of the connecting lead wires is usually not required because the total sensor resistance value is large compared to that of the lead wires.

It can be seen from Figure 7–1 that the highest point of helix shorting is located at the "actuation depth" (AD), below the liquid or slurry surface (because of built-in helix and jacket spring rate), and that all helix contacts below this point are shorted in parallel to provide a very reliable return current path. The resistance element remains unshorted above the actuation depth, and this portion of the helical winding is measured and related to the level quantity.

A resistance-tape sensor can also be used in closed-tank applications. Process pressure is equalized directly by venting the inner sensor cavity to process pressure using a specially designed capillary breather/equalizer. The static process pressure is applied to both sides of the total jacket system, which acts as a pressure-receiving diaphragm. The process environment is excluded by the breather from the helical cavity so that use in corrosive conditions is possible. Figures 7–2 and 7–3 show the construction and operation of resistance-tape level sensors.

Readout instrumentation is relatively simple because any resistance-measuring circuit can be used. A DC resistance Wheatstone bridge circuit (Figure 7–4) is one choice for the readout device; the electrical components of the detector circuit can be located remotely from the process vessel. Another common practice is to excite the sensor with a constant-current source, thereby converting sensor helix resistance directly to voltage.

The bridge circuit can operate from a null balance or unbalanced condition. A usual circumstance would be to adjust R_3 to balance the bridge for a value of R_X, which represents a maximum value when the process is at a zero-level condition. As the level increases, the value of R_X decreases, causing the bridge to be unbalanced. The bridge detector would be calibrated in level units for direct indications of level change.

The circuit shown is certainly an oversimplification and is not meant to represent an actual measuring circuit. In reality, an amplifier would probably be used to increase the unbalance signal to keep this signal level small—a few milliamps for an entire measurement range. This improves the overall linearity; the amplifier can also be used for scaling purposes. Refer to Figure 4-9 to review this operation.

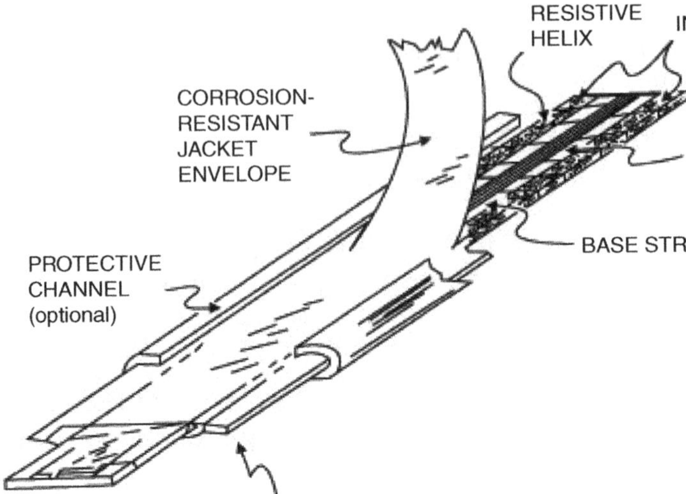

Figure 7–2 Resistance Tape Sensor Construction

For initial design considerations, the bridge balance is defined by the following resistor value relationships:

$$\frac{R_1}{R_2} = \frac{R_3}{R_x} \tag{7.1}$$

Also, for a balanced bridge the voltage across the bridge detecting points is zero ($E_X = 0$). By using conventional equivalent circuit theorems, the bridge unbalance voltage can be expressed as

$$E_x = E_{\text{Source}}\left[\frac{R_1}{R_1 + R_2} + \frac{R_2}{R_2 + R_y}\right] \tag{7.2}$$

The unbalance voltage, E_X, is a result of changes in R_X as the level changes from 0 to 100%. The total change in R_X is determined by the resistance/length relationship of the level sensor. The values of R_1 and R_2 are selected for the desired bridge sensitivity. The following example helps to explain the operation of the bridge circuit as readout for tape-type sensors.

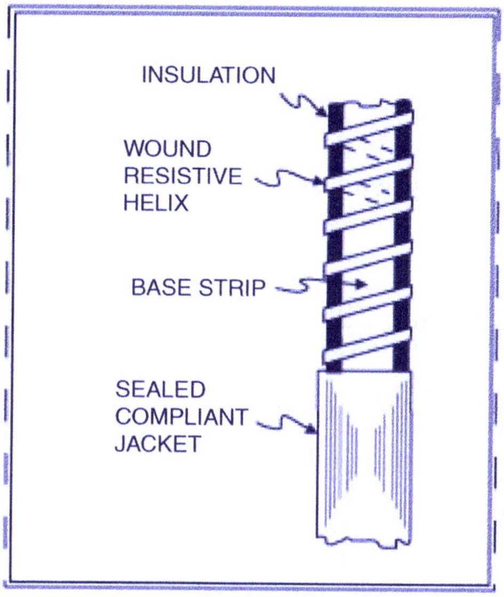

Figure 7–3 Resistance Tape Sensor Operation

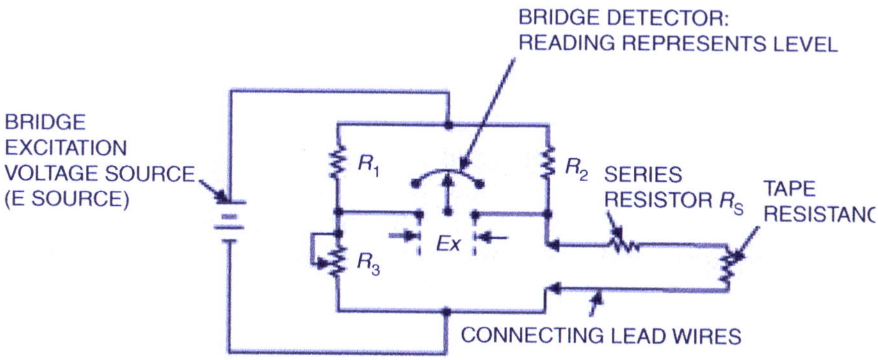

Figure 7–4 A DC Resistance Bridge Circuit

EXAMPLE 7-1

Problem: A bridge circuit similar to that in Figure 7–4 is used with a tape level sensor. The sensor slide wire has a resistance value of 1 Ω/in. and is 50 in. long. The bridge source voltage is 24 V, and the resistance values are $R_1 = R_2 = 10$ kΩ; $R_3 =$ 0-to-100 Ω pot = 55 Ω for initial balance; $R_s = 5$ Ω. Calculate the value of E_x as the level changes from 0 to 100%.

Solution: When the tank is empty, the value of Rx is

$$(50 \text{ in.})(1\ \Omega/\text{in.}) = 50\ \Omega$$

When the tank is full, the value of R_X approaches zero, but the series resistor in the connecting lead wire must be considered in the determination of the unbalance voltage. From Equation 7.2, when $R_X = 50\ \Omega$,

$$\begin{aligned} E_x &= (24\text{V})\left[\frac{10\ \text{k}\Omega}{10\ \text{k}\Omega + 55\Omega} - \frac{10\ \text{k}\Omega}{10\ \text{k}\Omega + 55\Omega}\right] \\ &= 24[(0.9945) - (0.9945)] \\ &= 0\ \text{V} \end{aligned}$$

This represents the initial balance condition when the tank is empty; when the tank is full, $R_X = 0$ and

$$\begin{aligned} E_x &= 24\text{V}\frac{10\ \text{k}\Omega}{10\ \text{k}\Omega + 55\Omega} - \frac{10\ \text{k}\Omega}{10\ \text{k}\Omega + 55\Omega} \\ &= 24(-0.005) \\ &= -0.120\ \text{V} \end{aligned}$$

The fact that the unbalance voltage is negative merely shows the direction of unbalance and is of little significance. The voltage determined in Example 7-1 can be used for control, switching or alarm purposes. When desired, it can be converted to a corresponding digital signal to provide any required digital function. The bridge unbalance voltage can also be scaled for conventional 4 to 20 mA, 1 to 5 V, or other ranges for signal transmissions.

An important feature of the resistance-tape level sensor is that its cross-sectional area is small. It can be suspended in small pipes, 2 to 3 inches in diameter, for water well, sump and tank applications with small access openings. The pipe can also serve as a stilling well in areas where process turbulence may create problems.

Calibration of a resistance-tape gage can be of an analytical nature, using a resistance decade box to simulate sensor values calculated for any defined liquid level, without liquid having to be present or cycled in the tank. By determining the resistance/length relationship of the resistance element, values of resistance related to corresponding level values can be substituted for the tape input to the measuring circuit. This procedure will check the accuracy of the readout device; to check the accuracy of the overall system, however, the tape should be included in the calibration procedure. The level can be gauged at any value, and the accuracy of the entire system verified or the instrument zero (offset) adjusted to achieve agreement.

Resistance-tape sensors are generally immune to errors caused by moderate variations in temperature, static pressure, dielectric constant, conductivity, viscosity or solids content. Although these sensors are generally unaffected by specific gravity, certain qualifying remarks are in order. When such devices are suspended in the liquid or slurry, the hydrostatic head of the liquid compresses the jacket and shorts the helix winding at a point below the actual level. This offset actuation depth (AD in Figure 7–1) between the apparent and actual level values is constant and compensated for during initial calibration. If the material specific gravity changes from the value at the time of calibration, error can result because of the variance in head pressure caused by the specific-gravity variations. If compensation for the offset is neglected, the error will be constant and the response to level changes will not be affected.

As an example, if actuation depth (AD) is 100 mm for water (specific gravity = 1.00), then a doubling of specific gravity would reduce AD to 50 mm on average, changing indicated output by 1 mm for a 2% change in specific gravity. Unlike pressure-type level gages, a change in specific gravity affects only actuation depth and is not related to the liquid level.

7.3 CAPACITANCE LEVEL MEASUREMENT

Capacitance is the property of a circuit that stores electrons and thus opposes a change in voltage in a circuit. A capacitor is an electrical component that consists of two conductors separated by a dielectric, or insulator. The value of a capacitor is measured in farads (F) and is determined by the area of the conductors (usually called plates), the distance between the plates and the dielectric constant of the insulator between the plates.

Generally, one thinks of a capacitor as having two small parallel plates separated by air or another type of dielectric. For capacitance level-measuring applications, however, one plate is typically a probe while the other plate comprises the side of the level vessel, or a reference probe or plate if the vessel is made of insulating material. The dielectric is the material in the vessel. The relationship between capacitance, plate area, dielectric constant, and the distance between the plates is

$$C = \frac{KA}{D} \tag{7.3}$$

where C is capacitance in microfarads, K is the dielectric constant pF/ft, A is the area of the plates, and D is the distance between the plates. With respect to Figure 7–5A, when the switch is closed, a current flows in the circuit; however, because the plates are separated by an insulator, current cannot flow between the plates and a buildup of electrical charge occurs on the plates. An electrical field or electrostatic charge will exist between the plates. In accordance with Equation 7.3, the charge C will vary directly with K and A and inversely with D. This explains the DC characteristics of capacitance and the instantaneous current flow.

By reversing the polarity of the DC source, the current flows in the opposite direction, the electrostatic charge collapses, and the current continues to flow until the negative plate is charged positively and the positive plate receives a negative charge. The current flow changes directions each time the polarity is reversed and flows only from the source to the plate and back to the source. Because of the insulating properties of the dielectric, there is no current flow between the plates.

When an AC voltage source replaces the DC source (Figure 7–5B), the voltage will reverse every half-cycle and a continuous current flow will result, although the term continuous implies a reversal of the current direction. This explains the AC characteristic of current. It may be seen, therefore, that a capacitor blocks DC current flow and passes AC. For AC capacitance applications, the frequency of the applied voltage affects the AC opposition (impedance) and current flow and is a consideration for AC capacitance circuits. As the frequency increases, the impedance decreases and the current flow increases.

A capacitance level-measurement probe in a tank is shown in Figure 7–6. It is apparent that for a particular application the area of the plates and the distance between them must both be fixed values. The dielectric constant varies when the level changes from A to B in the vessel, and this variation is used to produce a signal proportional to level. Level values are then inferred from a measurement of the capacitance resulting from a change in dielectric constant.

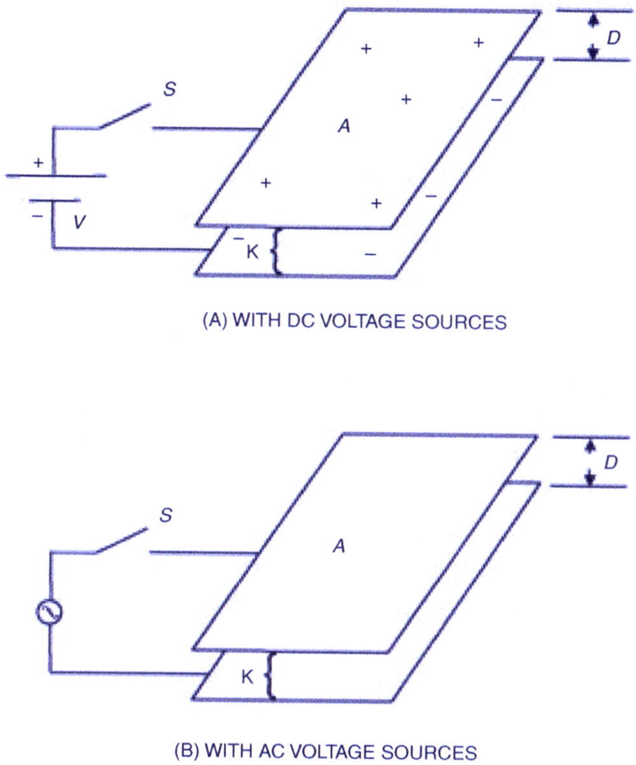

Figure 7–5 Capacitor Formed by Two Plates
(*Courtesy of Arjay Engineering Ltd.*)

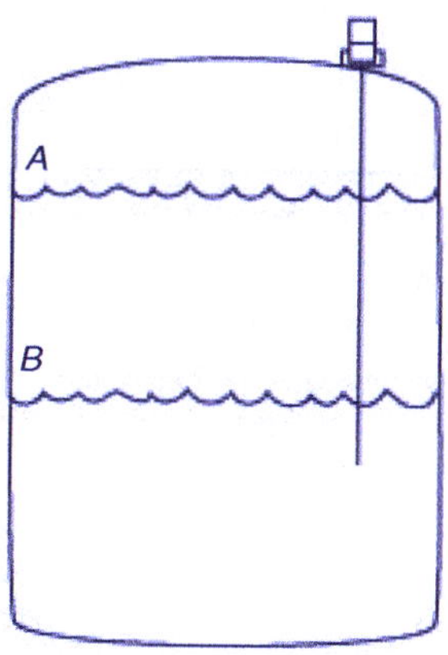

Figure 7–6 Capacitance Probe in a Closed Tank

Capacitance level-measurement probes are used in on/off applications for alarms or switches and control functions. For such applications, the probes are usually mounted horizontally, parallel to the level surfaces, so that a large plate area is effectively used. For liquid surfaces, small round rods are sufficiently large, whereas some solid measurement applications may require probes with larger plate areas. In any case, the best measurements are obtained when capacitance differences are high—a condition that results when the difference between the dielectric constants of the two media is relatively high. This is because the change in capacitance with level variations is a direct function of the dielectric constant and is greater as the level of the measured material (high K value) replaces the air or vapor (low K value) in the empty portion of the level vessel. The dielectric constant of air and most gases is 1. Values for solids are listed in Table 7–1, and those for granular and powdery materials are given in Table 7–2.

The dielectric constants of liquids (Table 7–3) vary with temperature. An increase in a material's temperature tends to decrease its dielectric constant. Obviously, wide changes in material temperature will result in dielectric changes that can cause significant measurement errors; therefore, temperature compensation is required if temperature variations are expected.

When measuring the level of a nonconductive material, a bare metal probe can be used. Figure 7–7 illustrates this application. The total system capacitance, C_E, is determined by

$$C_E = C_1 + C_2 + C_3 \tag{7.4}$$

where

$$C_2 = \frac{0.614(K_a)(L - l)}{\log_{10}\frac{D}{d}}$$

$$C_3 = \frac{0.614(K_p)(l)}{\log_{10}\frac{D}{d}}$$

and C_1 is gland capacitance[1], C_2 is vapor phase capacitance, C_3 is liquid phase capacitance, K_a is the dielectric constant of the vapor phase, K_p is the dielectric constant of the liquid phase, L is vessel height, l is level height, D is the diameter of the vessel, and d is probe diameter. When

Table 7–1 Dielectric Constants of Solids (*From Applied Instrumentation in the Process Industries, Vol. 1, 2nd Ed., by William G. Andrew and H. B. Williams, © 1979, Gulf Publishing Company, Houston, TX. Used with permission, all rights reserved.*)

Material	Dielectric Constant	Material	Dielectric Constant
Acetic acid (36°F)	4.1	Paper	45.0
Aluminum phosphate	6.1	Phenol (50°F)	2.0
Asbestos	4.8	Polyethylene	4–5.0
Asphalt	2.7	Polypropylene	1.5
Bakelite	5.0	Porcelain	5–7.0
Barium sulfate (60°F)	11.4	Potassium carbonate (60°F)	5.6
Calcium carbonate	9.1		
Cellulose	3.9	Quartz	4.3
Cereals	3–5.0	Rice	3.5
Ferrous oxide (60°F)	14.2	Rubber (hard)	3.0
Glass	3.7	Sand (silicon dioxide)	3–5.0
Lead oxide	25.9		
Lead sulfate	14.3	Sulfur	3.4
Magnesium oxide	9.7	Sugar	3.0
Mica	7.0	Urea	3.5
Napthalene	2.5	Teflon®	2.0
Nylon	45.0	Zinc sulfide	8.2

the dielectric constants of the vapor and liquid materials are constant, the system capacitance is a function of l, the level height.

For conductive material applications, insulated probes must be inserted into the vessel (Figure 7–8). The system capacitance is determined by

$$C_E = C_1 + \frac{C_2 C_4}{C_2 + C_4} + \frac{C_3 C_5}{C_3 + C_5} \tag{7.5}$$

1. Gland capacitance is the capacitance of the insulator between the probe and the tank for bare probes used with nonconductive fluids; for insulated probes, it is the capacitance of the probe insulation.

Table 7–2 Dielectric Constants of Granular and Powdery Materials
(Courtesy of Gulf Publishing Company)

Material	Dielectric Constant	
	Loose	**Packed**
Ash (fly)	1.7	2.0
Coke	65.3	70.0
Gerber® Oatmeal	1.47	Not tested
Linde 5A molecular sieve		
Dry	1.8	Not tested
20% moisture	10.1	Not tested
Polyethylene	2.2	Not tested
Polyethylene powder	1.25	Not tested
Sand, reclaimed foundry	4.8	4.8
Cheer®	1.7	Not tested
Fab® (10.9% moisture)	1.3	1.3
Tide®	1.55	
VEL® (0.8% moisture)	1.25	1.25

where the values are as shown in Figure 7–8.

Probe sizing is important in capacitance level applications. The differential capacitance over the range of level measurement determines the input span of the capacitance-measurement instrument. A differential capacitance of 10 pF or greater is generally desirable for good resolution and accuracy. For on/off measurement used in alarm and switching relays, spans as small as 2 pF are acceptable. Another important consideration is that the ratio between differential capacitance and terminal capacitance should be kept in the range of 0.25 to 4. Terminal capacitance is the measurement when the material in the vessel is at zero level.

The selection and installation of a capacitance probe are illustrated in Example 7-2.

EXAMPLE 7-2

Problem: An insulated capacitance probe coated with Teflon is to be used to measure propane. The gland capacitance, C_g, is assumed to be

Table 7–3 Dielectric Constants of Liquids (*Courtesy of Gulf Publishing Company*)

Material	Temp (°F)	Dielectric Constant	Material	Temp (°F)	Dielectric Constant
Acetone	71	21.4	Hexane	68	1.9
Ammonia	–30	22.0	Hydrogen chloride	82	4.6
Ammonia	68	15.5	Hydrogen sulfide	48	5.8
Aniline	32	7.8	Isobutyl alcohol	68	18.7
Aniline	68	7.3	Kerosene	70	1.8
Benzene	68	2.3	Methyl alcohol	32	37.5
Bromine	68	3.1	Methyl alcohol	68	33.1
Butane	30	1.4	Methyl ether	78	5.0
Carbon dioxide	32	1.6	Naththalene	68	2.5
Carbon tetrachloride	68	2.2	Octane	68	1.96
Caster oil	60	4.7	Oil, transformer	68	2.2
Chlorine	32	2.0	Pentane	68	1.8
Chloro-cyclohexane	76	7.6	Phenol	118	9.9
Chloroform	32	5.5	Phenol	104	15.0
Cumene	68	2.4	Phosphorus	93	4.1
Cyclohexane	68	2.0	Propane	32	1.6
Dibromobenzene	68	8.8	Styrene (phenylethene)	77	2.4
Dibromohexane	76	5.0	Sulfur	752	3.4
Dowtherm®	70	3.36	Sulfuric acid	68	84.0
Ethaol	77	24.3	Tetrachloro-ethylene	70	2.5
Ethyl acetate	68	6.4	Toluene	68	2.4
Ethylene chloride	68	10.5	Trichloroethylene	61	3.4
Ethyl ether	–40	5.7	Urea	71	3.5
Ethyl ether	68	4.3	Vinyl ether	68	3.9
Formic acid	60	58.5	Water	32	88.0
Freon-12	70	2.4	Water	68	80.0
Glycerin	68	47.0	Water	212	48.0
Glycol	68	41.2	Xylene	68	2.4
Heptane	68	1.9			

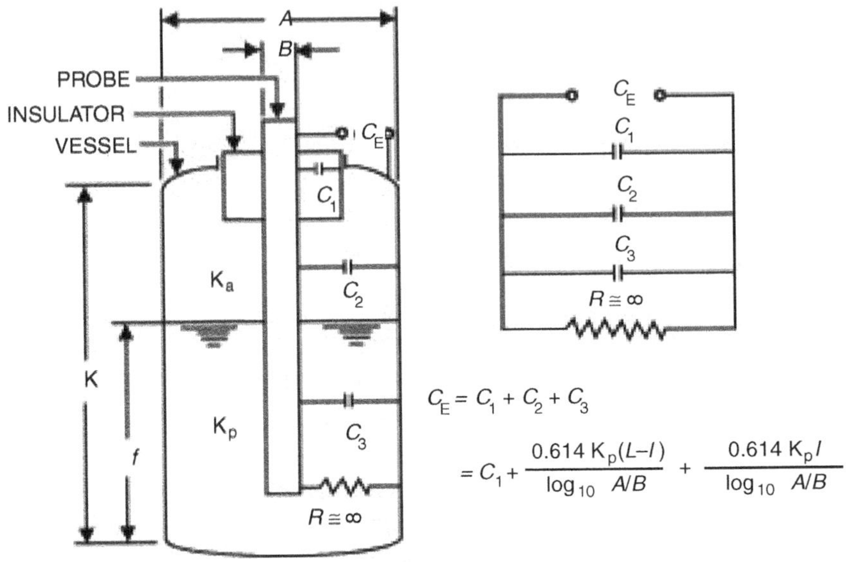

(A) CAPACITOR PROBE IN A TANK (B) EQUIVALENT CAPACITANCE

Figure 7–7 Bare Probe in Nonconductive Fluid

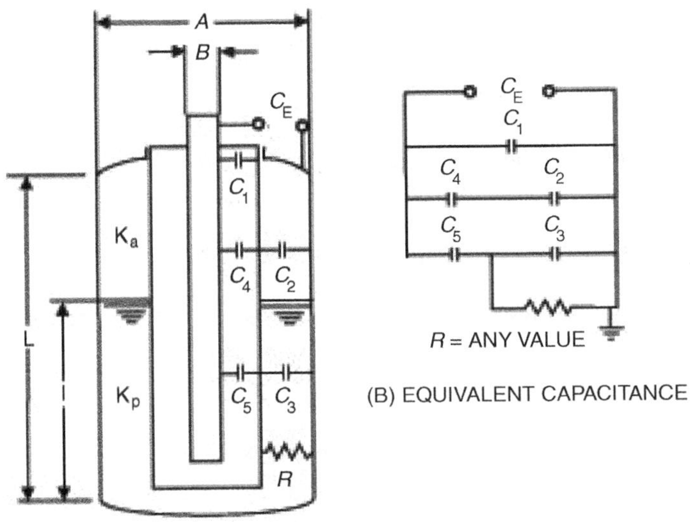

(A) CAPACITANCE PROBE IN CONDUCTIVE FLUID

Figure 7–8 Insulated Probe in Conductive Fluid

12 pF, the effective probe range is 5 ft, the dielectric constant of the vapor space above the liquid is 1, and the vessel depth is 7 ft with a capacitance C_E = 112 when the vessel is empty. Determine whether a No. 3 concentric probe is suitable for the application.

Solution: The value of C_E is determined by using a value of pF = 1 for the value of K and by finding the intersect point of the No. 3 concentric probe curve with the L\C axis in Figure 7–9. The value at the intersect point is 15 pF.

$$\begin{aligned} TC &= C_g + C_E \\ &= 12 + 112 \\ &= 124\text{pF} \end{aligned} \tag{7.6}$$

By again entering the chart in Figure 7–9 for the dielectric constant of propane (1.6), the differential capacitance is 22. The net ΔC is found by subtracting 15 pF/ft from 22 pF/ft = 7 pF/ft.

The total change in capacitance when the vessel is filled is

$$\Delta C = \frac{7\text{ pF}}{\text{ft}}(5\text{ ft}) = 35\text{ pF} \tag{7.7}$$

The established guidelines for a good installation are

$$\Delta C > 10\text{ pF} \tag{7.8}$$

and

$$4 < \frac{TC}{\Delta C} < 0.25$$

The value determined for the example is within the acceptable range of the selection guidelines:

$$\Delta C = 35\text{ pF and } \frac{TC}{\Delta C} = \frac{124}{35} = 3.54\text{ pF} \tag{7.9}$$

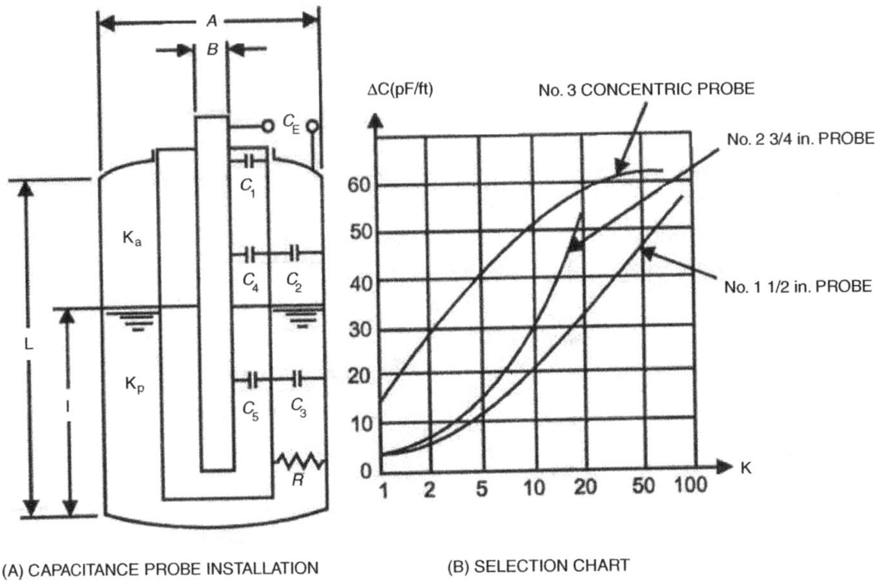

Figure 7–9 Capacitance Probe Selection Chart

7.4 CAPACITANCE MEASUREMENT TECHNIQUES

Various methods of capacitance measurement and detection are used in capacitance probe applications. The variable probe capacitance can be connected to one leg of an AC bridge circuit, and level variations will result in a bridge unbalance. The amount of unbalance can be measured and related to a corresponding level value. A general AC bridge is shown in Figure 7–10.

The operation of the bridge depends on the fact that when certain specific circuit conditions apply, the detector current becomes zero. This is known as the null or balance condition. Since zero current means there is no voltage difference across the detector, the voltage from point *a* to point *b* and from point *a* to point *c* must be equal. It then follows that

$$I_1 Z_1 = I_2 Z_3 \tag{7.10}$$

Also, the voltage from point *d* to point *b* and from point *d* to point *c* must also be equal

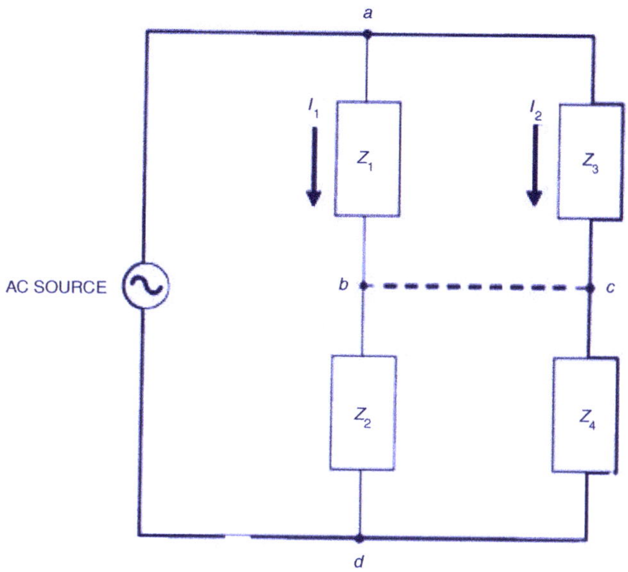

Figure 7–10 General AC Bridge Circuit

$$I_1 Z_2 = I_2 Z_4 \tag{7.11}$$

Dividing Equation 7.10 by Equation 7.11,

$$\frac{Z_1}{Z_2} = \frac{Z_3}{Z_4} \tag{7.12}$$

Equation 7.12 defines a bridge balance condition and applies to any four-arm bridge network, regardless of whether the branches are pure resistance or combinations of resistance, capacitance, or inductance. It should be noted that the ratios of impedances are not affected by the magnitude of the AC voltage source or the actual values of branch currents. The impedance values are complex, however, and are frequency-dependent.

A null or balance condition can be obtained only when both the magnitude and the phase angle of each of the four impedance branches satisfy Equation 7.12; that is, if the bridge is to be balanced, both the real components and imaginary (*j*) components of impedance must be balanced simultaneously. This is the reason for two variable resistance values in the capacitance bridge in Figure 7–11. When

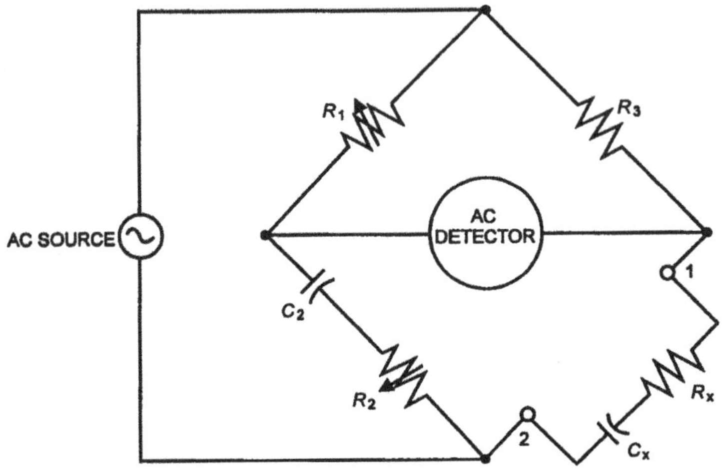

Figure 7–11 Capacitance Bridge

the bridge is not balanced, Equation 7.12 is not valid, and conventional equivalent circuit theorems or node equations must be used to solve for the voltage and current.

The simple form of AC bridge circuit shown in Figure 7–11 is used for capacitance measurement and is known as a capacitance comparison or series resistance capacitance bridge. The impedance arms of the bridge can be written as

$$Z_1 = R_1 \tag{7.13}$$

$$Z_2 = R_2 - jX_{C_2} \tag{7.14}$$

$$Z_3 = R_3 \tag{7.15}$$

$$Z_4 = R_x - jX_{C_x} \tag{7.16}$$

By substituting these values into Equation 7.13 and rearranging,

$$R_1(R_x jX_{C_x}) = (R_2 - jX_{C_2})R_3 \tag{7.17}$$

Equation 7.17 can be simplified by multiplying through and then grouping the real and imaginary terms:

$$R_1R_x - jR_1X_{C_x} = R_2R_3 - jR_3X_{C_2}$$

The real and imaginary terms must be equal at the same time for a balanced condition:

$$R_1R_x = R_2R_3 \tag{7.18}$$

$$(-jR_1)\left(\frac{1}{wC_x}\right) = (jR_3)\left(\frac{1}{wC_2}\right) \tag{7.19}$$

By further simplification,

$$R_1C_2 = R_3C_x \tag{7.20}$$

By solving Equations 7.19 and 7.21 for the unknown quantities R_X and C_X,

$$R_x = \frac{R_2}{R_1}R_3 \tag{7.21}$$

and

$$C_x = \frac{R_1}{3}C_2 \tag{7.22}$$

In the above expressions, the frequency dependence mentioned earlier has canceled out of the equations, and the bridge is not dependent on either the frequency or the applied voltage for balanced situations.

In practice, the C_X bridged component is a capacitance probe with a defined capacitance/level relationship. The R_X term, of little significance in capacitance probe applications, is the inherent DC resistance of the capacitor. When operating from null conditions for a particular level value, the bridge is balanced and the C_X value for a balanced condition is determined from Equation 7.22. By knowing the capacitance/level relationship of the probe, the level value can be found.

EXAMPLE 7-3

Problem: Find the level on a capacitance probe with a 5 pF/ft characteristic. The bridge component values for a particular application are $C_2 = 100\ \mu F$, $R_1 = 10\ k\Omega$, $R_2 = 100\ k\Omega$ and $R_3 = 50\ k\Omega$.

Solution: From Equation 7.22,

$$C_x = \frac{R_1}{R_3} C_2$$

$$= \frac{(10 \times 10^3)(100 \times 10^{-6})}{5 \times 10^3} = 20\ \text{pF}$$

and for the given probe characteristics,

$$\frac{20\ \text{pF}}{5 \frac{\text{pF}}{\text{ft}}} = 4\ \text{ft}$$

Another type of capacitance-measuring circuit utilizes a crystal-controlled oscillator that generates a 200-kHz signal used in a switching network. This switching circuit generates square waves that are applied alternately to charge and discharge two capacitors. One capacitor is the level capacitance probe, and the other is a reference or zero-adjust capacitor. The capacitors are discharged into a detecting network, and the sum of their capacitance discharge currents is output current that is applied to a DC amplifier. The gain of the amplifier sets the span of the instrument and typifies the negative feedback principle; the amplifier gain is controlled by regulating the amount of negative feedback.

Such a circuit is shown in the block diagram in Figure 7–12. By proper adjustment of the balance and zero controls, a wide range of values for capacitance probes can be selected. When the output current is at the zero reference level, the zero and balance capacitance values are equal to the total capacitance of the probe. This establishes a balanced condition; when the probe capacitance changes, the output current changes accordingly.

Capacitance level measurement devices play an important role in the manufacturing and processing industries, but under some operating

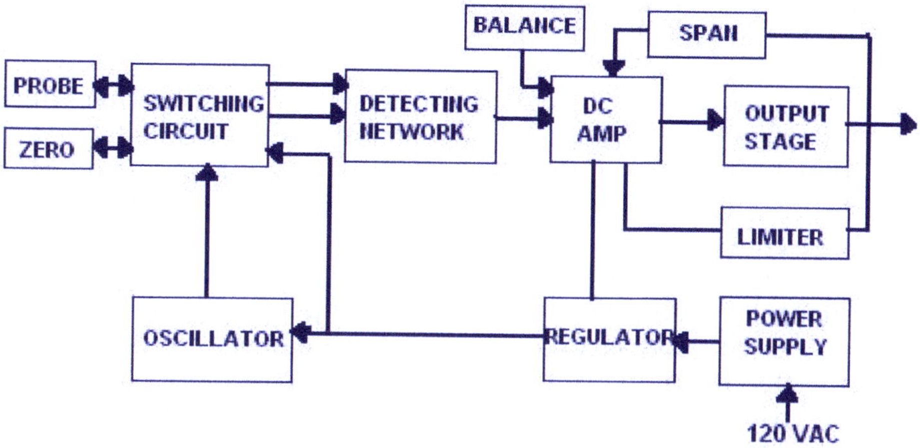

Figure 7–12 Block Diagram of a Capacitance Detecting Circuit

parameters they require special consideration. Generally, the capacitance values of a probe should be high compared to the capacitance of the lead wires connecting it to the measuring circuit. The wire capacitance should be swamped by the probe capacitance variations. Vendor-supplied extension wires for connecting the probe to remote mounted electronics are recommended to minimize the effect of the wiring on the measuring circuit; otherwise, instability and erratic operation may result. At other times, however, especially where maximum resolution is desired, the bridge must respond to very small values of capacitance change. Such operation requires special bridge design to negate the effect of stray capacitance.

Capacitance level devices offer a number of advantages:

- They contain no moving parts.
- Simple and rugged design is possible.
- They can be designed for corrosive applications.
- They are generally easy to clean.
- Sanitary design is available for the food processing industry.
- Extreme temperature and pressure requirements can be made possible by careful design.
- They can be designed for explosion-proof service.

As is true for most measuring systems, capacitance level techniques do have certain limitations. The following are some of the most notable:

- Measurement is subject to error caused by variables that affect the dielectric constant of the material to be measured. For a specific application the dielectric may be stable, but changes in process temperature, chemical or physical composition, moisture content, particle size and packing density can change the dielectric constant, causing a substantial error in the measurement. For example, an increase in temperature can cause a decrease in dielectric constant by as much as 0.1% per degree Celsius. To compensate for changes in dielectric, correction can be provided by a special probe design with two elements or with two probes. One of the probes actually measures the dielectric constant of the process material and a microprocessor-calculated algorithm uses the most recently measured dielectric constant to determine the level.
- Coating of the probe by conductive material causes error in measurement. An example of this is the condensation of water vapor in the vapor phase of a vessel, forming water droplets on the probe.
- Empirical calibration techniques are usually required.
- In solids measurement, variations in particle size and packing affect the dielectric constant.

7.5 APPLICATION CONSIDERATIONS

Many standard probes are coated with Teflon and thus have high chemical resistance, yet may be damaged by some highly corrosive materials. Erosive materials may cause mechanical abrasion, puncture the insulating material or otherwise limit use. For dry material such as powders and pellets, a bare probe is acceptable, and insulation material is not a major consideration. Active probe material can be stainless steel. Since the theory of capacitance level measurement depends on a material with a constant dielectric, any composition or process change will cause erroneous measurement and the instrument will need to be recalibrated.

When the tank is used as one of the plates, the tank material must be conductive; a reference capacitance shield is not needed unless the probe is installed too far from the tank wall and the dielectric is low. If tank dimensions result in a nonlinear reference (e.g., an oval tank) or if the tank is made of plastic, fiberglass or another electrically insulating material, a ground reference or shield must be used.

Tank pressure and temperature extremes can result in measurement error. Humidity variations can also cause a zero shift and inaccurate measurement. Circumstances to be avoided include material agitation, liquid splashing on the probe, condensation of vapor in the vapor space that causes droplets to form on the probe, and liquid draining over the probe from a fill line. A time delay to suppress the effect of liquid splashing and rippling is recommended, and a splash guard or stilling well may be required. Long probes may need to be secured to prevent movement, which would affect the capacitance reading by changing the distance between the plates.

The two basic types of capacitance level measurements are single point (on/off) and continuous. The single-point method of measurement is used for alarm points and for on/off control to start and stop pumps, open and close valves, and so on. Continuous level measurement is usually provided by a 4–20 mA output signal proportional to some specified level range.

This type of measurement is used for inventory control, reservoir monitoring, continuous process control and some batching operations (although single-point measurement is often adequate for batch control). The measurement span is dictated by the probe length inserted in the vessel. The measurement is the percent of level against the vertical probe and not the percent of tank volume. With proper signal conditioning and a given tank geometric configuration, a readout for percent tank volume may be obtained. Figure 7–13 illustrates various probe configurations.

7.6 INSTALLATION CONSIDERATIONS

As with most sensors, to improve performance and for a more successful operating system, special precautions in installation practice should be taken into account. The capacitance probe should be installed in such a way that material entering or leaving the vessel does not hinder the probe's operation. The impact of material striking the probe can cause measurement error and possible damage to the probe and insulator. In low dielectric material applications, it is important that the entire probe be covered, which is also a possible consideration in mounting.

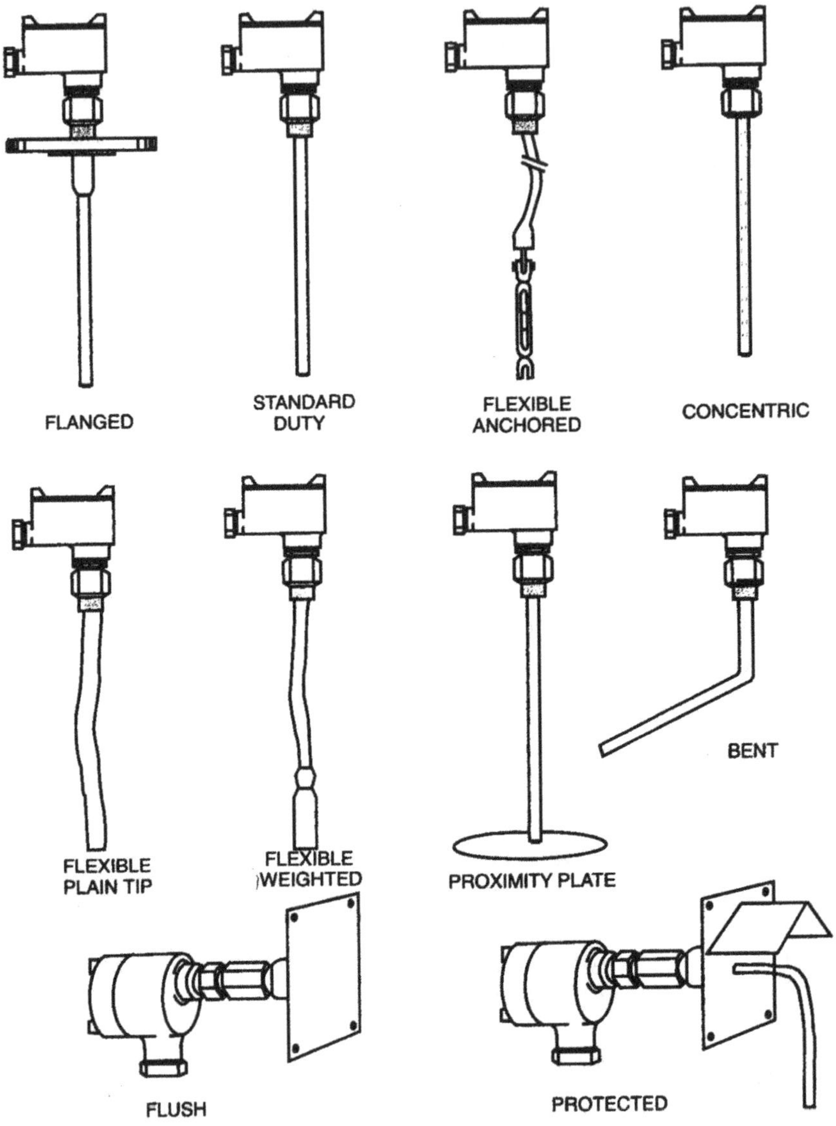

Figure 7–13 Various Probe Configurations (*Courtesy of Arjay Engineering Ltd.*)

The probe should be installed such that contact is not made with the vessel walls or any other structure in the vessel. When cable extensions are used, care should be taken so the probe does not contact the vessel. Also, the material should not form a bridge between the probe and vessel wall.

Some applications have more than one probe; in such cases, the distance between the probes should be no less than about 18 inches to reduce the possibility of interference between the electrostatic fields. When probes are installed through the sides of the vessel, protective covers should be installed to protect the probe from the weight of the process material when the weight may be excessive.

7.7 PROCESS CONSIDERATIONS

The success of a capacitance level system will depend in large measure on the overall process features and characteristics for a specific application. The specification and selection of capacitance level instrumentation must adhere very closely to process constraints and conditions. An application data sheet, ISA's specification form ISA-20.27, or some other check sheet should be used. Such forms will list the important criteria to be considered; many of the topics included on a specification sheet for capacitance level instruments are listed in the sample specification data sheet shown in Table 7–4.

Capacitance transmitters are typically calibrated by raising and lowering the level of the process fluid when the fluid is at normal operating conditions. After the electronics are successfully calibrated, the electronics can be connected to a variable capacitor and the settings for 4 mA and 20 mA outputs can be recorded. If the electronics require replacement, the new electronics can be calibrated using the variable capacitor, eliminating the need for process level changes to calibrate the transmitter.

7.8 MATERIAL TO BE MEASURED

The physical and chemical description of the process material must be considered for equipment selection. Most probes used in standard applications are Teflon-coated and have high chemical resistance to withstand corrosive chemicals. It is always recommended, however, that the process material be checked against the manufacturer's application data requirements to ensure that highly corrosive chemicals will not damage the probe. Physical abrasion and corrosive materials may erode or puncture the probe. Guards or other protective measures may be necessary.

Since the theory of capacitance level applications relies on consistent detection, any composition change of the process fluid that results in a dielectric change will affect instrument calibration and produce erro-

Table 7–4 Capacitance Specification Sheet Data Requirements

Data for Specification of Capacitance Level Instruments
Specify the following: • Stream description and/or pipe size and number or vessel number in which probe is installed • Solids level, liquid level, interface, foam detection, or other special material properly (chemical and physical). Is composition stable? • Probe model if known • Probe orientation: horizontal axis, vertical axis, other • Probe material (316 SS, etc.) • Sheath, if required (1/4-in. Teflon®, etc.) • Total immersion in inches, or in feet and inches • Whether electrodes are mounted at probe or remotely located • Length of special compensation cable to be furnished with probe • Upper fluid by name (state as liquid or vapor) • Dielectric constant of upper fluid • Lower fluid by name if interface application • Dielectric constant of lower fluid if interface application • Percentage moisture content of solids • Whether material is expected to build up on probe • Vibration environment of probe (mild, severe, etc.) • Tank construction material • Process fittings: construction type and material • Instrument enclosure • Explosion proof requirement of probe head • Linear ground reference availability and/or requirement

neous measurements. Recalibration is required when the material dielectric changes.

7.9 TANK CONSTRUCTION MATERIAL

Most probes require a ground reference in order to operate effectively and free of outside interference. When tanks are constructed of material that is not electrically conductive, a ground reference or shield must be used. The ground reference must be installed on either side of the nonconductive tank wall. Even for tanks constructed of electrically conducting material, reference shields are needed when the probe is installed too far from the wall and the process dielectric is low. Shields also should be used if the tank wall does not act as a linear reference (e.g., tanks with nonuniform sides).

7.10 TANK PRESSURE AND OPERATING TEMPERATURE

When temperature and pressure variations are great enough to cause significant changes in the dielectric, erroneous readings can result.

7.11 HUMIDITY CHANGES

Humidity changes can cause zero shift. For such conditions, the probe area exposed should be held to a minimum to reduce the amount of zero calibration error.

7.12 MATERIAL AGITATION

When tank agitation causes a liquid rise around the probe, the probe will detect this level instead of the true tank level. Liquid splashing and draining over the probe from the inlet or by other means also should be eliminated. Time-delay techniques can suppress the effects of such interferences.

Although capacitance level measurement is used in many applications, the market share of these installations is decreasing, due to the use other technologies that are relatively immune to changes in the dielectric constant.

7.13 RADIO FREQUENCY ADMITTANCE LEVEL MEASUREMENT

When using a capacitance probe to measure level, significant error can result from material coating the probe. Radio-frequency (RF) probes have been developed and used to improve the performance of capacitance level measurement (Figure 7–14). Although capacitance level applications have been available for many years, RF technology—which depends on the dielectric constant of the process material—can be used for point and continuous level measurement in both conductive and nonconductive process materials. Applications of RF technology include liquids, slurries, granular material and interfaces. The physical properties and chemical composition of the process material have little or no effect on measurement results. These improved operation features were brought about by anti-coating circuitry or driven shield technology, phase shifting, and other advanced characteristics.

To solve coating problems, the electrode is designed to ignore the effect of conductive material buildup. This is accomplished by adding a second element to the electrode that is electronically maintained or

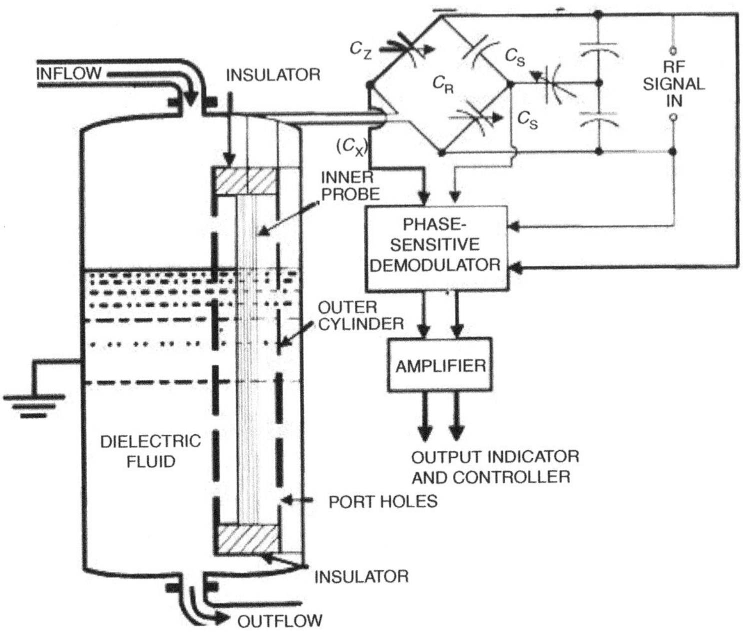

Figure 7–14 Capacitance Level Measurement System

driven ("driven shield" technology) at the same voltage and frequency as the measuring element. A driven shield element for point measurement in conductive material is shown in Figure 7–15.

With conventional probes, a buildup of material causes a current phase shift between the capacitance plates, but because there is no potential difference across the two sections, no current can flow through the coating to the vessel wall. When the conducting process fluid reaches the bare electrode detecting section, current passes through the material being measured to the vessel wall, completing the detection circuit. Without the second element in the probe, current can flow through the coated section to the vessel wall and cause a short circuit of the measuring section of the electrode, as would be the case in conductivity technology.

For continuous applications, measurement corresponds to the length of the portion of the probe that is submerged in the liquid or granular material. A signal is generated that is proportional to the rise of level on the detecting element. Methods used to minimize coating error include proper electrode selection, higher frequency measurements, phase shifting and conductive component subtracting circuits.

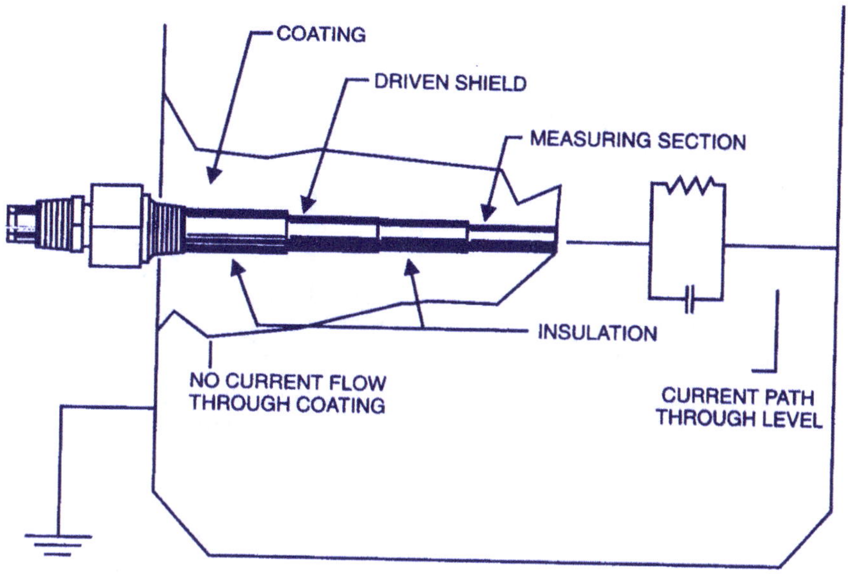

Figure 7–15 Driven Shield Element in Conductive Material
(*Courtesy of Great Lakes Instruments, Inc.*)

While the technology just described for point measurement relies on the special design of the probe, continuous level measurement incorporates an anti-coating circuit in the transmitter. RF capacitance admittance-type transmitters use a phase shifting circuit based on conductive component removal in the error signal to allow a pure capacitive measurement. Coating error is illustrated in Figure 7–16. The submerged portion of the electrode generates a nearly perfect capacitance or susceptance component. The conductive component is eliminated because the electrode is insulated. However, the upper section of the electrode is coated with conductive material and generates an error signal with a capacitive and conductive component. Readings resulting from the electrode are an admittance component that is 45° out of phase with a pure capacitance level signal. An equivalent circuit for the coated section is shown as a ladder network producing the phase-shifted error signal.

Two methods have been used to cancel the error signal. In method A (Figure 7–16), by measuring the conductance component *C* and realizing the 45° phase relationship, the capacitive error component *E* is of the same magnitude as *C* and can be subtracted from the total output from the electrode. The error caused by the conductive component has

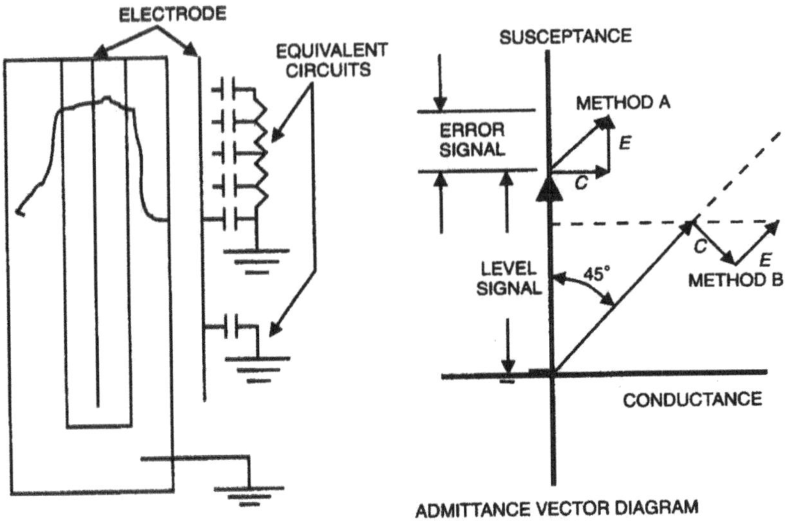

Figure 7–16 Coating Error (*Courtesy of Great Lakes Instruments, Inc.*)

been canceled. In method B the probe coating error is nullified by introducing a 45° phase shift to the entire measurement. This is possible because the conductance component *C* has the same magnitude as the capacitive error component *E*.

The vector diagram in Figure 7–16 illustrates the inverse of impedance for the coating error components. The total error seen by the probe will consist of a pure capacitance and a pure resistive component. Equal combinations of the two components are shown in admittance vector diagrams.

RF admittance technology can be used to measure the level of material with nearly any dielectric constant. If the material is insulating, the dielectric constant should be stable. If it is semiconducting in nature, can it be made to seem conducting by lowering the measurement frequency and thereby raising the capacitive reactance of the probe saturation capacitance? These issues will determine which of two types of RF admittance transmitters to use: level measurement only or a dual channel type to measure both level and material properties. The dual channel type can be used in all continuous level systems, but is needed in about 10% of all such applications.

Table 7–5 can be used to select the proper type of sensor. Prior testing and properly noting the characteristic properties of a particular mate-

Table 7–5 Single Versus Dual Measurement
(*Courtesy of Great Lakes Instruments Inc.*)

Material Characteristics	RF Admittance Transmitter Type
Insulating (stable dielectric)	Single measurement
Insulating (variable dielectric)	Dual measurement
Semiconducting (stable)	Single (low-frequency) measurement
Semiconducting (variable)	Dual measurement
Conducting (stable or variable)	Single measurement
Unknown (e.g., nondedicated vessels)	Dual measurement

rial to be measured will facilitate instrument selection. When the material types are unknown with respect to dielectric constant, tests may be required. When in doubt, use a dual-type transmitter.

As with other capacitance probes, RF admittance probes must be selected in accordance with process conditions. The primary considerations are temperature, pressure, mechanical strength and chemical compatibility. Rod diameter and insulation material determine the strength and process compatibility of a particular probe.

RF admittance technology can be used for extreme temperature ranges—300 to 1000°F—and most electronic transmitters can be used in environments with temperature ranges of –40 to 140°F. The dielectric constant of some material varies substantially with temperature, and this variation can affect the capacitance generated by the probe. These temperature effects are often given in appropriate tables and should be considered. Generally, materials with higher dielectric constants are more stable with respect to temperature change. The moisture content of granular material can also affect the dielectric constant.

Mounting positions must be carefully considered. They must be clear of flow into or out of the process vessel, as splashing can cause serious fluctuations of electrode capacitance. Side-mounted electrodes for point level applications should be mounted at a downward angle to allow any accumulation of liquid to drain from the electrode surface. Vertically mounted electrodes should be clear of all obstructions and should be far enough from the vessel wall to prevent material buildup to the wall.

Material such as air-conveyed granular solids and dust can cause a static charge to build up, which can damage circuit components. Although most manufacturers have designed a static-charge discharge capability to protect their equipment, a ground gage is sometimes required.

Foaming is generally undesirable in all capacitance applications and can cause problems. Sometimes the foam should be measured, while other times it may be better to ignore foaming effects. RF technology will support either or both, but consideration should be given to frequency selection and probe design.

Explosion hazards are always an important factor in any instrument selection and application. While many materials contain explosive gases or vapors in the vapor portion of a vessel, the low energy requirements of RF admittance usually make this instrument suitable for most explosive concentrations.

The selection of an electrode for a particular application will be based on all of the listed criteria. Finally, the capacitance versus dielectric graph provided by the manufacturer should be consulted to make an electrode selection that will yield satisfactory results for the desired range of level measurement.

The RF admittance probe is designed to overcome some of the limitations of capacitance probes and can be used for a wider range of level-measuring applications. RF admittance technology preserves many of the advantages of capacitance probes (e.g., no moving parts, immunity to many process fluids and mounting constraints, and so on) and offers additional advantages as well.

RF level-measurement technology has been used successfully in the following applications [Ref. 1]:

- Lined vessels with grounded shell
- Plastic vessels
- Differential level control
- The presence of high electrostatic charge
- Interface measurement

7.14 CONDUCTANCE LEVEL MEASUREMENT

The ability of a liquid to conduct electricity is a property that can be used in level measurement. The use of conductance-probe level detection is limited to water-based liquids of relatively high conductivity, such as brine solutions, acids, caustic solutions, steam drums and certain types of beverages. Conductance level measurement is usually limited to point level applications, such as alarm devices and on/off control.

A conductance level controller supplies a low voltage, usually less than 20 V, to a partially insulated probe. The operation is similar to an electrode circuit, with the probe acting as one electrode while another electrode is mounted on a wall of the process vessel. The resistance between the electrodes is nearly infinite when the liquid is below the probes, and a measurable value when the liquid rises to form a conductive path between the electrodes. The resistance of the process fluid covering the electrodes is detected by a DC Wheatstone bridge circuit. The unbalance bridge current or voltage is amplified and used to operate a relay or to forward-bias a transistor into saturation. The material being measured must maintain its conductance above the threshold level required to produce the bridge current for detection.

Consideration must be given to characteristics of the measured fluid that cause foaming, splashing, turbulence and solid particles. Protection cages, stilling wells and other devices may be needed to produce damping effects and to prevent splashing. The probes or electrodes should always be mounted so that the process fluid will clear or run off in low-level conditions.

Conductance level measurement and control provide a simple, low-cost means for point level control and detection for conductive materials.

7.15 SONIC AND ULTRASONIC LEVEL MEASUREMENT

In level-measuring applications where contact of the measuring instrument with the liquid in the process is undesirable, a sonic or ultrasonic device may be an option. These types of level-measurement instruments really measure the distance from one point in the vessel, usually a reference point, to the level interface with another fluid. The general operating principle of both sonic and ultrasonic devices is similar, and distinction between the two will not be made other than to define the frequency range of ultrasonic instruments to be around 20 kHz and that of sonic level instruments to be around 10 kHz and below.

7.16 PRINCIPLE OF OPERATION

Vibration of a device causes nearby objects to vibrate. This transference of vibration or motion throughout a medium is known as sound. Sound travels in the form of a wave with a characteristic frequency and velocity that are constant in a given medium. Sound waves also reflect at the angle of the incident wave.

An ultrasonic level measurement takes advantage of these properties of sound. Because sound travels at a constant, known velocity at a given temperature, the time between the transmit burst and the detection of the return echo is proportional to the distance between the sensor and the reflecting object. From this time, the distance between the two can be calculated by the relationship

$$\text{Distance} = \text{Rate} \times \text{Time} \tag{7.23}$$

where Rate is the speed of sound in air, and Time is half the time from transmit until echo detection.

The speed of sound through air varies proportionally with temperature. The temperature sensors used in ultrasonic level monitoring are usually located in the transducer and indicate the temperature of the air through which the ultrasonic sound wave is traveling. This temperature information is used to compensate for the variation in the speed of sound in the air, 0.17%/°C, which is in turn used in the distance calculation.

The minimum amount of change or increment that can be determined is called the resolution. Resolution in ultrasonic measurements is dependent on the incident frequency; the greater the frequency, the smaller the increment that can be resolved. Resolutions for various frequencies are given in Table 7–6.

An ultrasonic sensor generates ultrasonic energy that radiates outward. The sensor, however, has an area where the energy is higher in intensity. The boundary of this high-intensity area is defined as the point where the field intensity is reduced by one-half, or 3 decibels (dB), which is the unit of the log-arithmetic expression of a ratio. This is usually expressed in degrees of angle. The user must be aware of how

Table 7–6 Resolution at Various Frequencies

Frequency, kHz	Resolution, mm
23	14.4
40	8.2
100	3.3

the ultrasonic beam is spreading so that nearby objects or surfaces are not inadvertently detected instead of the intended target.

The sensors used for ultrasonic tank-level monitors pulse at high frequencies several times a second. After the initial transmit burst, the transducer sustains a mechanical vibration that decays exponentially. This is referred to as "ringing." The length of time required to allow for this ringing results in a measurement dead band. The transmit burst travels through the air until it encounters an obstruction, the "reflector," which reflects part of the signal back toward the sensor. The part of the reflected transmit burst that is detected by the sensor is referred to as the "echo." Between the decay of the ringing and the reception of the echo, the receiver signal is not completely quiet. There are relatively small, random fluctuations referred to as "noise."

In order for the instrument to accurately measure level, it must be able to determine the absolute center of the echo. The echo must be detected and interpreted as valid data and to do this, the instrument must determine whether the echo meets the following qualifications:

- It must occur within the instrument's set zero and span.
- It must meet or exceed a certain minimum duration.
- It must be greater in amplitude than the set "threshold."

These qualifications help to ensure that only the "real" echo is processed as valid data (Figure 7–17).

Ideal conditions are not always present during ultrasonic level measurement. Several types of problems are usually encountered: foaming, liquid turbulence, blocking of the signal by other objects, excessive ringing of the transducer and extreme electrical noise (line noise, arcing contacts, etc.). Although each condition presents a

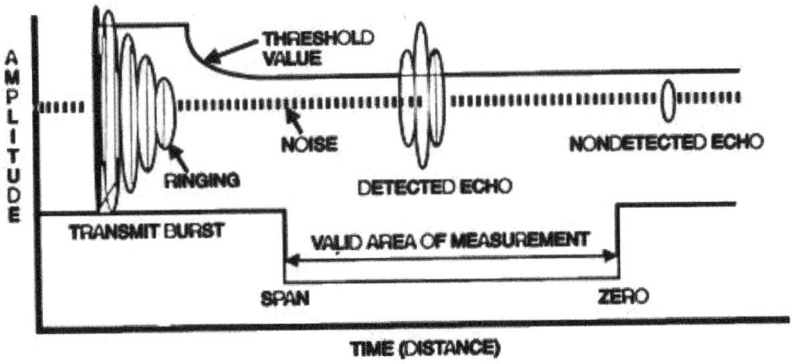

Figure 7–17 Ultrasonic Signal Composite

unique problem to the instrument, special signal interpretation and application techniques can create solutions.

Because the instrument depends on a reflected sound wave, it is important that the transducer be mounted perpendicular to the surface of the liquid or solid whose level is to be measured. Even a slight misalignment can significantly degrade instrument performance. Some instruments display the strength of the echo in the form of a digital number. The first few digits represent the amplitude of the echo, and the digits after the decimal point are related to the number of pulses counted or the return echo's pulse width. By maximizing the echo strength amplitude, proper alignment is achieved. The echo width is used to determine whether the instrument is receiving the exact number of echo pulses that it is sending. It should be noted that highly absorbent surfaces, such as heavy foams, may sufficiently absorb the sound waves to the point of eliminating or restricting the use of ultrasonic level-measurement techniques.

A standpipe can be used as a waveguide to make an ultrasonic measurement under adverse conditions, such as extreme turbulence, intermittent foaming, a partially obstructed echo path, or mounting of the sensor near a tank wall. The wave travels down the standpipe, reflecting off its walls at the angle of incidence. These internally reflected waves will not be detected as false level echoes because the wave's center will travel faster, down and back up the standpipe. If a break in the standpipe is present or the standpipe has a jagged internal surface, the reflected wave from these imperfections can return before the target wave. The false reflected wave can be detected incorrectly as a valid echo, and the instrument can lock up at this point. The use of a smooth standpipe eliminates this problem.

Foaming can have unpredictable effects on an ultrasonic measurement. Depending on the properties of the foam, the echo may be reflected off the top of the foam; reflected from somewhere within the foam; absorbed by the foam completely, resulting in echo loss; or be unaffected by the foam.

Generally, foaming is undesirable for reliable and accurate ultrasonic measurement and is best avoided. In an application with foaming, the solution may be as simple as relocating the sensor to a portion of the process that is not subjected to the presence of foam. In other cases, a standpipe can be used to shield the measurement from the foam. In some instances, however, foaming may restrict or eliminate the effective use of the ultrasonic technique for measurement.

Liquid turbulence is another problem that may cause fluctuating readings or weakened echoes, and reduce the usable range. In applications with turbulence, adjustment of the instrument's damping and echo loss time constant (ELOSS TC) settings will help.

The damping setting selects the amount of digital damping to which the measurement will be subjected. Increased damping will result in a slower, more heavily damped response, which will tend to integrate fluctuations in the reading. Conversely, decreased damping will result in a faster response, making the reading more susceptible to fluctuations with a turbulent surface. Since increased damping will slow the response, it is important to set damping no higher than it needs to be, or the response may be too slow to monitor the level satisfactorily.

If the instrument is sporadically indicating echo loss due to the poor and varying reflectivity of a turbulent surface, increasing the ELOSS TC setting will help. Another reason to increase the ELOSS TC is to prevent interference due to objects that periodically cross into the beam, such as tank impellers or objects floating on the liquid surface. ELOSS TC selects how often or how long a valid echo must be received in order to avoid echo loss indication. If ELOSS TC is set for 7s, then echo loss indication will not occur as long as a valid echo is received at least once every 7s. Conversely, a lower ELOSS TC will mean that valid echoes must be received more often to avoid echo loss indication, thereby making the measurement more stringent. Since a turbulent surface is already a poor reflector, proper sensor alignment is especially critical for applications with turbulence.

Two problems that are integral to the level instrument are noise and ringing, which are controlled by internal settings. These settings can be used to overcome problems in the field that may result from extended sensor ringing or an unusually noisy operating environment.

The ringing setting controls the rate at which the threshold decays from its maximum value at transmit-to-the-target threshold value. This setting helps to ensure that the threshold decays above the ringing illustrated in Figure 7–17.

If normal ringing becomes extended and crosses the threshold, it can be detected and interpreted as a very near echo. An indication of this would be if the instrument always indicated a 100% (or higher) level measurement even though the level was not at 100%. Increasing the ringing setting will prolong the decay time to prevent the detection of the extended ringing. This has a trade-off. As the threshold decay is stretched out, near echoes will have to be stronger to be detected. This is because at any given point, the threshold will now be higher than it was before the ringing setting was increased. If the ringing setting is set very high and the reflectivity of the surface is poor, the reflection from a near surface may not be sufficiently strong to be detected because the threshold will now be high soon after transmit. For this reason, it is important to set the ringing setting no higher than it really needs to be.

The noise setting limits the minimum value the threshold can assume. This setting helps to ensure that the threshold stays above the noise illustrated in Figure 7–17. Noise is usually generated by one of four possible sources:

- *Circuit noise.* The circuitry itself generates some electrical impulses that are detected in the receiver and result in noise.
- *External electrical noise.* Other electrical activities in the proximity of the instrument (pumps, motors, arcing contacts, etc.) can generate electrical fields that are detected by the receiver and result in noise.
- *Acoustically induced noise.* The sensor is a crystal device with a resonant frequency of 40 kHz. When impacted physically, it will tend to generate an electrical signal at its resonant frequency. Very loud acoustical noises may sufficiently impact the sensor to cause it to generate weak signals that will appear as noise.

- *Mechanically induced noise.* If the sensor is impacted mechanically, or is mechanically in contact with a vibrating surface, the sensor's crystal will generate 40 kHz of noise.

Noise is always present, but if the sensor is mounted properly its amplitude is usually sufficiently small with respect to the echo that it does not interfere with the measurement. If the amplitude of the noise at the receiver input becomes unusually high and crosses the threshold, it can be detected and interpreted as an echo. An indication of this would be if sporadic random readings were occurring, possibly accompanied by momentary echo losses, or if the reading were floating near or above the 100% reading. Increasing the noise setting will raise the minimum value the threshold can assume. By raising it sufficiently high, the noise will not be able to cross the threshold and thus will no longer be detected.

The trade-off in this case is that as the noise setting is increased, the instrument becomes less able to detect weaker echoes. This diminishes the effective range of the instrument, and the echoes from a poor reflector at moderate distances may be too weak to be detected if the noise setting is too high.

If the noise setting is very high and the reflectivity of the surface is poor, or the surface is relatively distant, the echo may not be sufficiently strong to be detected because the threshold can no longer be dropped low enough to intercept it. If the instrument is operating over a relatively short range and the reflectivity of the surface is good, the noise setting can be raised substantially to achieve much improved noise immunity. The user is cautioned, however, that if the noise setting is increased too much, frequent echo losses may occur and the required range may be unattainable.

Some instruments can evaluate how high the noise setting may be set by displaying the noise level on their digital readout. This is used to set the strength of the echoes being detected under the worst conditions anticipated in the application (longest distance to surface, etc.). The noise setting can then be lowered a bit to optimize noise immunity while retaining sufficient sensitivity to make the measurement.

The method of data presentation is very important in level monitoring. Microprocessors have greatly improved the ability of instruments to manipulate and condition data and have increased the flexibility of modern instrumentation.

The most common need for data conversion is to display the distance or level information in standard engineering units. This is easily accomplished for most common tank shapes through the instrument's software using simple geometrical formulas. Conversion of level to volume, however, is not as easy with irregularly shaped tanks, with nonstandard engineering units, or when the physical dimensions of the tank change as a function of the fill. Most major instrument manufacturers have included in their software a user-generated table that allows for a continuous distance-to-level conversion.

This table also has other uses. For example, one can be made that compensates for the atmosphere through which the ultrasound is traveling. Most instruments base their distance calculations on the time of flight through air. Often the atmosphere is not air; tanks sometimes contain atmospheres of inert gas or organics that have different speeds of sound. A table can be used to compensate for the variation in sound speed through different media. The sides of very large tanks tend to bulge as the level in the tank increases, making the volume measurement based on distance inaccurate. Again, the user-generated table can become a strapping table and compensate for the bulging effect.

Data collection systems that multiplex between a large number of tanks using a computer and special software to control the sequencing are another way to collect and present level data. These systems typically employ the use of a "blind" (no local display) ultrasonic level transmitter that feeds its analog input directly into the controlling computer via an A/D converter. These systems can handle many monitoring points and store the data on computer disks. This type of total level package provides simpler operation, effective and efficient data archiving, and a lower cost per monitoring point [Ref. 2].

7.17 PARASITIC ECHOES

In addition to the true target echo, many other undesirable parasitic echoes are present in a tank. When these echoes appear within the expected echo time frame, they can cause erroneous readings. As mentioned, the obvious parasitic echoes are reflections from obstacles such as pipes, ladders or even rough weld seams on tank walls or standpipes. Other, less obvious parasitic echo sources are described below. They are related to secondary reflected echoes that interact with weak target echoes.

An ultrasonic signal in air is attenuated exponentially as a function of distance from the transmitting transducer. The echo, however, is also a function of the reflecting surface. Wavy surfaces tend to divert part or most of the reflected echo away from the receiving transducer; thus, at times, weak target echoes may be below receiver threshold level and consequently ignored, while secondary parasitic echoes are accepted. Since this is usually an intermittent phenomenon, such parasitic echoes will appear as "spikes" at the instrument level output. Two specific cases based on actual field installations are considered here.

7.18 TRANSDUCER-RELATED PARASITIC ECHOES

In one installation, unexplained intermittent output spikes were observed whenever the level surface was wavy. Investigation of the received signal showed two echoes on an oscilloscope screen: a primary echo, time-wise equal to reflection from the water level, and a secondary echo with a time distance equal to twice that of the primary one. At times, depending on water surface waves, the primary echo appeared to be weaker than the secondary one, resulting in a secondary echo trigger that caused a downscale spike in the instrument output.

Since there were no overhead reflecting surfaces, the transducer itself became the only suspect. How is it possible for an echo to be strong enough to bounce back from the transducer and cause a secondary triggering echo without first being recognized as a primary echo? The answer is based on Snell's law of refraction. In this case, an ultrasonic echo directed toward the transducer face at some angle is refracted within the silicon rubber face, and again within the aluminum horn, toward the transducer wall rather than toward the sandwiched ceramic piezoelectric elements (Figure 7–18). For example, an ultrasonic echo traveling at an incident angle of 3° in air (point *A* to point *B*) will bend to a 9° angle in the face layer (point *B* to point *C*), because sound velocity in silicon rubber is three times faster than that in air. When this ultrasonic wave encounters the aluminum horn (point *C*), it is again refracted, at a 70° angle. (Aluminum sound velocity is about six times faster than silicon rubber sound velocity.) This signal eventually ends up at point *D*, does not reach the ceramic element, and hence is not recognized. At the same time, another component of echo *A–B* is reflected back to the water-wave surface at point *E* and, under some water-wave conditions, will be reflected back to the transducer and recognized as a primary lower liquid level signal, producing a downscale spike.

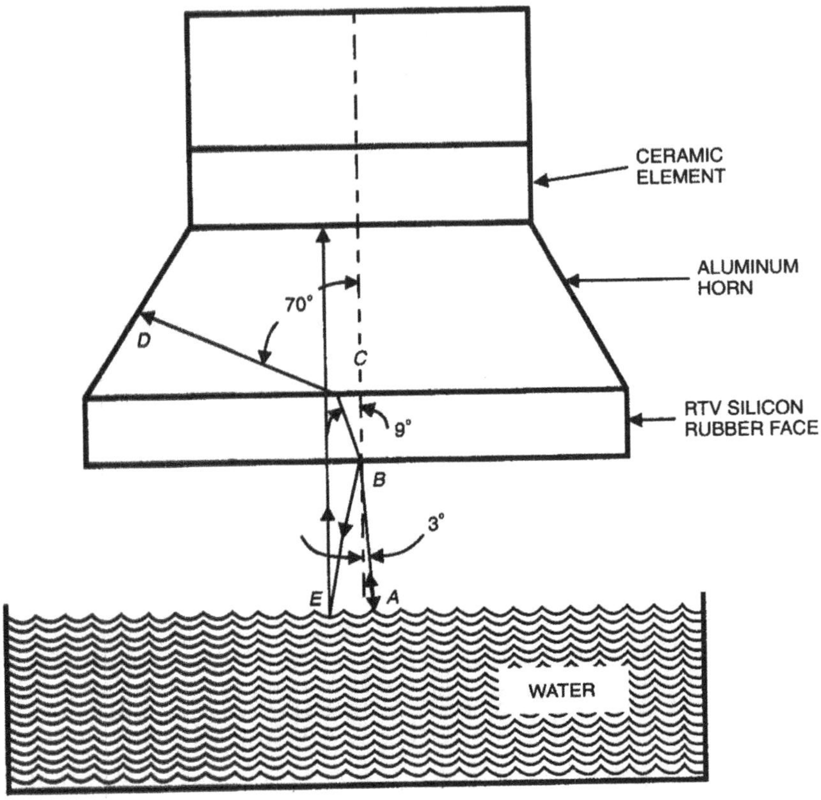

Figure 7–18 Refraction of Ultrasonic Beam

7.19 SECONDARY ECHOES IN COVERED TANKS

Covered tanks are another source of parasitic echoes. This is similar to the previous case, except that here, secondary echoes were reflected from a tank ceiling, as shown in Figure 7–19. In both cases, a wavy surface caused the phenomenon. It is interesting that in one installation of a covered tank, spikes were observed even without waves caused by the filling or emptying of the tank. Here, falling drops of ceiling condensation were enough to cause spike-yielding secondary echoes.

To reject parasitic echoes, they must be distinguished from true target echoes, even though both appear within the expected echo-allocated time slot. The solution is based on "majority vote." True target signals are present most of the time, whereas parasitic echoes appear only

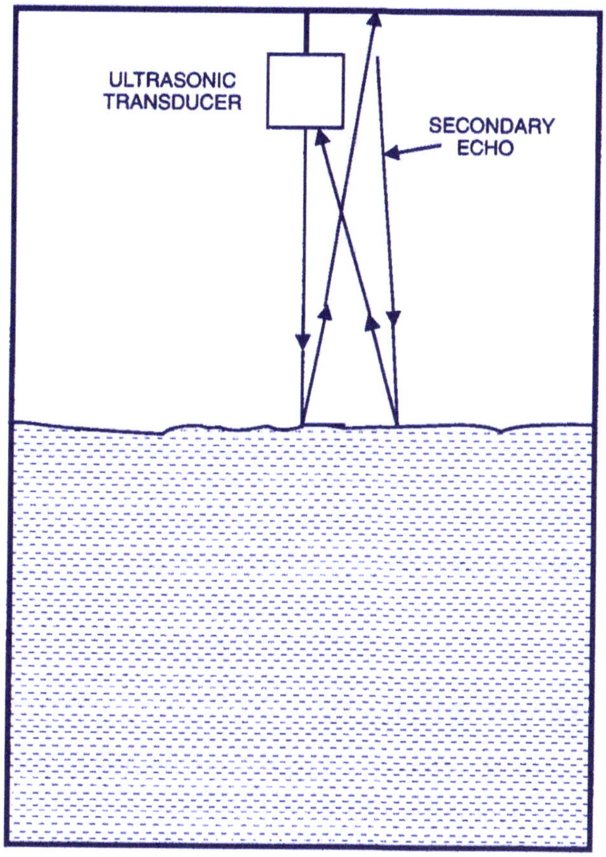

Figure 7–19 Secondary Echo Reflected From Tank Ceiling

intermittently; in this case, because the standard software's running-average filter was insufficient, a new algorithm was developed. This algorithm rejects any new level reading that is a selected percentage different from the established average level. However, to respond to true level changes, consecutive rejected readings are counted; if they exceed a given count, they are then accepted as true level change readings. Any reading below the selected deviation will reset the rejected readings counter and start a new cycle of consecutive rejected readings [Ref. 3].

For further clarification of the operation of continuous measurement of level by ultrasonic methods, Figure 7–20 shows a method of signal generation.

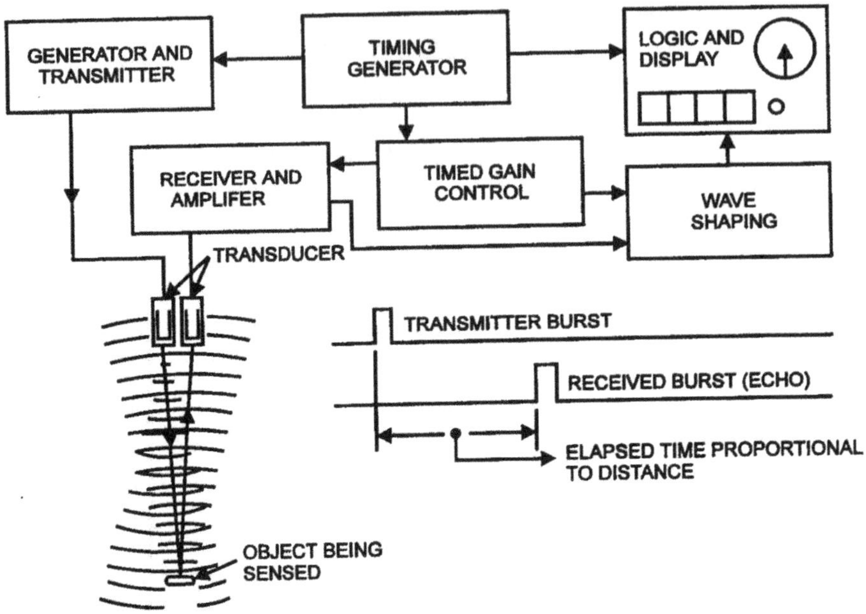

Figure 7–20 Ultrasonic Level Measurement Principles

7.20 POINT MEASUREMENT

Point measurement can be accomplished by two-sensor systems or single-sensor elements. Operation of the former type is such that one sensor emits a signal that travels from one side, through the normal vapor area, to another sensor on the other side. When the level increases to the point where the wave path is interrupted, the strength of the received wave decreases, resulting in the generation of a signal. For dirty fluids and long signal paths, this method normally yields unreliable results.

Operation and reliability can be improved by mounting the transmitter and the receiver of the two-sensor system on the same probe. The distance of the signal path is fixed, regardless of the dimensions of the process vessel. The two-sensor single probe system requires only one vessel entrance and normally consists of piezoelectric transducers in the sensor separated from the transmitter by a 4-in. gap. When the low-energy ultrasonic beam is interrupted by the surface of the process liquid, a control relay is operated for alarm or control action. Two-element systems are shown in Figure 7–21; a two-sensor system mounted on a single probe is shown at the right.

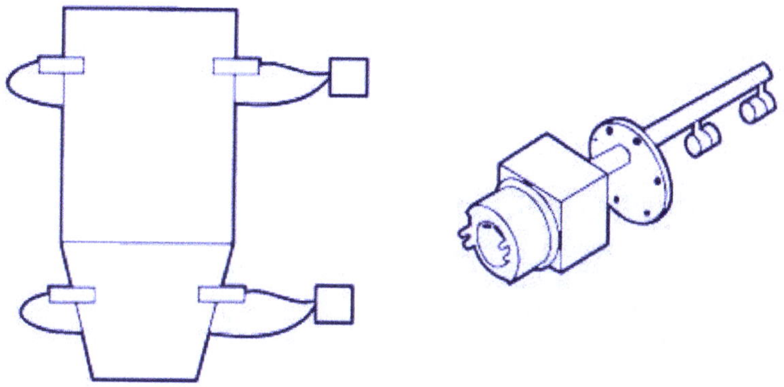

Figure 7–21 Two-Element Systems

One type of single-element sonic system uses an air gap to separate the transmitter from the receiver portion, while the other type presents only one face for material contact. A control unit generates an electrical signal that is converted to an ultrasonic signal by the transmitter crystal. When the gap is filled with liquid, as would be the case when the process level rises to that point, the ultrasonic signal that reaches the receiver element, being stronger than when air or vapor filled the gap between transmitter and receiver, is amplified and operates a control relay. In the absence of liquid, the received signal is too weak to energize the relay. More than one of the signal-element systems can be arranged as shown in Figure 7–22C to represent a progression of level in a process vessel. An interface arrangement is shown in Figure 7–22A, and the probe is shown in Figure 7–22B.

The sensor face of the single-element system shown in Figure 7–23 is energized by an ultrasonic signal that is damped when a liquid contacts the surface area. When the face is dry, a control relay will be energized, and de-energized when the presence of liquid results in damping.

7.21 NONINVASIVE ULTRASONIC SENSORS

Conditions may preclude installation of a level-sensing device within a process vessel. Sensors have been designed to mount on the outside wall of the vessel and can be adapted to various types of applications. These devices transmit ultrasonic signals through the walls or chamber of the vessel. A receiver is also located on the outside wall. The control unit generates and transmits the signal, which will reach the receiver

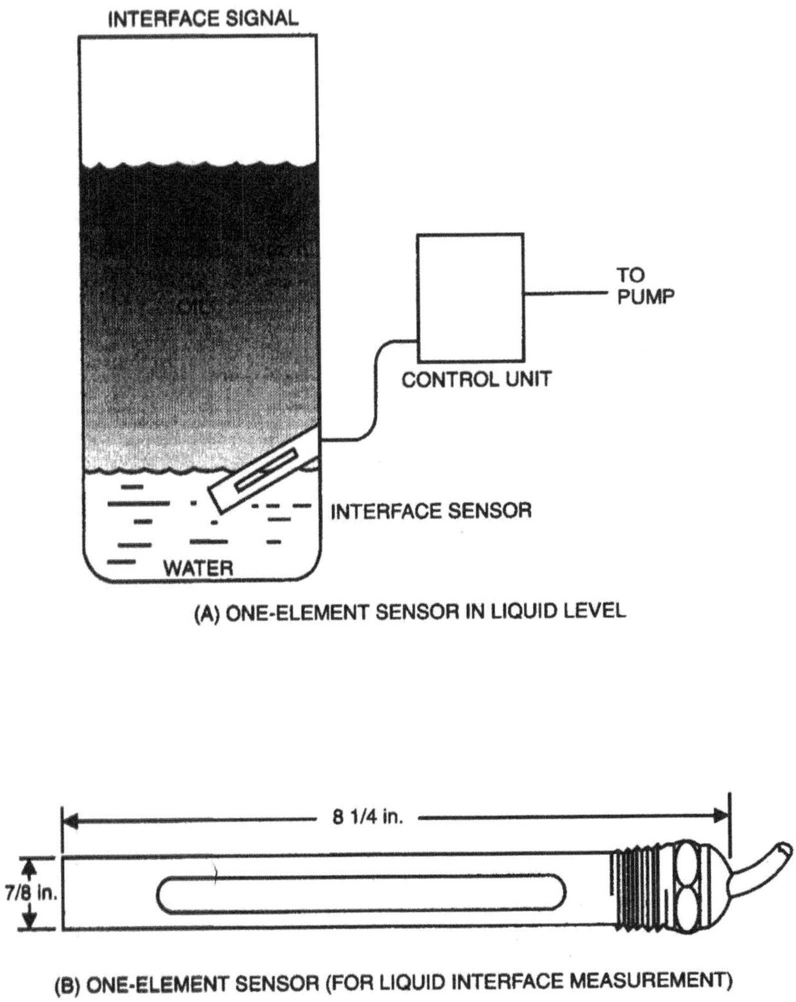

Figure 7–22 A and B One-Element Sensors

when a liquid is present in the wave path. The amplified signal will be sufficient to operate a control relay. When the medium between the transmitter and the receiver is air or vapor, the signal will not be strong enough to operate the relay. Noninvasive devices cannot be used when the vessel walls are of a material that would absorb the ultrasonic signal.

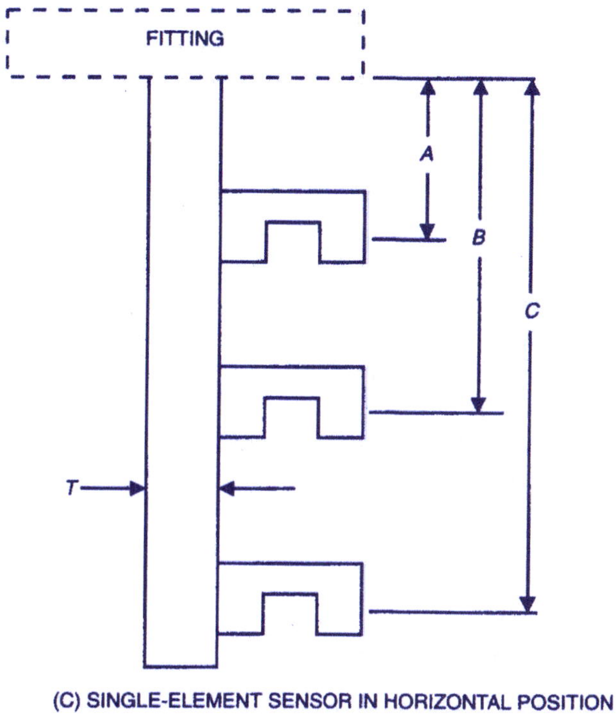

Figure 7–22C One-Element Sensors

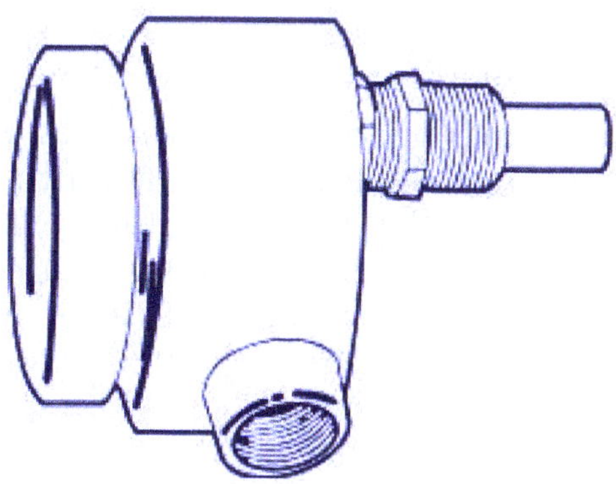

Figure 7–23 A Single-Face One-Element System

Most current ultrasonic instruments include temperature compensation, signal conditioning, self-calibration and other features that can improve the accuracy to better than 0.1%. Intelligent units can perform calculations for various tank geometric patterns and one unit can be configured, through multiplexing, for different applications.

7.22 SUMMARY—ULTRASONIC MEASUREMENT

Ultrasonic level-measurement techniques can be expensive, though the cost is decreasing. They are sometimes more sophisticated than conventional measuring techniques and are therefore generally used only when the application of other techniques would be more difficult and yield less successful results. When the transmitter and sensing probes can be kept clean, they are reliable and relatively maintenance-free because of the lack of moving parts and the fact that the solid-state electronic components are in sealed housings. Except for noninvasive types, and unless isolation chambers are used, maintenance may require the emptying and evacuation of the process vessel. Calibration is usually accomplished by empirical means.

7.23 RADAR LEVEL DETECTION

In 1925, A. Hoyt Taylor and Leo Young used Navy radar (Radio Detection and Ranging) to measure the height of the earth's atmosphere. In the 1930s radar was used for Navy ships and later for aircraft. The first radar sensors were developed in 1976, and in the mid-1980s radar became economically feasible for industrial applications. Radar signals transmit at the speed of light at frequencies in the range of 1 to 300 GHz and at an approximate power level of 0.1 to 0.5 mW/cm^2.

The function of a radar level gage can be described as in Figure 7–24, where the gage and its environment are divided into five parts: microwave electronic module, antenna, tank atmosphere, additional sensors (mainly temperature sensors) and a remote (or local) display unit. The display may include some further data processing, such as calculation of the mass. Normally, the transmitter is located at the top of the vessel and the solid state oscillator transmits an electronic wave at a selected carrier frequency and wave form that is aimed downward at the surface of the process fluid in the vessel. The standard frequency is 10 GHz. The signal is radiated by a dish or horn-type antenna that can take various forms depending on the needs of the specific application. A portion of the wave is reflected back to the antenna, where it is

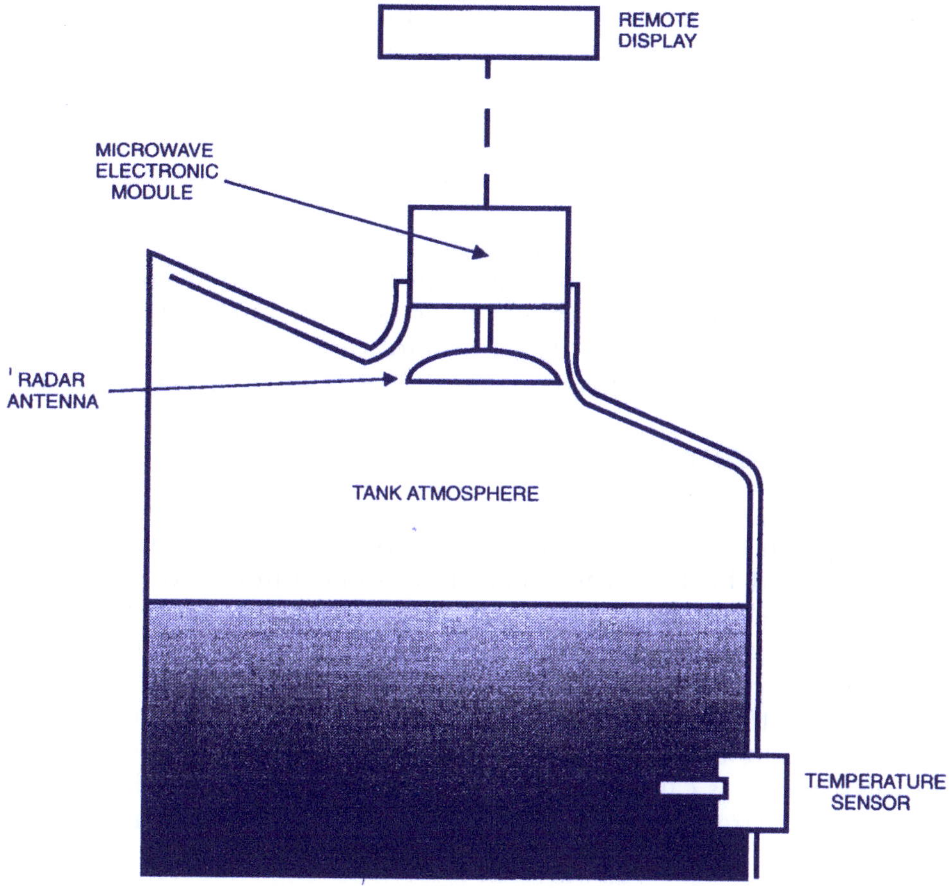

Figure 7–24 Radar Level System

collected and sent to the receiver; there a microprocessor determines the time of flight for the transmitted and reflected wave. Knowing the speed of the wave and its travel time, the distance from the transmitter to the process fluid surface can be calculated. The detector output is based on this difference.

Radar (microwaves) is currently the most widely used method of level gaging onboard tankers, primarily because of the simplified maintenance offered by a noncontact gaging method. In the early 1980s this method was introduced for land-based installations. The two main differences between marine and land-based radar tank gaging systems are that land-based applications require much higher accuracy and must be suitable for retrofit installations. This means it should be possible to utilize existing plant cabling and to mount the radar transmitters without taking the tanks out of service.

7.24 MICROWAVE PRINCIPLE

Radar altimeters have been in use for fifty years. They work on the same principle as a radar tank gage. When the method was adapted for level gaging, however, a number of modifications had to be made to meet special requirements. One major modification involved accuracy, which for a standard aircraft altimeter is 0.5% but is 0.01% for a custody transfer level gage. Also, the hardware had to be suited to a tank and its atmosphere. A radar level system is shown in Figure 7–25.

7.25 PULSE RADAR

In this system, a pulsed radar signal is emitted from the antenna, travels through the vapor portion of the vessel to the reflecting layer, and is reflected back to the receiver in some interval time t_1. The distance to the reflecting layer, which is independent of the radar method used, is determined by the transit time t of the microwave signal. The microwaves travel a distance of twice the distance from the transmitter to the surface of the level. The waves travel approximately 1 ft. in 2 ns. The total time from transmit to receive is divided by two and the distance is calculated, which is the distance of the level from the transmitter/receiver (sometimes referred to as the transponder) (Figure 7–26).

A technical difficulty in pulse radar is measuring the short travel time of the pulse; it can be as short as about 6 ps for a resolution of 1 mm, which is required for custody transfer applications. An extended time scale is normally used by sequential sampling of a periodic signal at continuously delayed points of time so that the signal evaluation can be performed in the range of lower frequencies. The requirement of time measurement accuracies in the picosecond range remains the same.

An illustration of an extended time scale circuit and signal diagram is shown in Figure 7–27. The circuit shown consists of two oscillators that operate on a slight frequency shift, Δf, stabilized by a phase-locked loop (PLL) circuit which sets the frequency at the exact time and sequence. The signal received by a directional coupler is sampled at every following pulse with an exact time delay Δt:

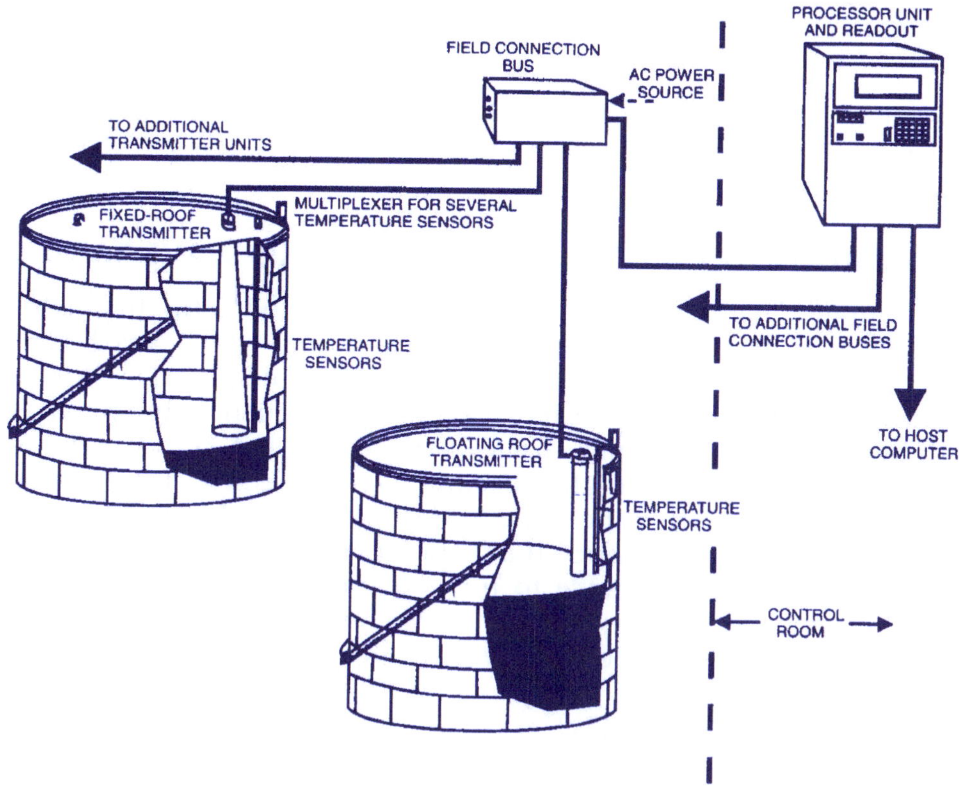

Figure 7–25 System Layout

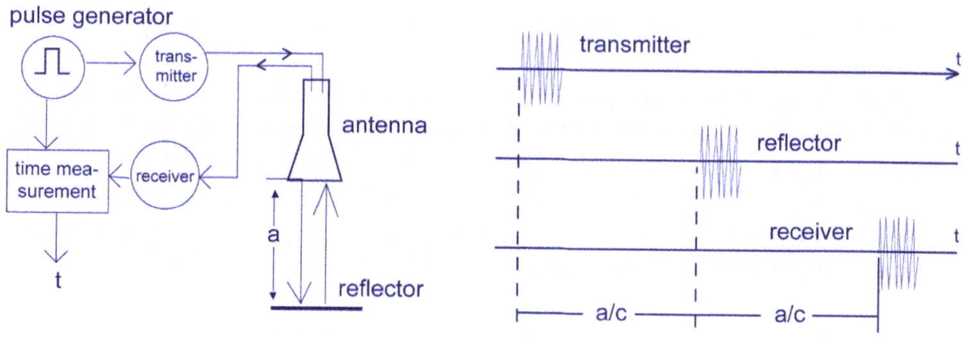

Figure 7–26 Pulse Radar Measuring System (*Courtesy of KROHNE*)

$$\Delta t = \frac{1}{f_0 - \Delta f} - \frac{1}{f_0} \approx \frac{\Delta f}{f_0^2} \quad \text{and} \quad \Delta f \ll f_0 \tag{7.24}$$

The time scale factor is then $f_0/\Delta f$; other terms are noted in Figure 7–27, and Δt is the time shift between the transmitted and received signal.

7.26 FREQUENCY MODULATED CONTINUOUS WAVE RADAR

For accurate radar gaging, the most suitable method is the frequency modulated continuous wave (FMCW) technique, which has higher precision than the better known pulse radar principle. The FMCW method's working principle is based on a radar frequency sweep that is reflected against the liquid surface. The echo that comes back is delayed in proportion to the distance. The reflected signal has a frequency different from the signal transmitted, and the two frequencies mix into a new frequency that is proportional to the distance. As the conversion from distance to frequency can be made with great precision (possibly with an undetermined conversion factor) a calculation of the distance can be made from the converted data with virtually no loss of precision.

This method is illustrated in Figure 7–28, which shows the output frequency of a linear radar sweep.

7.27 SIGNAL EVALUATION FOR FMCW RADAR

In FMCW radar, the information that reveals the measured distance for the level is contained in the frequency of the low-frequency signal of the mixer circuit, which beats the transmitted and received signal together. A counting method to detect the frequency is shown in Figure 7–29. The time interval is between two zero crossings of oscillations, $t_1 - t_0$, $t_2 - t_1$, etc. The signal-to-noise ratio must be reasonably good for best results in the counting method because large interference components can render this method unsuitable. More elaborate methods of frequency measurement are

- *Fourier Transform*—the signal is digitized with sampling at constant intervals followed by a Fourier transformation into the frequency range.

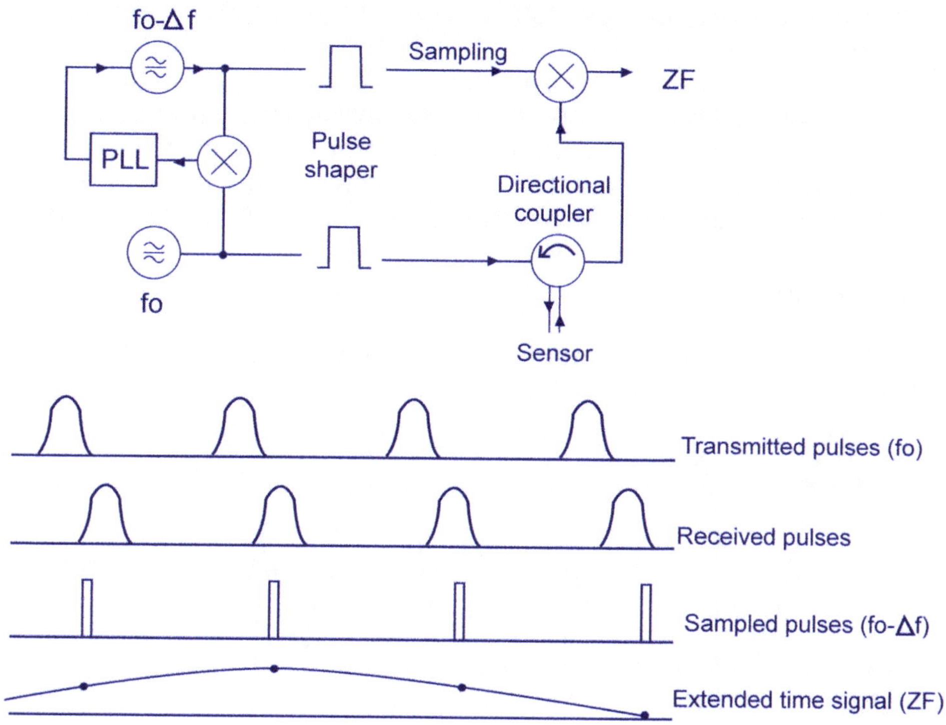

Figure 7–27 Block Diagram of an Extended Time Scale Circuit and Signal Diagram (*Courtesy of KROHNE*)

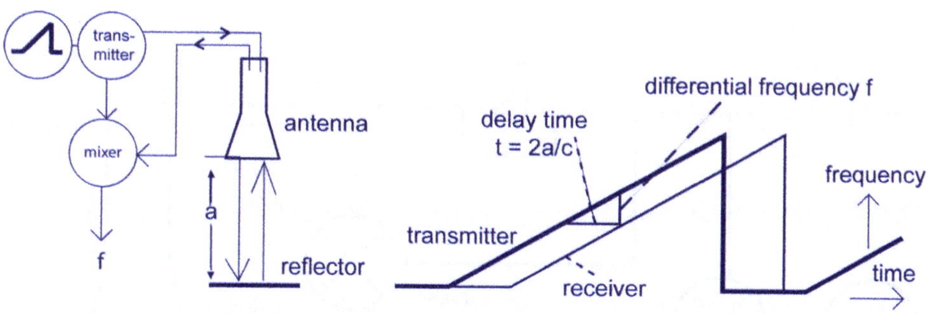

Figure 7–28 FMCW Principle (*Courtesy of KROHNE*)

- *Phase gradient*—phase information related to the transmitted and received signal frequency shift and a special transform are used to determine the time shift. This method yields improved accuracy but a large signal-to-noise ratio is required.

- *Tracking*—the signal is digitized and the frequency is determined by a series of steps that include estimating the frequency then synthesizing a second signal of the estimated signal frequency. This is compared to the measured frequency to create an error from which the deviation of the estimated from the real value is calculated. The corrected frequency can be used as a starting value for the next measurement. The change in frequency and thus the change in level can be accurately tracked.

Figure 7–30 shows a basic diagram of an FMCW system [Ref. 4].

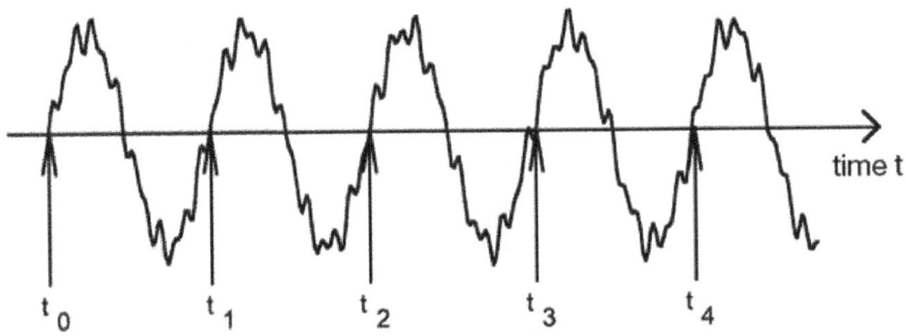

Figure 7–29 Counting Method to Determine Frequency (*Courtesy of KROHNE*)

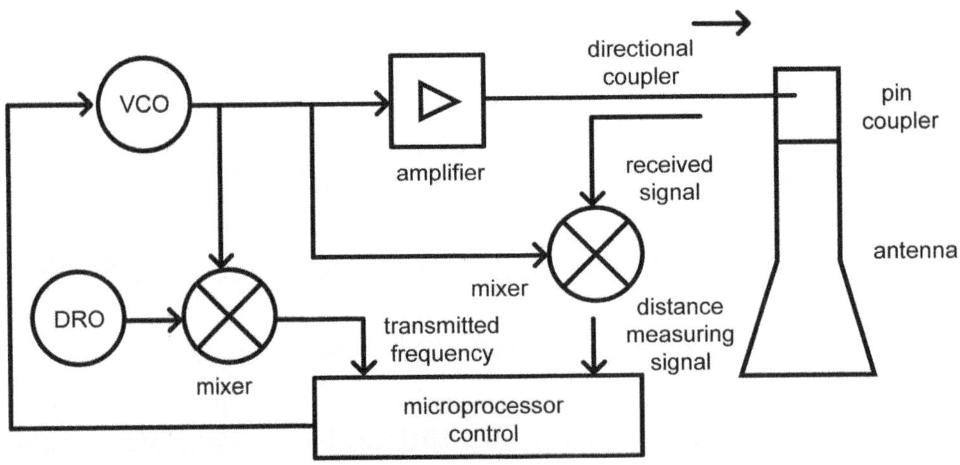

Figure 7–30 Basic Diagram of an FMCW Microwave Circuit (*Courtesy of KROHNE*)

7.28 MICROWAVE ANTENNA

The microwave signal is emitted by an antenna that directs the signal perpendicularly toward the liquid surface. The width of the signal created by the antenna is inversely proportional to the diameter of the antenna. A rather large antenna is often required to produce a beam that is narrow enough to pass between pipes and other obstacles inside the tank. A radar wave length in the "x-band" range (3 cm) and an antenna diameter of 0.2 to 0.4 m give a free-space requirement of a cone with about a 10° opening angle.

The antenna is the only part actually exposed to the internal atmosphere of the tank. Its design is of utmost importance to ensure performance despite contamination from viscous or condensing deposits. Many large crude oil tanks use a still-pipe through a floating roof for level gaging. Obviously, such tanks do not permit a radar beam to pass down to the surface, but the still-pipe can be utilized as a waveguide. The still-pipe has holes or slots, and its inner surface is often covered with rust and deposits. With special precautions, it is possible to avoid the influence of these conditions. The same method is generally used in tanks for liquefied petroleum gas (LPG) and liquefied natural gas (LNG), although such tanks do not use floating roofs.

Several types of antennas are used in radar level measurement (five types of antennas are shown in Figure 7–31). The horn antenna is designed for mounting on small tank nozzles near the tank wall of fixed-roof tanks. A process of signal polarization developed by the manufacturer eliminates the effect of false radar echoes on measurement quality. All liquids can be measured with a horn antenna radar gage, but highly viscous products like asphalt are less suitable for measurement.

A still-pipe antenna is designed to be used on tanks with a still-pipe attached, whether of the fixed or floating roof type. Such tanks are used for petroleum products such as crude oil and gasoline. Even with rusty and deposit-coated still-pipes, the accuracy required for custody transfer applications can be obtained. This antenna has a removable wave guide to facilitate any process sampling that may be required.

The parabolic antenna is large and is constructed from stainless steel and PTFE (Teflon). It is somewhat immune to problems caused by sticky and condensing products and can be used in applications ranging from very clean products to asphalt and dusty products.

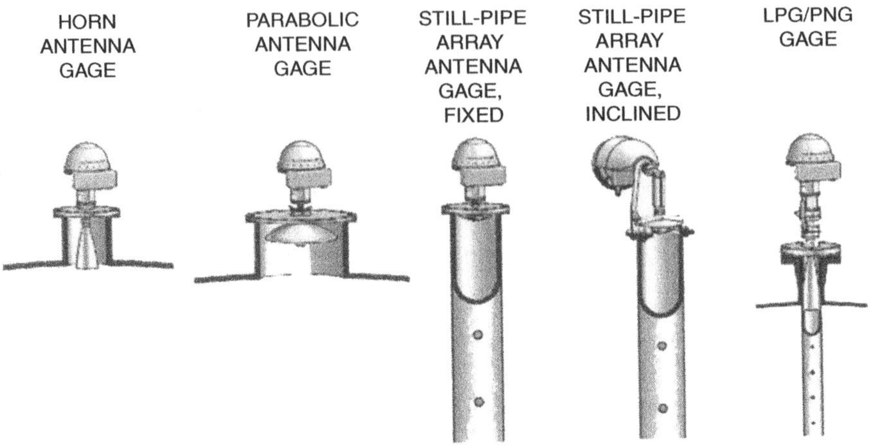

Figure 7–31 Microwave Antennas (*Courtesy of Emerson Process Management*)

The LPG/LNG antenna is used in pressurized tanks with liquefied gas. The signal is sufficiently strong for use with boiling surfaces.

Manufacturers that supply all of the antennas described may use the same transmitter head that contains the electronic module and is interchangeable with other gage heads regardless of the antenna used.

For both antenna and still-pipe installations, the mechanical design must be adapted for installation on tanks without disturbing normal operation. No hot-work will generally be allowed, and existing openings or still-pipes must be used.

7.29 CONTACT AND NON-CONTACT OPERATION

Non-contact radar detectors operate by using pulsed radar waves or frequency-modulated continuous waves (FMCW). In pulsed wave operation, short-duration radar pulses are transmitted and the target distance is calculated using the transit time. An FMCW sensor sends out continuous frequency modulated signals, usually in successive (linear) ramps. The frequency difference caused by the time delay between transmission and reception indicates the distance.

Radar beams can penetrate plastic and fiberglass so non-contact radar gages can be isolated from the process vapors by a seal which can be above the parabolic disc or can be totally isolated from the sensor. The low power of the beam permits safe installation in both metallic and non-metallic vessels. Radar sensors can be used when the process material is flammable and when the composition or temperature of the material in the vapor space varies.

Contact radar measuring devices send a pulse down a wire, cable or probe to a vapor-liquid interface where a sudden change in the dielectric of the materials causes the signal to be partially reflected. The time of flight is measured and the distance traversed by the signal is calculated. The non-reflected portion of the signal travels to the end of the probe and gives a signal for a zero reference point. Contact radar can be used for liquids and small granular bulk solids. Because the radar pulse travels down a wire, cable or probe, it is typically called "Guided Wave Radar."

Reflection-type microwave switches measure the change in amplitude of a reflected signal. Air and vapors return a small portion of the signal because of their low dielectric constants, while materials with high dielectric constants, such as water, return nearly all the signal. More sensitive switches can distinguish between liquid-liquid or liquid-solid interfaces with as little as a 0.1 dielectric constant difference. Plastic pellets and other materials can be measured when particle sizes are less than about 0.8 in.

A beam breaker switch can be used to send a microwave beam from a transmitter to a receiver located on the opposite side of a vessel. As the beam is blocked, the signal becomes weaker, a change that is used to detect material blocking the beam. This beam breaker method can be used for distances up to about 100 ft. Both reflection and beam breaker detection techniques are used in applications where it is not desirable to penetrate the tank, and can be used in plastic, fiberglass, ceramic and plastic-lined well tanks.

In radar applications, the reflective properties of the process material will affect the reflected signal strength. Liquids have good reflective qualities but solids usually do not. When heavy concentrations of dust particles or other such foreign materials are present, these materials will be measured instead of the liquid.

7.30 TANK ATMOSPHERE

The radar signal must be reflected directly from the liquid surface to obtain an accurate level measurement. Any dust or mist particles present usually have no significant influence as the diameters of such particles are typically much smaller than the 3-cm radar wavelength. For optical systems with shorter wavelengths, this is not the case. For comparison, when navigating with radar aboard ships, a substantial reduction of the possible measuring range is experienced, but even with a heavy tropical rain the range will be around 1 km, which is large compared to a tank.

There can be slight measurement errors with a few specific products in the vapor space of the tank. This is especially true when the composition may vary between no vapor and fully saturated conditions. For these specific products, pressure and temperature measurement may be required for compensation. Such compensation is made by the software incorporated in the tank intelligence system provided by the equipment manufacturer.

Following is a partial list of vapors and gases that are known to affect the transmission of radar waves:

- Acetaldehyde
- Acetone
- Ethyleter
- Isobutyraldehyde
- Methanol
- Propionaldehyde
- Propylene oxide
- Propylether
- 2-Butanone

Ammonia vapors attenuate the radar signal level to the extent that a technical evaluation is recommended to be made by the manufacturer for such applications [Ref 5].

7.31 TEMPERATURE SENSORS AND DISPLAY EQUIPMENT

Tank gaging systems used for inventory purposes must incorporate temperature measurement. In order to reduce installation cost, the signals from a number of temperature sensors can be multiplexed in the same cable as the level signals. Sensors for oil/water interface measurements and other sensors also can be connected.

Data display is provided at the control room as well as locally. Existing plant cabling is generally used in order to reduce costs. In many cases, an advanced PC-based inventory system with net volume calculations and various other functions is used.

7.32 APPLICATIONS

Various liquids have been successfully measured with a radar gaging system, including hydrocarbon products, crude oil, asphalt, chemicals, LPG and liquid sulfur. Solids such as ore pellets or coal can also be measured.

7.33 FLOATING-ROOF TANK INSTALLATIONS (PIPE INSTALLATIONS)

A still-pipe (typical diameter of 6 or 8 in.) connected to a radar level transmitter can be used. A cone is used to transform the still-pipe to a standard size waveguide that enters the level transmitter. The transmitter covers the still-pipe but can be folded aside for sampling or hand-dipping. The hardware can be adapted to existing still-pipes.

7.34 FIXED-ROOF TANK INSTALLATIONS

The parabolic antenna has a diameter of 0.4 m and can be mounted on a fixed-roof tank at a convenient location (e.g., on an existing manhole cover).

7.35 LIQUEFIED GAS INSTALLATIONS

Figure 7–32 shows an installation for pressurized LPG and LNG tanks. The pressure may be up to 600 psi, and a pressure sensor is included to provide the internal pressure in the tank. The measurement is done through a 4-in. still-pipe installed for this purpose and thus is insensitive to boiling or turbulence. Free-space restrictions are also avoided. The still-pipe has a built-in test echo for periodic verification without opening the tank [Ref. 6].

7.36 TANK FARM STORAGE AND WASTE CHEMICALS

Tank farm and waste chemical vessels must accommodate a wide variety and mix of chemicals, oils, sludge and other products. Level-measuring devices for each application must be unaffected by major

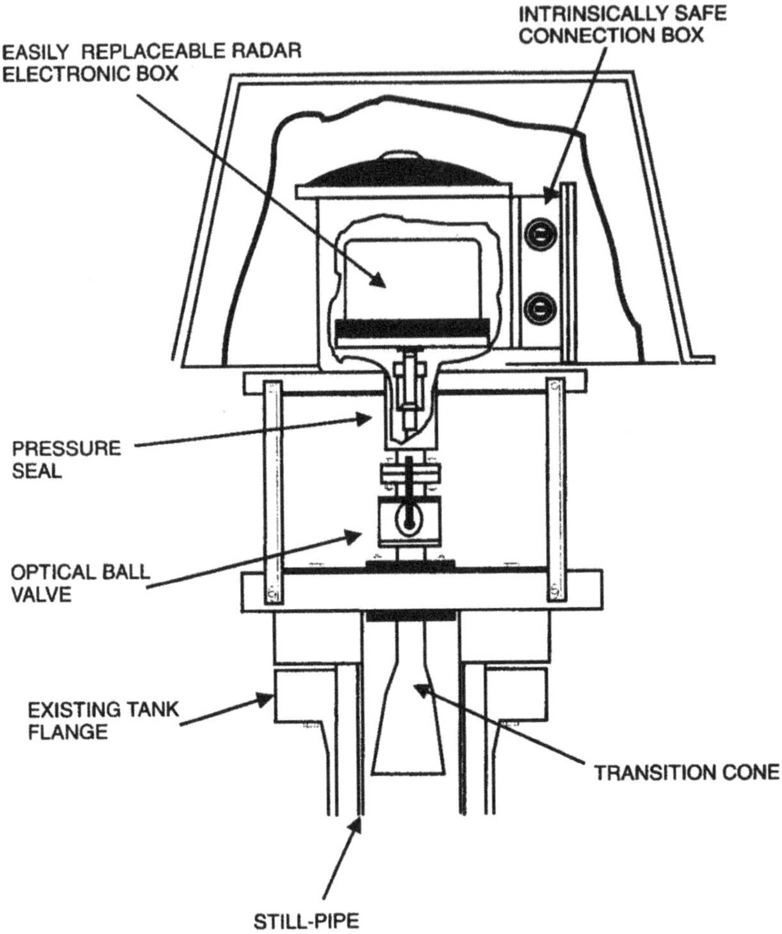

Figure 7–32 Microwave Transmitter for LNG and LPG Tanks

variations in liquid composition, density, dielectric constant, conductivity, vessel wall coating, corrosive environment, foaming, turbulence, condensation, mist, fill pipe and countless other factors. The obvious limitations of other level technologies with respect to these adverse conditions render most measurement schemes inoperative. Radar level techniques, however, have proved successful.

7.37 FOOD INDUSTRY

Many food products, including fruit juices, milk, and so on, are stored at cooler temperatures and pressurized with a 2.5-psi nitrogen or another inert gas blanket to prevent contaminant leakage into the vessel. Sanitation requires that containers be frequently washed and sprayed with caustic or other agents, which may be harmful to wetted-type level sensors. Rapid temperature variations can also result in calibration and zero shift in some measuring instruments. Because radar devices use electromagnetic wave transmissions that are independent of environment, composition, foam, specific gravity and other such variables, they are used in such applications with considerable success.

7.38 HEAVY HYDROCARBON STORAGE VESSELS

Asphalt and other heavy hydrocarbons are stored at elevated temperatures ranging to 400°F. Level-measuring instruments for such applications should be immune to high temperatures, coatings and buildups, agitation, vapor density gradients, and other undesirable situations. Radar level measurement is unaffected by these conditions.

7.39 POSSIBLE INTERFERENCE ISSUES IN RADAR LEVEL MEASUREMENT

Consideration should be given to all mounting procedures to avoid possible interference with radar transmission, as depicted in Figure 7–33. Some possible interference sources are discussed below.

Multiple Reflections

These occur, for example, when the signal is reflected from the surface of the medium, then strikes the tank cover or some other "good" reflector, and is again reflected from the medium before being received by the antenna. Since multiple reflections occur at periodic intervals, they can be detected and taken into account in the signal processing. A better solution is to change the mounting position so as to eliminate multiple reflections altogether.

Multipath Propagation

If, for example, the signal is reflected from the tank wall, its propagation path is lengthened; the reflection signal is thus broadened in time and the measuring accuracy reduced. The antenna should be moved further away from the wall.

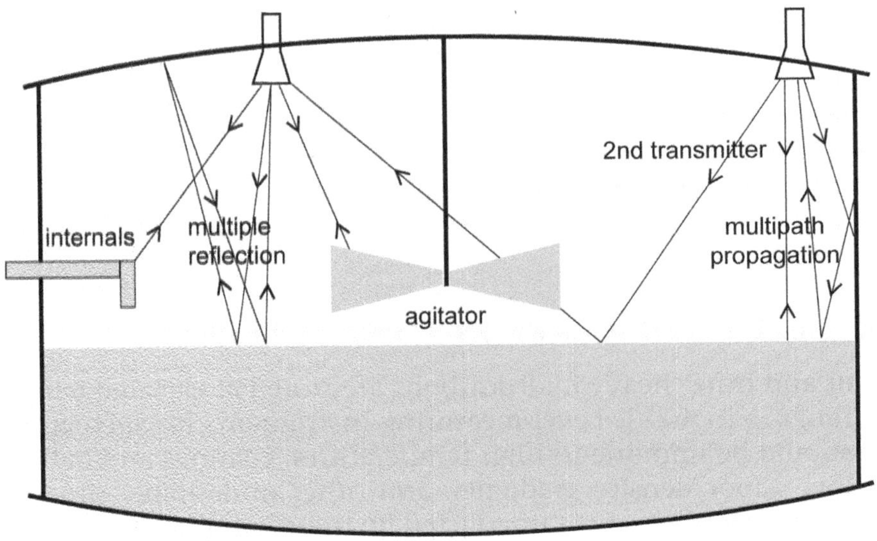

Figure 7–33 Possible Sources of Interference in Radar Level Measurement (*Courtesy of KROHNE*)

Other Microwave Transmitters

Several radar systems that are installed in one tank can mutually influence one another. With FMCW radar, however, this probability is normally very low because the systems would have to operate in synchronism down to fractions of microseconds in order to generate an additional differential frequency portion within the processing bandwidth of a few kHz. In pulse radar with a high pulse repetition rate, interference can, however, easily occur when the signals from several transmitters are interpreted as being the total reflection pattern [Ref. 4].

7.40 END-OF-THE-PROBE ALGORITHM

An end-of-the-probe algorithm can be used in guided wave radar when there is no reflection coming back from the product. This new technology innovation provides a downward-looking time-of-flight situation which allows the guided wave radar system to measure the distance from the probe mounting to the material level. An electromagnetic pulse is transmitted and guided down a metal cable or rod which acts as a surface wave transmission line. When the surface wave meets a discontinuity in the surrounding medium, such as a sudden change in dielectric constant, some of the signal is reflected back to the source, where it is detected and timed. The portion of the signal that is not reflected travels on and is reflected at the end of the probe (Figure 7–34). This concept is well understood, but what does the end-of-the-probe signal reflection mean and how can this enhance the application?

The dielectric constant of every material is different. The dielectric constant of a vacuum is 1, and all other materials are related to this. The speed of the electromagnetic wave changes as it travels through other media, and the speed varies inversely with each medium's dielectric constant (often referred to simply as dielectric). A hydrocarbon has a dielectric of around 2, which may seem low, but it is twice that of a vacuum. This difference in the dielectrics of various materials enables the end-of-the-probe technology to be useful in measurement.

Each sensor has a fixed cable length and the end of the cable gives a large reflection as the wave grid ends and all energy travels back to the electronics unit. As was stated, that portion of energy that travels to the end of the probe will be reflected as an end-of-the-probe signal. There will be a time shift as the electromagnetic wave speed is decreased by traveling through product. This time delay between the signal traveling to the end of the probe through the vapor of an empty tank and the signal traveling through the liquid in the tank is proportional to the level and the dielectric of the material. This is shown in Figure 7–35.

The end-of-the-probe algorithm can be used to calculate the dielectric of the material under process conditions, using the shift in propagation speed and the surface reflection. The level can be calculated by the end-of-the-probe time shift and the best measured dielectric in cases where the surface reflection may be intermittently lost.

In level measurement applications based on the constant dielectric of the material, error can result because the material's dielectric can vary with temperature, humidity and sometimes pressure. When the measurement system fails to compensate for changing dielectric, the measurement is compromised. With end-of-the-probe technology, the dielectric is determined with each surface reflection, which usually averages two times per second.

In solids level measurement like filling a silo, dust resulting from the material being blown in can create much disturbance to the measurement signal. An end-of-the-probe algorithm determines the level by two methods, surface reflection and end-of-the-probe technology; the noise generated in the surface signal can be lower than the noise created and only the end-of-the-probe signal is used to determine the level.

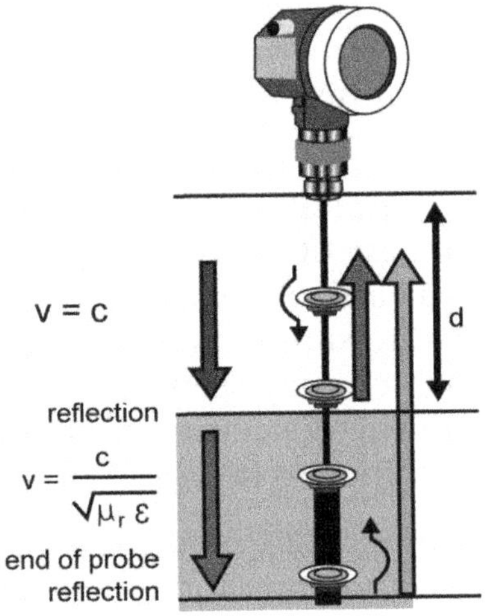

Figure 7–34 Signal Propagation Through Different Media (*Courtesy of* Flow Control *Magazine*)

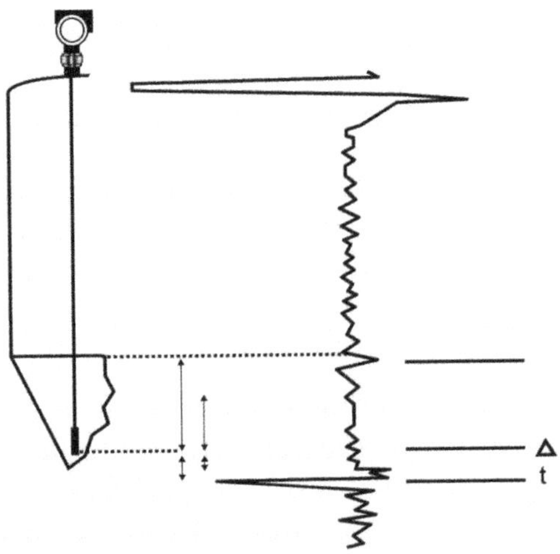

Figure 7–35 Signal Time Shift Through Different Media (*Courtesy of* Flow Control *Magazine*)

The same disturbance difficulties can occur in liquid applications in cases of mist, agitation and waves. In such applications, and when the medium's dielectric is low, the surface reflection can be lost. The end-of-the-probe signal will still give a reliable measurement.

Many manufacturers offer end-of-the-probe guided radar systems. Each may have different set-up requirements and give various degrees of accuracy. Careful research and analysis should first be exercised to select an end-of-the-probe system for a given application [Ref. 7].

7.41 INTERFACE DETECTION BY TIME DOMAIN REFLECTOMETRY (TDR)

Liquid/liquid measurement can be provided by guided wave radar, also called time domain reflectometry (TDR), which is similar to pulse radar but is conductor-based and used with electrical pulses without a carrier frequency. This technology is particularly suited for interface measurement because there is less signal weakening and reduced interference from tang geometry situations.

Figure 7–36 shows that a generated electrical pulse starting at t_0 is guided by a two-wire line as an electromagnetic wave. At each position where the relative permittivity[1] changes, some of the wave is reflected back to the sensor. When the wave travels to the surface of the top fluid, part of the wave is reflected back to the sensor at t_1. The wave continues to the interface between the two liquids and is reflected back to the sensor at t_2. Propagation of the wave continues until it reaches the end of the wave guide and part is reflected back to the sensor a third time, at t_3. The received signal delay times ($2t_1$, $2t_2$ and $2t_3$) indicate the position of the top interface (the top liquid and tank vapor), the interface between the two liquids, and the end of the wave guide. The polarity of the signal is reversed at each reflection point from lower to higher permittivity. The difference in relative permittivities will determine the reflected power. Typical interface applications for guided wave radar require that the liquid with the higher dielectric constant be the lower liquid, and the difference between the dielectric constants of the two liquids be greater than approximately 13.

For successful operation, some considerations should be evaluated for signal processing:

- Deposits of materials along the wave guide that may cause additional reflections.

---

1. Relative permittivity (ε_r) is a unitless constant describing the physical behavior of a material in an electric field. For water, it is about 80, for a vacuum 1, for gases slightly greater than that, and for most liquids other than water, greater than or equal to 2.

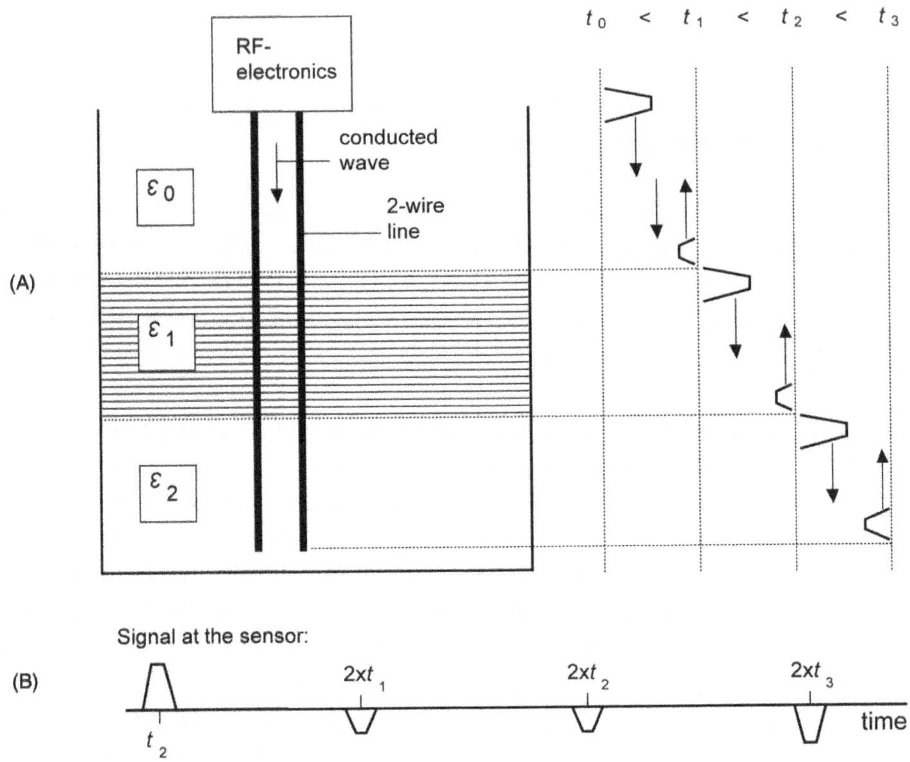

Figure 7–36 Principle Operation of a Wire-conducting TDR Level Gage (*Courtesy of KROHNE*)

- Changes in permittivity values as a result of temperature or product composition.
- Excessive absorption or attenuation of the microwaves.
- Conditions where there is a lack of sufficient discrimination between the two fluids.
- No abrupt transitions within the process material, as is the case when emulsions form.

7.42 SUMMARY—RADAR LEVEL MEASUREMENT

The radar level gaging technique used on tanker ships for many years has been used in refineries and tank farms in recent years. Its high degree of integrity against virtually all environmental influences has resulted in high level-measurement accuracy; one example is the approval of a radar gage for 1/16-in. accuracy. Nearly all level gaging in a petroleum storage environment can be done with radar level gages adapted for that purpose.

Although radar level technology is a relatively recent introduction in the process and manufacturing industries, it is gaining respect for its reliability and accuracy. While the time-of-flight principle of sonic and ultrasonic level measurement systems is similar to radar, there are distinct differences. The primary difference is that sound waves produced by ultrasonic units are mechanical and transmit sound by the expansion of a material medium. Since the transmission of sonic waves requires a medium, changes in the medium can affect the propagation. The resulting change in velocity will affect the level measurement. Other factors can also affect the transmitted or reflected signal, including dust, vapors, foam, mist and turbulence. Radar waves do not require a medium for propagation and are inherently immune to the factors which confuse sonic-type devices [Ref. 8].

Advantages and other characteristics of radar level measurement include

- Generally immune to vapor space material and conditions.
- Immune to obstructions in the tank.
- Improved accuracy and stability by microprocessor generation and computation of algorithms—can be used for custody transfer applications.
- Two-wire designs.
- Little affected by temperature and pressure, for which both can be compensated.
- Relatively immune to errors caused by foam; however, thick, dense foam can impair measurement.
- Relatively easy to install and configure.
- Up to 300 ft measured span.
- Used with guide wires to measure a liquid/liquid interface.
- Some units are battery powered.
- Some units are wireless.

- Used extensively in petroleum-based industries.
- Limited use in the pharmaceutical and food industries.

Although radar level measurement is a relatively new technology, first used in 1976 and perfected to a compact gage in 1989, antenna mounting and design as well as software development have enhanced the number and success of applications. Blind spots, how measurement is affected by multiple transmissions, and other limitations caused by hardware or process parameters can be eliminated [Refs. 4 and 5].

Following are some considerations for radar level measurement.

- Degradation of signal strength can result
    - In the vapor space.
    - At the antenna.
    - At the surface of the material.
- Dirt or coating on the antenna can reduce the signal strength.
- Materials with very low dielectric constants should be avoided or measured with caution.
- Sufficiently thick, dense foam can attenuate the reflected signal or result in spurious reflected signals.

7.43 FIBER-OPTIC LIQUID LEVEL MEASUREMENT

The theory of operation of a prism-type fiber-optic liquid level sensor can be simply stated. A low-loss fiber-optic probe or cable is interfaced with a quartz prism. One leg of a bifurcate (split) probe is used to transmit light from an LED source; the other leg is attached to a photodetector and processing electronics. The common end of the probe is attached to the prism.

In air, light transmitted from the LED source is internally reflected at the prism/air interface; the air (index of refraction = 1) acts as a cladding material around the prism. Light is reflected in the prism and is sent back to the receiving electronics. If the prism is allowed to contact a liquid with a higher index of refraction than air, the light trapped in the prism will be refracted into the liquid, causing a dark condition at the photodetector. The amount of light absorbed by the liquid will depend on the index of refraction of the liquid being sensed. If the corresponding electronics are properly configured and no other interfering conditions are present, the control electronics will detect this drop in light intensity and switch the state of their output.

Electronic controls can vary in operation and provide a number of switching functions. Some common types are capable of handling input voltages from 5 to 28 V DC and can sink up to 300 mA through a normally "on" or normally "off" open collector output. Other DC models can provide transistor/transistor logic (TTL) or CMOS (complementary metal oxide semiconductor) logic level signals directly to a computer.

AC-powered devices can operate from a 110- or 220-V AC source and can switch loads up to 8 A through a dry contact relay. Some models have solid-state outputs capable of handling up to 2 A or to provide a signal directly to a process controller. In particularly hazardous areas, telecommunications-grade fiber can be used to remotely locate the electronics several hundred feet from the hazardous area.

7.44 APPLICATIONS

The most common use of this type of system is a high- or low-level indication, such as "Is the liquid present at this level?" This question reflects a simple "yes" or "no" condition. There is a very specific need for improved single-point level sensors in industry. Liquid level sensors using optical fibers hold particular promise as replacements for many existing technologies, especially in harsh environments. A prismatic sensor coupled to fiber can provide a safe, reliable (and low-cost) means of level detection in most non-coating liquids. The use of fiber allows the liquid level sensor to be mounted hundreds of feet away from the corresponding electronics, thus providing safe detection and signal transmission from a hazardous area to the control room.

7.45 LEVEL SENSORS FOR REFINERY VESSELS

Level sensors for refinery vessels can be fairly costly—several hundred dollars per switch. The most commonly used devices are float-type sensors that mechanically actuate a switch and/or alarm. Other techniques include resistance temperature detectors (RTDs) and differential-pressure and ultrasonic instruments. All work well in liquids and slurries, but are usually more expensive than float or displacer devices. Typical problems with float-type switches include gumming up or freezing up.

7.46 SYSTEM CONFIGURATIONS

Figure 7–37 depicts a typical liquid level installation. Not included in the diagram are various other sensors for temperature and pressure, and the electrical boxes for the level sensors and other instruments. In all installations, the electronics are enclosed in explosion proof boxes. The electrical code states that in class I, groups B, C, and D, divisions 1 and 2, all electrical circuitry must be sealed from the volatile liquids being sensed (Table 7–7). The use of fiber provides electrical isolation and an inherently safe barrier between the process and electronics. The fiber does not provide a path for the conduction of a charge into explosive or hazardous areas.

Special attention must be paid to the packaging of the sensor and electronics to meet electrical codes and the mechanical and environmental constraints of refinery or other outside installations. Figure 7–38 shows the mechanical outline of a fiber-optic level sensor. The switch has a 1/2-in. NPT thread to mount directly into a vessel or fitting. It may be desirable in most installations to place a shutoff valve between the vessel and the sensor to provide a bypass in case of failure or for ease of maintenance.

Note that if the sensor does require maintenance, the work can be performed without the risk of electrical hazard, unlike conventional sensors where the electronics housing is in close proximity to the switch. The back of the sensor has a 3/4-in. pipe thread to provide a conduit mount. An 18-in. fiber-optic cable is permanently attached to the sensor, sealing and protecting the sensor body and allowing it to withstand up to 1500 psi continuous pressure.

The electro-optic interface can be remotely located by the use of a fiber-optic extension cord up to 1000 ft away from the sensing location. The fiber extension cord is of a duplex oval construction and can be cost prohibitive, in some cases several dollars per foot.

An alternative is to use simplex fiber cable to transmit an optical signal from the processing electronics to the control room at a significantly lower cost. This will ensure safe, noise-free transmission of data from the electronics to the control room. Figure 7–39 shows a typical tank installation with two sensors to control level. Multipoint level sensing can be accomplished by installing several sensors in the vessel at various levels.

Table 7–7 Hazardous Locations, Classes, Groups, and Divisions (Abbreviated)

Classification	Description
Class I	Locations where flammable vapors or gases may be present.
Group A	Acetylene and similar atmospheres per National Electric Code (NEC) Table 500-2c
Group B	Hydrogen and similar atmospheres per NEC Table 500-2c
Group C	Ethyl-ether and similar atmospheres per NEC Table 500-2c
Group D	Propane, most solvents, and similar atmospheres per NEC Table 500-2c
Class II	Locations where combustible dust may be present
Group E	Atmospheres containing metal dust and similar atmospheres per NEC Table 500-2
Group F	Atmospheres containing carbon black, coal dust, and similar atmospheres per NEC Table 500-2
Group G	Atmospheres containing flour, grain, and similar atmospheres per NECT Table 500-2
Class III	Locations where ignitable fibers and flyings may be present
Division 1	Locations where hazardous agents are present under normal operating conditions
Division 2	Locations where hazardous agents may be present only in case of accidental rupture or breakdown

For complete definitions, classifications, and installation requirements of hazardous locations, see NEC Articles 500-503.

7.47 OTHER APPLICATIONS

The same liquid level sensor described in the preceding section can be used to distinguish the difference between two liquids by responding to typical losses or changes in light intensity. For example, a properly configured electronic system can detect the presence of water in gasoline. In this case, the sensor can be placed at the bottom of the tank. If water is detected, it should trigger an alarm, shut off the pump, and allow the system to be drained. This would be especially useful in refueling trucks for fuel-storage tanks and in oil/water separators.

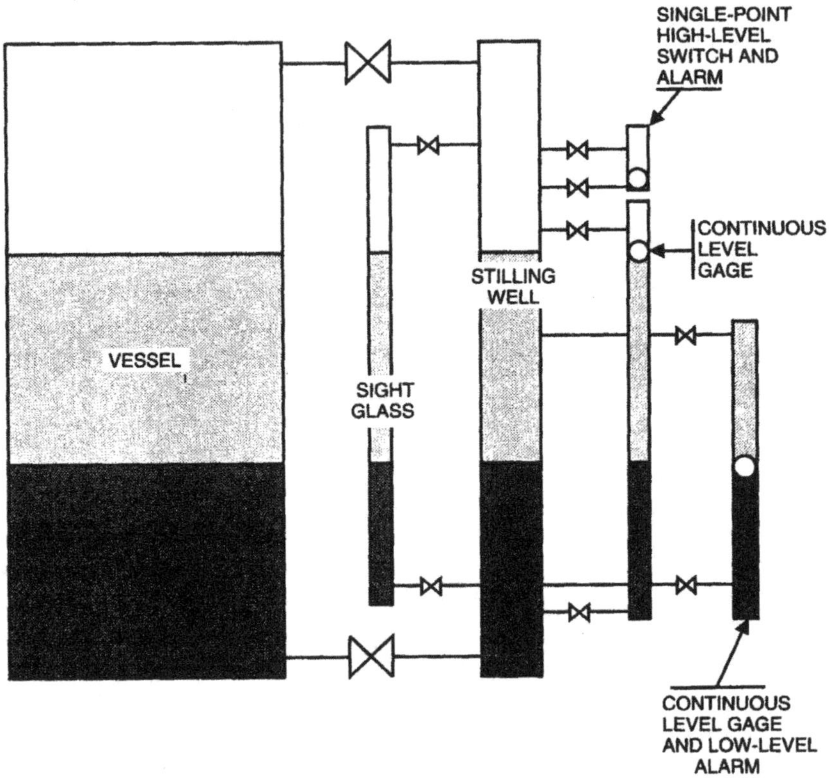

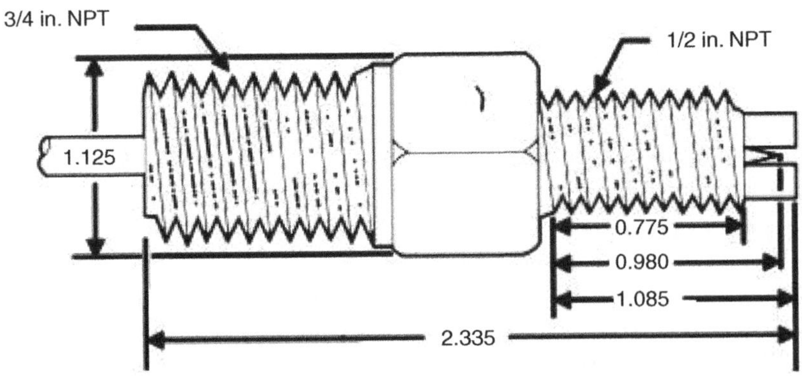

Figure 7–38 Mounting Drawing for Fiber Optic Level Sensor

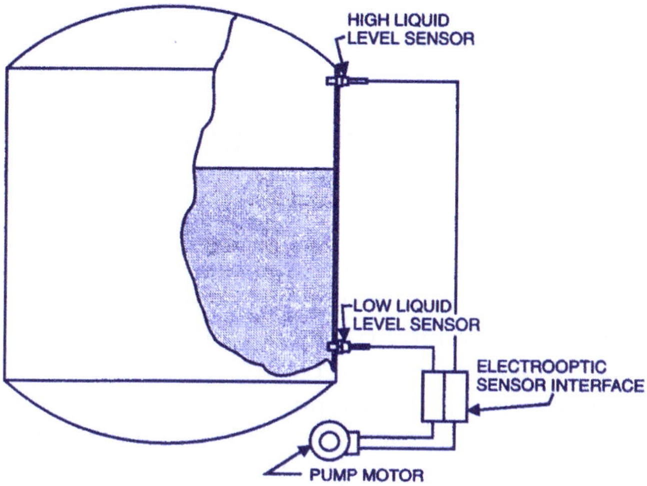

Figure 7–39 Application Drawing for Multiple Fiber Optic Level Measurement and Control

7.48 FACTORS AFFECTING INDEX MEASUREMENTS

Because of its inherent design, a prism can also be injected with light from outside sources. In a switch application, the signal differential from air to liquid is great enough to overcome the effect of light being reflected back to the prism. When the sensor is used as a liquid discriminator, light leaving the prism can be reflected back into it by outside sources of reflection. Turbid liquids containing suspended solids also can result in false readings. This type of sensor is limited to clear liquids.

Fiber-optic level switches are inherently safe and provide a cost-effective approach to level detection where hazardous liquids are involved. Their small size makes them easily adaptable to a variety of applications, allowing the original equipment manufacturer (OEM) designer greater flexibility in system design.

Fiber-optic cable connection has become routine and user-friendly. The installation cost is lower than that of other conductors and less routine maintenance is required [Ref. 9].

7.49 OTHER TYPES OF LEVEL MEASUREMENT

The level devices described thus far constitute the major types used in most industrial applications, ranging from simple visual gages and sight glasses to more complex and costly capacitance and ultrasonic instruments. Certain measurement applications require still other approaches.

7.50 MAGNETOSTRICTIVE LEVEL MEASUREMENT

The method of measuring level by hydrostatic principles includes correcting for density, pressure, temperature and vessel volume. This procedure, known as hydrostatic tank gaging (HTG), is covered in detail in Chapter 9 and has become popular because of its improved level-measurement accuracy. Magnetostrictive level measurement, covered here, is a precise, repeatable and accurate technique that has enabled level to be measured within a tolerance of ±1/16 in. or ±0.025% FS. Magnetostrictive technology has been used in HTG applications.

One primary device for level measurement is an inductive sensor/encoder that initiates its lineal measurement from the tank bottom, thereby circumventing the potential error introduced by tank-roof movement. A float, without mechanical linkage but with an integral energy transponder, is designed to float at the vapor/liquid and/or liquid/liquid interface, and to travel up or down as the liquid level rises or falls (Figure 7–40). A measurement element extending the full height of the storage vessel becomes an electronic digital dipstick. This element is a flexible tape-like assembly that consists of a strip of metallic material with high magnetic properties, to which is bonded an excitation loop running the full length of the element. Also bonded to the element is a Gray-code pattern of wires; each increment of level has its own unique pattern (Figure 7–41). A highly precise silicon piezoresistive pressure sensor is integrally mounted at the bottom of the element. The assembly, except for the pressure sensor, is encased in a thin-wall Teflon jacket. The toroidal float, with a hermetically sealed integral transponder, is kept close to the measurement element. The excitation loop in the element is excited with a low-voltage, high-frequency AC signal and may be likened to the primary of a transformer.

The transponder's solid-state electrical circuit is powered by the excitation loop, and the transponder intermittently induces a small burst of energy back into the measurement element's Gray-coded pattern at the point of liquid level. Because a unique pattern exists for each increment along the measurement element, the position of the float is known precisely, anywhere in the operating range for that element. The measurement is therefore absolute and discrete. At this point, the

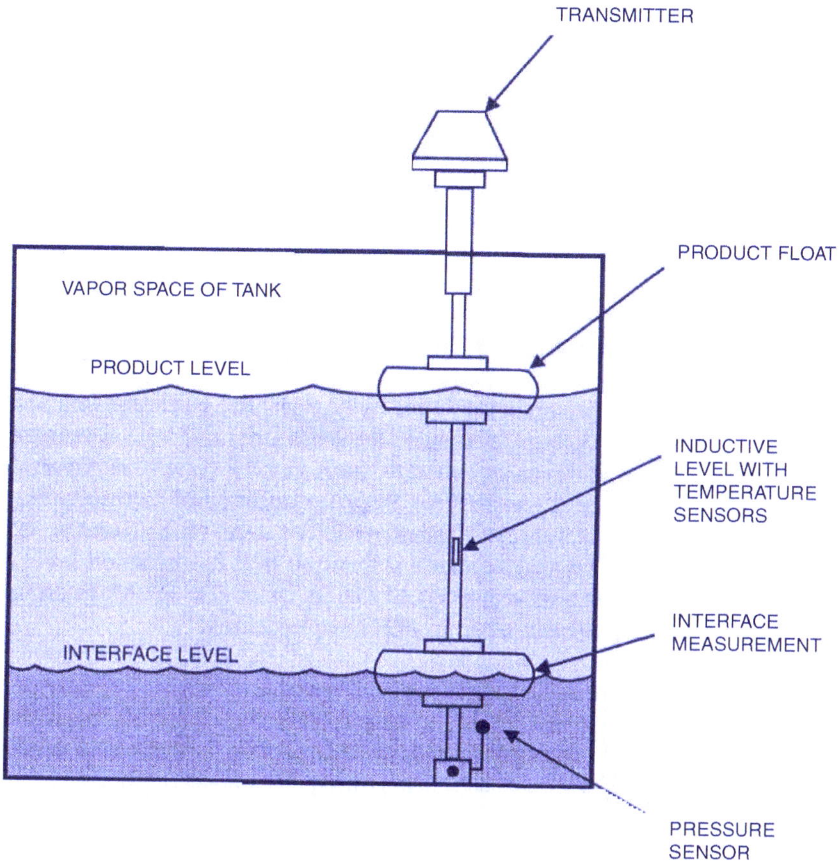

Figure 7–40 Floats and Measuring Element for Magnetostrictive Level Measuring

level measurement is digitally encoded, whereas the hydrostatic pressure sensor produces an analog signal proportional to level. A transmitter unit, wired to the upper termination of the measurement element, provides intrinsically safe power to the excitation loop, while simultaneously storing float position in its on-board memory and monitoring the hydrostatic pressure signals. By means of a time multiplexing technique, two floats can be used on the same measurement element. With proper ballasting, the second float will position itself at the interface between two known liquids in the tank.

A microprocessor-based processing unit and transmitter memory located on the tank roof will store the position of both floats as well as hydrostatic pressure data. To complete the measurement and to compensate

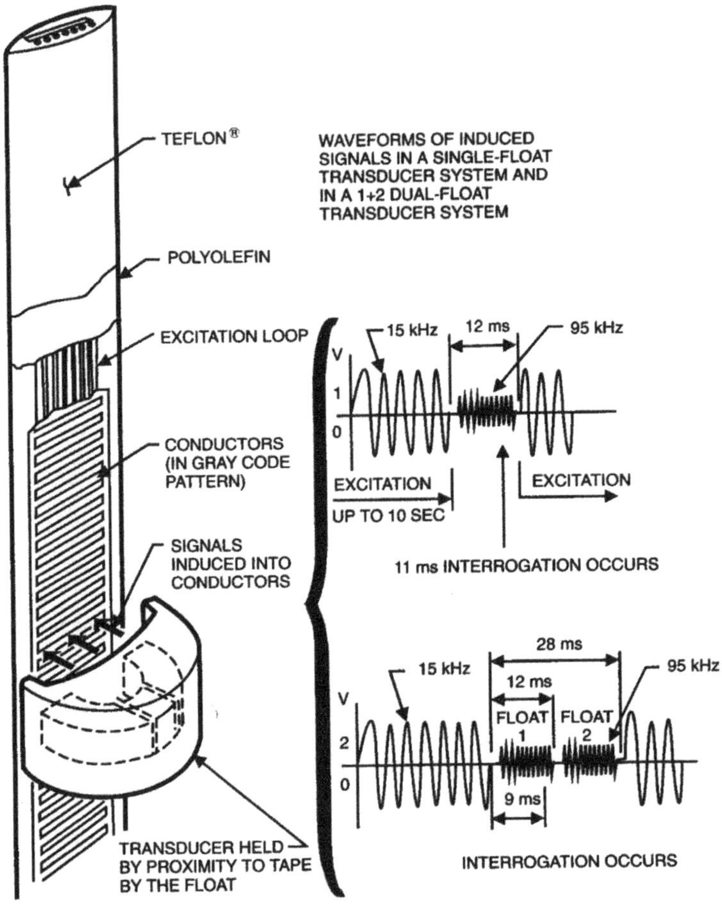

Figure 7–41 Measurement Element Signals

for temperature variations, platinum-film RTD sensors may be embedded directly into the measurement element to compute the average temperature of the product as well as to provide a profile of vertical temperature stratification in the tank. The addressable transmitter formats the level, temperature and pressure data, and transmits them to a local computer. Mass and density can be computed from the data [Ref. 10].

A magnetostrictive transmitter/probe assembly can be attached externally to a magnetic float gage for improved measurement accuracy and signal transmission. The magnetic float position in the level gage is detected by the magnetostrictive transmitter.

7.51 NUCLEAR RADIATION DEVICES

Nuclear radiation devices were first used in instrumentation and related industries in the early 1950s. These devices are used where others prove unsuccessful. Adverse process conditions (such as extreme temperatures; thick, abrasive or corrosive materials; and highly pressurized vessels) represent typically difficult measurement applications. Conventional measurement systems are limited in such applications, and nuclear radiation devices are an ideal choice.

7.52 THEORY OF OPERATION

Radioisotopes used for level measurement emit energy at a fairly constant rate and in a random fashion. The disintegration of the radioactive source results in the formation and emission of alpha and beta particles and gamma radiation. Alpha particles have found no significant use in industrial applications because they have no penetrating power or nature. Beta particles exhibit modest penetrating power and have been used in the thickness testing of thin materials. Gamma radiation is present in high-energy, short wavelengths that produce great penetrating power and have found wide acceptance for level gaging applications and other measurement techniques.

The basic unit of measurement of radiation intensity is the curie, defined as that source intensity which undergoes 3.70×10^{10} disintegrations per second. The more practical unit of radiation measurement related to level applications is the millicurie.

The two most common radioactive isotopes used in level-measurement applications are cobalt 60 and cesium 137. The half-life of each of these isotopes is 5.2 years and 33 years, respectively. An analysis of their relative strengths shows that 0.88 mg of cobalt 60 and 11.5 mg of cesium 137 have about the same power of radioactivity generation. The choice of radioactive source will depend largely on the penetrating power needed.

For industrial applications, radiation field intensity is normally measured in milliroentgens per hour. The radiation field intensity in air can be calculated from

$$D = 1000\ (KM_c/d^2) \qquad (7.25)$$

where D is the radiation intensity in milliroentgens per hour (mR/hr), M_C is the source strength in millicuries (mCi), d is the distance to the source in inches and K is the source constant (0.6 for cesium 137, 2.0 for cobalt 60).

The radiation from the source penetrates through the vessel wall and through the process liquid. For a practical application design, it is necessary to know the intensity of the field available at the detector. The following example illustrates the procedure used to determine the source size.

EXAMPLE 7-4

Problem: The minimum field intensity at the detector of a measuring system in an empty steel tank (described later) is

$$D = 2.0 \text{ mR/h}$$

A 50% change in intensity is desired when the tank is full. The source material is cesium 137 (K = 0.6), M_C – 10mCi, and d = 24 in. The source is housed in a 1/4-in.-thick steel tube, and the tank wall is 1/2-in.-thick steel. Determine whether the installation guidelines are met.

Solution: From Equation 7.24,

$$D = 1000\frac{KM_c}{d^2}$$

$$= 1000\frac{(0.6)(10)}{24^2} = 10.4 \text{ mR/h}$$

In order to find the amount of absorption lost in the source housing and the tank, or the percent transmission through the metal, a plot of percent transmission versus material and thickness is needed. Such a curve is found in Vol. 1 of the *Instrument Engineer's Handbook* (CRC Press), which shows the transmission to be 67% for the 3/4-in. combined thickness of the steel tank wall and protective source housing.

The field radiation intensity at the detector for the minimum level condition is

$$(0.67)(10.4) = 7 \text{ mR/h}$$

This value is greater than the 2.0 radiation intensity required.

When the tank is full, the radiation must penetrate a 24-in. wall of water in addition to the steel wall and probe source housing. From the transmission curve for cesium 137, the transmission through the water is 7.5%. The field intensity at the detector is

$$(0.075)(7) = 0.525 \text{ mR/h}$$

The installation guidelines are met because the ratio between the radiation intensity for empty and full tank conditions is greater than 50%:

$$\frac{7}{0.525} > 0.5$$

If the desired installation requirements in the preceding example had not been met, other radiation sources could have been selected.

7.53 APPLICATIONS OF NUCLEAR RADIATION LEVEL MEASUREMENT

Like many of the other level-measurement devices discussed, nuclear radiation measurement can be used for either point or continuous applications. The measuring principle is the same for both applications; the hardware and its arrangement are different.

Point measurement uses two basic components: a radiation source and a detector with an associated amplifier. Generally, the source is located on one side of the process vessel, and the detector is located on the opposite side. As illustrated in Figure 7–42, more than one detector can be used for level indication at various points. It should be noted, however, that the ability to use more than one detector depends on several factors, including source size, distance between the level points, and vessel diameter. The radioactive source and the holder will have a lead shield to prevent radioactive emission during shipping. Shielding is also provided as a safety precaution during periods of use. The detector is usually a Geiger-Mueller tube with a solid-state amplifier and associated power supplies.

Gamma rays are emitted by the source and pass through the walls and the material in the tank. The variation in transmission of the gamma rays will be a function of the medium through which the rays travel on their path from the source to the detector. When the level is below the

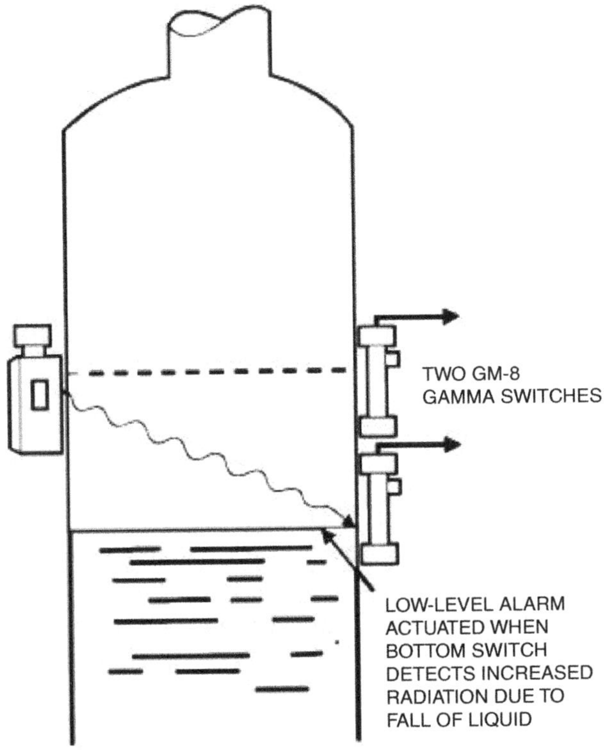

Figure 7–42 Point Measurement with a Radioactive Device (*Courtesy of Ohmart/VEGA Corporation*)

elevation point of the detector, the intensity of the field will be greater than when the level reaches the detector. When the liquid interrupts the path of the gamma rays, the percent of transmission will decrease, and the output of the detector amplifier will decrease sharply. This will activate a control relay, signifying the presence of a liquid level at the elevation point of the detector. Although satisfactory for level detection at one point, when more than one or a few measuring points are desired, a continuous measurement should be used.

Continuous level measurement is depicted in Figure 7–43. In Figure 7–43A, a point source is shown with a detector covering the entire desired measuring span of the instrument. The radiation of gamma rays occurs in all directions, and the percent of transmission decreases as the liquid level increases, interrupting more of the nuclear energy of the gamma rays. As the level continues to rise, fewer of the gamma rays are transmitted or absorbed, and radiation slowly decreases to a low value as the source for that detector or portion of

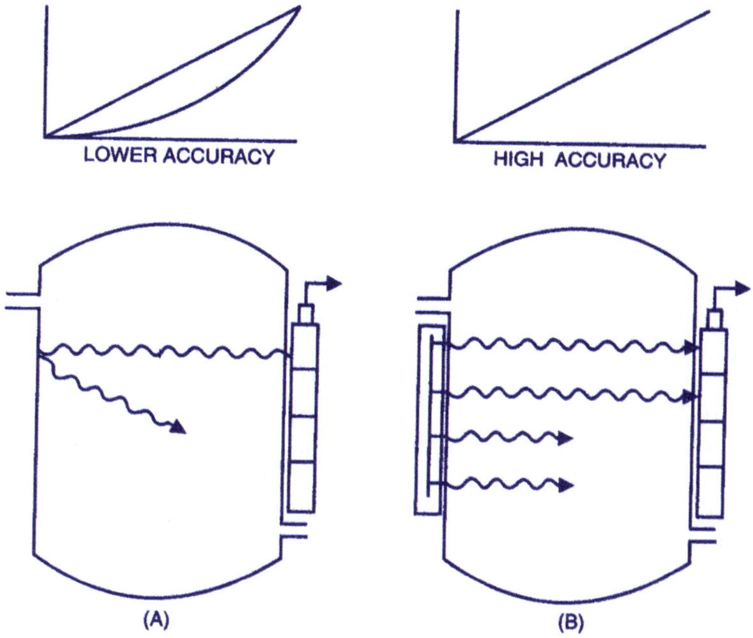

Figure 7–43 Continuous Level Measurement with a Radioactive Device (*Courtesy of Ohmart/VEGA Corporation*)

the detector is completely covered. This procedure continues for the next detector until the entire measurement span has been covered. As shown in the curve in Figure 7–43A, this method of measurement yields nonlinear results.

A better method of continuous measurement is shown in Figure 7–43B. In this technique, each measuring cell receives radiation from each source along the span or from all points of a strip source. The detectors at the center of the vessel have a shorter mean distance from every source than do those at the end, so a small amount of nonlinearity exists, which can usually be reduced to a tolerable level. Ion chambers or scintillation detectors are used for continuous level measurement. These detectors generate a small current caused by ionization that is amplified to produce a current that is proportional to level.

Radiation level measurement is expensive, but provides the most reliable operation for applications where other level-measurement means would be difficult and questionable. When the walls of the vessel are very thick, as is the case in high-pressure processes, error may be significant when the ratio of wall and probe absorption to material absorption is high. As Example 7-4 illustrates, components must be properly selected to keep the expressed ratio low.

By placing the source within the vessel, absorption by a vessel wall can be eliminated. The disadvantages of this procedure are obvious. For vessels with large diameters, the path of gamma rays can be on a chord of the vessel instead of a diameter. Since absorption is not merely a function of process material, all other factors (such as density) that govern transmission and absorption must be constant [Ref. 11].

Radiation level detectors are considered by some as a sort of last option device; they may be used where other technologies fail to be satisfactory or when process penetrations required by other devices present a risk to human life or to the environment. Liquids and bulk solids that are very corrosive and applications of high pressure are some examples that are good candidates for nuclear level applications. The installation and operation of nuclear gages are required to be done by NRC (Nuclear Regulatory Commission) licensed technicians. There are also requirements for periodic testing and associated record keeping. (There are some devices with sources that have low enough radiation levels that they do not have to meet these stringent requirements). All such design and applications are guaranteed safe. When this is compared to the risk of having untested or unreliable technology, the perceived and real disadvantages of nuclear detectors seem to fade. As detectors become more sensitive with the aid of computers, there will be a reduction in source size. The associated reduction in radiation levels and the associated improvement in safety will likely result in an expansion of the use of this technology.

The NRC limits the radiation intensity to a maximum of 5 milliroentgens per hour (mr/hr) at a distance of 12 in. from the source. When the level is higher than this, the area requires Radiation Area posting. The distance of 12 in. is critical because radiation intensity decreases by the inverse of the square of the distance. Nuclear gages are sized to provide radiation intensity at the detector that exceeds the minimum required but is under 5 mr/hr maximum. The required level is determined by the detector type. For ion chamber detectors, the level is 1 mr/hr, for the Geiger-Mueller tube it is 0.5 mr/hr and for scintillation detectors it is 0.01 to 0.02 mr/hr. Detector operation is such that as the vapor space above the liquid decreases (a rise in level) the intensity is essentially zero when the tank is full.

Criteria other than the dose limit can be used for radiation safety to avoid the need for area posting to the NRC by keeping the average or yearly dosage level under 5 rems (roentgen + equivalent + person). For example, should a person be exposed to radiation throughout the year, such a dosage will result from exposure to radiation at an intensity of 0.57 mr/hr while for a worker that is exposed for only 40 hours per week, 5 rems/year will correspond to what that person would receive if exposed to 2.4 mr/hr in the work area. It is the lifetime dosage received that matters (maximum 250 rems). The acceptability level of 5 rems/year is also a function of age and total time of exposure. It should be considered, however, that the radiation at the detector must still be sufficient to produce a usable change in detector output when the level changes [Ref. 12].

7.54 ROTATING PADDLE

A paddle wheel or rotating paddle is a good means of point level measuring for solids and slurries. The paddle (Figure 7–44) is mounted inside the process vessel, and rotation is powered by a small electric motor (~1/100 hp). The paddle is connected to the motor by a torsion spring and turns continuously as long as it is free to rotate. When the level increases to the point where the process material touches the paddle, rotation ceases while the motor continues to turn. The rotation is slow, about 10 rpm, to keep paddle momentum from digging into the process material. The continued rotation of the motor causes the torsion spring to expand, striking and activating a limit switch. This action stops the motor and initiates control relay action that empties the tank, activates an alarm or provides some other function.

When the material in the vessel falls away from the paddle, rotation is again established as the torsion spring unwinds and collapses to its original position, releasing the limit switch. Such operation is convenient in blending and batch control applications.

The paddle must be located away from areas of the tank where falling solids and dust particles contact the wheel, causing fouling and erroneous operation of the sensing device. Failure to do so will hamper reliable operation of the measuring system. To prevent the buildup of material in close proximity to the paddle area, baffles above the paddle may be required. Fine powder also has a tendency to enter the paddle shaft bearing and cause shaft seizure and false stops of the motor. Dust

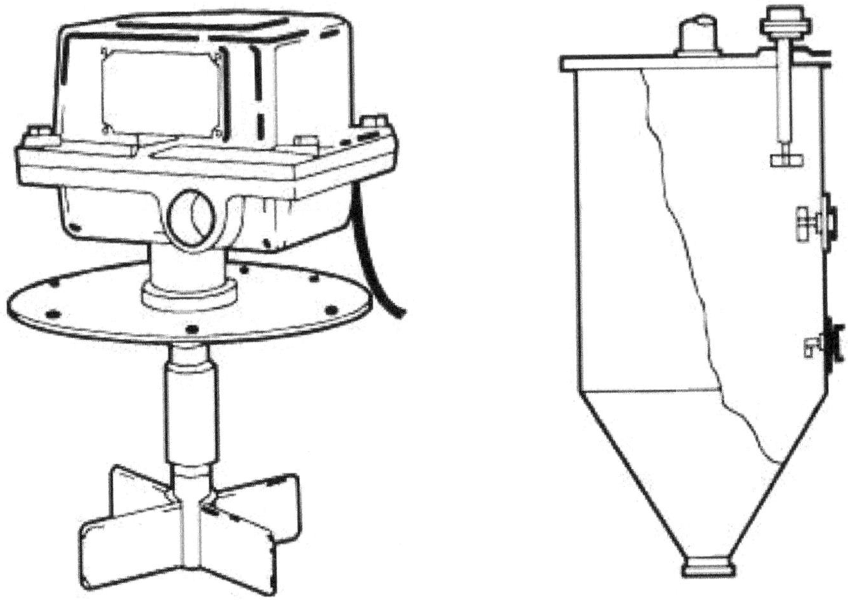

Figure 7–44 Rotating Paddle (*Courtesy of Bindicator*)

seals or small repetitive bursts of air, focused on the bearing to keep it clean, may be helpful.

Paddle switches are available in both general purpose and explosion proof housings. Generally, they are designed for low-temperature and low-pressure service. The paddles can be mounted at various levels on the side of the tank for multipoint level applications (Figure 7–44) or at the top if desired.

7.55 VIBRATION-TYPE LEVEL MEASUREMENT

The vibration-type level detector is similar in operation to the rotating paddle. It is a bit more sophisticated, however, and consists of a vibrating paddle that is used for high or low liquid level detection. The vibrating paddle is driven into a 120-Hz mechanical oscillation by a driver coil that is connected across a 120-V, 60-Hz line. A second coil with a permanent magnetic stator is located in the pickup end, and produces a 120-Hz voltage signal proportional to the vibration amplitude. When the process liquid contacts and stops the paddle, the voltage amplitude drops significantly because of the decrease in vibrations. This reduction in voltage operates a control relay. Some designs are two-wire devices that produce an 8- or 16-mA output depending on whether or not a liquid is present. The unit also performs diagnostics and will output 4 or 20 mA if a fault is detected.

The unit is available for explosion proof operation (class I, group D, division I classification). An advantage of the vibrating paddle over the rotating paddle is that it can be used for much higher pressure and temperature environments. Pressure limits as high as 3000 psig with temperatures as high as 300°F are possible.

7.56 THERMAL LEVEL MEASUREMENT

Thermal level technology is based on the difference between the thermal conductivity of a liquid to be measured and that of vapor in the space above the liquid. The detector for thermal level measurement is an RTD with a specific resistance/temperature relationship. Equation 7.26 explains the relationship of the various resistance values used in thermal level measurement:

$$R_s + R_n + R_H = R_t \tag{7.26}$$

where

R_t = the total resistance

R_n = the nominal resistance value of the RTD

R_s = the self-heating value of the RTD resulting from the bridge current

R_H = resistance value of the RTD when immersed in the process fluid

RTDs have a positive coefficient of thermal resistance. The resistance varies directly with the temperature.

Figure 7–45 shows a temperature sensor (R_m) connected in a bridge circuit. The bridge current is selected by values of R_1 and R_2 to heat the temperature element to a safe level of power dissipation. The bridge is balanced for either high- or low-level conditions by R_3 and R_4. When the level moves past the sensing element, the sensing element will be either cooled or heated, depending on the relative temperature of the process liquid with respect to the vapor. The resulting value of R_t with the change in temperature will unbalance the bridge, producing an

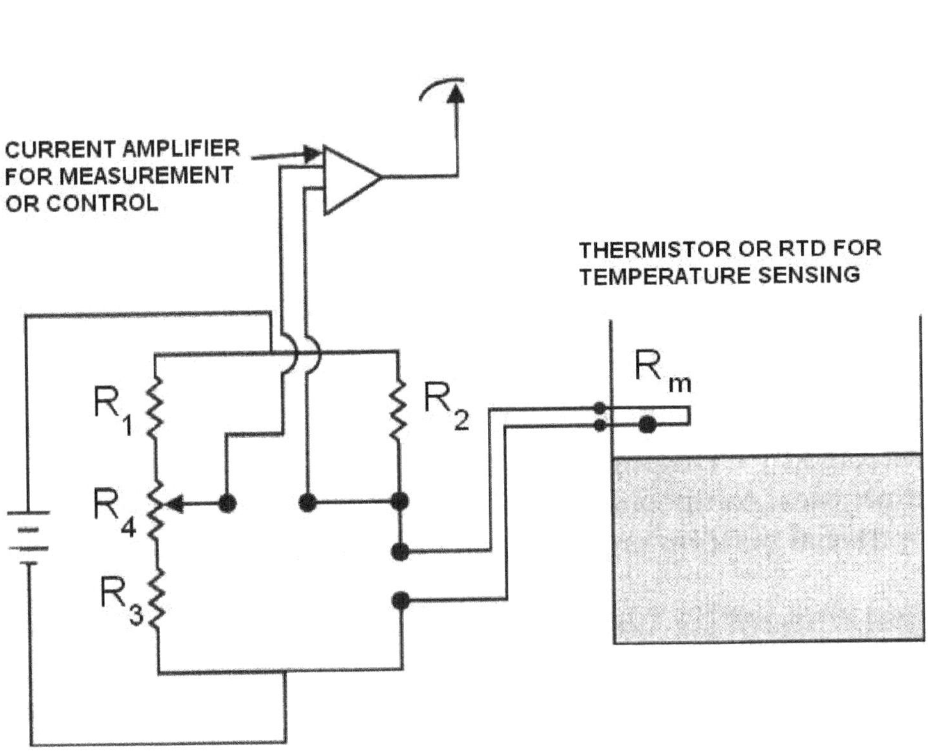

Figure 7–45 Temperature Element and Bridge Circuit

amplified current capable of operating a control relay for alarm or control functions, as desired.

Being much more sensitive to temperature changes, thermistors can be used as sensing elements instead of RTDs when small changes in temperature between the liquid and the vapor space are expected. The resistance of a thermistor varies inversely with temperature. The bridge circuit should be designed to be compatible with either thermistors or RTDs.

To provide protection for the sensing device, it is usually installed in a thermowell; however, this increases the response time of the sensor. The measuring circuit can be designed to be very accurate and sensitive to small level variations. Common usages are in liquid/liquid and liquid/air combinations, liquid/vapor and slurries. When temperature changes in the vapor section of a tank are great enough to cause erroneous responses, a second sensing element can be mounted at a point above maximum level and connected to the bridge circuit, replacing the resistor (R_3 in Figure 7–45). Static tank temperature variations will then cause equal changes of resistance in opposite bridge components, thus maintaining bridge stability.

The thermal level measurement just described is actually a level switch used for contact or point indication. By a different arrangement, level detection can also be used for continuous level measurement. This technique uses two RTDs instead of the single element employed in contact measurement. The difference between the thermal conductivity (heat transfer) of the liquid and vapor above the liquid bases the concept of a continuous temperature level measurement device.

A thermal-differential probe is the sensing element and consists of a cylindrical tube containing twin RTDs that extend the active length of the probe in a parallel configuration. When energized with a current for heating, the resistance of the two elements is proportional to the average temperature of the liquid and the vapor space above the liquid. Current applied to one of the elements causes it to heat to a temperature of its dry surface to a few degrees above the process temperature in the vapor space. When the sensors extended in the tank over the measurement span are exposed to both the liquid and the vapor, a thermal differential is created in the heated sensor due to the cooling effect of the greater thermal conductivity of the liquid. The change in resistance of the heated sensor is linearly related to the length cooled by the liquid which has a greater thermal conductivity than the vapor, and the level can be determined. The parallel unheated sensor experiences the same process conditions as the heated sensor; process temperature effects are thus eliminated.

Thermal-differential sensors are viable for any two media with different thermal conductivities. It is therefore possible to measure liquid/foam, liquid/liquid and solid/liquid interfaces. A DC Wheatstone resistance bridge circuit is commonly used to measure the resistance of the detectors. A signal-conditioning circuit can provide a 4–20 mA current linearly proportional to level over a specified range, usually up to about 20 ft. Thermal level detectors are unaffected by changing dielectric constants, moisture content, or chemical and physical composition changes and are less susceptible to error caused by material buildup on the probe.

The two sensors (RTDs) can be installed inside a stilling well or basin and can be used to measure in the temperature range of –24° to 28°F.

7.57 LASER LEVEL MEASUREMENT

The operating principle of laser level measurement is similar to ultrasonic and radar, where a transmitter mounted on the top of the vessel sends a signal toward the level surface. The measured level reflects a portion of the signal back to a receiver also mounted on the top of the vessel. The time measured between the transmitted and reflected beam with the known velocity of the traveling wave is used to calculate the distance from the transmitter and receiver to the level surface. The ultrasonic unit uses pulses of acoustic energy, the radar method uses electromagnetic waves from the microwave spectrum, and laser technology uses infrared light. This light can be formed as electromagnetic radiation or corpuscular current. The physically different types of wave lengths behave differently during propagation and reflection. The laser uses a very short wavelength, as short as 1 micrometer. This short wavelength reduces angular divergence even over long distances. The speed is 0.2998 meters per nanosecond (10^{-9} seconds) in a vacuum and slightly less, 0.2997 meters per nanosecond in ambient air. Even with the high speed and short time measurements involved, by using fast electronic components and various software algorithms the accuracy and resolution required for level measurement can be realized.

Because the speed of light is nearly constant regardless of the material in the vapor portion of the vessel and because a medium is not needed for propagation (which is not true for ultrasonic applications), the type of material in the vapor is insignificant. However, the presence of dust, fog and steam does interfere with laser level measurement and care should be exercised when such are present as these can preclude the use of laser technology.

The type and condition of the process surface should also be considered. When measuring the level of solids, the particle size should be greater than the wave length of the measurement signal, otherwise interference signals such as indirect reflection from vessel walls and obstructions can result. This situation is not likely, however, because of the very short wavelength of the laser and because laser beams are nearly parallel so interfering reflections from braces, material build-up, vessel, joints, etc. are usually not a problem.

The characteristics of the process material and liquid surface can determine the success of laser level applications. Static and still transparent liquids such as clear water are not good candidates for laser measurement because the laser beam may not be reflected by the surface but will travel to the bottom of the vessel before being reflected. Ideal choices for surfaces are bright solids and cloudy liquids. Black surfaces produce a weak reflected signal and glossy surfaces can cause the signal to be reflected laterally.

Some laser level applications that are known to be successful include wood pulp (even when steam is present) green and brown colored silo fill material during non-filling times, brown tobacco, beer without thick foam, and high-temperature applications like molten aluminum and liquid polyethylene. Unsuccessful applications are some polyurethanes, fermenters with transparent liquids, some cleaning agents, and liquid copper.

Factors to be considered in laser level applications are the possible degradation of beam strength at the transmitter or receiver (transponder), dirt or coating on the transponder, beam attenuation through a sight glass and material in the vapor space, poor reflective qualities of the process surface and possible reflection from surface foam. An air purge with dust guard can help prevent coating of the optics.

Laser level detectors are good for many difficult situations. Specific applications should be reviewed with the manufacturer and tested under actual conditions. A laser level measuring system is shown in Figure 7–46 [Ref. 13].

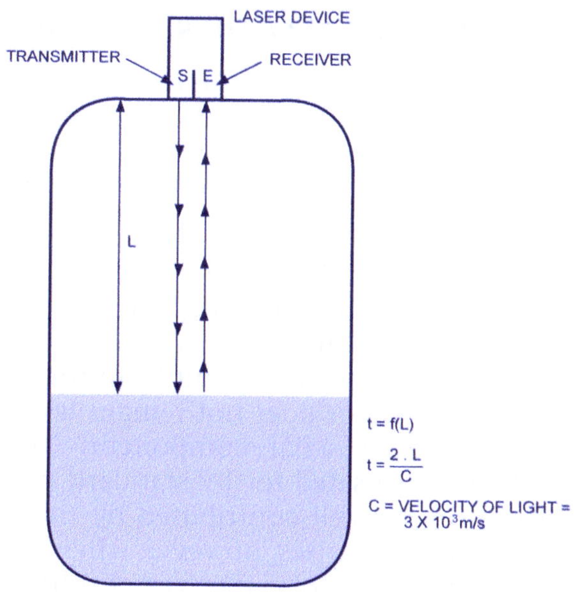

Figure 7–46 Laser Level Measuring System (*Courtesy of CRC Press*)

7.58 LEVEL MEASUREMENT BY WEIGHT

Occasionally, level measurement in a process vessel presents such a challenge that none of the previously mentioned level measurement techniques is satisfactory. A reactor-type container, for example, may be heated to a fairly high temperature and contain a highly viscous slurry and an agitator so that the interface point is not constant enough for reliable measurement. A "last-ditch" effort to quantify the vessel's contents may involve a weight measurement system. This really is a mass or volumetric measurement, one that is often lumped together with level measurement principles. Caution: successful level measurement by weight is usually obtained by designing the system (vessel, piping, supports, etc.) for weight measurement. Load cells retrofitted to vessels that are not designed and installed for their use may produce unsatisfactory results.

The arrangement in Figure 7–47 shows a tank supported on three load cells. The empty weight of the tank is nulled to give a zero reading on the indicator; additions of material into the tank result in a response at the indicator. The load cells may be hydraulic or electrical strain gages.

7.59 MOUNTING AND INSTALLATION OF LOAD CELLS

Most load cell data sheets provided by the developer or manufacturer will state such factors as input-output relationships, non-linearity, hysteresis, drift and temperature sensitivity. Such parameters are based on ideal or axial loads which are normal to the cell. The load must act on the cell at a 90° angle between load and cell.

Angular loads, those applied to the cell at other than a 90° angle, will act on the cell at some point other than that intended. These loads can originate from two conditions. First, they can result from an axial load plus a side load, and second, they can result from the load acting on a cell that is not mounted level or does not remain level, since the cell responds only to the load's axial component. This situation is common and can be compensated for by standard calibration procedures when the force on the cell contributed by the angular load is small and constant. A problem arises, however, when the angular loads shift during operation, the result of cell or vessel shifts or other conditions that can be caused by poor mounting procedures and practices.

Eccentric loads are applied parallel with the axial load but are not concentric nor at the target point. It is desirable for the cell to respond

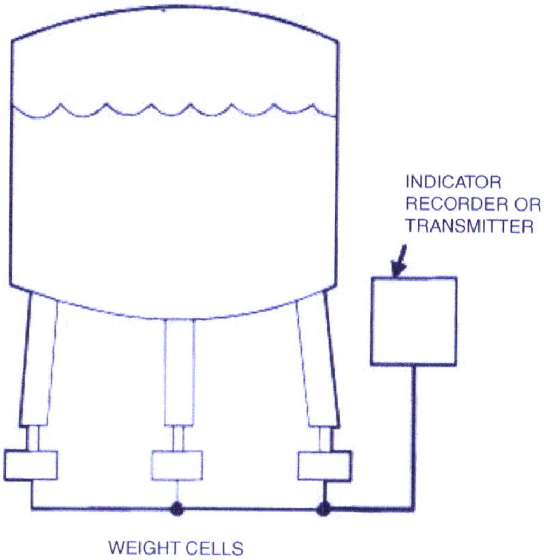

Figure 7–47 Level Measurement by Weight Method

to an eccentric load with the same sensitivity as to an axial load, but this is not normally the case. Just as a constant angular load can be compensated for by calibration, an eccentric load with constant eccentricity is not a source of error when calibration-compensated. It is when the load shifts position during operation that an error source exists. Side loads act on the cell 90° from the primary axis of the axial load. Side loads are somewhat common and are introduced by shifting loads, expansion and contraction of the structure caused by pressure and temperature, and other possible infractions. The total force on the cell contributed by side loads is generally small and is difficult to generate in a laboratory environment. As with other non-axial loads, constant side loads can be compensated for in calibration.

To reduce error and malfunction caused by the non-axial forces mentioned, good standard practices should be considered when installing and mounting load cells. The mounting surface should be level and remain so for all values in the measuring range. Piping should not hinder the movement of the vessel that is caused by changes in weight. The mounting surface should be flat and the mounting bolts should keep the cell firmly attached to the foundation. The mounting bolts should be grade 8 or equivalent high tensile quality so they do not stretch under load, causing the cell to become loose in its contact with the foundation. The bolts should be tightened to the recommended torque [Ref. 12].

7.60 HYDRAULIC LOAD CELLS

When hydraulic load cells are used, each cell is filled with hydraulic fluid until the piston floats to a reference mark, indicating that the chamber and associated components are full. The output of each cell goes to a summer or averaging relay, and that output is fed to a readout device. Almost any of the conventional types of pressure elements commonly used can serve as the recording or indicating component with the scales calibrated, usually by empirical means, to read weight units. A pressure transmitter can also be used, and the weight values can be converted to any desired unit for transmission to a recorder, controller, or other device.

Hydraulic load cells are relatively trouble-free for extended periods and require no outside power source for indicators only. They respond quickly to load changes and within certain limits are unaffected by small variations in fill material. Leaks in the system, however, will cause it to become inoperative.

7.61 STRAIN GAGES

A strain gage is a passive transducer that uses electrical resistance variations in wires to sense a strain (response to a force). Most strain gages are of the bonded type shown in Figure 7–48. When a load is applied to the supporting column, the column is compressed, causing the wires in the gage bonded to the sides to decrease in length, increase in cross-sectional area, and decrease in resistance. The strain-gage support in Figure 7–48 is designed so that the gages on the Y sides are unaffected by force on the support. They are to null the effect of temperature variations. The unbalance of the four-element bridge shown in Figure 7–49 is proportional to the force on the load cell that contains the strain gage. The output is linear with load changes on the cell.

The cell operation explained is that of a compression type, but Young's modulus of elasticity, which relates wire tension and length variations to cross-sectional area and temperature changes, is true for tension or compression. The load cell support can be designed for tension as well as compression. To help in the linearization of the relationship of bridge response to force, the bridge components are designed to produce relatively small output signals for corresponding variations in force. The overall bridge sensitivity is established in initial component design and is determined by the bridge excitation voltage, compensating resistors, strain-gage sensitivity, and detector

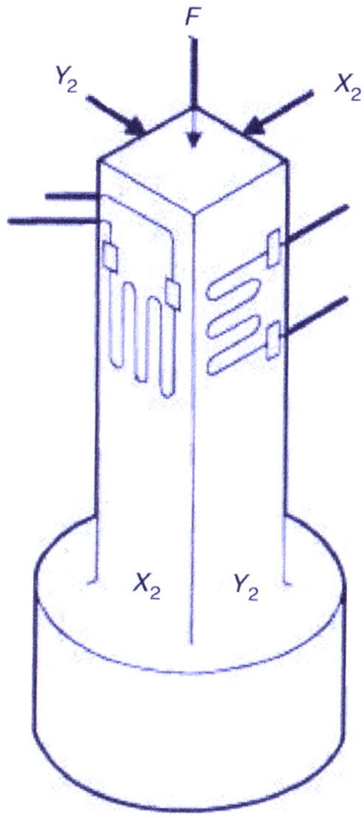

Figure 7–48 Level Measurement by Weight Method Using a Bonded Strain Gage

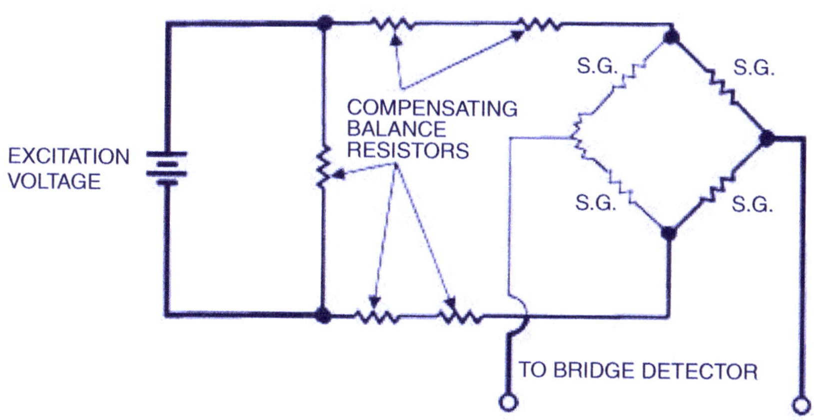

Figure 7–49 A Bridge Circuit for Strain Gage Measurement

sensitivity. The excitation voltage may be either AC or DC, but to avoid reactive component compensation, a bridge voltage of 10 to 20 V DC is common.

The force on the load cells should be normal to the cell; angular or nonaxial loads can produce error, because the strain gage cannot discriminate between the forces resulting from axial or bending loads. Without special consideration, overloads should not exceed 125% of rated capacity.

Strain-gage load cells are compact and can measure up to 250,000 lbs, but the size does increase with the load rating. They can measure up to their rated capacity with deflections ranging from 0.005 to 0.01 in. As is true of hydraulic cells, the output of several cells can be added electrically to obtain the total force or weight that is exerted on more than one cell.

By knowing the force/resistance relationship of the strain gages, equivalent bridge circuits can be used to calculate the unbalance voltage relating to different weights. However, the calculation would require the determination of many variables and would at best produce "ballpark" results; therefore, analytical methods of calibration are generally not suitable. Empirical calibration methods are generally preferred.

EXERCISES

7.1 Explain the principle of level measurement by the following methods:

- Resistance tape
- Capacitance
- Conductance
- Ultrasonic
- Radar
- Pulse radar
- FMCW

7.2 Explain the difference between point and continuous level measurement and list applications for each.

7.3 List three contact and three noncontact methods of level measurement.

7.4 List two radioactive isotopes used in nuclear level measurement and the forms of radiation emitted by the source. Which of these is applicable to level measurement?

7.5 Describe the principles of nuclear radiation level measurement and explain the effect of other process conditions on measurement by this method.

7.6 What level-measurement technique can be applied to the measurement of solids and slurries?

7.7 State the effect of temperature variations on sonic level measurement.

7.8 How can radar technology be used to measure a liquid/liquid interface?

7.9 List an advantage of the "end-of-the-probe" application.

7.10 List three possible sources of interference in radar level measurement.

7.11 What is the primary device used in magnetostrictive level measurement?

7.12 What is the advantage of a lineal measurement that is tank-bottom based as opposed to that which is tank-top based?

7.13 Can interface measurement be provided by magnetostrictive technology? If so, how?

7.14 Why may reactor level measurement be suitable for a weight-based level measurement application?

7.15 Why is laser level measurement more immune to the characteristics of the material in the vapor section of a vessel than ultrasonic and radar level technologies?

7.16 Why is the type and condition of the liquid surface a consideration in laser level applications?

7.17 Why is interference by reflected signals normally not a problem with laser level measurement?

REFERENCES

1. Schuler, Edward L., "Sensors," *Measuring Level with RF Admittance*, 1989.
2. Groetsch, John S., *Application of Ultrasonic Level Monitoring,* Research Triangle Park, NC: ISA, 1989.
3. Soltz, Dan, *Improved Ultrasonic Level Measurement,* Research Triangle Park, NC: ISA, 1989.
4. KROHNE, *Fundamentals of Radar Technology for Level Gauging,* 2003.
5. SAAB Rosemount Tank Radar Technical Literature.
6. Westerlind, Hans G., *Level Gaging by Radar,* Research Triangle Park, NC: ISA, 1989.
7. Vermeulen, Bob, "The End-of-Probe Algorithm", *Flow Control* magazine, Sept. 2004.
8. Radar Technology, 4th Edition, Krohne Messtechnik Ganlett and Co.
9. Buffone, Louie, *Fiber Optic Liquid Level Sensor Application.* Research Triangle Park, NC: ISA, 1989.
10. Piccone, Roland, *Combining Inductive Liquid Level Technology with a Pressure Sensor to Produce a Highly Accurate Hybrid System,* Research Triangle Park, NC: ISA, 1992.
11. Hendrix, W.L., *Industrial Applications of Radar Technology for Continuous Level Measurement,* Research Triangle Park, NC: ISA, 1992.
12. Flow Level and Environmental Handbook, Omega Company.
13. Lipták, Béla M., *Process Measurement and Analysis,* 4th Edition, Volume 1, CRC Press, 2003.

8

Liquid Density Measurement

8.1 INTRODUCTION

Density is defined as mass per unit volume. The density of a substance is an important characteristic that is often used to determine such properties as concentration, composition, and the BTU content of some fuels, and for mass flow and volumetric applications. Liquid properties are often measured in terms of density and are expressed as specific gravity, which is the ratio of the density of a substance to that of water. Most liquids are only slightly compressible, and pressure effects in density and specific gravity applications are usually neglected. Most industrial density instruments operate on the principle of density changes with respect to weight.

8.2 UNITS AND DEFINITIONS RELATED TO DENSITY

Relative density, or specific gravity, is defined as the ratio between the density of a process material to that of water for liquid applications. Being a ratio, this property has no associated units. Both density and specific gravity express the same physical property of a material and they are meaningful only if defined at a stated temperature. In the case of specific gravity, the temperature can be different for the process and the reference fluid. This is acceptable but must be clearly stated. A specific gravity table, for example, might list a process fluid as having a specific gravity of $0.92^{80/40}$. This means that the liquid at 80°F (27°C) will have a density of 0.92 times that of water at 40°F (4.4°C).

Since density is expressed as mass per unit volume, these units must be expressed. Commonly used units are grams per cubic centimeter (g/cm^3), kilograms per cubic meter (kg/m^3) and pounds per cubic foot (lbs/ft^3).

Also, specific gravity can be expressed in different scales. Following are some scale conversion factors:

- For petroleum products, the American Petroleum Institute (API) introduced the unit of API degrees:

$$°API = \frac{141.5}{SG@60°F} - 131.5 \tag{8.1}$$

- The brewing and sugar industries use Balling degrees to express the weight percentage of dissolved solids at 60°F (17.5°C):

$$°Ba = \%\text{ by weight of dissolved solids} \tag{8.2}$$

- The sugar industry also uses Brix degrees which represent the weight percentage of pure sucrose in a 60°F (17.5°C) solution:

$$°Br = \text{weight }\%\text{ of sucrose} \tag{8.3}$$

- The tanning and tanning-extract industry expresses specific gravity in Barkometer degrees. Each degree is 0.001 SG above or below SG = 1.0; therefore, Barkometer degrees can be negative or positive (+10 °Bk = 1.01 SG):

$$°Bk = (SG - 1.0)(1000) \tag{8.4}$$

- The density of acids and of light and heavy syrups is expressed as Baumé degrees. This scale was introduced by Antoine Baumé and consists of two scales, one for "heavy" liquids (above SG = 1), the other for "light" (below SG = 1) liquids at 60°F (17.5°C):

$$°Be\text{ (light)} = (140 \div SG @ 60°F) - 130 \tag{8.5}$$

$$°Be\text{ (heavy)} = 145 - (145 \div SG @ 60°F) \tag{8.6}$$

- The dairy industry uses Quevenne degrees in testing the fat content of milk. Each degree corresponds to 0.001 SG in excess of SG = 1.0, so 40°Q = 1.04 SG:

$$°Q = (SG - 1)(1000) \tag{8.7}$$

- The alcohol industry uses Sikes, Richter, or Tralles scales on their alcohol meters. They each use the volumetric percentage of ethyl alcohol in water:

$$°S, °R, °T = \text{volume \% of ethyl alcohol} \tag{8.8}$$

  The alcohol industry also uses the unit of proof which is °S, °R, or °T multiplied by two:

$$\text{Proof} = 2\ (\text{volume \% of ethyl alcohol}) \tag{8.9}$$

- The sugar, tanning, acid industries and other industries working with liquids that are heavier than water all use Twaddell degrees. Twaddell divided the SG range between SG = 1.0 and 2.0 into 200 equal parts, each representing 0.005 SG [Ref. 1]:

$$°T_W = 200\ (SG - 1.0) \tag{8.10}$$

A density meter is an instrument that automatically measures the density of a liquid. As used in this context, "density" is the degree of the mass of substances that are dissolved or suspended in a liquid such as water, and is expressed as percent weight.

For example, when 10 grams of starch are mixed into 100 grams of water, the density of the mixture is 9.1% as calculated in the following expression:

$$\frac{10\ g}{(100 + 10)\ g} \times 100\% = \frac{10}{110} \times 100\% = 9.1\% \tag{8.11}$$

Density, also called consistency or concentration, is a very important index of control in such processes as paper making and sludge treatment. "Consistency" as used in the paper industry designates the concentration of weight of air-dry pulp in the liquid (mixture of water and solids including pulp).

$$\text{Air-dry consistency} = \frac{\text{Air-dry weight of solids}}{\text{Weight of water and solids}} \times 100\% \tag{8.12}$$

Also, consistency is used to mean the total solids density, which is the same as the terms bone-dry, oven-dry and moisture-free consistency.

The bone-dry method is explained as the dry weight method or manual analysis. This drying method is primarily used in laboratory analysis when air-dry consistency is considered to contain 10% moisture in the solids, so that

$$\text{Bone-dry consistency} = 0.9 \times \text{air-dry consistency} \tag{8.13}$$

The solids which are suspended or dissolved in water are measured by the dry weight method, which is used as a reference or a standard value. In the dry weight method, a weighed amount of sample is heated to about 105°C to evaporate the water content; the remaining solid weight is measured to calculate the density. Because measuring is done manually using a weight balance, this method is also called "manual analysis."

Following are examples of manual methods of density measurement.

When a 50 g sample is dried to 2 g of solids, the density is

$$2/50 = 0.04 = 4\%$$

- Total solids (TS) density and suspended solids (SS) density:

  TS (Total Solids) density—The density of total solids, which is the sum of suspended solids and dissolved solids in water. Total solids density = suspended solids density + dissolved solids density.

  SS (Suspended Solids) density—The density of suspended solids in water.

  Dissolved solids (DS) density (or total dissolved solids, TDS density)—The density of dissolved solids in water.

- Density (consistency) in the case of pulp:

  This method is berst illustrated using a measurement example.

  A sample of 100 g is placed on filter paper to filter out water. The solids remaining are dried in an oven together with the paper, then are weighed. Since the dissolved solids are filtered out with water, dissolved solids will not remain on the paper and cannot be measured. The measured density obtained by this method is the Suspended Solids density.

- Measurement example in sewage sludge application:

  A 10 to 50 g sample is dried on an evaporating dish and weighed again. The density given in this method is a Total Solids density. Measurement in this procedure takes more than two hours.

  A simplified test is often performed by using a commercially available water content meter which has an infrared lamp and a balance. With this method, for a sample of about 10 g, one can measure and get results in less than 30 minutes. The density obtained by this method is Total Solids density.

  A 50 mL sample is placed in a centrifuge to separate precipitated solids and supernatant liquid, which is drained and evaporated. The precipitated dry hard solids are weighed to obtain the density [Ref. 2].

These methods are accurate enough for most applications but are considered laboratory applications requiring manual measurements and calculations. In most process applications, however, continuous and unattended density measurement methods are required. Several such methods are discussed in this chapter.

8.3 DENSITY MEASUREMENT BY HYDROSTATIC HEAD

In Chapter 6, methods to infer a level value from a hydrostatic head measurement were discussed. Recall that pressure measured at some reference point on the vessel is converted to liquid level by the formula

$$P = 0.433\frac{\text{psi}}{\text{ft}}(G)(H) \tag{8.14}$$

where P is the measured pressure in pounds per square inch, G is specific gravity, and H is the level in feet measured from the reference point. When the level from a reference point is a constant, the hydrostatic pressure becomes a function of specific gravity. Figure 8–1 illustrates this principle. As H is a constant value, specific gravity changes will cause changes in hydrostatic pressure.

Several arrangements can be used for hydrostatic head density measurement; a flange-mounted pressure or differential-pressure instrument is often preferred. Dip-tube instruments (Figure 8–2) are also used. Figure 8–1 shows that with a constant height of liquid

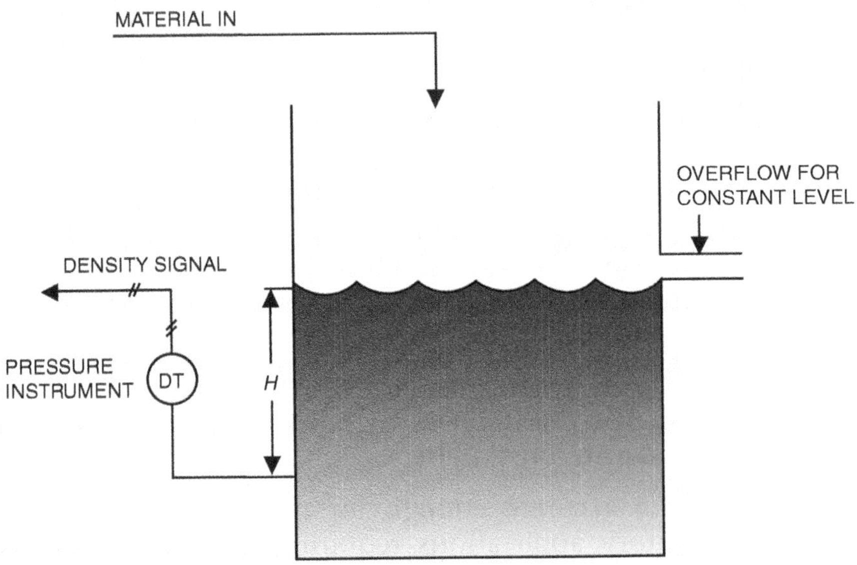

Figure 8–1 Constant Level Application for Density Measurement

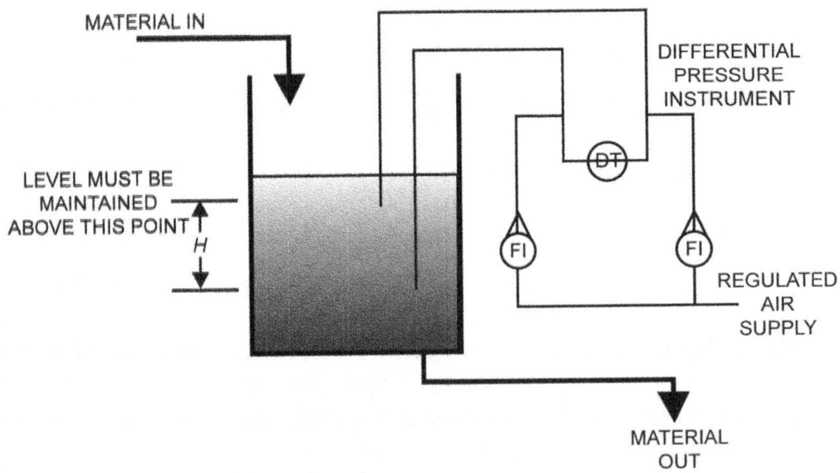

Figure 8–2 A Dip Tube Arrangement for Density Measurement

exerting a pressure on the pressure instrument, the output will change as the weight of the liquid changes with specific gravity. To measure the change in head resulting from a change in specific gravity from G_1 to G_2, H is multiplied by the change in specific gravity:

$$P = H(G_2 - G_1) \tag{8.15}$$

where P is the difference in pressure (usually expressed in inches or meters of water), G_1 is the minimum specific gravity, G_2 is the maximum specific gravity, and H is the distance between the high and low pressure points. Instead of measuring actual specific gravities G_1 and G_2, it is usual practice to measure only the span of change. The initial measurement zero is suppressed, or the lower range value is elevated to the minimum head pressure to be measured. The entire instrument span becomes a measure of the pressure differential caused by changes in the density of the fluid to be measured. Examples will help to explain the suppression principle with respect to density measurement.

EXAMPLE 8-1

Problem: Referring to Figure 8–1, H = 100 in. and G_1 = 1. Find the amount of suppression needed.

Solution: In order for the measurement to start from a zero reference point when the specific gravity starts to change from the minimum value of 1, the instrument zero must be suppressed by an amount equal to $H(G_1)$, which is 100 in. The zero suppression is therefore equal to 100 in. of H_2O.

EXAMPLE 8-2

Problem: Again referring to Figure 8–1, H = 2 m and G_1 = 0.5. Find the amount of zero suppression required.

Solution: In order for the measured response of the instrument to start at a zero reference point as the specific gravity increases from G_1, the instrument's zero must be suppressed by an amount equal to $H(G_1)$. Therefore, zero should be suppressed by 2(0.5) = 1 m H_2O.

For density and specific-gravity applications using hydrostatic measurement principles, the following relationships are important:

$$\text{Span} = H(G_2 - G_1) \tag{8.16}$$

$$\begin{aligned}\text{Zero suppression} &= H \times G_1 \\ &= 2\text{m}(0.5) \\ &= 1\text{m}\end{aligned} \tag{8.17}$$

It is evident that low-gravity-span applications will require instruments with low measurement spans. Vessels with greater H dimensions may also be required.

Flange-mounted pressure and differential-pressure instruments are sometimes preferred for density measurement in closed tanks and pressure vessels. Such arrangements are shown in Figure 8–3B. The wet leg shown in Figure 8–3A may require a seal fluid if the process fluid should be isolated from the instrument. The specific gravity (G_S) of the seal fluid should be greater than that of the process fluid (G_1), and zero elevation will be required. Equation 8.18 is used to calculate the amount of elevation required:

$$\text{Zero elevation} = H(G_s - G_1) \tag{8.18}$$

When the use of a wet leg is undesirable, a pneumatic repeater can be used (Figure 8–3B). The upper transmitter, the repeater, simply transmits the measured pressure to the low side of the lower transmitter. The process pressure at the point of the repeater is subtracted from the pressure at the lower transmitter, and the output of the lower transmitter becomes a function of the specific gravity of the process fluid. Equations 8.19 and 8.20 relate to the lower specific gravity transmitter, and Equation 8.21 relates to the repeater:

$$\text{Span} = H(G_2 - G_1) \tag{8.19}$$

$$\text{Zero Suppression} = (H)(G_1) \tag{8.20}$$

$$\text{Repeater output (max.)} = [D_1(\text{max})][G_1][P(\text{max})] \tag{8.21}$$

where D_l is the level measured from the repeater position and P is static pressure.

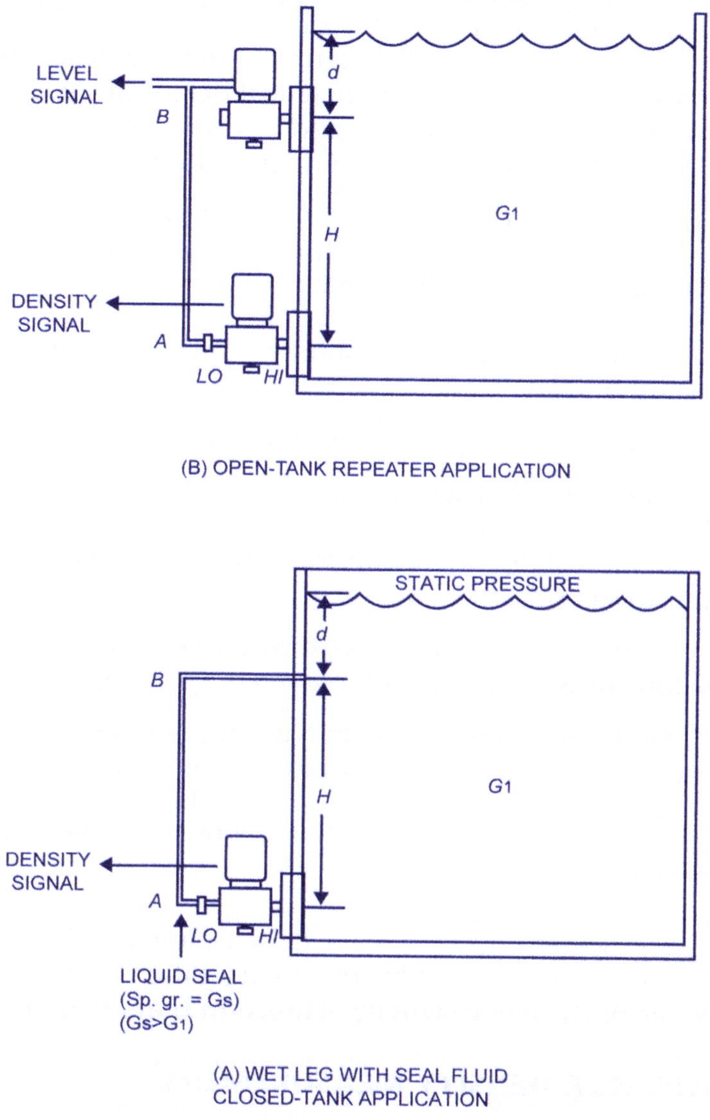

Figure 8–3 Flange-Mounted Instrument for Density Measurement

Because of the expansion of fluids with temperature and the resulting changes in density and specific gravity, it may be necessary either to correct for temperature variations in the process fluid or to control the temperature at a constant value. Necessary corrections to a desired base temperature can be part of the density instrument's calibration procedure.

Conventional pressure and differential-pressure instruments can be used in hydrostatic density and specific-gravity applications. The nature of the process fluid will determine the practical design and range to be selected. The maximum practical height of the liquid vessel and the minimum span of gravity and density changes will set the limitations of hydrostatic measurement applications.

Head type density measurement is low cost, simple, and easy to install, and can be dry calibrated. There are, however, several points that may be limitations and must be considered:

- All wet legs must be at a constant height, with seal material of constant density.
- The stratification of process fluids can result in error.
- Small differences in specific gravities or a limitation in height will result in a small-span requirement for the transmitter, although most digital transmitters can accommodate low range applications.
- Level height control is critical, or the level must be constant. The low pressure tap must be covered by the process fluid.
- The condensation of vapors in a dry leg will result in error. A means of drainage may be required when vapor condensation is expected, or a wet leg with a seal fluid may be used as required. Also, a repeater can be used to avoid the wet leg as in Figure 8–3A.

The multipurpose tank gage discussed in Chapter 9 for hydrostatic tank gaging measures density by the hydrostatic method. The operating principle of the device will be detailed in that chapter.

8.4 DISPLACER DENSITY MEASUREMENT

A detailed description of the principle and operating features of displacer level measurement was given in Chapter 5. This principle also relates to density measurement; a review of the earlier material will aid in the understanding of density measurement by displacers.

The fundamental difference between level and density displacer applications is that in the case of level measurement, the density of the process liquid remains constant and the movement of the displacer is a result of level changes. In density applications, the displacer is always fully submerged and the change in buoyant force is a result of process density variations only.

Figure 8–4 shows the basic components of a torque-tube displacer instrument used for density measurement. The process fluid enters the chamber around the center portion through a piezometer ring, which eliminates the velocity effects of the fluid. In applications where the fluid velocity is less than about 2 ft/min, the fluid should be stable enough so that velocity effects will not affect instrument operation. When it is desired to measure the density of material flowing in a pipe, a sample is routed through the chamber. When density detection in a vessel is desired, the displacer is mounted in an appropriate flange connection, as it is suspended in the process fluid.

Displacers for density measurement are sized in accordance with the amount of buoyant force required for a particular measurement range:

$$\text{Displacer volume (in}^3\text{)} = \frac{\text{Torque tube force lb}}{(G_2 - G_1)(0.036\ \text{lb/in}^3)} \tag{8.22}$$

where the torque tube force is that required for a particular instrument and application, $(G_2 - G_1)$ is the range of measurement desired, and 0.036 lb is the weight of 1 in^3 of water.

EXAMPLE 8-3

Problem: A measurement application requires a torque tube force of 0.36 lb. The desired specific gravity range is 0.975 to 1.000. Find the volume of a displacer with a standard diameter and length for the application.

Solution: From Equation 8.22,

$$V = \frac{0.36\ \text{lb}}{(0.025)(0.036)\text{lb/in}^3} = 400\ \text{in}^3 \tag{8.23}$$

From Table 8–1, a 402 in^3 displacer for that specific gravity range corresponds approximately to one that has a diameter of 4 in. and is 32 in. long.

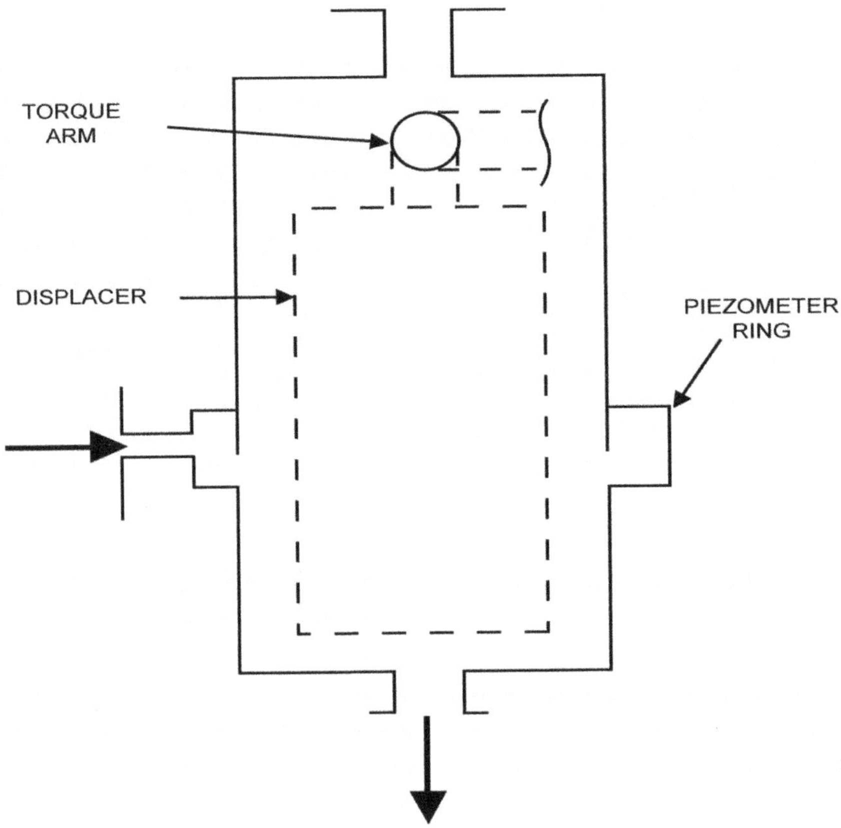

Figure 8–4 Displacer Type Density Sensor with Torque Tube

For density transmitters, displacer instruments can be pneumatic, analog or "smart" electronic. Temperature compensation can usually be provided as an option. Displacer density instruments are compatible with most fluid characteristics, with temperature and pressure ratings high enough for most applications. Displacers are best suited for clean, nonviscous fluids; a buildup of solids resulting from dirty liquids and slurries would result in errors caused by a change in the weight of the displacer.

8.5 RADIATION DENSITY MEASUREMENT

The principle of radiation density measurement is much like that of radiation level detection (Chapter 7). The obvious difference between

Table 8–1 Displacer Specifications (*Courtesy of CRC Press*)

Displacer Float			Minimum Specific Gravity Span	
Diameter in. (mm)	**Length in. (mm)**	**Volume in^3 (cm^3)**	**Standard Tube**	**Thin-Well Tube**
3 (76)	14 (356)	99 (1622)	0.202	0.101
3 (76)	32 (813)	226 (3703)	0.088	0.044
3 (76)	48 (1219)	340 (5572)	0.059	0.030
3 (76)	60 (1524)	425 (6964)	0.047	0.024
4 (102)	14 (356)	176 (2884)	0.114	0.057
4 (102)	32 (813)	402 (6588)	0.050	0.025
4 (102)	48 (1219)	602 (9865)	0.033	0.017
4 (102)	60 (1524)	753 (12,339)	0.027	0.014
6 (152)	14 (356)	396 (6489)	0.051	0.026
6 (152)	32 (813)	905 (14,830)	0.021	0.011
6 (152)	48 (1219)	1360 (22,286)	0.015	0.008
6 (152)	60 (1524)	1700 (27,858)	0.012	0.006

radiation level measurement and density measurement is that in level measurement, the process fluid density is constant or compensated for, and the amount of absorption of the transmitted gamma rays through the measured medium is a function of liquid level. Level measurement is based on the presence or absence of material. In radiation density measurement, the path length through the process fluid is constant and density is a variable. When gamma rays pass through a process fluid, they are absorbed in proportion to the density of the material. An increase in process fluid density will result in more of the radiation being absorbed by the fluid. This resulting attenuation is a function of density only, as the distance between the source and detector is fixed. The radiation is attenuated in accordance with the following:

$$I = I_0 e^{\mu \rho d} \tag{8.24}$$

where I is radiation striking the detector, I_0 is unattenuated radiation from the source, μ is the mass attenuation coefficient (absorption coefficient) (cm^2/g), ρ is the density of the process material (g/cm^3), and d is the thickness of the sample chamber (cm). The exponential function of the absorption is converted to an output that is linearly proportional to density.

A radiation density instrument will generally include the following major components:

- The radioactive source
- A shielding container for the source
- The detector
- The signal-conditioning circuit

8.6 RADIATION SOURCE

Atoms that have the same chemical behavior but a different number of electrons are called isotopes. Many elements have naturally occurring stable isotopes, while many others have unstable isotopes. These unstable isotopes disintegrate to form lighter elements. Radioactive disintegration is accompanied by the emission of three kinds of rays: alpha, beta, and gamma. Alpha and beta rays consist of electrically charged particles, are deflected by an electric or magnetic field, and have reduced penetrating power compared to nondeflecting gamma rays. Gamma radiation sources, because of their greater penetration, are used in radiation density measurement.

Two common gamma radiation sources are cobalt 60 and cesium 37. The radioactive disintegration of these elements causes cobalt 60 to produce nickel, and cesium 37 to form barium. It should be noted that the decay process produces electromagnetic energy, which cannot induce other materials to become radioactive. Gamma sources are thus safe for use around food processes. Another point of interest with respect to these sources is that they lose their strength as they decay. The rate of decay is expressed as half-life, which is the period of time during which the source strength decays 50%. Cobalt 60 has a half-life of 5.3 years and will decay about 23% per year. The half-life of cesium 37 is 30 years, with a decay rate of 2.3% per year. The size of the radiation source will be in accordance with the type and thickness of the vessel walls or the pipe size and material, the process material, and the amount of process material through which the radiation must pass. The size of the source will also depend on the specific application with respect to desired precision and the desired measurement system response time.

8.7 SHIELDING

Because gamma sources radiate energy in all directions, individuals within the vicinity of a radiation source could be exposed. Short-term exposure to high-intensity gamma radiation or long-term accumulated exposure to lower radiation should be avoided because of health hazards. Radiation sources are shielded to prevent these hazards.

The (source head) shielding can consist of a lead-filled steel pipe; the radioactive source field is allowed to exit the shield through a guide tube. The radiation exit channel is normally an angle of 12°. The radiation exit channel must always be closed during installation and is usually equipped with a blocking shutter with a safety lock. Some special designs have a pneumatic lock, with switch contacts indicating its position.

8.8 RADIATION DETECTORS

There are three basic types of gamma detectors: the Geiger tube, the ionization chamber, and the scintillation detector. The Geiger tube is a low-accuracy device that measures radiation through the ionization of a halogen gas sensitized to a potential of about 500 V DC. Ionization chambers (also called ionization cells) operate by the ionization of pressurized gas between two dissimilar metals under incident radiation that generates a very low current (on the order of a nanoampere). The scintillation detector senses the light photons resulting from gamma rays incident on certain crystal materials. For density applications, both scintillation and ionization detectors are used. When the ionization chamber detector is used, it is temperature-controlled to maintain stable environmental conditions. The scintillation detector is advantageous because of its increased sensitivity; it usually allows a smaller source to be used than would be permissible with an ionization chamber detector.

8.9 SIGNAL CONDITIONING

The detector output signal requires amplification and scaling, which is done by both DC and AC amplifiers. AC amplifiers were once preferred because of better stabilization. High-gain stable DC amplification by linear operational amplifiers and digital circuits is gaining prominence and is now more common. Drift caused by component aging, temperature shifts and other factors is compensated for electronically in the signal conditioning circuitry.

Minimum full-scale span is about 0.01 g/cm^3 (or another density unit), and depends on the thickness of the material to be measured. Corresponding resolution as great as 0.0001 g/cm^3 is possible. When measuring small spans, the zero drift caused by source decay can be an important consideration, however. A means of source decay compensation, incorporated into digital signal conditioners, is now available.

8.10 DENSITY GAGE APPLICATIONS

As with any measuring instrument, careful initial selection and installation are important for reliable and precise operation of a radiation density detector. When selecting the site for installation, several criteria are critical. The pipeline or sample chamber must be completely filled with the process fluid or sample at the detecting point. Corrosion, abrasions and material deposits on pipeline walls and sample chambers must be eliminated when possible. Installation on vertical pipes is preferred. Entrained gas-forming bubbles in the product will also cause errors and should be eliminated (the product sample chamber can be operated at higher pressure to reduce bubbles). Mechanical vibration of the pipe resulting from cavitation and so forth should be reduced to avoid error and damage to the detector.

The maximum operating temperature of some detectors and related conditioning equipment is about 50°C. When conditions warrant, cooling of the detector will be required. Cooling jackets are provided by some manufacturers and should be used when needed. Water is usually the best cooling medium in high-temperature applications because of its greater cooling capacity. Careful mounting and insulation can help reduce problems caused by temperature.

Microprocessors have significantly expanded the capabilities of these density gages as well as simplified their operation. All application data can now be entered into nonvolatile memory at the factory; thus field startup and calibration consist simply of a one-step standardization procedure: the gage's response to a known material (e.g., water) is measured. Multiple readouts, digital and analog, are available in any engineering units. High- or low-relay outputs can be easily adjusted by the operator to alarm, and the time constant of the system can be adjusted to optimize process control loops.

A typical density gage is mounted externally to a process pipe as shown in Figure 8–5. Recall the radiation equation given earlier (Equation 8.24). In the density gage configuration, the material thick-

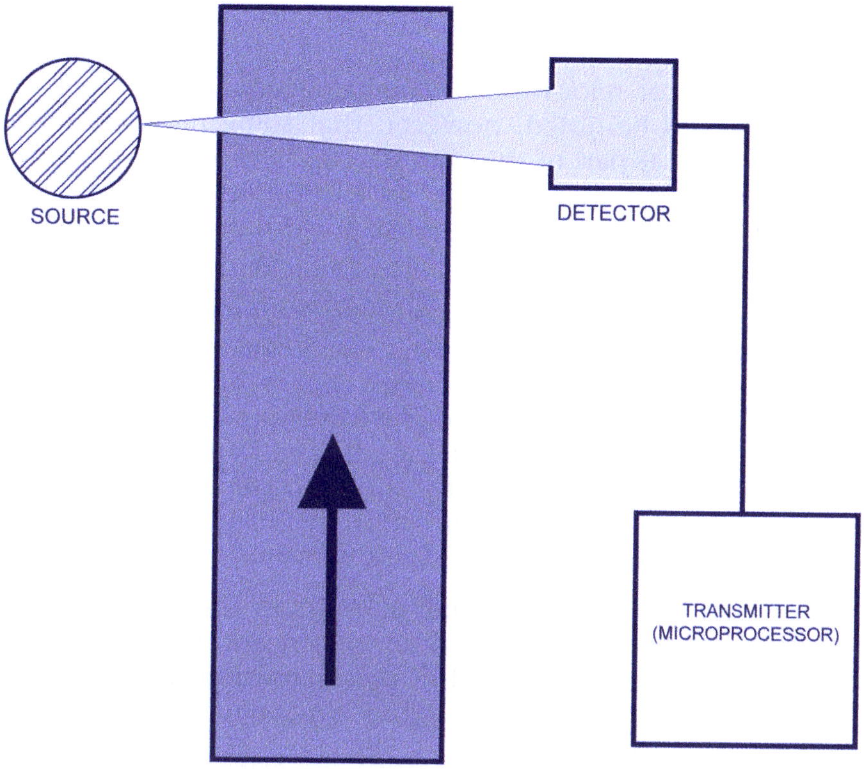

Figure 8–5 A Radiation Density Detector

ness *(d)* does not vary because the source and detector are fixed to the process pipe. Since the mass absorption coefficient (μ) is assumed to be constant, any change in the radiation intensity at the detector occurs only when the density (ρ) of the process material varies.

Nuclear density gages are employed for gases, liquids, slurries, solutions and emulsions. Because no contact is ever made with the process material, density measurement can be performed easily in caustic, corrosive and abrasive processes.

If density is to be measured inside a large vessel, it may not be practical to place both source and detector outside the vessel, as the source required would be too large. For this type of application, an insertion source can be used.

Since a density gage measures the total density of a process material, air or other entrained gases in the material may be a problem in some applications. The user may not know that the bubbles are present until the gage is installed. Entrained gas will cause the gage to read too low and possibly cause erratic fluctuations in the output. Replacement of faulty pump seals or relocation of the gage generally will solve this problem. It should be noted, however, that some products have air injected into them as part of the manufacturing process. Air, then, is a critical part of the product, and in such instances a nuclear density gage, which measures the bulk density, cannot ensure consistent quality control.

The buildup of solids on the inner pipe wall will also introduce error into the density reading. The worst case occurs when the material builds up gradually and then suddenly flakes off. This problem is usually solved by the installation of a process pipe section lined with glass, Teflon or some other material unaffected by buildup. For slurry applications, the source and detector should be mounted on a vertical pipe section if possible. This configuration ensures a full pipe and prevents the stratification that can occur with horizontal mounting.

In most process streams, very small changes in the mass absorption coefficient of the process material are insignificant. For elements with atomic numbers 2 through 50, the mass absorption coefficient for cesium 37 radiation varies only slightly and may be assumed to be 0.077. The mass absorption coefficient increases as the atomic number goes above 50, due primarily to increased photoelectric absorption. Hydrogen, because of its high electron density, has an absorption coefficient of 0.1537. This is not a problem for solutions or slurries with a water carrier, as it is simply figured into the mathematics used to calibrate the system. However, for organic materials with varying hydrogen contents, this large difference in the absorption coefficient may introduce error into the density measurement.

The application of microprocessor-based electronics to density measurements has solved or greatly reduced the impact of many problems with analog systems that were quite formidable in the past. For example, it is well known that the density of a substance may vary significantly with fluctuations in temperature. However, the intent of the nuclear gaging system may be the measurement of density variations associated only with product makeup. It is thus essential to provide compensation for variations in process temperature. Analog systems offer limited capability for temperature compensation, but they cannot allow for the fact that the temperature coefficient of most materials is itself a function of temperature; hence, analog temperature compensation is effective over only a very narrow temperature range. Today's microprocessor-based systems offer true temperature compensation over a wide temperature range and can compensate for both components of two-component streams, such as slurries or emulsions.

Of all the specifications pertaining to nuclear density systems, two of them—time constant and precision—are probably not considered fully by the average user. Since the two are interrelated, they will be discussed together. Precision is simply repeatability. However, recall that the amount of radiation emitted by a gamma source is a statistical phenomenon; hence, a nuclear density gage will not produce a given repeatability 100% of the time. To be meaningful, then, a statement of precision must be accompanied by a statement of confidence level. For example, if precision at one standard deviation is quoted, the confidence level is 66% and the system can be expected to measure at the quoted precision approximately two-thirds of the time. At a two-standard-deviation level, the system will measure density within the stated precision about 95% of the time. If no confidence level is given, let the user beware, because no meaningful statement of gage performance has been made.[1]

The statistical variations that affect precision appear as "noise" on the output. A filtering time constant is used to damp out the noise (statistical variations in radiation reaching the detector). The time constant is the time required for the system to provide about 63% response to a step change in process density. A 95% response requires three time constants. There are two ways to improve the precision (reduce the noise) of a given nuclear density gage configuration: increase the source size or increase the filtering time constant. Increasing the source size increases system cost, and there are limits to how large a source can be supplied. Increasing the time constant costs nothing, but this must be balanced by the response requirements of the application. Some microprocessor-based density systems provide a solution to this problem of trade-off between response time and precision, which is known as dynamic process tracking. These systems operate with a relatively long time constant when the process density is stable, thus providing good precision. When the process density changes, these systems automatically switch to a short time constant, thus providing good dynamic response.

---

1. The concepts of precision and confidence level apply to all types of physical measurement instrumentation, regardless of the technology used, and should always be considered when specifying such equipment.

When density changes must be measured in a small process pipe, the change in radiation may be so small that measurement becomes extremely difficult. This problem is best handled by utilizing a Z-section pipe and a scintillation detector (Figure 8–6). The Z-section is used to produce an effective measurement path of approximately 8 in., even though the actual pipe diameter may range from 1 to 4 in. Increasing the process material thickness increases the change in radiation for a given change in density and makes a more precise measurement possible. When an ionization chamber is used on a Z-section or other small-diameter pipe, the performance is degraded because only a small part of the ionization chamber is actually being irradiated. This is illustrated in Figure 8–6A. Since a scintillation detector is totally illuminated by the radiation beam (Figure 8–6B), it can make the measurement with a smaller source. A general rule is that scintillation detectors improve density gage performance for applications using very small pipes (<4 in. diameter) or very large pipes (>30 in.). Manufacturers offering both systems should always evaluate the requirements of the specific application to ensure that the optimum system is provided.

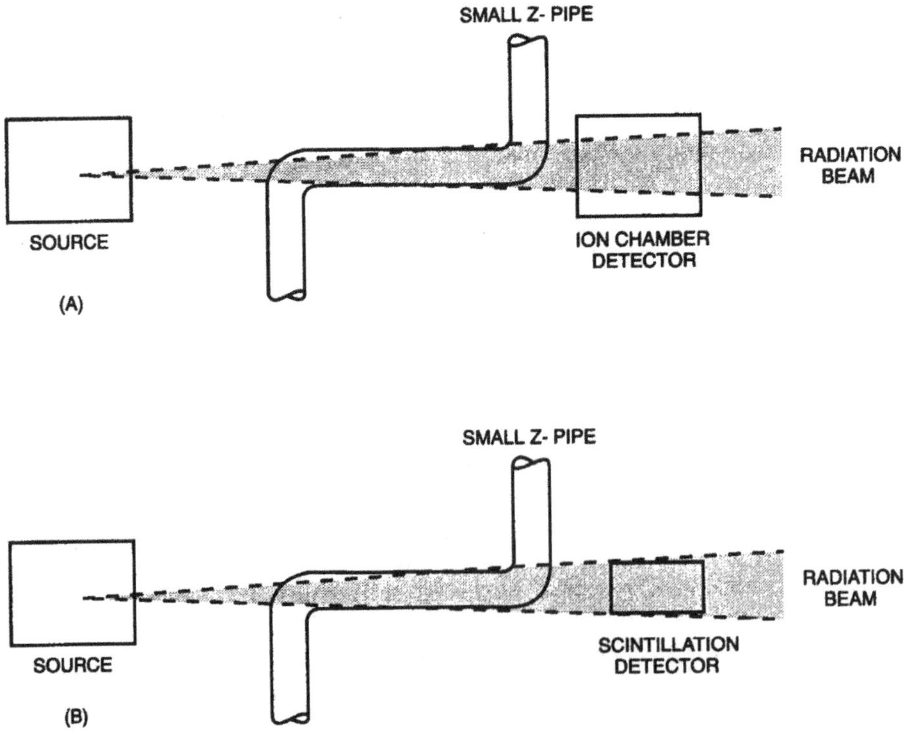

Figure 8–6 Pipe Section for Density Measurement

As previously discussed, the availability of microprocessor-based systems and scintillation detectors has significantly expanded the capability of nuclear density systems. With the many options now available, good communication between the user and the supplier is essential [Refs. 3 and 4].

8.11 OSCILLATING CORIOLIS DENSITY MEASUREMENT

A distinct feature of a coriolis meter is that it is truly a mass measuring device for flow and density. The density of a process fluid can be detected even when the measurement of flow is of no interest. A typical coriolis instrument is shown in Figure 8–7. The sensor is comprised of either one or two flow U-tubes. The double tube version is used to negate the error effect that can result from mechanical vibrations.

The instrument consists of the tube(s), a driving mechanism to force the free (closed) end of the tube to oscillate, detectors at the vibrating end, and (usually) a temperature detector for compensation. The process fluid is contained in the tube; the coil and magnetic driver cause the tube to oscillate in opposition to its natural frequency, which is a function of total tube mass. The free end of the tube moves in a twisting motion caused by the coriolis forces on the tube. The twist of the tube is proportional to flow rate and is measured by proximity detectors. The detectors are usually electromagnetic, optical, or capacitance sensors. The detector output is a sinusoidal voltage with a shift in frequency or phase that is a function of the amount of twist on the tube. The tube temperature is required because the natural frequency of the tube is affected by temperature. The tube temperature measurement is used to correct the natural frequency shift caused by changes in the elastic modulus of the tube material.

The spring constant (k) in the general equations (8.25 and 8.26) is a function of geometry and material properties and is determined from various physical parameters relating to tube material, modulus of elasticity, and initial forces. These quantities are used to determine the natural frequency, the oscillation frequency and/or the period of oscillation. Finally, Equation 8.27 is used to determine density:

$$f = \frac{1}{2\pi}\sqrt{\frac{k}{m}} \tag{8.25}$$

$$\rho = \frac{k}{4\pi^2 V f^2} \tag{8.26}$$

$$\rho = K_1 C_T T^2 - K_2 \tag{8.27}$$

where

f = frequency of oscillation

k = a spring constant

m = mass

V = volume

ρ = density in g/cc

K_1, K_2 = calibration constants

C_T = a function of tube material and temperature that is empirically determined by the manufacturer

T = the period of tube oscillation

Values of the calibration constants are determined by measuring T at two known fluid densities. By knowing T at two density values, K_1 and K_2 are derived from simultaneous equations. Any two fluids of known density can be used to find the calibration constants; fluids selected for calibration should have significantly different density values to minimize the amount of error associated with K_1 and K_2. Air and water at given temperatures are commonly used as calibration fluids. The success and accuracy of coriolis density measurement to even determine the standard density values of air and water at given physical conditions, with the aid of real-time self-diagnostics and compensation, are significant developments in this technology.

8.12 SUMMARY OF CORIOLIS MEASUREMENT

The coriolis instrument can measure process liquid density and true mass flow rate for various applications. In addition to this, through available configurations, the percentage of one or more components in a mixture can be determined. This can be accomplished by microprocessor programming for various complex calculations. Common applications include the determination of percent of solids by mass volume, degrees Brix, percent fructose, percent alcohol and percent of solids in black liquor. Because many of these measurements are temperature dependent, the measurement from the tube temperature detector is used for temperature compensation. Since coriolis flow measurement can also be provided, the net flow rate of one or more components in a process mixture can also be determined. For example, the amount of oil in an emulsion of oil and water can be measured [Ref 1]. When only density measurement is required, a small coriolis meter can be installed in a bypass configuration to reduce transmitter cost. The achievable accuracy using a coriolis meter to measure density is ± 0.0002 g/cc.

8.13 BALL-TYPE DENSITY METER

The ball-type density meter in Figure 8–8 consists of a number of opaque silica balls about 1/4 in. in diameter and enclosed in a glass tube that is immersed in the process fluid. Each ball has a different density in the range of specific-gravity values to be measured. For example, for the measurement range of 0.7 to 0.9 specific gravity, 20 balls graded in increments of 0.01 specific-gravity units are used.

A light source projects light onto one side of the tube containing the balls immersed in the process fluid, and a series of photodetectors are arranged on the side of the tube opposite the light source. Glass-fiber probes conduct light through the tube, and the location of the height of the highest ball—the last one to float—is detected. This height corresponds to the density of the ball and relates to the density of the process fluid. Applications are limited to transparent or translucent process fluids, but a distinct advantage is that the full intrinsic safety feature enables use with explosive mixtures. The device is also immune to electric or magnetic fields and has an accuracy of about 0.01% (FS).

8.14 CAPACITANCE-TYPE DENSITY MEASUREMENT

The process fluid flows through a capacitance cell containing two electrodes. A bridge circuit or some other appropriate capacitance-detecting circuit is used to measure the capacitance of the process fluid with respect to density variations. The capacitance is proportional to the dielectric constant (a function of the force of attraction between two opposite charges separated by a distance in a uniform medium), which is in turn proportional to density. This principle is similar to capacitance level measurement (Chapter 7).

This method of measurement is limited to nonconductive fluids, and with appropriate electrode arrangement can be used for level and mass

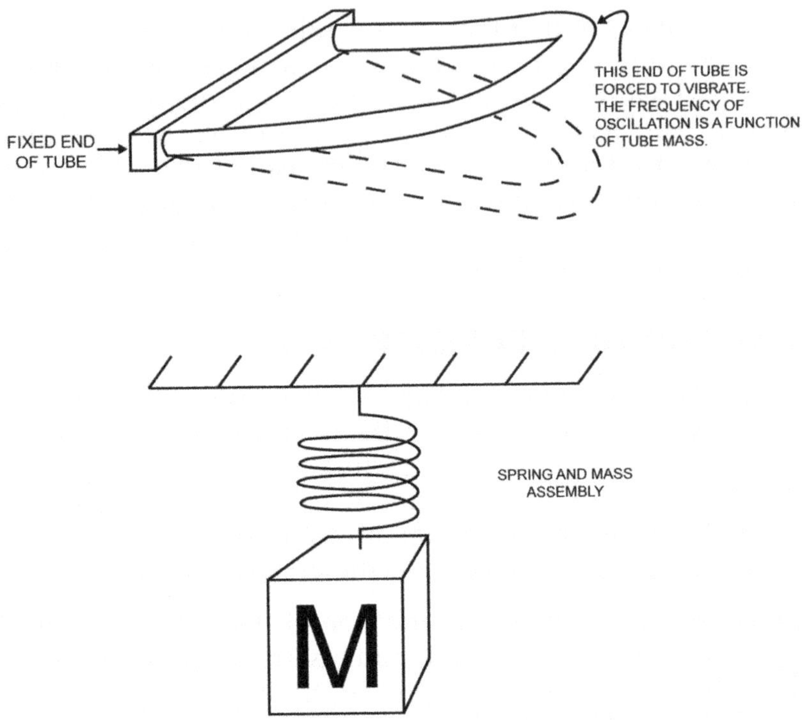

Figure 8–7 A Coriolis Density Meter (*Courtesy of Micromotion Inc.*)

measurement. A common application is for fuel tanks and storage vessels. Accuracy is on the order of 1% FS. A capacitance density detector is shown in Figure 8–9.

8.15 HYDROMETER

One of the simplest in-line density measurement devices is the hydrometer, which is a weighted float with a small-diameter indicator stem that has graduations of various specific-gravity ranges and units. The process fluid enters the bottom of a glass that contains the hydrometer and a glass thermometer. The sample flow rate is maintained at about 1 gal/min to avoid velocity and turbulence effects. The density of the fluid is read directly from the graduations on the stem and temperature values from the thermometer are used to make corrections as required.

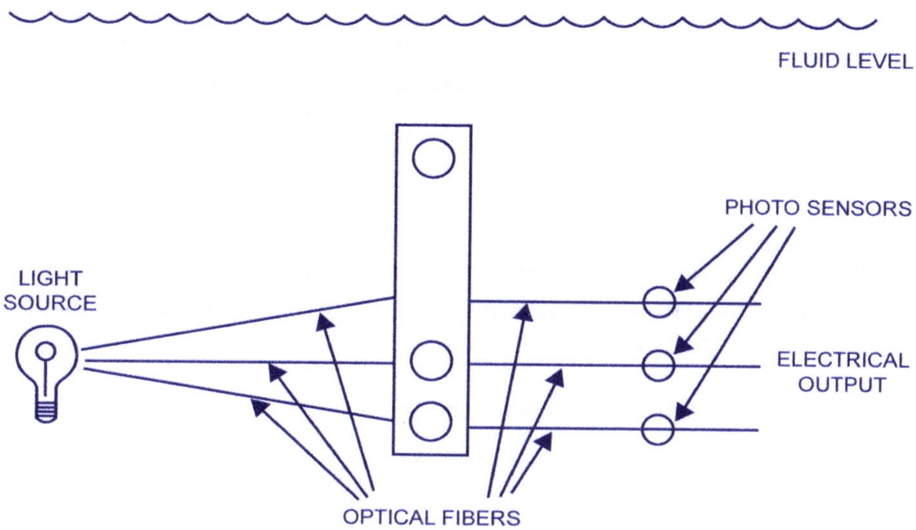

Figure 8–8 A Ball Type Density Meter

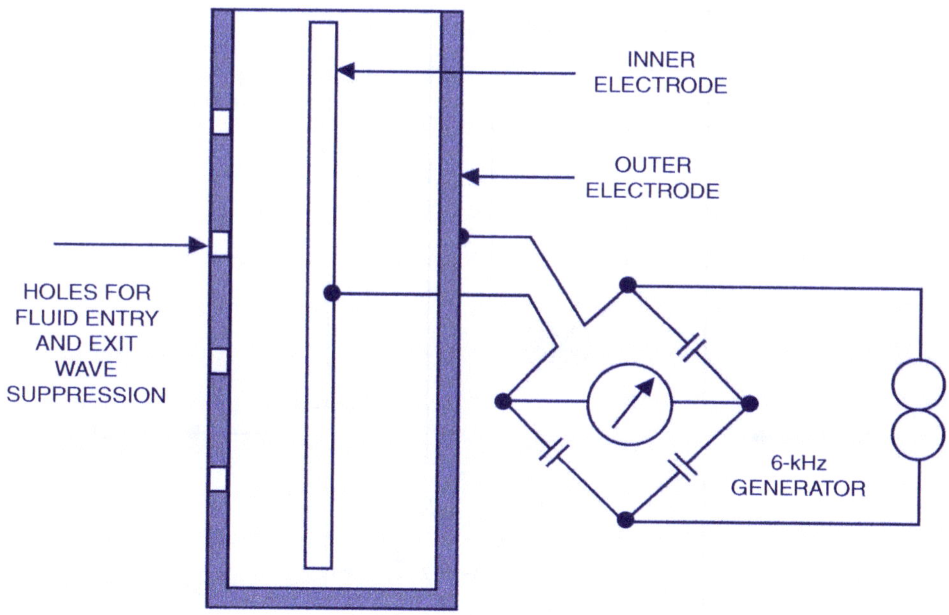

Figure 8–9 A Capacitance Density Detector

The element can also be mounted inside a rotameter housing, with overflow pipes used to maintain a constant level. The position of the hydrometer stem has been used with various sorts of position transducers to constitute a density transmitter. Automatic temperature compensation can be provided with appropriate temperature measurement and signal conditioning.

Density measurements with hydrometers have about 1% error and can be direct indicating, without mechanical linkages or electrical sources. Because of error resulting from material coating, they are limited to clean, nonviscous fluids. Figures 8–10A and B show the application of hydrometers for density measurement.

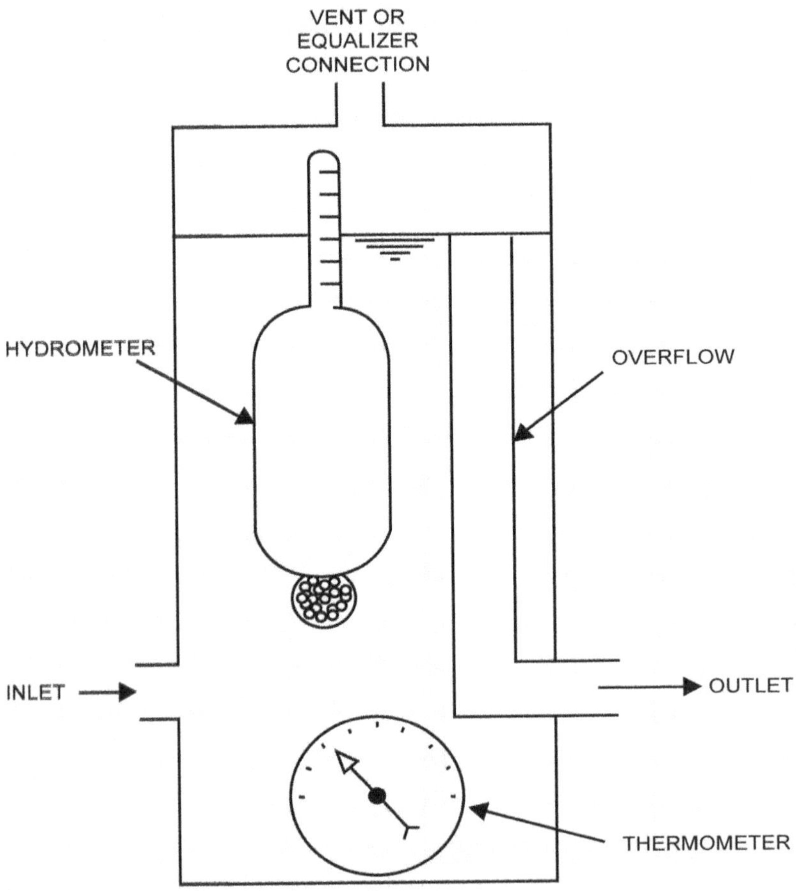

Figure 8–10A Hydrometer Density Measurement

8.16 VIBRATING SPOOL DENSITY MEASUREMENT

When a cylindrical spool is immersed in a fluid and caused to oscillate about its longitudinal axis, the spool will vibrate at a frequency that is a function of its stiffness and the combined oscillating mass. The fluid around the spool is also caused to oscillate, and the combined mass of the system consists of the oscillating process fluid and the spool. For a spool of fixed mass and stiffness, fluctuations in the oscillation frequency are a result of variations in fluid density. Figure 8–11 shows a vibrating spool for density measurement.

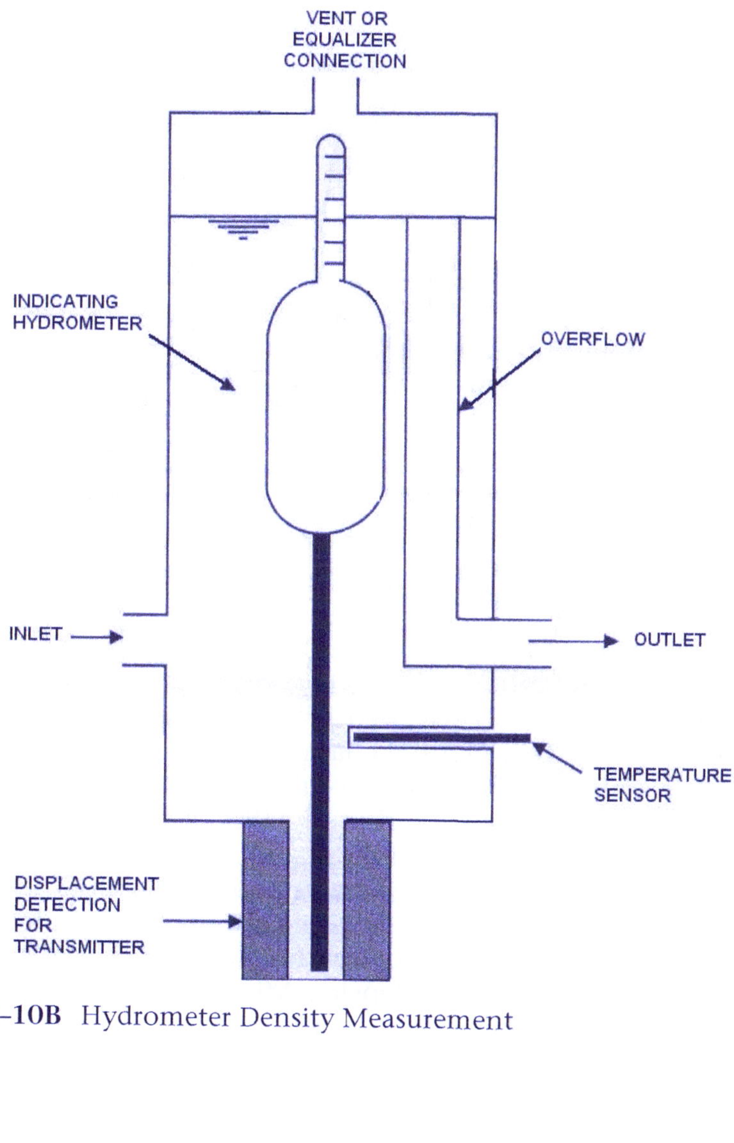

Figure 8–10B Hydrometer Density Measurement

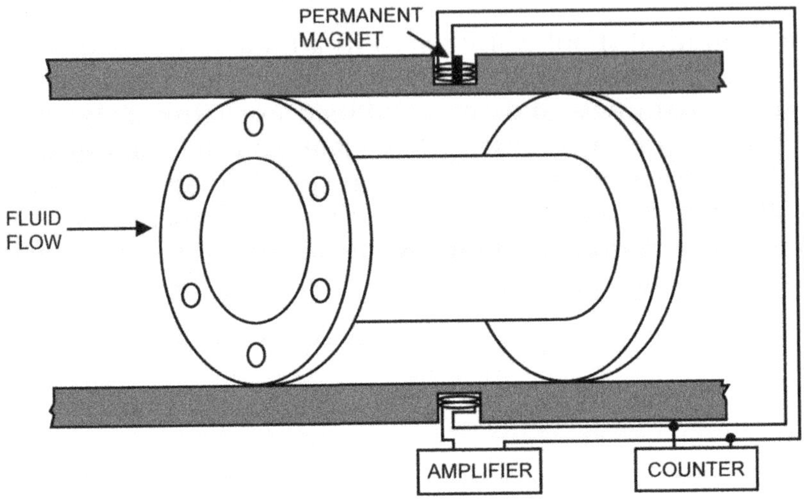

Figure 8–11 A Vibrating Spool Density Detector

Oscillations are normally induced by a feedback amplifier, and a predetermined number of oscillations are counted. The time of that number of oscillations is determined by a timing circuit. The frequency is then determined from the measured time and set number of oscillations. The fluid density is determined from the associated frequency. Errors of about 0.1% are common for a range of 0.5–1.0 specific gravity.

8.17 WEIGHT OF FIXED VOLUME

Density is defined as mass per unit volume; therefore, the most direct means of density measurement is to measure the weight of a fixed volume of a substance and to determine the density in any desired unit. Most weight/volume methods consist of a fixed-volume flow-through chamber that is continuously weighed. These devices are generally classified by the design of the weighing chamber, which is usually a section of pipe that is either straight or a U-tube. A small tank or another volume can also be used. Small volumes will yield a greater degree of accuracy and precision; holdup time is also less which shortens the response time for a true measured value. Load cells and strain-gage methods have been discussed as level-measurement techniques. The same principle can be applied to determine density by converting the weight of a known volume to density. Microprocessors are used to perform the calculations and to express the data in the unit of choice.

8.18 U-TUBE DENSITY GAGE

This design involves a U-tube section of pipe supported near the closed end by a load sensor, equipped with pivoting flexible connections on the open ends (Figure 8–12). Process fluid flows through the tube, and any increase in the weight of the tube causes an increase in force at the closed end of the tube. A flapper-nozzle system can be used to produce a pneumatic signal proportional to the increase in weight as the density increases. A conventional nozzle relay and feedback mechanism is used in a force-balance arrangement to produce a scaled signal representing a range of density or specific-gravity units. The total movement of the flapper is but a few thousandths of an inch. Electronic proximity detectors can be used for measuring the tube deflection for electronic instruments, which are probably more common.

Any electrical method of displacement detection can also be used to measure the deflection of the closed end of the tube, including capacitance, inductance, strain gage, and optical methods. Feedback stabilization methods are used.

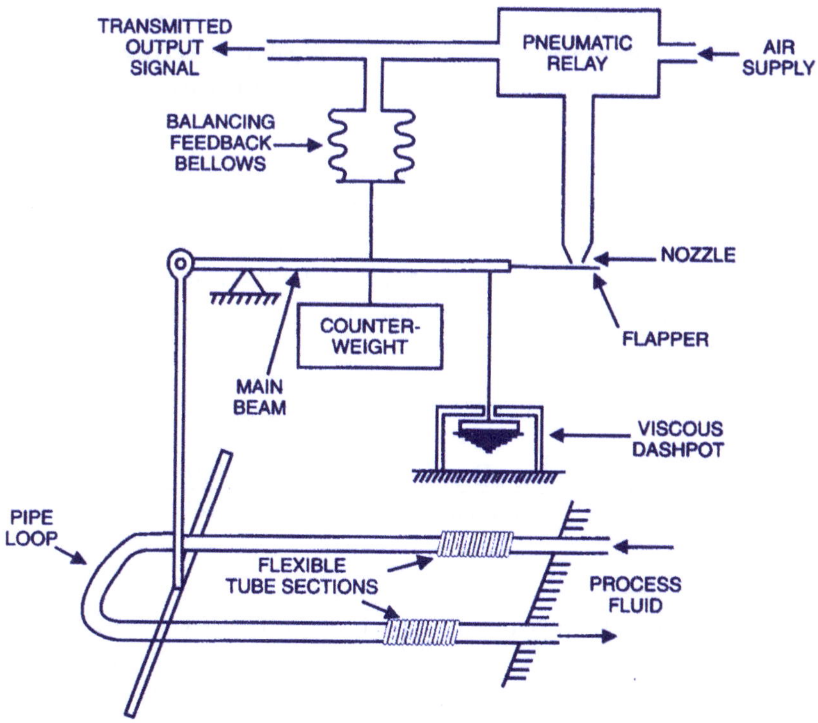

Figure 8–12 A U-Tube Density Transmitter

Some U-tube designs have experienced problems associated with process pressure and flow-rate variations. Any changes in tube deflection or movement resulting from the stress and forces on the tube can produce erroneous results. Careful design of pivot location and mounting have helped to stabilize the horizontal movement of the tube, and viscous dashpots have been used to eliminate vertical force effects.

This design can be used for slurry applications when the flow velocity is great enough to prevent settling. If gas bubbles are present in the process fluid and the density without the gas is desired, a trap should be used to remove the bubbles or the process pressure must be great enough to eliminate them.

Small variations in temperature can usually be compensated for by the density instrument if it has been factory calibrated for a particular process fluid at the operating temperature. When large variations are expected (more than ±20°F), a temperature transmitter should be used to correct the output of the density transmitter. The maximum span of the temperature transmitter is a function of the expected change in process density (density span). The following example helps to illustrate the method of determining the span of the temperature transmitter used with a density transmitter for temperature compensation.

EXAMPLE 8-4

Problem: A temperature compensator is to be used for a density range of 0.70 to 0.78 and a span of 0.08. The coefficient of thermal expansion for the process fluid is 0.0005 SG/°F. If the density measurement is based on 60°F, find the range and span of a temperature transmitter used for compensation.

Solution:

$$\text{Temperature span} = \frac{\text{Density span}}{\text{Expansion coefficient}} = \frac{0.08}{0.0005} = 160^\circ\text{F}$$

$$\text{Temperature range} = 60 \text{ to } 220^\circ\text{F}$$

Most density gages are limited in use to clean, nonviscous fluids; the selection of gage type is normally a function of accuracy and economics. Radiation detectors have gained popularity in recent years because of advances in computers and electronic circuits for compensation and calculations. These detectors are also preferred for slurries, viscous fluids, high-pressure applications and applications that require non-contact measurement. U-tubes are limited to moderate pressure and temperature ratings. Hydrostatic head-type measurements are best suited to moderate to high ranges because low-range applications require a correspondingly high standpipe or liquid volume. Another limitation or disadvantage of head-type density measurement is that when such devices are used for corrosive or viscous fluids, repeaters or purge fluids are often required [Ref. 1].

8.19 INSERTION LIQUID DENSITY MEASUREMENT

An insertion transducer with a tuning fork sensor can be used to measure most liquids in tank and pipeline applications. The tuning fork vibrates at a resonant frequency, which changes with fluid density. This change in frequency is continuously monitored and corresponds to the change in density. The sensor frequency can be used as a sensor input to signal converters or used with a transmitter with integral signal conditioning circuitry for a scaled output. A long stem version that is available from some manufacturers can measure to an insertion depth of up to ten feet or more. By selection of the standard or long stem version of the tuning fork density instrument and by appropriate calculations with available temperature measurement, the following can be measured or calculated:

- Interface location
- Specific gravity
- Percent solids in a slurry
- Percent concentration
    - Degrees Brix

The following applications are suitable for insertion density measurement:

- Solvent separation
- Wort gravity for brewing
- Evaporation control
- Product mixing
- End point detection

Insertion density measurement offers several advantages over other measurement methods, with the following features:

- Direct insertion sensor
- High pressure in line service
- Various types of sensor material for corrosive service
- Low or no regular maintenance required
- Integral temperature measurement
- Hygienic service
- Tolerant of solids and bubbles
- Vibration tolerant
- Suitable for open or closed tanks
- Approved for hazard areas
- Factory calibrated
- No moving parts
- Does not require sample transport

Insertion density measurement is limited to accuracy on the order of 0.001 g/cc, whereas coriolis type detectors can have accuracies of 0.00015 g/cc to 0.00035 g/cc [Ref. 5].

8.20 MICROWAVE DENSITY METER

A microwave is an electromagnetic wave or radio wave of a frequency between 300 MHz and 30 GHz. These waves are used in communication radar, microwave ovens and cellular phones. The power of the microwaves used in a density meter is about one millionth of the power used in a microwave oven, and is too weak to generate heat; therefore, there are no safety concerns.

Radio waves travel in air at 3×10^{10} cm/s, fast enough to go around the earth seven and a half times a second. When they travel in water, their speed becomes slower, to about one ninth of that in air, or 3.33×10^9 cm/s. If the water contains suspended substances or dissolved substances, the speed of the waves varies slightly by the density of the substances.

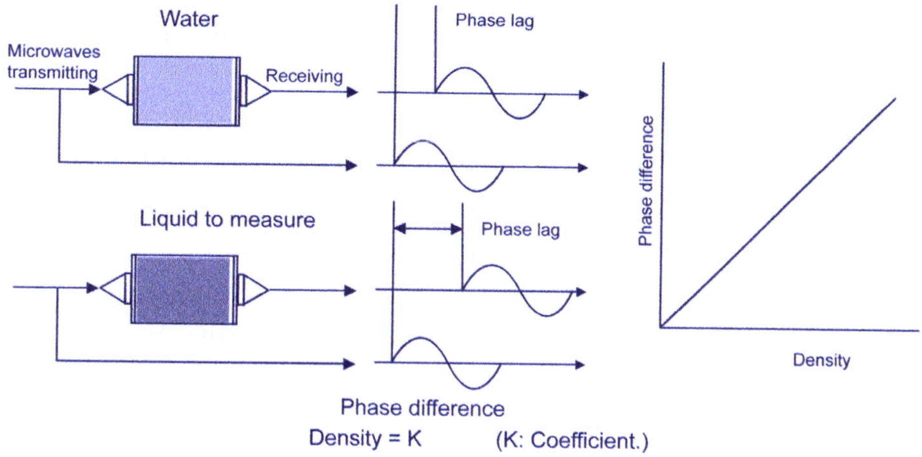

Figure 8–13 Principle of Measuring by Microwave Phase Difference (*Courtesy of Toshiba*)

A microwave density meter has a microwave transmitter on one side of the detector tube and a receiver on the other side to let the waves pass through the liquid. The meter measures the phase difference of the waves and calculates the difference of the speed of the waves. The phase difference gives the density as shown in Figure 8–13. If the phase lag in water (density 0%) is θ_1, and the phase lag in a certain density liquid is θ_2, the difference $\Delta\theta = \theta_2 - \theta_1$ has a good linear relationship with density. Density can therefore be obtained by measuring the change in phase of the waves in a liquid.

A microwave density meter can detect, in principle, both the suspended solids and the dissolved solids, which makes the meter good for measuring the total solids density.

The features of the microwave density meter are

1. Stable density measuring. The microwave density meter measures the phase of waves and is not susceptible to the effects of bubbles and solids buildups.
2. Wide range measurement, from low to high density. The meter has good linearity in measuring density that ranges from as low as 1% or less to as high as about 30%, and allows measuring in varieties of ranges by simply changing the range through the keypad.

3. Continuous measuring. The density meter has no sample-taking process and is suitable for continuous in-line measurement; thus it is applicable to process control.
4. Highly reliable, with simple maintenance. It has no moving mechanical parts to wear out, eliminating periodic parts changing.
5. No velocity compensation necessary. The microwave density meter is free from the effects of flow velocity.
6. Not affected by pulp materials. In paper-making processes, the pulp materials and the freeness (the rate at which water will drain through the pulp) have very little affect on pulp density measurement.
7. Additives compensation function for the paper-making process. In the paper-making process, various additives (in accordance with the color and gloss required for paper) are added to the pulp. The microwave density meter has an optional function to correct the difference in sensitivity of additives in reference to the pulp, and gives true density.

8.21 DENSITY APPLICATIONS OF MICROWAVE MEASUREMENT

Microwave measurement is applicable to measuring the density of substances suspended or dissolved in a liquid. The following conditions apply:

1. *Suspended*—A state in which the particles of one or more substances are distributed in a liquid. When static, the particles settle to the bottom of the vessel. An example of the suspended state is starch and water.
2. *Dissolved*—A state in which the substance is dissolved in a liquid. The substance does not settle in a static state. An example of this state is salt or sugar dissolved in water. In a supersaturated solution, however, the solute can settle out under certain conditions.
3. *Mixture*—A state of two or more suspended and/or dissolved substances.

Substances of an identical density can show different responses in some density meters so it is necessary to calibrate the instrument in the substance to be measured to obtain a density correction factor. An accurate measurement of a mixture of two or more substances is possible when the ratio of the mixture's components is constant or known at the time of measurement.

Microwave density instruments can generally be used to measure materials with dielectric constants that differ by more than 0.05 between the solvent and the substance to be measured. As stated earlier, dielectric constant is a function of the force of attraction between two opposite charges separated by a distance in a uniform medium. Relative dielectric constant describes the electric nature of a substance and is specific to that material. When the dielectric constant of a vacuum is assumed to be 1 and a dielectric constant is expressed relative to a vacuum, it is called a relative dielectric constant. The relative dielectric constant of water is about 80. There is no simple way to know the relative dielectric constant of a liquid that contains a dissolved or suspended substance. This difference can be measured by microwave and can be done with the microwave density meter. Generally, if the solvent is water and the measured substance is inorganic, density can be measured. With organic substances it can also be measured but the sensitivity is lower. This is because the microwave density instrument is sensitive to the conductivity as well as the density of the material to be measured.

When a liquid contains electrically conductive particles, such as powders of carbon or metal, to 0.1% density or more (as is the case when sludge in a water plant contains activated charcoal) the instrument will not measure accurately. In such a case, a microwave density meter should not be used.

Both density and the dielectric constant, content ratio of the conductive particles, can be measured with the microwave density instrument, but either the density or the dielectric must be known and constant. When the fluid contains conductive particles, the dielectric is inversely related to the density and directly related to the density for fluids with nonconductive particles. This interaction between dielectric and density must be considered. Therefore, a microwave density meter cannot make correct measurement of a liquid that contains conductive particles whose content ratio is unknown. However, by utilizing the capability to measure either density or dielectric, it is possible to measure the content ratio of carbon in a slurry of known constant density [Ref. 2].

EXERCISES

8.1 List four scales that are used to measure specific gravity.

8.2 List the similarities between hydrostatic head, displacer, and nuclear radiation level measurement and density measurement. Explain the differences in application.

8.3 Describe a wet leg application for density measurement and a means by which the wet leg can be eliminated.

8.4 Explain the principle of operation of the following density-measurement methods:

- Oscillating coriolis
- Ball type
- Vibrating spool
- U-tube
- Tuning fork
- Microwave

REFERENCES

1. Lipták, Béla M., *Process Measurement and Analysis,* 4th edition, CRC Press, 2003.
2. Toshiba Technical Paper LQ-G-7075.
3. Hendrix, William G., and Nelson, John B., *Industrial Gamma Ray Gage Gaging Applications and Pitfalls,* Research Triangle Park, NC: ISA, 1985.
4. Kay-Ray Model 3680 Density Transmitter Technical Literature.
5. Rosemount Technical Literature 78265004_US, Rev. C, October 2007.

9

Hydrostatic Tank Gaging

9.1 INTRODUCTION

Level-measurement techniques discussed in Chapters 5, 6, and 7 considered the position of air/liquid or liquid/liquid interfaces. Level units are normally linear to express the height of liquid in a vessel, or in percent to express the extent to which the vessel is full or empty. Recently, however, the ability to infer a volumetric quantity from a level measurement has been shown to result in more accurate level measurement. Hydrostatic tank gaging (HTG) was developed in response to the desire for increased accuracy.

9.2 HTG PRINCIPLES

At one time the most that was required of many level-measurement applications was to determine whether a level value was changing or steady, or to control the level in a process within certain limits to prevent a vessel from overflowing or being emptied. Except for certain specialized processes, level measurement within ±1% was adequate.

With the rising cost of raw material and finished products, tank inventory became increasingly important. Also, considering that the accuracy of process measurement instruments can approach ±0.1% of measured value, it seemed advantageous to combine other variables with level quantities to provide a volumetric measurement system. This technique had been used for decades, but not until the 1980s was the means of determining an actual tank volume from instrument-measured level values a reality.

The concept of automated tank gaging (ATG) is reviewed here to aid in the understanding of HTG principles. There are three major reasons for measuring the level in large storage tanks. Production personnel need to know the exact liquid level and volume

- To determine safe high and low limits.
- To safely transfer product into or out of the tank.
- To determine the tank inventory for other job functions.

As mentioned, these concerns have been met by conventional level-measurement methods. These techniques, however, are limited if the goal of tank-level measurement is to achieve an accuracy of ±1/8 in. or to provide accurate calculations of mass and volume. The accuracy of most level-measurement devices or systems is sufficient for operating production units, but when true mass balance or inventory controls are required, such systems may be inadequate.

The procedure of calculating mass and volume from level could yield data sufficient for inventory control if other variables were not involved. True volumetric measurement must account for the change in volume with variations in temperature and, to a lesser extent, in pressure. The average temperature of the product is required because mass is calculated from net standard volume. Under certain conditions, level values vary with pressure (drum swell and shrink in boilers, for example) and thus pressure measurements must be included in the calculation of mass. The API (American Petroleum Institute) manual of *Petroleum Measurement Standards* provides equations used to calculate a net standard volume at 60°F or an observed volume at an operating temperature. These volumes have been determined from tank-capacity tables derived from tank-calibration data. For some petroleum products, temperature variations of 1 to 3°F can account for level errors as great as 0.2%. This can result in very significant errors in volume. Also, if true mass measurement is required, density variations caused by temperature changes must be compensated for unless all measurements are made at 60°F or an adopted standard value.

9.3 TANK CALIBRATION

The level/volume relationship of a tank depends on several variables, each of which must be considered for volumetric measurement. Tank calibration is one means for determining this relationship. Once data have been accumulated (usually by empirical means) to relate volume to level, the data are then used in all volumetric calculations. The capacity table is a record that is kept on file for a particular tank and is used to verify actual storage volume.

9.4 TANK CALIBRATION METHODS

For more than half a century, vertical cylindrical storage tanks have been calibrated by the manual strapping method. In this method, calibrated measuring tapes are used to measure the circumference of the tank at several elevations. During the past decade, however, new technologies for tank calibration have emerged; many are in the final stages of standardization by national and international standards organizations. These new technologies offer advantages in terms of improved safety, accuracy and efficiency in the overall spectrum of tank calibration. Guidelines are presented here for tanks with both internal and external floating roofs. Specific construction details are considered for these guidelines.

The manual strapping method, the oldest method of tank calibration, was introduced as a standard in the early 1960s by organizations such as API and the Institute of Petroleum (IP). The optical methods developed during the early 1970s are now commercially available for routine application. These new methods have attained a level of maturity comparable to, and in many instances better than, the manual method. Table 9–1 summarizes the available methods and the appropriate industry standards for their application.

The brief overview presented here of the different methods of calibration will help to promote better understanding of the guidelines for their application. Only those methods that are already standardized or that are currently in the process of standardization are discussed, although other methods, such as photogrammetric and laser imaging technology, are also commercially available. These methods may well be considered for standardization later.

The manual strapping method involves strapping each ring of a tank with a certified tape measure. As tank diameter and tank height increase, it becomes difficult to tape the tank uniformly due to tape sag, tank indentations, and the limited length of tapes. Nonetheless, it is a popular method and is cost effective in many applications.

Liquid calibration is considered the most accurate and preferred method of calibration, and should be used whenever possible. This

Table 9–1 Tank Calibration Standards (*Courtesy of Oil and Gas Journal*)

Standard Number	Title	Remarks
API 2550	Method for measurement and calibration of vertical cylindrical tanks.	Issued by API in 1965; currently undergoing revision.
API 2555	Method for liquid calibration of tanks.	Issued by API in 1966; currently undergoing revision.
API Chapter 2, section 2B	Calibration of vertical cylindrical tanks using optical reference line methods	New standard issued by API in 1989.
ISO standards	Petroleum and liquid petroleum products; volumetric calibration of vertical cylindrical storage tanks	ISO drafts international standards; technical work in the preparation of standards completed.
ISO DIS 7507-1	Strapping method.	Standards being processed for industry publication
ISO DIS 7507-2	Optical reference line method (ORLM).	
ISO DIS 7507-3	Optical triangulation method (OTM)	
...	Electrooptical distance ranging (EODR) method.	ISO work group in the process of preparing a draft proposal.

method consists of metering a known quantity of liquid and gaging the tank at regular height intervals to develop a capacity table. It is time consuming and may not lend itself to easy and quick application. However, under certain circumstances it may be the only available option for calibration.

The optical reference line method (ORLM), which can be applied internally or externally, uses an optical theodolite (an instrument used for accurate angle measurement) to establish a perpendicular ray in a vertical plane. With this optical ray, the deviations in tank diameter at various course heights are measured with respect to a reference offset at the bottom course. In addition, the reference circumference is measured at the bottom course. These deviations, together with the reference offset and the reference circumference, are then translated into appropriate tank diameters.

The optical triangulation method (OTM) can be applied internally or externally. The principle of this method is that a tank profile can be defined by triangulation with two theodolites and a target point on the tank, or with one theodolite and two target points on the tank at a number of stations around the tank. For external application, depending on the tank height, a minimum space around the tank is required. Already in use in some countries, OTM is in the final stages of standardization within ISO.

Electrooptical distance ranging (EODR), currently under development toward standardization within ISO, is intended primarily for internal calibration. With a single optical laser, a distance-ranging device and an on-line computer, circumferential target points on the tank shell wall for any given course are mathematically and statistically analyzed almost instantaneously to give the required course diameter. The calibration can be accomplished from the ground level by one person. This method is presently in use in Europe.

Selection of a calibration method often depends on a number of factors, which can be broadly grouped as follows:

- Type of tank: floating or fixed roof
- Operational constraints: entry or no entry
- Insulation or no insulation
- Riveted or welded
- Other parameters: e.g., number/size of wind girders

Selection of specific methods based on these factors is presented in the form of decision charts in Figures 9–1 to 9–3. These guidelines have been developed to determine the best technology for calibrating specific tanks used for vertical cylindrical storage [Ref. 1]. In the development of these guidelines, it is assumed that the insulation requirements pertain only to fixed-roof tanks. Also, the term "entry," as applied to floating-roof tanks, refers to access onto the top of the floating roof, with the roof resting on its legs. The bottom calibration requirements are not considered separately, because they would belong to the same category where entry is required or permitted.

For each category, technology selections are presented in a prioritized order. The priority recommended is based on the most expedient method of calibration for a given set of conditions, ensuring overall accuracy. However, the recommended priority is not intended necessarily to optimize the overall cost of calibration. The cost associated with any given method is dependent on many factors.

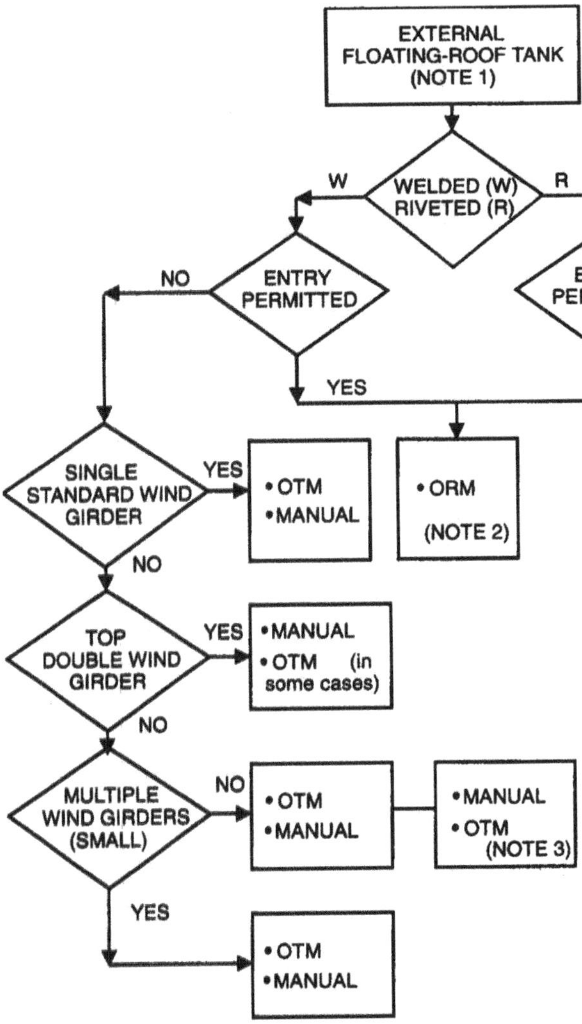

Figure 9–1 External Floating-Roof Tank Decision Chart
(*Courtesy of Oil and Gas Journal*)
Note 1: Liquid step calibration is preferred whenever possible and thus is not listed as an option where other methods lend themselves for applications.
Note 2: Assumes the existence of a standard wind girder at the top.
Note 3: Possible choice.

Generally, the on-site cost of calibration of a tank using a new technology may be high when compared to the cost of a conventional manual method. However, with routine application and with the support and encouragement of the petrochemical and refining indus-

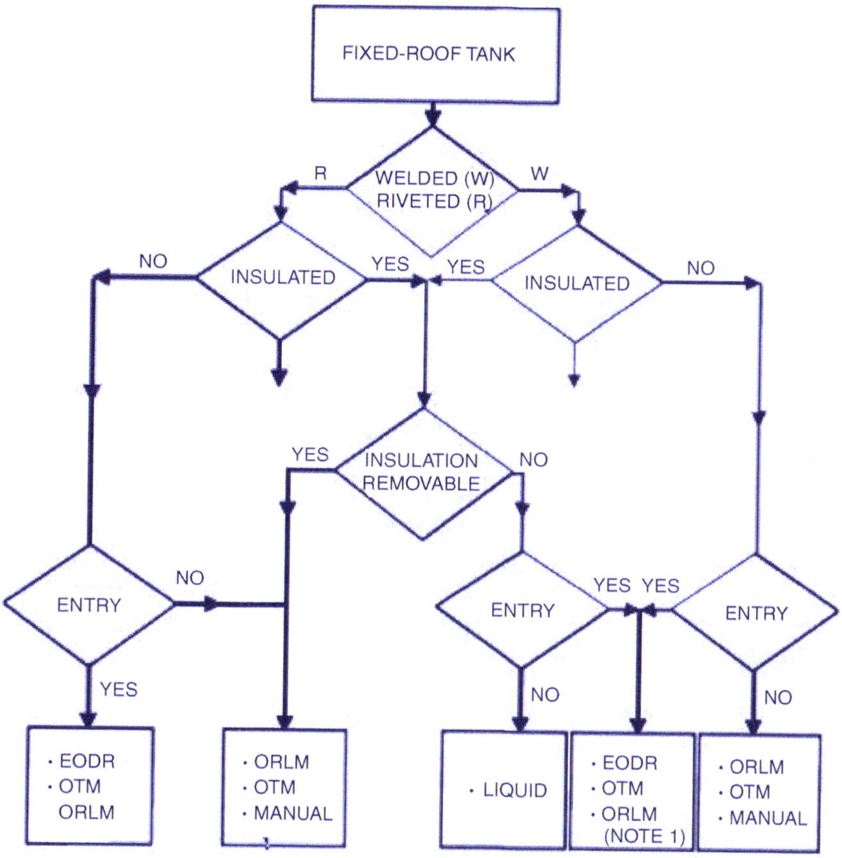

Figure 9–2 Fixed-Roof Tank Decision Chart
(*Courtesy of Oil and Gas Journal*)
Note 1: Reference circumference, where required, must still be measured externally on the tank bottom course with appropriate sections of insulation removed.

tries, the new technologies can be cost effective in the long run. In general, the following factors can be expected to affect the cost of calibration:

- Number of tanks available at any given time
- Size of tanks: diameter and height
- Cost of necessary auxiliary and support services, such as power supplies and scaffolding

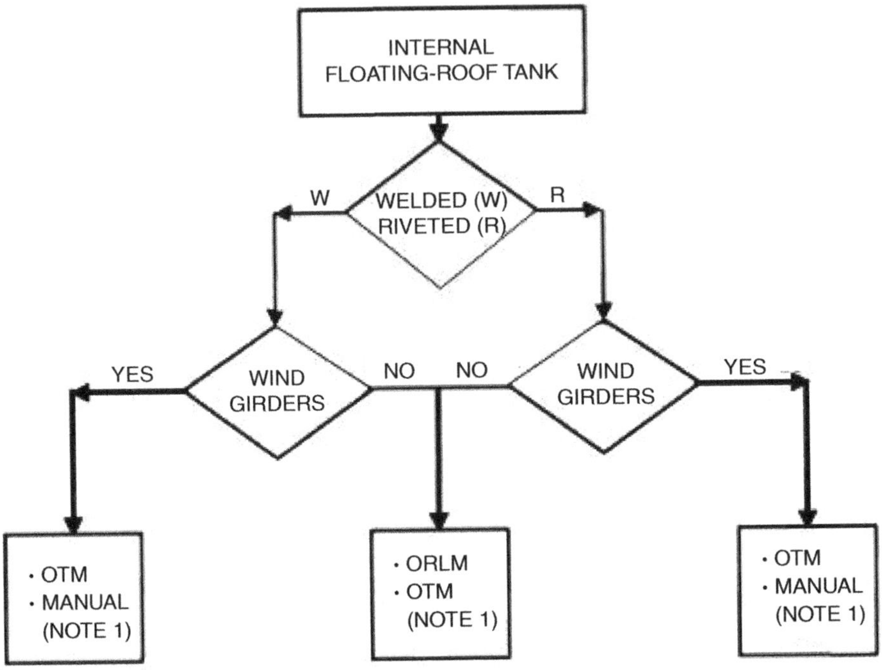

Figure 9–3 Chart for Internal Floating-Roof Tanks
(*Courtesy of Oil and Gas Journal*)
Note 1: All methods are external.

- Cost of stripping and reinstallation of sections of insulation where and when required
- Ultrasonic thickness measurements, if required, as part of calibration
- Cost of bottom surveys where and when specified
- Proximity of calibration contractor to site
- Calibration work done on a straight-time basis or on an overtime basis

In addition, indirect costs are involved, such as downtime and the cost of preparing a tank for safe entry. The cost of tank calibration often becomes an issue for intense and lengthy debate. It is important to relate the cost to the frequency of calibration. While the cost factor is no doubt important, it may not necessarily be the most decisive factor in the selection of a particular method. Other advantages, such as safety, accuracy, expediency, and efficiency, should also be considered. New technologies using optical devices provide quicker, safer methods with accuracy comparable to, and most likely better than, the manual method. Tank owners now have a much wider choice and greater flexibility in calibration selection.

When calibration tables are constructed from manual level measurements, data so accumulated can result in errors as great as 1/8 in. Bottom movement, which is a function of construction material, foundation, and so on, can cause tank volume errors as great as 0.1% of full tank volume.

In addition to errors caused by product temperature variations, expansion and contraction of the tank shell will cause errors similar to tank dynamics resulting from pressure changes. If product temperatures are different from those at which tank calibration was performed, error will result.

Some tanks have water accumulated in the bottom, which will be measured as product level. For increased accuracy, some systems compensate for the water level by actually measuring the water/product interface level.

When the conditions that affect the calibration of the tank for volumetric measurement change, recalibration may become necessary. Three sets of variables, each of which affects the calibration data, must be considered:

- Measurement variables, including tank diameter, tank plate thickness, and tilt
- Structural variables, including tank deadwood (foreign material, sludge, etc., that lies permanently on the tank bottom), reference height, tank structure, and repairs to the tank
- Operating variables, including product temperature, ambient temperature, specific gravity of the product, and reference height

Usually, only changes in structural and measurement variables require recalibrating; changes in operating variables necessitate recomputation of the capacity table.

Once a set of options is chosen using the decision charts, a calibration contractor may be requested to bid on each of the options. Such bids should be based on sound and well-documented technical specifications. Depending on specific site conditions and local requirements, any one method could prove to be cost effective.

It is hoped that these guidelines to the selection of the proper technology will promote a technology application in its entirety and will help to avoid the combining of different methods, which could lead to erroneous applications and compromise on recommended procedures. The final selection of a particular method, of course, may eventually be dictated by internal policies, local regulations, and/or budget constraints [Ref. 2].

9.5 TANK RECALIBRATION AND RECOMPUTATION

Tank calibration and the establishment of calibration tables should be considered prime factors in implementing HTG technology. When conditions that set the tank calibration factors change, recalibration may be necessary. Guidelines can be followed to help determine the frequency and necessity of recalibration. Tank parameters such as bottom course diameter, bottom course plate thickness, and tilt should be observed and verified to maintain tank calibration tables. The physical properties of the material stored should also be checked and updated as required. When HTG measurements are in error and after research has revealed this error as resulting from erroneous tank calibration data, further evaluation should be considered. Recalibration or recomputation could be required. Five specific tank parameters and operating variables can constitute the need for recalibration and/or recomputation:

- Tank bottom course inside diameter variations
- Tank bottom course plate thickness variations
- Volume correction for tank tilt
- Tank shell temperature variations
- Product specific-gravity variations

Based on API Standard 2550 and generally accepted custody transfer stipulations, an error of ±0.01% of tank volume is significant. For each of the parameters listed, tolerable limits can be reached before recalibration or recomputation may be required, and the overall range of variability may approach ±0.05%. Criteria have been established to determine when recalibration or recomputation is needed. Tables 9–2 to 9–4 can be used to ascertain when the measurement variables are considered to be out of tolerance and in need of recalibration. Tables 9–5 and 9–6 specify conditions where recomputation is required.

9.6 RECALIBRATION GUIDELINES

The tank conditions and parameters given in Tables 9–2 to 9–4 will change with time because, as in most physical structures, settling and shape distortion will occur based on soil type, foundation, tank design, climatic conditions and other factors that affect structural stability. Tank owners should maintain an operating history and data to indicate possible performance criteria. Such records have indicated that a recalibration will be required in the first five to ten years of operation, with the second accumulation of settling occurring in the following ten to twenty years.

The following guidelines should be observed. For tanks in custody transfer service, verification of the bottom course diameter, bottom course plate thickness, and tank tilt is suggested once every five years. If any of these parameters exceeds the criteria for a predetermined variation in volume, a total recalibration should be undertaken. These variations in tank inside diameter should be computed after de-stressing the measured diameter to a zero stress condition using correlations presented in API Standard 2550.

For tanks not in custody transfer service, a total recalibration should be undertaken at least once every fifteen years, even if a five-year verification shows no variations or variations well within the limits presented in the criteria tables. Verification of the physical parameters (diameter, thickness of the bottom course and tank tilt) of such tanks should be considered once every five to ten years. If any of the individual parameters exceeds the criteria for a predetermined variation in volume (Tables 9–2 to 9–4), a total recalibration should be undertaken.

If at any time there is a change in the deadwood that could affect the tank volume by a preset value (0.01 to 0.05%), a recalibration may be required. If the reference height is altered by modifications to the datum plate or by the rewelding or resupporting of the datum plate, a recalibration may be required. Any variations in the floating-roof dead weight that could alter the tank volume by a predetermined value (0.01 to 0.05%) mean that a recalibration should be considered. It is

Table 9–2 Tank Bottom Course Inside Diameter Variations (*Courtesy of Oil and Gas Journal*)

Approximate Variation in Basic Volume, %	Nominal Tank Diameter,* ft, Up To:					
	50	100	150	200	250	300
	Allowable Variation in Diameter, mm					
0.01–0.02	3	4	4	5	6	7
0.02–0.03	4	5	7	9	10	12
0.03–0.04	4	7	10	12	15	18
0.04–0.05	5	9	12	17	20	24
0.05–0.06	6	10	15	20	25	30

*Tank diameter is the zero stress inside diameter.

Table 9–3 Tank Bottom Course Plate Thickness (*Courtesy of Oil and Gas Journal*)

Nominal Tank Diameter, ft	Plate Thickness Variations,* mm
50-300	1.5-3

*Plate thickness measured at eight points circumferentially on the bottom course and averaged.

Table 9–4 Volume Correction for Tank Tilt (*Courtesy of Oil and Gas Journal*)

Tilt, ft/100 ft	Volume Correction Factor, %	Remarks
1.4	0.010	Meaure tilt at the same location.
1.6	0.013	Compute variability in volume based on the initial and final tilt.
1.8	0.016	
2.0	0.020	Maximum tilt variation 0.024% allowable, vol%.
2.2	0.024	
2.4	0.029	A variation of 0.005% due to tilt should be considered significant and should warrant recalibration.
2.6	0.034	

assumed that the variations in dead weight are caused by physical modifications, such as pontoon repairs, seal additions, and so on. If major repair work to the tank bottom plate is undertaken, a total recalibration may be required, including a bottom survey. A bottom survey should also be undertaken regardless of bottom repair work if other measurements indicate that the bottom volume has changed by more than 0.01%.

Table 9–5 Tank Shell Temperature Variations (*Courtesy of Oil and Gas Journal*)

Variation in Shell Temperature (T_S),°F	Variation in Ambient Temperature (T_A), °F	Variation in Liquid Temperature (T_L), °F	Approximate Variation in Volume, %
10	20	20	0.01
20	40	40	0.03
30	60	60	0.04
40	80	80	0.05

*Shell temperature (T_S) can vary due to variation in liquid temperature (T_L), variation in ambient temperature (T_A), or both. The values presented correspond to variations in either T_A or T_L for uninsulated tanks. For insulated tanks, variation in T_S will be equal to the variation in T_L. Ambient temperature has no effect on shell temperature for insulated tanks.

Table 9–6 Product Specific-Gravity Variations (*Courtesy of Oil and Gas Journal*)

Variation in Specific Gravity, %	Approximate Variation in Volume,* %
10	0.008–0.015
20	0.015–0.030
30	0.030–0.040
40	0.040–0.050
50	0.050–0.065

*Actual variation in hydrostatic head correction volume could be higher than specified, depending on tank plate thicknesses.

9.7 RECOMPUTATION GUIDELINES

Two major parameters affect recomputation of a tank's capacity table: the tank shell temperature and the specific gravity of the stored product. The reference height marker may also affect recomputation under certain circumstances.

In API Standard 2550, the tank shell temperature (T_S) is computed as the mean of the liquid (product) temperature (T_L) and the ambient temperature (T_A). For insulated tanks, the shell temperature is assumed to be equal to the liquid temperature. A variation in liquid temperature or ambient temperature, or both, will affect the net tank shell temperature and thus the shell volume correction factor. Variations in liquid and ambient temperatures can be in both directions.

For a preselected allowable volume variation, Table 9–5 provides allowable shell temperature variation and corresponding variation in liquid or ambient temperature. Under such circumstances, a recomputation of the capacity table should be considered to account for varying temperature conditions. With the increasing use of on-line computers and microprocessors, it becomes fairly easy to instantaneously develop correction factors for shell temperature variation.

Likewise, Table 9–6 provides criteria for product specific gravity variation and its effect on tank volume. Depending on the allowable volume variation, the allowable specific gravity variation can be determined and the need for recomputation due to specific gravity changes ascertained.

A 10% variation in specific gravity (not API gravity) will affect the hydrostatic head correction volume by approximately 0.01%. The actual effect could be higher or lower, depending on design considerations. Again, with the application of on-line computers, such recomputations can be accomplished easily.

The floating-roof correction is affected by variations in product specific gravity, but the effect will be insignificant as long as the floating roof is in a floating mode at all times. If, however, the floating roof is likely to be resting, then the effect of variations in product specific gravity could be significant and should be considered.

To facilitate recomputation, it is suggested that the tank capacity table be developed in four distinct parts:

- *Part 1*—basic capacity table at 60°F for zero stress conditions, including internal deadwood.
- *Part 2*—shell temperature expansion factor table in increments of 5°F.
- *Part 3*—hydrostatic head correction table for varying specific gravities (up to ±50% variation, or as required).
- *Part 4*—floating-roof correction table.

Finally, recomputation of the ullage table may sometimes be necessary when the reference height marker (or point) on top of the tank is relocated to facilitate easy verification of the reference height.

Operating variables such as specific gravity and temperature, therefore, affect recomputation, but they do not affect the basic capacity table at 60°F at zero stress conditions.

9.8 HTG MEASUREMENTS

Figure 9–4 shows a typical vertical cylindrical tank equipped with two precise pressure probes, P_2 and P_3. (P_1 is used with pressurized vessels.) These pressure probes interface digitally with the microprocessor-based HTG processor unit.

With the temperature of the pressure probe also known, a temperature-compensated pressure is derived for each pressure point. Temperature-induced variations caused by the probe are thus minimized. An example of this type of pressure transmitter is a smart transmitter operating in the digital mode of communication with an HTG processor unit.

The digital transmitter also has the unique feature of performing all of the characterization itself, thus relieving the HTG processor of this task. No calibration constants have to be entered into the database of the processor unit. This "decoupling" results in fewer maintenance problems, since there is no need to match pressure probes to the processor unit.

For most products, an RTD (resistance temperature device) is used to measure the actual product temperature. Although only one spot probe is shown in Figure 9–4, multiple spot probes, or an averaging temperature RTD strip, would probably be used.

With these pressure and temperature values, the HTG processor unit calculates a substantial amount of information about the product within the tank. These values include:

- Product level
- Product density
- Product mass
- Product temperature

The density of the product can be calculated for both the actual product temperature and a standard reference temperature, if the particular process application so demands. The reference value can be derived, for example, through a standard API algorithm or using a standard or customized look-up table.

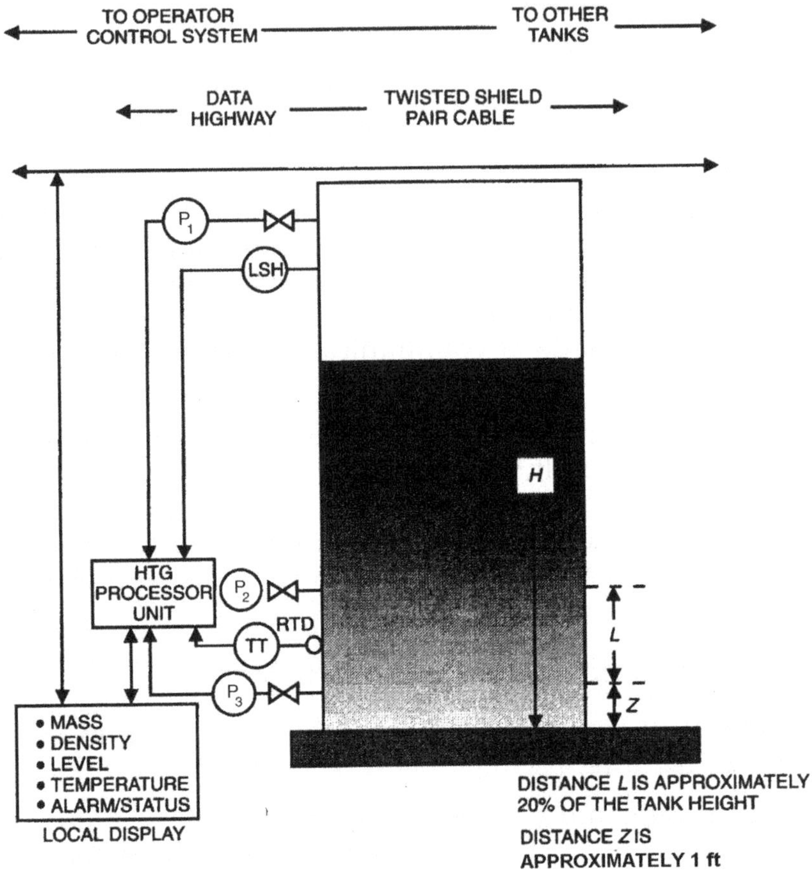

Figure 9–4 Vertical Tank with Hydrostatic Field Equipment

To exemplify the HTG calculations performed by the HTG processor unit, product density is derived by the following:

$$\text{Density} = \frac{(P_3 - P_2) \times (G_c)}{L} \tag{9.1}$$

where P_3 and P_2 are pressure values in engineering units, G_c is the local gravitational constant, and L is the distance between pressure probes P_3 and P_2. Hydrostatic density measurement was discussed in Chapter 8.

In addition to generated product information, the HTG processor unit can provide alarm and status information, including such values as:

- Independent high level alarm (discrete input to the HTG processor unit)
- High- and low-temperature alarm limits
- High- and low-level alarm limits
- Drastic density change
- Rate of density change
- Drastic mass change (leaking tank)
- Rate of mass change
- Drastic level change
- Pressure probe P_2 exposed
- Water high/low limits (if tank is equipped with water detection)

If this information were used only at the tank, it would be of limited value. The maximum usefulness of HTG technology is realized when it is used in conjunction with other information that is usually concentrated at higher levels of plant operations. The operator interface at this higher level may be a console on a DCS (distributed control system), an intermediate-level system designed specifically for use with management of the tank farm, or a programmable controller (PLC), which provides batch processing functions based on the HTG-generated information.

Whatever the type of operator interface, there are two requirements for the HTG processor units in a tank farm:

- They must have network connectivity to these upper-level systems, either directly or through an intermediate operator interface and data converter.
- They must be able to provide the HTG-generated information in an intelligible format for these upper layers.

For instance, the typical DCS looks at all process information in the form of some type of 12-bit analog value. Thus, the HTG processor unit must have the ability to format the rather large (e.g., 32-bit) HTG-derived values, such as mass, into one or more analogs before transmission to the DCS.

9.9 APPLICATIONS OF HTG TECHNOLOGY

The following examples describe some typical problems solved using HTG technology and the benefits derived.

9.10 VERTICAL CYLINDRICAL TANKS

One of the most popular applications of HTG is in the replacement of traditional float/tape or servogaging systems on vertical cylindrical tanks (Figure 9–4) used in almost every type of refinery and marketing terminal, as well as in chemical plants. The products stored in this type of tank can range from crudes and asphalts to refined gasolines.

In some areas of the food industry, it is important to be able to measure Brix, a value correlated to the sugar concentration (and hence density) of the stored liquid product. Hydrostatic tank gaging has been used to perform this function and to provide information on product quality equal to a typical laboratory analysis. The HTG measurements are carried out in "real time," so that no dead time is lost waiting for an analysis to be performed.

Some issues for HTG application in the food industry differ from those facing petroleum operations. For example, the pressure probes normally have to fulfill a "clean in place" requirement. Secondly, the tanks usually are significantly smaller in capacity (25,000 gal versus 250,000 barrels, typically). This latter fact allows HTG to provide a much more accurate mass (or weight) value, since strapping errors are typically not encountered with a very small tank.

By using other inputs to the HTG processor unit, additional benefits can be derived. For instance, by using a separate sensor at the bottom of the tank, water bottom measurement is facilitated. A differential-pressure instrument can be used to measure the density within the tank bottom area (approximately the first 0.3 to 0.46 m above ground plane). This density is then compared to the product density calculated by Equation 9.1, and an equivalent water level at tank bottom is derived.

9.11 LIQUID PETROLEUM GAS (LPG) TANKS

Liquid petroleum gas tanks are notoriously difficult to gage, particularly using a mechanical system. These tanks are under pressure, making it impossible to service the gage mechanism. If the gage fails, the tank may have no reliable means of being measured for product content until it is next taken out of service. Unfortunately, this length of time can be very long.

Several costs are either explicitly or implicitly incurred as the result of gage failure: (1) the cost of repairing the tank's gage; (2) the cost of not knowing very precisely the quantity of the tank's contents while the gage is out of operation (resulting in inaccurate inventory information, etc.); and (3) the potential loss of product inventory capacity while the tank's gaging system is out of service.

HTG technology is a natural fit for these types of tanks. Figures 9–5 and 9–6 depict how the pressure points of P_3, P_2, and P_1 are applied for bullet and sphere tanks, respectively. Points P_3 and P_1 will normally exist for retrofitting the tank with HTG technology. Since P_2 is unique to HTG, it would have to be installed during the tank's out-of-service state. However, to measure product level, which has been seen as the most immediate requirement, P_2 need not be installed if the product density can be characterized over the product's temperature and pressure operating ranges.

The respective characterization curve for the product is conveniently downloaded to the HTG processor unit, which uses points from the curve in calculating product level. Product level for an unpressurized tank is calculated by

$$\text{Level} = \frac{P_3}{\text{Density}} + Z \tag{9.2}$$

where Z is the distance between P_3 and the tank bottom.

Another point to be made is that, from a console, the operator can download an appropriate density value to the HTG processor unit. This procedure may be used if characterization is difficult.

9.12 BATCH PROCESSING

A final example illustrates how HTG can provide very accurate batch processing using a programmable controller (Figure 9–7). A network of HTG processor units interfaces with a programmable controller that has its own local and dedicated operator interface. The mass and density information from each tank is used in conjunction with the other recipe information residing in the PLC to perform the necessary batch processing. The database of the PLC is also made available to a

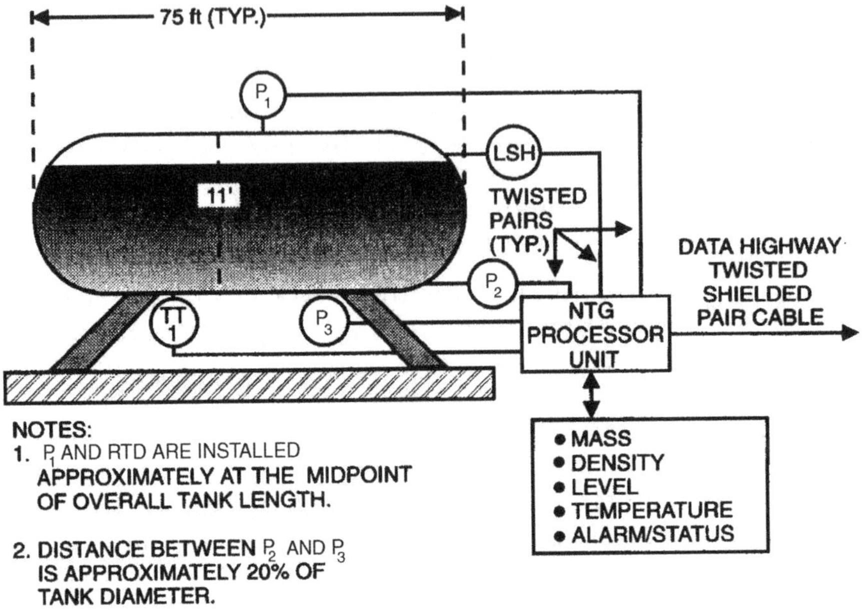

Figure 9–5 Typical LPG Bullet Tank with HTG Technology

higher-level system, such as a DCS or a mainframe system, for status, alarm, and accounting purposes.

The products within these tanks could be anything—from fine chemicals to orange juice. Tank size is not a primary factor in deciding when (or when not) to use HTG technology: The value of the tank's product and the possibility of unaccounted-for loss in inventory are more important considerations.

The integration of very precise pressure probes with intelligent microprocessors has created a new dimension to hydrostatic tank gaging. Some of its unique advantages include:

- *Improved accuracy*—using pressure probes with very high accuracy (up to 0.02% FS [full scale]) and high repeatability (up to 0.005% FS).
- *Lower maintenance*—no moving parts, in addition to virtually no calibration requirements.
- *Higher reliability*—due to solid-state electronics and a high level of self-diagnostic capability.

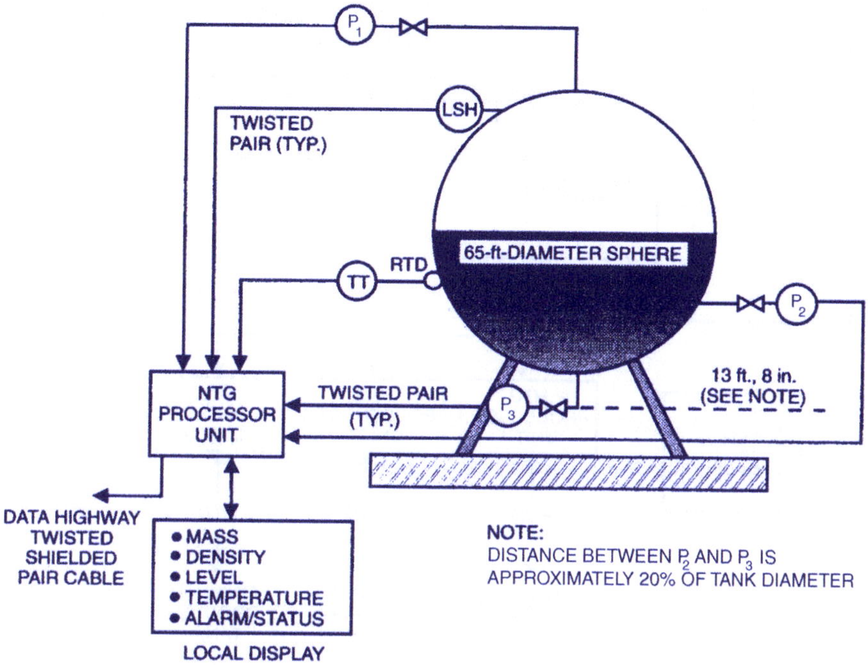

Figure 9–6 Typical LPG Sphere with HTG Technology

- *Ease of installation*—typically while the tank is in service.
- *More information*—real-time product density and mass, in addition to tank level and volume.
- *Lower costs*—overall operational costs for the plant or refinery reduced [Ref. 3].

9.13 HTG CALCULATIONS

The basic equations that the HTG system uses to perform its calculations are for density at temperature, level, gross volume, mass and net volume. As mentioned, these calculations are made by taking accurate pressure measurements at two or three points in a tank, and a point temperature of the tank's product between the two pressure probes. Figure 9–4 shows an HTG system installation and is reviewed here. A pressure probe is placed at a point near the bottom of the tank (P_3). The distance from the bottom of the probe to an internal reference point at the bottom of the tank is referred to as the *Z* dimension in the HTG calculations. Another pressure probe is placed a known distance above

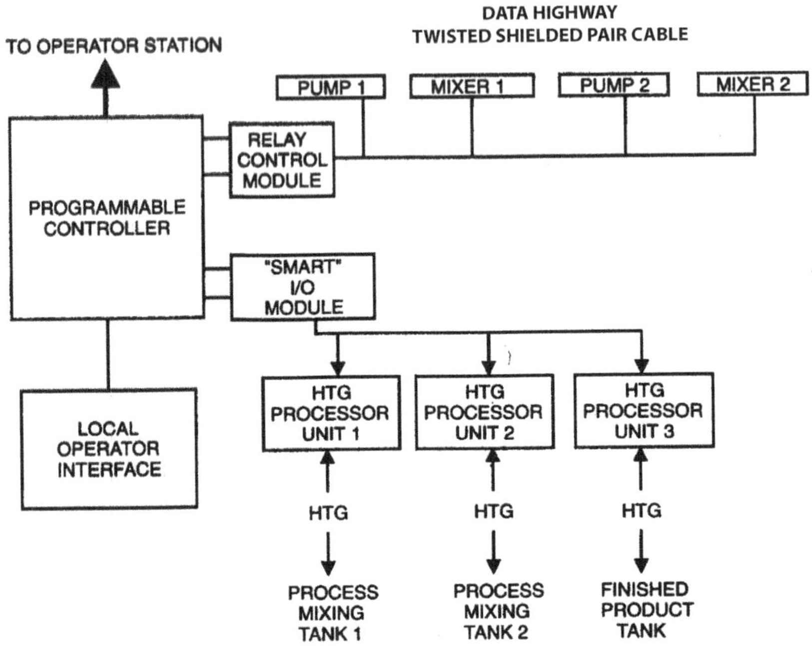

Figure 9–7 Typical Use of HTG Information for Batch Processing

the bottom probe (P_2). This distance, which is normally a function of the height of the tank, is referred to as the *L* dimension. Typical probe separation distance is normally recommended to be from 4 to 8 ft, or 20% of the tank height. This value should be as large as is practical to minimize the magnitude of the density error, although good test results have been obtained on small tanks with separations of less than 4 ft. Generally, the probes are placed no more than 8 ft apart to provide for ease of maintenance. Also, the density can be calculated only when the middle probe is covered, so minimizing the separation distance gives maximum level range in which density may be calculated.

The temperature probe is placed between the two pressure probes at a point that is the best estimate of the average temperature of the region between the two probes. Typically, this point is 3 ft from the bottom and 3 ft into the tank for unheated tanks. However, for heated tanks the RTD should be placed at a point that reflects average temperature of the region between the probes. This may involve moving the RTD lower or higher in the tank, depending on the location and configuration of the heaters in the tank.

The top pressure probe (P_1) is used only in applications in which the head pressure in the vapor space of a tank is not atmospheric pressure. When used, the probe must be placed above the maximum possible product level.

The sequence of calculations is similar for most of the HTG systems available today:

$$\text{Density@}T = \frac{(P_{\text{bot}} - P_{\text{mid}})(G_c)(144)}{L} \text{ (English)} \tag{9.3}$$

$$\text{Density@}T = \frac{(P_{\text{bot}} - P_{\text{mid}})(G_c)(101.97)}{L} \text{ (SI)} \tag{9.3a}$$

$$\text{Level} = \frac{(P_{\text{bot}} - P_{\text{top}})(G_c)(144)}{\text{Density@}L} \text{ (English)} \tag{9.4}$$

$$\text{Level} = \frac{(P_{\text{bot}} - P_{\text{top}})(G_c)(101.97)}{\text{Density@}L} \text{ (SI)} \tag{9.4a}$$

$$\text{Gross volume} = f(\text{Level})$$

$$\text{Units} = \text{gallons or barrels} \tag{9.5}$$

$$\text{Mass} = (P_{\text{bot}} - P_{\text{top}})(\text{Equiv. area})(G_c)(144) \tag{9.6}$$

$$\text{Equiv. area} = \frac{\text{Volume}}{(\text{Level})7.48} \text{ (English)}$$

$$\text{Mass} = (P_{\text{bot}} - P_{\text{top}})(\text{Equiv. area})(G_c)(101.97) \tag{9.6a}$$

$$\text{Equiv. area} = \frac{\text{Volume}}{(\text{Level})7.48} \text{ (SI)}$$

$$\text{Net volume} = \frac{\text{Mass}}{\text{Density@Ref}} \text{ (English)} \tag{9.7}$$

$$\text{Net volume} = \frac{\text{Mass}}{\text{Density@Ref}} \text{ (SI)} \tag{9.7a}$$

In these equations, the constant G_C is the gravitational correction factor, which permits the measured force to be interpreted as mass. The constant is dimensionless and has a value of approximately 1.0. The actual value depends on the measurement of the constant in the proximity of the location where the HTG system is used. If a site-specific value for G_C is not known, then this is defaulted to a value of 1.0. Acceptance of HTG as a standard for mass measurement by such standards organizations as API and ISO will provide further incentive to use this new and exciting technology.

9.14 CALCULATING HTG ACCURACY

Hydrostatic tank gaging is gaining more acceptance by the user community as its benefits become better understood. An important issue to users is the accuracy that can be expected from an HTG system. This section presents a method for calculating the typical accuracy of an HTG system. The equations used in the accuracy analysis are based on a mathematical analysis of HTG equations. The calculations used in an HTG system are based on a set of assumptions that relate to how the equipment is installed and to the actual operation of the tank. These assumptions are discussed along with other subjects, including equivalent area, the relationship between tank geometry and equivalent area, and HTG characteristic accuracies for various tank geometries.

In an HTG system, the density calculated is a density measurement of tank contents between the bottom and middle pressure probes. HTG systems assume that the product contained in a tank is homogeneous (uniform in composition throughout its volume). The HTG system must assume that the density of the material above the middle probe is of the same density as was calculated.

Likewise, it is assumed that the density of the material below the bottom probe is of the same density as was calculated. When a product stratifies (i.e., becomes layered), the density in the region between the bottom and middle probe can be different from the density of the material above the middle probe. The assumption about the density outside the measured region is then invalid.

The temperature of the tank contents (measured with an RTD) is assumed to be the average temperature of the region between the lower and the middle probe. The density at this measured temperature is called the density at temperature.

The RTD is positioned to obtain the average tank temperature of the region between the probes. If the temperature stratifies in the tank, and the temperature reading is not the average temperature of the region, the HTG system will not determine the correct density at reference. If the density at reference is inaccurate, then the net volume calculation will also be in error. However, if the temperature stratifies and the RTD indicates an accurate average temperature, the correct density at reference for the entire tank is obtained, and an accurate net volume will also be calculated.

9.15 HTG ASSUMPTIONS AND LEVEL CALCULATION

The level accuracy is dependent on pressure measurements and the accurate measurement of the *L* and *Z* constants. It is often difficult to determine *Z* (the distance from the bottom reference to the bottom probe) by making physical measurements. An HTG system must be calibrated for the tank on which it is installed. This calibration involves taking a reference measurement and comparing the measured level of the tank to the level displayed by the HTG system. The level measurement that is accepted by API for the calibration of an ATG is a hand gage. The level height is obtained by making an ullage measurement of level (tank roof to product level) and subtracting it from the tank reference height. The reference height is the distance from a calibration point on the roof to the bottom reference plate (Figure 9–4). The hand measurement of level is thus referenced to the internal reference plate. Quite often it is difficult to determine the exact location of the reference plate. One approach is to estimate the *L* and *Z* values initially, and then use a tuning procedure to tune the constants for best level accuracy.

The HTG level measurement assumes that an accurate density at temperature has been previously calculated. If the product in the tank has become stratified, then the density calculated by the HTG system will be higher than the actual average density in the tank. This is due to conditions of stratification, in which the heavier product migrates to lower regions of the tank, and the lighter product migrates to the higher regions of the tank. The level will be lower than expected, because it divides pressure by density. This means that the HTG system will calculate a level less than the actual level.

9.16 HTG ASSUMPTIONS AND GROSS VOLUME CALCULATION

The gross volume is calculated with a polynomial that is a function of level. The coefficients for the polynomial are obtained by performing a nonlinear regression on the tank strapping table data. This approach is valid for all tank geometries, although several polynomials are often required for reasonable volume accuracies. The accuracy of the polynomial will be no better than the strapping table used to derive the polynomial. Error usually derives from the fact that a tank table is either unavailable or is more than ten years old. When a tank table is unavailable, a perfect cylinder is often assumed. The lack of a strapping table is a serious problem if accurate volumes are desired using any gaging technology.

9.17 HTG ASSUMPTIONS AND MASS CALCULATION

The mass calculation is dependent on an accurate measurement of the average surface area of the tank product at all product levels within the tank. This area is often called equivalent area because it is not calculated directly. The equivalent area is derived by dividing the volume by the level. The assumption is that the equivalent area is an average area that is independent of level. While this assumption is true for a vertical cylindrical tank, it is not true for spherical, horizontal cylindrical or bullet-shaped tanks.

What this means is that HTG is most accurate in applications where tank geometry contributes no inaccuracy to the mass calculation. The fact that the liquid level surface area of a sphere changes with tank height means that a significant contribution to error results from tank geometry. The equivalent area of the liquid surface in a sphere is the surface area of a vertical cylinder that meets the following requirements:

- The volume of the cylinder is equal to the volume of the sphere to the current level.
- The height of the cylinder is equal to the level in the sphere.

The equivalent radius for a sphere is a percentage of the sphere height (radius) compared to percent sphere height (level). The radius of the sphere will differ from the radius of the equivalent area by more than 10%. In fact, the surface area of the liquid in the sphere and the equivalent area of the cylinder coincide at only one point, where the level is 75% of the sphere height. This applies to all spherical tanks. For tanks with complex shapes, the equivalent area, which changes with tank height, is very different from surface area. The error in measuring equivalent area is magnified by the sensitivity of equivalent area to error in the level measurement. This error in equivalent area makes a significant contribution to error in the mass calculation.

9.18 HTG ASSUMPTIONS AND NET VOLUME CALCULATION

The net volume calculation makes use of the previously calculated mass and density at reference. On a percentage basis, the mass accuracy is almost an order of magnitude smaller than the density accuracy. The main limitation of net volume accuracy is the accuracy of density at reference. The HTG gage, for the purposes of comparison, does not rely on the same techniques to derive net volume as used by traditional level gages. Net volume derived by a conventional float/tape or a servogage requires that the level and temperature of the tank be measured. The level is applied to a tank capacity table to determine a gross volume. Using a product sample and the tank temperature, the density of the product is determined and a volume correction factor (VCF) derived. The VCF is multiplied by the gross volume to obtain net volume. When an HTG system is used to calculate net volume, the HTG system requires a mass volume and a density at reference. An accurate density at reference depends on four factors:

- The density at temperature measurement must be accurate.
- The product in the tank must not be stratified.
- The temperature measurement in the region between the probes must be the average temperature of that region.
- The algorithm that converts density at temperature to density at reference must be accurate to within 0.1 API gravity [Ref. 4].

9.19 EFFECT OF TANK GEOMETRY

Equations have been derived that are used for accuracy analysis and are obtained by taking the partial differential of the HTG equations (Equations 9.3 to 9.7a). The resulting terms of the error equations are then added together using root mean square (rms) methodology. This technique is used to ensure that the error estimates exceed normal conditions where some of the errors cancel. Data obtained from error calculations were used in the curves in Figures 9–8 to 9–10.

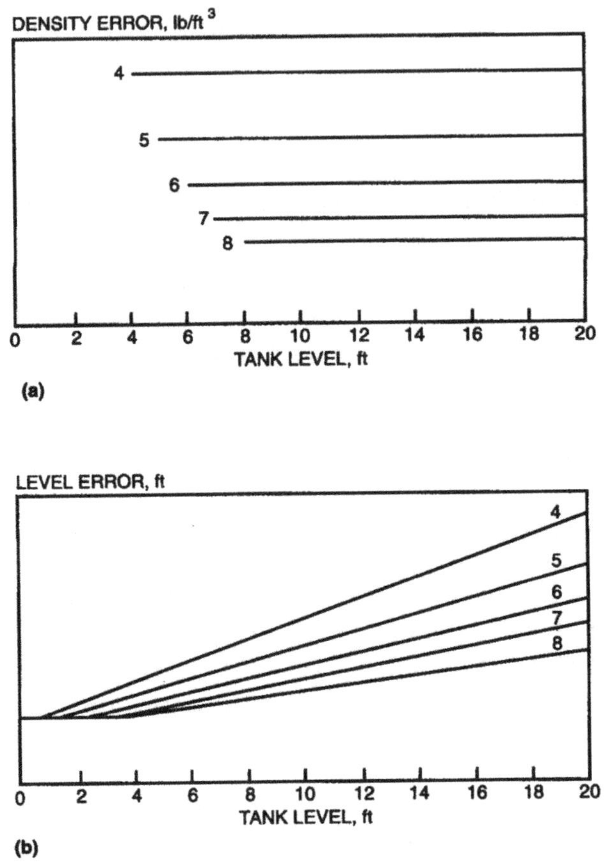

Figure 9–8 Density and Level Error Curves for All Tank Shapes

9.20 ADVANTAGES AND LIMITATIONS OF HTG

Several advantages have led to the increasing use of HTG technology. The accuracy of the method is consistent with the requirements of the application. Its typical level measurement accuracy of ±1 or ±2 in. is perfectly adequate for operational needs. Although this degree of accuracy would be unacceptable for a level gage that takes measurements for subsequent inventory calculation, it is perfectly adequate for HTG. This is because the HTG inventory measurements of mass and net standard volume are based on precise pressure measurements, rather than on level measurements.

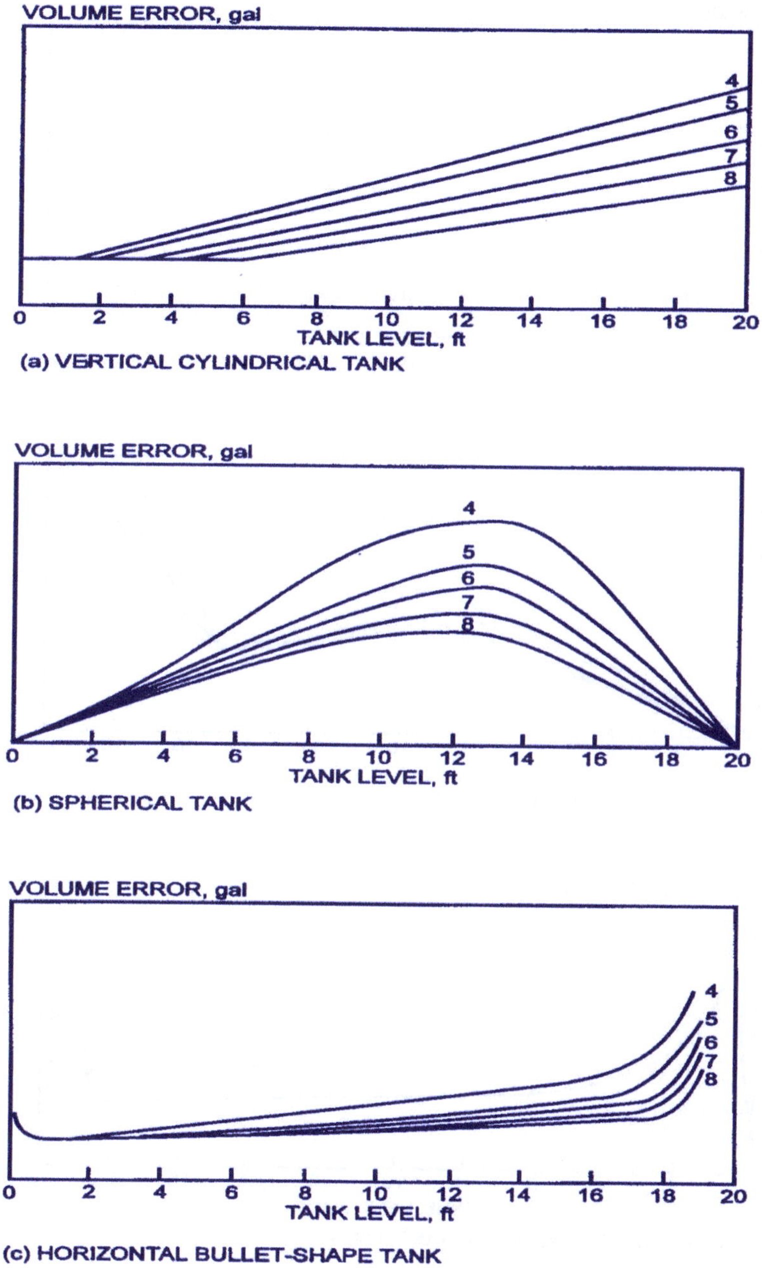

Figure 9–9 Volume Error Versus Level for Different Tank Shapes

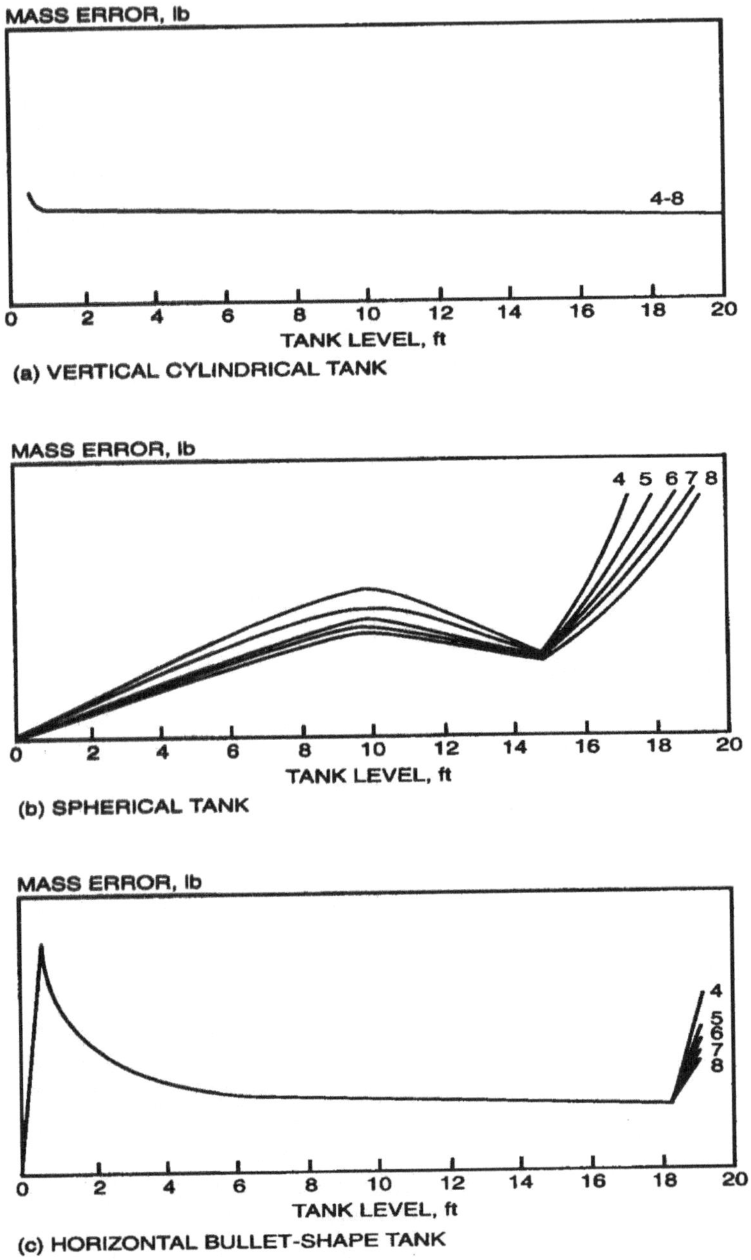

Figure 9–10 Mass Error Versus Level for Different Tank Shapes

The lack of in-tank moving parts not only increases the likelihood of high reliability on clean service, but also allows HTG to be applied to heavy and viscous products such as asphalt and lubrication oils.

Although HTG sensors must be mounted on a stable reference to avoid measurement errors due to tank shell movement, it is much easier and less costly to properly anchor the sensors than it is to install an in-tank stilling well. Also, the sealed nature of the connection of the sensor to the tank allows a tank design that can better contain vapors than can a slotted stilling well.

Another advantage is on-line density measurement. This allows the composition of the product to be monitored and provides an early warning of potential "off-spec" product or contamination due to the improper addition of an incorrect product to a tank.

One of the key benefits of HTG is that it is a mass-based measurement system. This feature provides not only accurate inventory measurement, but also leak detection and movement control. Both are difficult with level-based systems, which are affected by expansion and contraction with temperature.

9.21 NEW DEVELOPMENTS AND TRENDS

Since the first major application of HTG in 1986, acceptance of this technology has increased because of its ability to provide direct static mass measurement by continuous on-line measurement and computations. HTG is now considered to be a mature technology.

Recently, Ijkwezen, the Dutch weights and measures authority, reasoned that a true static mass meter using HTG principles could meet the performance criteria applicable to legal product transfer and inventory measurements. After conducting a series of rigorous tests, it was concluded that the term "static mass meter" should be used to differentiate between HTG and ATG measurement.

Work group ISO/TC28/SC3/WG4 of the International Organization for Standardization finalized a draft standard for mass measurement by hydrostatic tank gaging. The Ijkwezen work complements that of ISO and establishes guidelines for HTG applications by adopting performance criteria and operating conditions to be met for custody transfer applications. The security that must be provided to indicate malfunction or unauthorized access is also stipulated. The support of HTG by international standards clearly substantiates this technology as a viable method for tank farm and storage tank operators to determine extensive parameters such as mass, volume, level, temperature, product density and true product quantity.

The success of a high-accuracy installation test that led to the mass measurement standard will be used to adopt a volume measurement standard. The American National Standards Institute (ANSI) has contributed to the ISO research and is working to adopt the ISO standards.

The American Petroleum Institute revised their standard dealing with the use of automatic tank gages for level measurement. This standard differentiates between ATGs for custody transfer and for internal inventory control by setting different performance criteria for each application. It establishes acceptable accuracy limits for each application and stipulates that any ATG that can be shown to meet these requirements is suitable for the application. Float, servo, HTG, radar and several other types are discussed. One of the limitations of the standard is that it deals only with the measurement of level. Even the calculations of the desired inventory values of standard volume and mass are left to other standards. However, a recommendation has been made within API to consider adopting the ISO standard on static mass measurement, either as is or with API modifications. Hydrostatic tank gaging is now a fairly well-established technology, with at least four of the world's largest oil companies having purchased many hundreds of tank sets from all of the major suppliers and many more oil and petrochemical companies having purchased significant quantities for small- to medium-size projects. With this maturation, attention has logically turned from evaluation of the gage to focus on applications in order to better solve user problems and better meet user needs.

One application where HTG is being considered as an alternative to level-based gages is on tanks where the gaging system is expected to double as both an inventory system and a first-line leak-detection system. Results of work done by several companies show that HTG has the capability of detecting leaks on a mass basis equivalent to a level change of approximately 1/16 in. over an 8-hr test period on a static tank. This is a good example of capitalizing on the inherent mass measurement accuracy of HTG. It would be very difficult to achieve this level of performance with a level-based system.

Users have also begun to turn to alternate technologies to solve long-standing tank gaging problems with pressurized spherical and bullet-shaped tanks. Tank farm operators were quick to see the potential benefits of nonintrusive radar and HTG gages for heavy and viscous products that cause problems for floats and servogages. For high pressures (>150 psi), differential-pressure HTG sensors offer greater accuracy than the standard gage-pressure types used for pressures up to 150 psi. HTG sensors with capillaries and seals have even been applied to high-pressure bullet tanks where specialized products were known to plug up small-diameter piping.

Another trend in the use of HTGs is the evolution of computer display software to more fully exploit the capabilities of the gage. One example is in the area of movement authorization. Monitoring mass has allowed much tighter control of tank product movements. Software is available to detect the rate of change of mass and to indicate a movement condition almost as soon as it begins. The programs allow the operator to declare a tank product movement acceptable for inbound, outbound, both or none. An unauthorized movement signals an alarm.

Similarly, the measurement of mass allows the net flow rate in or out of a tank to be accurately calculated and displayed. The operator can set high and low alarm limits that bracket the flow rate. This can be useful, for example, on an inbound movement where the flow rate from the pipeline or barge is known and any significant difference between the source flow rate and the receipt flow rate needs to be signaled and accounted for.

Another software feature used with HTG systems is alarms for both the absolute values of high and low density and the rate of change of density. This allows early detection of product contamination due to the introduction of improper product into a tank.

Remote monitoring of tank conditions is another development. Software enhancements make it easier for telecommunications techniques to be used to monitor inventory and prepare management reports at a central location remote from the tank farms. Also, alarm conditions can be monitored to increase the safety of operations by having a backup watch on tank levels and other tank measurements and alarms.

9.22 A MULTI-FUNCTION TANK GAGE

A single instrument has been designed to provide the information needed for an HTG application. This multi-function tank gage (MTG) can be used for ATG and HTG and to provide measurements needed in many related applications.

MTG Construction and Design

The MTG probe includes multiple 2.8" O.D. stainless steel pipe sections with three to six pressure and temperature modules in each section. The probe assembly is manufactured in sections of specific lengths designed for the particular application and is normally 5 to 8 ft long; the individual sections can be connected by threading or by a sleeve.

Each sensor module has a four wire RTD and a proprietary pressure sensor that is accurate from 0.01% to 0.02% URV (upper range value). The sensor assembly includes a dual microprocessor-based transmitter located on top of the probe. The transmitter polls all sensors and performs accurate A/D (analog to digital) conversions and data processing including correction algorithms and strapping tables. From this computation all measurements such as mass, level, temperature, volume, water content, interface, specific gravity and density are provided.

The MTG transmitter communicates with any standard Modbus® RTU host via IEEE/EIA RS-485 communication bus. Included also are communications via Bluetooth, spread-spectrum, radio, cellular, and satellite using the Modbus RTU protocol. The MTG has a solar power capability and an Enhanced Tank Gage Interface (ETGI).

Other features of the MTG include

- Intrinsic safety
- Various cabling/communication topologies
- Ethernet with 10Base-FL connections
- Web-enabled device
- GSI Bluefield™ Topology

MTG Theory of Operation

Measured values from the MTG are based on mass and density from the pressure and temperature measurements. The method of these measurements overcomes the disadvantages of some other measurement technologies, such as reference point movements and changes in temperature effects, problems encountered with moving parts and mechanisms, bubble tubes in enclosed vessels, various tank entry and measurement points, and uncompensated changes in the physical properties of the process fluid.

The MTG is included in the HTG section because that is where its application has found much success. It provides accurate inventory measurement, a real-time profiling of product within a liquid storage tank. It can determine product quality by measuring specific gravity and density, data that can be used to determine water content in a product at several points in the tank. This is an important criterion in petroleum products as water can often form emulsion layers along the tank strata. The MTG provides continuous knowledge of the interface point or points with liquids with as little as a 0.05 difference in specific gravities. This can show a real-time profiling of any type of density or temperature stratification within the tank. This continuous measurement along the tank strata can improve tank blending applications.

From Figure 9–11, it can be seen that density is determined from measuring points P_2 and P_3, and from points from P_2 and/or P_4, P_5 and P_1. Temperature from T_T is used in the calculation of density. A multifunction tank gage (MTG) has been developed to provide measurements at various points to improve the accuracy of HTG. The MTG shown in Figure 9–12 can produce measurements used for calculated process variables at six points. These units can be stacked to get data at many points along a desired measurement range. From the illustration shown, the following is provided:

1. Density at five strata levels
2. Free water at the bottom of the vessel
3. Percentage of water at selected interface (emulsion) points
4. Liquid level
5. Vapor pressure
6. Vapor temperature
7. Percentage of water in the product
8. Mass
9. Multi-spot temperature points
10. Free water level
11. Mass sensitivity for leak detection

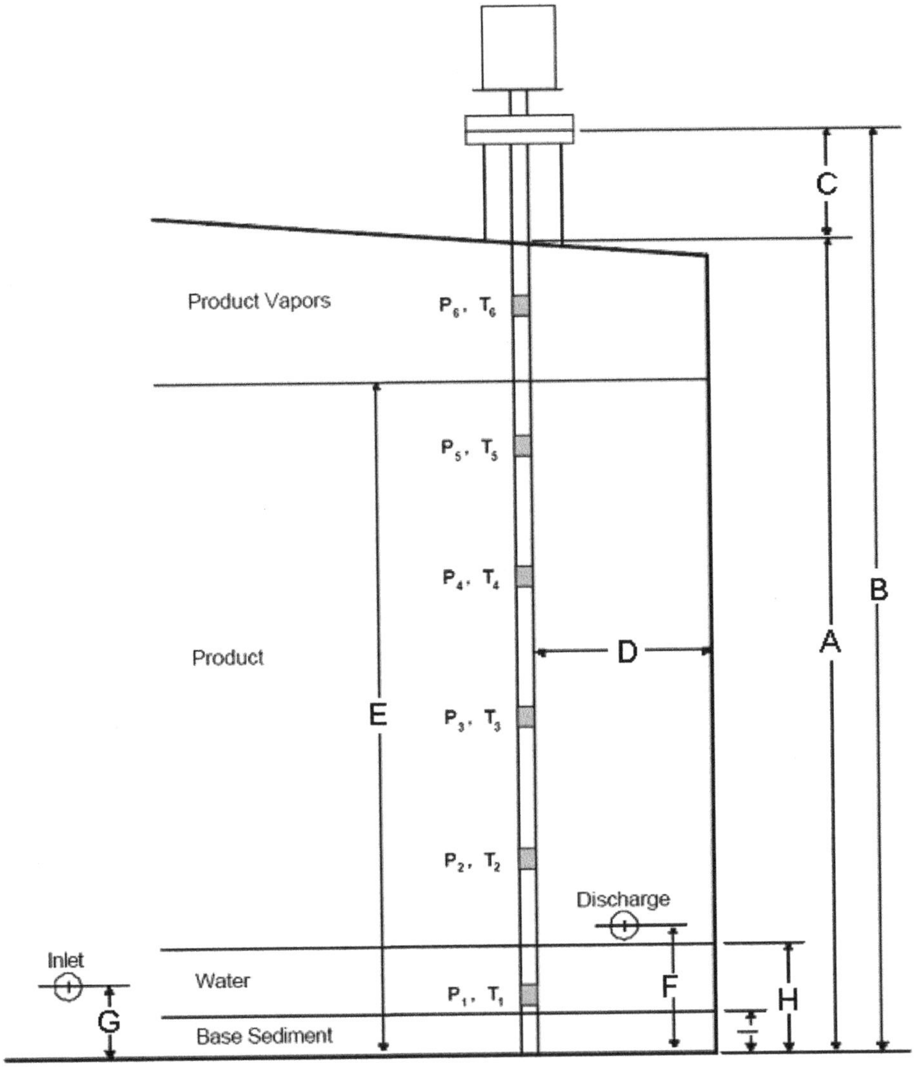

Figure 9–11 Measurement Points to Determine Density
(*Courtesy of Gauging Systems*)

Real-time calculations by a proprietary algorithm can give complete tank profile data, including net volume, by mass or liquid level at desired points within the vessel. RTDs are used for accurate and precise temperature, and level is calculated from pressure measurement at various points along the probe. Alarming is also available for the various measurements.

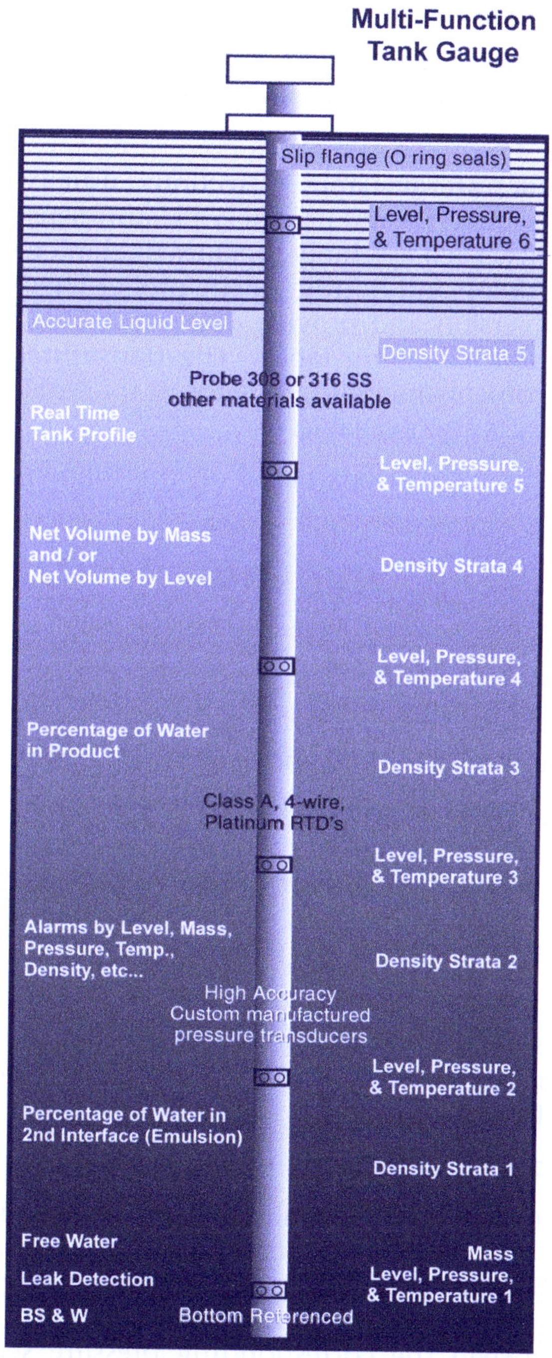

Figure 9–12 Measurement Provided by Multi-function Tank Gage (*Courtesy of Gauging Systems*)

Built-in redundancy is provided by the multiple sensor design.

True level measurement is independent of tank top movement because the probe is mounted on the bottom of the tank, with an O-ring seal slip flange at the top.

The sensor modules are available in sections, usually 5 ft or 8 ft in length, with four wire RTDs and propriety pressure sensors that measure ±0.01 to 0.02% URV. Each assembly includes dual microprocessor-based transmitters, at the top and bottom of the probe. They poll all sensors, perform A/D conversion and provide data processing with correction algorithms and tank strapping tables. The transmitters communicate with most bus protocols including Bluetooth for wireless communications. Local display is also available.

The MTG performs calculations based on the following relationships (Figure 9–11):

$$P_1 - P_6 = D \times H \times G \tag{9.8}$$

where

D = density of the liquid

H = the height of the liquid above the first sensor

G = the local gravity constant

The mass of the liquid above the first pressure tag is

$$M = (P_1 - P_6)\ (SG) = (D)(H)(S) = D(V) \tag{9.9}$$

where

S = the average cross sectional area of the vessel given by the strapping table

G = the local gravity constant

V = the volume of the liquid above the first pressure point

The pressure difference $DP_{12} = P_1 - P_2$ gives the average density value D_{12} in the area between the first and second sensors, provided that the distance H_{12} between the corresponding pressure sensing points is known. D_{12} is the density between P_1 and P_2. H_{12} represents the level between P_1 and P_2. The same reasoning applies for the remaining equations.

$$D_{12} = (P_1 - P_2)/G/H_{12} \tag{9.10}$$

The density layers between the rest of the sensors are calculated as before:

$$D_{23} = (P_2 - P_3)/G/H_{23} \tag{9.11}$$

$$D_{34} = (P_3 - P_4)/G/H_{34} \tag{9.12}$$

$$D_{45} = (P_4 - P_5)/G/H_{45} \tag{9.13}$$

The density values measured by formulas 3 through 6 and temperature measurements taken at the same point along the tank height enable a density distribution curve to be plotted.

The level of liquid L in the tank is calculated as follows:

$$L = H_1 + P_1/G/D_1 \tag{9.14}$$

where

H_1 = the location of the last sensor unit covered with liquid

P_1 = the pressure measurement taken from the last sensor covered with liquid

D_1 = the current density value

> All of the above calculations are from simplified formulas. More detailed expressions for some of the values can be found in ISO11223-1 and API *Standards* Chapter 16, Section 2.

When pressure sensors in the second to fifth units are not covered by liquid, the MTG system identifies these, and when stable conditions exist the uncovered sensors are recalibrated. A proprietary algorithm recalibrates the bottom sensor by using simultaneous measurement results from three to six pressure sensors.

The performance specifications of the MTG are:

1. Mass ±0.05% of full scale
2. Density ±0.0015 grams/cubic centimeter
3. Level ±0.15% of full scale between the sensor modules
4. Temperature ±0.1°C

The above criteria meet the requirements for custody transfer as set forth by API standard 3.18, API 3.6, and ANSI/ASQC.

Leak detection is measured in GPM. HTG installations that are accurate enough to meet the standards can be used in these applications [Ref. 5].

The following are some considerations to take into account when selecting the instruments and technology to be used (for HTG applications).

1. The system should accurately provide full information about the systems to include
   A. Level
   B. Multipoint measurement of
      i. Temperature
      ii. Density (may be calculated from other variables)
      iii. Mass
      iv. Water content
      v. Net volume
2. The system should
   A. Be accurate
   B. Be reliable
   C. Have few or no moving parts that need maintenance
   D. Be of rigid construction
3. The system should be easily configurable, self-calibrating and maintenance free.
4. The sensor should be bottom referenced to eliminate error caused by tank growth.

5. Only one tank entry should be required.
6. The system should be interoperable.
7. Communication should be based on an OSI model.
8. Various communication topologies should be available to meet the user's demands/desires.
9. Radar level technology can be used for level measurement in hydrostatic tank gaging operations.

Increased automation of tank farms improves productivity and quality and increases safety, and will no doubt lead to greater use of ATG equipment. The age-old technique of level-based gaging with both inexpensive float gages and more accurate level gages such as servo, nuclear, and radar types is certainly still viable. However, HTG will continue to capture an increasing share of the market as the advantages of a mass-based system become better known and accepted. The availability of more useful application software by suppliers and the establishment of static mass measurement standards by various standards organizations will help to accelerate the use of mass-based systems such as HTG.

EXERCISES

9.1 Explain the difference between ATG and HTG.

9.2 Explain a general procedure for tank calibration and the reason for it.

9.3 When is recomputation necessary?

9.4 What measurements are required for HTG applications?

9.5 What calculations are required for HTG applications?

9.6 Aside from instrument accuracy, list three possible sources of error in HTG applications.

9.7 Discuss the concept of HTG and custody transfer.

REFERENCES

1. Hampton, B.J., and Sivarman, 5., "Guidelines Set for Recalibration of Storage Tanks" *Oil and Gas Journal,* June 1989.

2. Sivarman, 5., "Guidelines Help Select Storage Tank Calibration Method." *Oil and Gas Journal,* Feb. 5, 1990.

3. Early, Paul 1., Jr., "Solving Old Tank Gaging Problems with the New Hydrostatic Tank Gaging System" ISA Paper 87-1019, Research Triangle Park, NC: ISA, 1987.

4. Adams, David, "Method for Calculating HTG System Accuracy" ISA Paper 89-0664, Research Triangle Park, NC: ISA, 1989.

5. Gauging Systems, Incorporated (GSI) Technical Literature, 2006.

10

Instrument Selection and Applications

10.1 INTRODUCTION

In the previous chapters, material was presented that detailed the operation of several types of pressure, level and density measurement technologies. Material in this chapter will give information on the selection of a measurement technology for a given application.

10.2 SUMMARY OF PRESSURE APPLICATIONS

Many times pressure instruments are categorized on the basis of the ranges of pressure they can measure and the design principle involved in their operation, or on the basis of a specific application. By listing a short summary of basic features applicable to a particular group of devices, the user can better determine which category of instruments is most suitable for a desired application. Taken together, the measurement technologies presented can be used to measure pressures ranging from about 10^{-12} mm Hg to several hundred thousand psig.

While the range of costs is dynamic in nature, it will be given as low, medium and high to represent a few hundred, a few thousand and several thousand dollars, respectively. The cost also depends on the specifics of the application, such as the use of a pressure gage, one or more disposable transmitters, or a multivariable transmitter with field control options. The designation of analog or digital has largely lost its significance because most transmitters are of a digital design, and analog only to the extent that they will transmit 4–20 mA signals. Most pressure and differential pressure instruments are designed with digital communication, calibration, re-ranging and two-way communication between field and control center for asset management application. The major distinction between digital transmitters is the selection of HART or FOUNDATION Fieldbus capabilities. Even this distinction will be fading now that HHCs (handheld communicators) are available to accommodate both communication protocols.

Although many devices are presented for the purpose of completeness, for a particular application the selection does not involve a great many choices. Table 10–2 will help in this selection and may be used for an orientation to pressure sensors. It generally notes the overall pressure range that the sensor is capable of measuring, the stated reference accuracy, and the approximate price range. When not listed, the accuracy is generally expressed as a percent of calibrated span, although the HART and FOUNDATION Fieldbus transmitters are probably listed using percent of rate. It should also be noted that the quality statement given is for ideal conditions, and installed accuracy will probably be somewhat less depending on the specific application, process connections, connecting lines, seal conditions, fill fluids, etc.

As slight as it may be, reference error can result in measurement error and is often overlooked. For absolute pressure measurement, if the vacuum reference chamber is not evacuated to absolute zero pressure (a perfect vacuum) but instead to a few thousandths of a millimeter of Hg (torr), the measurement accuracy will be diminished by the amount of error in the reference pressure. For gage pressure applications, changes in the referenced barometric pressure will also result in that fluctuation, resulting in a measurement error by the same amount. The error caused by a false reference pressure may be insignificant in relatively high pressure applications, but for low pressure and compound pressure applications, measuring above and below atmospheric pressure, the error should not be neglected.

The following listing (Table 10–1) will help in the selection of pressure sensors for given applications.

The data listed is for reference and may not be needed or used in the selection of a pressure instrument for a given application. The manufacturer has made the selection of a pressure element for a particular measurement range. Table 10–2 lists the characteristics of various types of transmitters. Industry standard specification forms available from ISA can be used as a checklist in the selection of instruments used in the processing industry. All items to be listed are explained on the specification form.

Table 10–1 Types of Pressure Elements

Type of Sensor	Possible Full Range Values
C tube	A few psi to a couple thousand psi
Bellows	About 1 psi to a couple thousand psi
Diaphragm	10^{-2} mm Hg to a thousand psi
Quartz helix	10^{-3} mm Hg to a thousand psi
Liquid and electronic manometer	10^{-3} mm Hg to a thousand psi
Capacitance diaphragm	10^{-5} mm Hg to a thousand psi
McLoud gage	10^{-6} mm Hg to about ten psi

When there is a possibility of the process liquid or condensate from gas or vapor freezing in the instrument and connecting lines, the instrument can be sealed for freeze protection as outlined in Chapter 3. To avoid errors caused by temperature fluctuations and density changes, insulation and even heat tracing may be required. Again, these errors may be neglected for high pressure applications and for differential pressure service, where density changes in the impulse lines will cancel each other, assuming the temperatures are equal. When heat tracing is used, care must be exercised to control the temperature to avoid density changes and the vaporization of the seal fluid. Problems encountered from the above considerations can cause zero drift or even erratic readings.

It is also possible for the seal fluid to become viscous over a period of time, which will result in a loss of sensitivity in the measurement. For this reason, a preventive maintenance schedule can be established for "blowdown" of the system. A configuration like that shown in Figure 10–1 is recommended for pressure and Figure 10–2 for differential pressure. For liquid and vapor service, the instrument will be located below the process line to maintain a liquid-full system. For vapor, especially steam service, care must be exercised to inform operating personnel of an impending zero shift of the reading until the impulse lines are liquid-full and at a stable temperature. Seal pots as shown in Figure 3–15 can help in this situation. A review of the instrument mounting section of Chapter 3 will help in understanding instrument maintenance.

Table 10–2 Types of Transmitters and Their Characteristics [Ref. 1]

Type	Range	Error (Percent of Rate)	Relative Cost
General-purpose bourdon-tube indicator	15–10,000 psig (1–690 bar)	±2%	Low
High accuracy process calibrator	Low vacuum to 3000 psi (low vacuum to 207 bar)	±0.25%	Medium to High
Bourdon/spiral case-mounted indicator/recorder	Low vacuum to 50,000 psig (low vacuum to 3450 bar)	±0.5%	Medium
Spring-and-bellows case-mounted recorder	Low vacuum to 50 psig (low vacuum to 3.5 bar)	±0.5%	Medium
Nested capsular case-mounted recorder	10–90 psig (0.7 to 6.2 bar)	±0.5%	Medium
Low pressure bell case-mounted indicator	–0.1 to 0.1 inches of water (–3 to 3 mm of water)	±2%	Medium
Beam-mounted strain gage, 4–20 mA DC output	0 – 1000 psig (0 to 69 bar)	±0.25%	Low
Piezoresistive and capacitance transducers, 4–20 mA DC output, HART and FOUNDATION Fieldbus compatible	0 to 6000 psig (0 to 414 bar)	±0.1%	High
Digital hand-held communicators for HART and FOUNDATION Fieldbus transmitters	Compatible with transmitter range	Better than ±0.1%	High

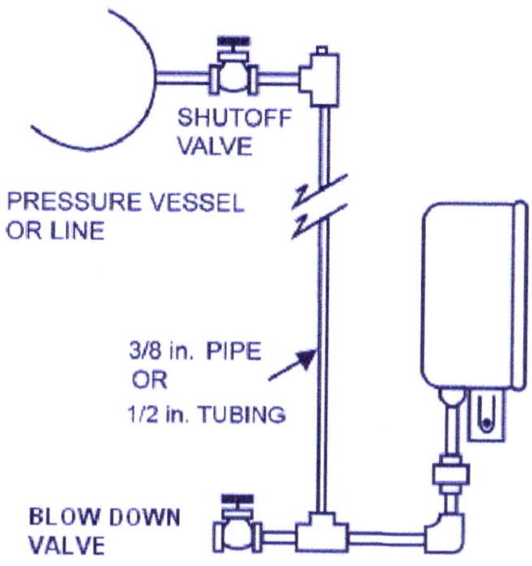

Figure 10–1 Blowdown Arrangement for a Pressure Instrument

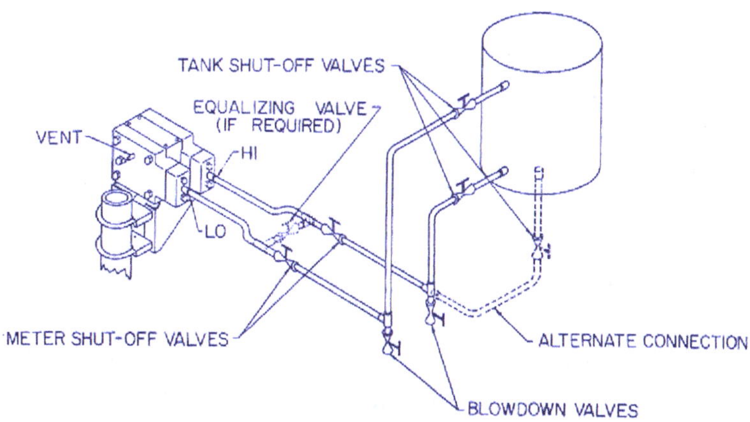

Figure 10–2 Blowdown Arrangement for a Differential Pressure Instrument

Instrument maintenance and calibration procedures have changed substantially with the introduction of digital transmitters. The HART and FOUNDATION Fieldbus protocols provide self-diagnostic capabilities to aid in troubleshooting, and really, most situations that require maintenance are not instrument-related but are problems encountered in interfacing the instrument with the process. As outlined in

Chapter 4, conventional calibration using a transfer standard is seldom needed. Calibration checks are really re-ranging procedures where a range is noted or changed with an HHC or host workstation. As microprocessor-based transmitters have become more accurate, it is becoming increasingly difficult to find calibration equipment that is as accurate as the transmitter. Calibrating with a standard that is less accurate than the transmitter can degrade the accuracy of the device.

10.3 SUMMARY OF LEVEL APPLICATIONS

Instrument Selection

There are many factors to be considered in selecting a level measuring technology for a specific application. Following is a list of some of these factors:

- Is the application for measurement or for control?
    - Measurement will be concerned with high accuracy for material balance or custody transfer.
    - Control will be more concerned with precision (repeatability).
- Interface
    - Liquid/gas
    - Liquid/liquid
    - Liquid/solid
    - Solid/gas
        - Measurements are generally made at one point in the vessel. For liquid/gas and liquid/liquid, the level arranges itself to be horizontal. This results in a well-defined level. For solids/gas, however, this may not be the case because particulate solids do not arrange themselves to be horizontal. Solids often form a conical hole where the material exits the vessel and a conical mound where the material enters the vessel. Sensing level at the mound or hole should be avoided when possible. The sensor should be positioned so the measurement represents the actual level in the vessel.

- Material surface condition
  - Static
  - Agitated
  - Characteristics
    - Reflectivity for light or RF energy
    - Presence of foam or a "rag" layer between interface fluids
- Physical constraints
  - Temperature
  - Pressure
  - Material consistency and characteristics
    - Viscosity
    - Density stability
- Can the level sensor be mounted in the vessel or should it be external?
- Continuous or point measurement?
- Contact or noncontact sensor application?
- Direct or inferred measurement—can hydrostatic head measurement be used?
- Is vessel depressurization or process shutdown required and acceptable for sensor maintenance or removal?

By evaluating the above criteria the list of sensors to be considered will be shortened. Sensor selection can further be narrowed by considering only the types and designs that are compatible with the process material, sensor material, construction, required accuracy, pressure, temperature and any constraints or characteristics. Tables 10–3 and 10–4 can be used in selecting the appropriate or acceptable measurement technology. Often a number of sensor and transmitter types could be used and in such cases, traditions and preferences will play a part in the selection process. User familiarity, spare parts inventory, availability, cost, ease of mounting and calibration are also considerations.

Table 10–3 Orientation Table for Selecting Level Sensors [Ref. 2][a]

Type	Max. Temp (°F)	Available as Noncontact	Inaccuracy (1 in. = 25.4mm)	Applications: Liquids: Clean	Viscous	Slurry/sludge	Interface	Applications: Solids: Foam	Powder	Chunky	Sticky	Limitations
Air Bubblers	UL	No	1–2% FS	G	F	P	F					Introduces foreign substance into process; high maintenance.
Capacitance	2,000	No	1–2% FS	G	F-G	F	G-L	P	F	F	P	Interface between conductive layers and detection of foam is a problem.
Conductivity Switch	1,800	No	1/8 in.	F	P	F	L	L	L	L	L	Can detect interface only between conductive and non-conductive liquids. Field effect design for solids.
Diaphragm	350	No	0.5% FS	G	F	F			F	F	P	Switches only for solids service.
Differential Pressure	1,200	No	0.1% AS	E	G-E	G	G					Only extended diaphragm seals or repeaters can eliminate plugging. Purging and sealing legs are also used.
Displacer	850	No	0.5% FS	E	P	P	G-E					Not recommended for sludge or slurry service.
Float	500	No	1% FS	G	P	P	F					Moving parts limit most designs to clean service. Only preset density floats can follow interfaces.
Laser	UL	Yes	0.5 in.	L	G	G		F	F	F	F	Limited to cloudy liquids or bright solids in tanks with transparent vapor space.
Level Gages	700	No	0.25 in.	G	F	P	F					Glass is not allowed in some processes.
Microwave Switches	400	Yes	0.5 in.	G	G	F	G		G	G	F	Thick coating.
Optical Switches	260	Yes	0.25 in.	G	F	E	F-G	F	F	P	F	Refraction-type for clean liquids only; reflection-type requires clean vapor space.
Radar/ Noncontact	450	Yes	0.12 in.	G	G	F	F		P	F	P	Interference from coating, agitator blades, spray or excessive turbulence.

Type	Max. Temp (°F)	Available as Noncontact	Inaccuracy (1 in. = 25.4mm)	Applications								Limitations
				Liquids				Solids				
				Clean	Viscous	Slurry/sludge	Interface	Foam	Powder	Chunky	Sticky	
Radar/Contact TDR/FMCW	221	No	0.062–0.125 in.	F	F	F	G		G-E	G	F	Limited performance on sticky process materials.
Radiation	UL	Yes	0.25 in.	G	E	E	G	F	G	E	E	Requires NRC license/reporting.
Resistance Tape	225	No	0.5 in.	G	G	G						Limited to liquids under near-atmospheric pressure and temperature conditions.
Rotating Paddle Switch	500	No	1 in.						G	F	P	Limited to detection of dry, non-corrosive, low-pressure solids.
Slip Tubes	200	No	0.5 in.	F	P	P						An unsafe manual device.
Tape-Type Level Sensors	300	No	0.1 in.	E	F	P	G		G	F	F	Only the inductively coupled float is suited for interface measurement. Float hang-up is a potential problem with most designs.
Thermal	850	No	0.5 in.	G	F	F	P	F				Foam and interface detection is limited by the thermal conductivities involved.
Ultrasonic	300	Yes	1% FS	F-G	G	G	F-G	F	F	F	G	Presence of dust, foam, dew in vapor space; sloping or fluffy process material interferes with performance. Susceptible to error caused by density and temperature variations in vapor space.
Vibrating Switches	300	No	0.2 in.	F	G	G	F		F	G	G	Excessive material buildup can prevent operation.

a. TDR = Time Domain Reflectometry, FMCW = Frequency Modulated Continuous Wave, AS = in % of actual span , FS = in % of full scale
E = Excellent, G = Good, L = Limited, P = Poor, UL = Unlimited

Table 10–4 Applicability of Level Sensors for Different Services [Ref. 2][a]

		Liquids		Liquid/ Liquid Interface		Foam		Slurry		Suspended Solids		Powdery Solids		Granular Solids		Chunky Solids		Sticky Moist Solids	
		P	**C**	**P**	**C**	**P**	**C**	**P**	**C**	**P**	**C**	**P**	**C**	**P**	**C**	**P**	**C**	**P**	**C**
Beam Breaker		.	.	.	.	2	.	.	.	.	.	1	.	1	.	3	.	1	.
Bubbler		1	1	.	.	.	.	3	2	.	.	.	.	.	.	.	.	.	.
Capacitance		1	1	1	1	1	2	1	2	.	.	2	2	1	2	2	2	1	2
Conductive		1	.	2	.	1	.	1	.	.	.	3	.	3	.	3	.	1	.
Differential Pressure		2	1	2	2	.	.	2	2	.	.	3	3	.	.	.	.	.	.
Electromechanical	Diaphragm	1	1	2	.	.	.	2	2	.	.	1	3	1	.	3	.	2	3
Electromechanical	Displacer	1	2	2	2	.	.	3	2	.	.	.	.	.	.	.	.	.	.
Electromechanical	Float	1	.	2	.	.	.	3	.	.	.	.	.	.	.	.	.	.	.
Electromechanical	Float/Tape	3	1	.	.	.	.	.	3	.	.	.	.	.	.	.	.	.	.
Electromechanical	Paddlewheel	.	.	.	.	.	.	3	.	.	.	2	.	1	.	3	.	2	.
Electromechanical	Weight/Cable	3	1	.	.	.	.	.	1	.	1	.	1	.	1	.	1	.	1
Gages	Glass	1	1	2	2	3	3	3	3	.	.	.	.	.	.	.	.	.	.
Gages	Magnetic	1	1	.	.	3	3	3	3	.	.	.	.	.	.	.	.	.	.
Inductive		.	.	.	.	.	.	2	.	.	.	2	2	2	2	2	2	3	3
Microwave		1	1	.	.	.	.	1	1	.	.	1	2	1	1	1	1	1	1
Radiation		1	1	.	.	.	.	1	1	.	.	1	1	1	1	1	1	1	1
Sonic Echo	Sonar	.	.	2	2	.	.	.	3	1	1	.	.	.	.	.	.	.	.
Sonic Echo	Sonic	1	1	3	3	.	.	1	1	2	2	.	3	1	1	1	1	2	1
Sonic Echo	Ultrasonic	1	2	2	2	.	.	1	2	1	1	.	3	2	2	1	2	2	2
Thermal		1	.	1	.	2	.	2	.	.	.	.	.	.	.	.	.	.	.
Vibration		2	.	3	.	.	.	2	.	1	.	1	.	1	.	2	.	1	.

a. C = Continuous, P = Point, 1 = Good, 2 = Fair, 3 = Poor, . = Not Applicable

In situations where the material is agitated, there may be no vapor space available to insert a probe-type sensor. Also, the liquid surface is not flat, so sonic, ultrasonic or radar usually cannot be used successfully. With displacers and head-type sensors, agitation can cause erratic or cyclic signals from the transmitter. Stilling basins can be used in such cases or filtering can be provided in most transmitters. Setting the filtering time should be in accordance with the response time of the process.

The relationship between level and volume is a function of the cross-sectional area of the tank shape and may be a factor in instrument selection (Figure 10–3). This relationship is linear for straight-sided vertical tanks and nonlinear for spherical, cone-shaped and horizontal tanks. For head measurement applications a multivariable transmitter can be used to compensate for process temperature and density. With the appropriate configuration and number of process and temperature measurements, volume and mass values can be determined.

For steam and other vapor service using head level measurement, an application technique is to reverse the connections to the differential pressure transmitter. The high side is connected to the wet leg, which is filled with condensate from the vapors in the vapor space of the vessel. The low side of the transmitter responds to the liquid level in the vessel. The output then increases as the level decreases. It is therefore necessary to invert the signal from the transmitter or use a transmitter configured for reverse acting. This method was necessary when DP (differential pressure) transmitters required the higher pressure to be applied to the H side. This is typically not required today, but is traditionally used in some industries. Another, perhaps more common, method to compensate the wet leg is to use the conventional arrangement of connecting the low side of the transmitter to the top of the vessel and elevating zero by the amount of calibration span or more depending on the location of the low side connection to the vessel.

When the process fluid is sludge, slurry, foam, molten metal, or an otherwise difficult to handle material where physical contact with the instrument is undesirable, non-contact sensor technology can be considered. Non-contact sensing can be accomplished by placing the sensor outside the vessel. Examples can include weighing or radiation. By using ultrasonic, radar or microwave, the sensor can be located in the vapor space of the vessel. For sludge or sticky, highly viscous, or otherwise difficult material that can result in the accumulation of solids in the dead-end cavities and lead lines (impulse lines), or cause

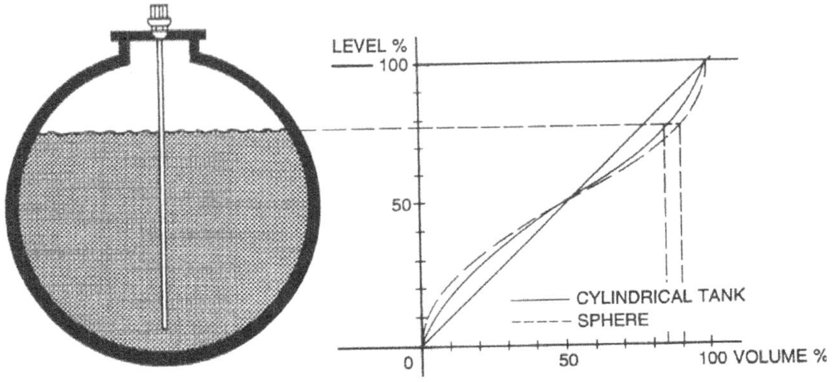

Figure 10–3 Relationship Between Level and Volume for Different Tanks (*Courtesy of Instrument Engineer's Handbook, 4th Edition, CRC Press*)

damage to the wetted portion of the instrument, Teflon-coated diaphragm seals or those using exotic materials can be used.

Where it is required to control foam level thickness below a maximum point and to control the liquid level below the foam, selected radiation detectors can be used as well as optical or thermal switches or capacitance or conductivity switches. Where applicable, a tuning fork or vibrating switch can be used. Automatic washers may be required in point applications when fouling of the sensor occurs when the level falls below the detectors.

For glass and molten metal and other such special applications, the following can be considered:

- Proximity capacitance—with about 0.1 mm sensitivity over a 150 mm range.
- Laser-based systems with better than 0.1 mm sensitivity over a 609.6 mm range.
- Floats constructed with refractory material attached to an LVDT, where high resolution is not required.
- Dip or bubble tubes constructed with a refractory material, with argon or nitrogen used as the bubble gas [Ref. 2 and 3]. Note: It is extremely important that any gas used for a bubbler in molten fluids be dry.

The characteristics of the material surface can affect most sensors.

- Must reflect the energy used for laser, ultrasonic or noncontact radar, i.e., must reflect light, sound or RF signals.
- Radar surfaces must reflect RF with dielectric constants over about 10.
- Dielectric constants below 10 require more radar energy to measure level; therefore, the effective range can be reduced by as much as 80% because of limited energy produced by the sensor. Radar cannot be used with dielectric constants below about 1.5.
- The condition of the surface can affect the range of level measurements. A calm surface is usually desired, and surfaces generally become more turbulent when the material is stirred, blended or boiled. To ease the burden of providing stable, level measurements under such conditions, filters can be incorporated to reduce the effects of turbulence, but as always with filtering, one must guard against a slow response time that can result in severe degradation of measurement quality. Also, in such applications, more energy is needed and the effective range of the sensor can be reduced. Stilling basins can be used to help in such situations.
- Foam on the surface area of the material generally requires more energy and can also reduce the range. At times, foaming conditions cannot be avoided but precautions such as modifying the process by relocating the lines where material enters or leaves the vessel, or the use of agents to reduce the foam, can help. Ultrasonic systems usually measure the top of the foam and not the actual level. Measurements by radar can detect the surface of the foam or some intermediate position within the depth of the foam depending on the dielectric of the foam. Dense, translucent foam will generally not reliably reflect a laser source, which will be reflected by the top of opaque foams and the bottom of transparent foams and at some intermediate position with translucent foams.
- Conditions in the environment near the sensor can affect the quality of level measurement. Dust and dirt coating the sensors can decrease the signal strength and render the measurement system inoperable. In harsh environments, etching of the

sensor surface can occur. Special chambers to enable purging or routine removal for cleaning may be required.

The characteristics of the material in the vapor space of the vessel can have many undesirable affects on the operation of level measurement systems. Since non-contact level systems including ultrasonic, radar and laser technologies must transmit energy through the vapor space and measure the reflected energy, attenuation, absorption or any other effect on the waves can have detrimental effects on measurement quality.

The range and operation of the system at reference conditions assume that the energy is not affected or altered in the vapor space. The energy produced by the sensors can be weakened when the vapor space is not transparent to the energy in the wave, thereby reducing the range of the sensor. Vapors that are opaque to the energy can even render the system inoperable, which is the case in laser systems with dust and smoke in the vapors. A higher powered sensor may be needed.

As the operation of ultrasonic systems depends on the speed of the wave, any characteristic or property of the vapor that changes the velocity will cause error in the measurement. Temperature, pressure, composition and density can affect the velocity. Compensation from temperature measurement at the sensor can help reduce error but may not reflect the true temperature of the vapor.

Consideration should be given to certain characteristics relating to the vessel. Internal structures near the sensor can reflect energy, apart from the reflection at the material surface. Many transmitters can discern and ignore spurious or parasitic reflections. Moving parts and agitators can cause turbulence, interference with detector mounting and false reflections. Again, special precautions must be taken, especially when transmitters cannot discern between actual and false reflections.

Sensor mounting and considerations include the following points:

- Sensors should be mounted perpendicular to the level surface. (One exception is point level detection using a capacitance probe. The installation is typically from the side of the vessel, with the probe in the horizontal position to ensure a repeatable activation point).
- Vapor condensation on the sensor should be avoided.

- Care should be exercised so as not to exceed pressure and temperature limits on the sensor.
- Sensors should be mounted to avoid sensor coating or damage by the process. Laser and radar sensors can be equipped with windows or a Teflon® shield to avoid contact with the process material and vapors.
- Radar and ultrasonic energy coupling from vessel linings should be considered because linings can act as waveguides and result in spurious reflections with more energy than those from the surface of the level.
- Some level sensors have a minimum distance that can be measured; therefore, the sensors can be mounted too close to the level to permit reliable measurement.
- Radar and ultrasonic sensors that are mounted on a nozzle or case in a vessel can cause multiple reflections. Also, to prevent multiple reflections these sensors should not be installed in the symmetrical center of the vessel.
    - Radar and ultrasonic can have beam angles wide enough so that the beam size can increase significantly as the distance from the sensor increases. When the sensor is mounted close to the vessel wall the beam can be reflected from the wall, causing multiple reflections. This is a concern for laser applications but they have small beam angles, reducing the likelihood of multiple reflections.

Accuracy Statements and Reference Performance

A great deal of consideration should be given to the real meaning of information given to specify the performance of a level measurement instrument or system. Major among these specifications are reference performance and installed or actual results. Quite often, the data listed on an instrument's specification sheet is for reference performance. This describes an instrument's accuracy under ideal conditions that differ from those of actual operation. Such data may not reveal actual performance, but is useful in comparing performance because reference operating conditions for most manufactures are generally similar. Actual test conditions, however, may vary because of the manner in which the test data was gathered. The operating effects previously listed can result in the degradation of an instrument's performance.

Accuracy Statements

There are several means to express the accuracy of a level measurement system. The user must understand the parameters described and the terms expressed by the statement. Following is a listing of several such terms:

- The percent of span describes the error in terms of full scale range, and span is the maximum measurement minus the minimum measurement. A level instrument that measures 20 to 120 in. ±1% of span has an accuracy of 1 in. As previously mentioned, when the level is at 120 in., the accuracy is expressed as 1% but at 20 in. the accuracy drops to 20 in. ±1 in. = 5%.

Note that like the accuracy statement expressed as a percent of full scale, the error is directly proportional to the measurement. As the measurement increases, the accuracy increases.

- The percent of rate specifies a constant error and is a better quality statement. As compared to the previous example, the instrument will measure level from 20 to 120 in. ±1% at every point on the scale. At 20 in., the measurement will be ±0.2 in. and at 100% the measurement will be ±1 in.
- The percent of absolute or fixed distance error describes an error that is also constant and independent of the calibration range and actual level in the vessel. For example, for a range of 20 to 120 in. ±10 in., the error is 50% on the low end of the range and 8.33% on the high end of the range.

From the foregoing, it can be seen that consideration should be given to ascertaining exactly what is meant by a given quality statement. One should assume that an accuracy statement expressed simply as a percent is probably best taken to mean percent of full scale or maximum measured value.

Pursuant to the above features and considerations, an additional summary of specifications for the application of level measuring technologies is given below.

- "Visual" includes gages and floats that are used for local indication, often to provide a back-up for transmission systems. Floats can also be used as primary devices for transmission.

Gage glasses can be equipped with magnetic gages and floats for magnetostrictive transmitters, guided wave radar or other technologies. Visual applications can be used for both open- and closed-tank measurement using tubular glass gages and armored designs. The following specifications are common:

- Temperature up to 200°F for tubular, 700°F for armored and 750°F for magnetic.
- Pressure—up to 15 psi for tubular, 4000 psi for armored reflex and 3000 psi for transparent armored. Magnetic gages are available for 3500 psi. Bull's-eye gages, common in boiler drum level measurement, can go up to 10,000 psi.
- Range—20 to 24 in. for armored reflex gages that can be ganged in four sections.
- Accuracy—scale markings are in 1/4-in. graduations. Foaming, boiling and temperature variations between the gages and process can cause errors.

- Pressure and differential pressure level measurement (head type)
  - Temperature—140°F for electronics in transmitters, up to 350°F for wetted parts of transmitters and up to 1200°F for filled systems.
  - Pressure—up to 10,000 psi.
  - Accuracy—from 0.1% to 0.2% of span for most transmitters. Accuracy can be compromised with fill fluids, wet legs, suppression and elevation. Repeaters that are often used to eliminate wet legs will increase the accuracy nearly to that provided by the transmitter, especially with temperature compensation.
  - Range—the rangeability of most digital transmitters is from 0 to 5 in. H_2O to 0 to about 430 ft H_2O.
- Head type with bubblers
  - These can be used for local indication and transmitters in corrosive fluids, slurries, viscous fluids, and when it is not convenient to mount the instrument at the zero reference point. Bubblers can be used in both open- and closed-tank applications but care must be exercised to maintain a con-

stant air flow because of variations in process pressure in closed-tank systems.

- Pressure—atmospheric and up to the supply pressure for the air or other supply for the bubblers.
- Accuracy—limited by the read-out device of the transmitter.
- Range—nearly unlimited.

- Capacitance and RF (radio frequency) admittance technology
  - Applications—point and continuous for conductive and non-conductive fluids.
  - Pressure—up to 4000 psi.
  - Temperature—300°F to 500°F for Teflon-insulated probes and 1800°F for bare probes.
  - Span of probe capacitance—0.25 to 4000 Picofarads, but preferred span is 10 Picofarads.
  - Accuracy—point measurement within 1/4-in. of set point and from 1 to 2% of full scale for continuous applications.
  - Sensitivity and drift—usually caused by temperature variations; are from 0.1 to 5 Picofarads.
  - Measurement range—about 0.1 in. to a few feet for proximity applications and up to about 20 ft for probes.
  - Deadband and time delay for capacitance switches is about 0.5% over the measurement range and from 0 to 25 s.
- Conductivity and field-effect switches
  - Applications—point and differential measurement of conductive fluids and slurries with dielectric constants of 20 and above and fluid resistivities up to 20,000 ohms/cm.
  - Pressure—up to 3000 psi for conductivity probes and 100 psi for field effect conductivity probes.
  - Temperature— –150°F to +140°F for integrally mounted transmitters and –15°F to +1800°F for units with remote electronics for conductivity. Field effect probes can operate at up to 212°F.

    - Range—0–6 ft up to 0–100 ft for conductivity and 0–8 in. for field effect.
    - Accuracy—±1/8 in.
- Displacers
    - Applications primarily for liquid/liquid interface control in bootlegs in the oil-related industries.
    - Pressure is usually set by the flange ratings and is usually up to 100 psi for flexible discs and up to 600 psi for diaphragm-sealed designs. Torque tube design up to 2500 psi and magnetically coupled up to 6000 psi.
    - Temperature—usually –50 to +450°F for standard, and from –350°F to +850°F for Inconel tubes. With electronic transmitters the upper limit is 180°F. Torque tubes should be insulated for a process temperature range of 0°F to 500°F.
    - Accuracy—±0.5% full scale (stable fluid densities are required).
    - Range—0–14 in. to 0–60 in. and ganged units up to 60 ft.
- Radar
    - Applications—used in high accuracy non-contact and guided wave radar service for liquids in larger tanks.
    - Pressure—up to 600 psi for suspended antennas and 300 psi for isolated antennas.
    - Temperature—up to 400°F for standard designs.
    - Range—up to 650 ft for large antenna sizes of 8-, 12- or 18-in. diameter and to 50 ft for small antenna sizes of 3-, 4- or 6-in. diameter used in free space, and up to 120 ft when used in a stilling basin.
    - Accuracy—±0.04 to ±0.125 in. for tank farm gages and ±1 in. to ±0.5% of full scale for process applications.
    - Tolerated coating—up to 1 in. of buildup in nonconductive oil, tar or wax and up to 1/8 in. if the buildup is conductive or water-based.
    - Zero shift up to 1/3 in. can be corrected with a wet calibration.

- Tolerated turbulence—up to 3-ft waves. For 1 GHz sensor bandwidth, large diameter sensors (12 in.) have no error, and small diameter sensors have about 3 in. Doppler effect can cause additional error for sensors with less bandwidth.
- Tolerated foam—up to 6 ft if conductive, up to 6 to 12 in. depending on density and conductivity for other applications.
- Mist and spray effects—no effect for mist; full falling curtain or spray will block the beam. Full spray that is horizontal will be measured as a level.
- Tank ceiling angular deflection limit—about 5° for large diameter (12 in.) sensors and 12° for small diameter (4 in.) sensors. This will result in less than 1/4 in. of error.

- Radiation
  - Temperature—for external detectors, –40°F to +460°F.
  - Radiation source—Cobalt 60, Cesium 137, Americium 241 and Radium. Cesium is most commonly used.
  - Accuracy—1/4 in. for point applications and 1/4 in. to 1% of span for continuous service.
  - Range—for external units, 1 in. to 50 ft. Traversing backscatter units can measure up to 150 ft when mounted in dry wells.
- Resistance Tapes
  - Application—liquids and slurries.
  - Resolution—1/8 in. or the distance between the turns of the helix.
  - Actuation Depth (AD)—the material depth required to short the tape varies with the specific gravity (SG) and is AD in inches = 4/SG. For example, when the SG = 0.5 AD = 4/.5 = 8 inches
  - Temperature effects—a 100°F change in temperature will change the unshorted resistance of the tape by 0.1%. Temperature compensation is normally provided.

- Accuracy—±0.5 in. when AD is corrected and AD temperature is constant. When SG changes, a zero shift based on 4/SG will result. Cold temperatures will also raise AD.
- Operating pressure—from 10 to 30 psia.
- Operating temperature— –20°F to +225°F.

- Rotating Paddle
  - Application—solids level switches.
  - Design Pressure— –14.7 psi to +100 psi.
  - Design temperature—up to 1600°F and air cooled extensions may be required.
  - Accuracy—±1 in. and repeatable using the same material.
- Tapes for ATG applications
  - Application—solids and level in tanks and tank farms.
  - Types—wire-guided float/displacers, inductively coupled float (magnetostrictive), wire-guided sensors, and surface detectors (plumb bob).
  - Design pressure—5 psi for solids, standard wire-guided 30 psi, special designs up to 300 psi.
  - Range—100 ft for standard wire-guided and up to 200 feet for plumb bob applications.
  - Accuracy—±0.2 ft for solids and ±1/16 in. for liquid service.
- Thermal sensors
  - Applications—liquids, interface and foam.
  - Design pressure—up to 3000 psi.
  - Design temperature— –100°F to +350°F and –325°F to +850°F by special design.
  - Response time—10 to 300 s for standard design and 1 to 150 s for fast response systems. Can be less than 1 s for very high temperature service.
  - Accuracy and repeatability—±0.25 in.

- Ultrasonic
  - Applications—wetted and non-contact switch and transmitter applications for liquid level, interface and solids.
  - Design pressure—for switches, up to 3000 psil; for transmitters, 0 to 150 psi.
  - Design temperatures—for switches, –100°F to +300°F; for transmitters, –30°F to +150°F.
  - Range—for tanks and silos, up to 200 ft for special design and up to 25 ft for standard design.
  - Accuracy—±1/8 in. for switches and 0.25 to 2% full scale depending on vapor space conditions.
- Vibrating Switch
  - Types—tuning fork, vibrating probe, vibrating reed.
  - Applications—liquids, solids, slurries.
  - Design pressure—liquids and solids up to 150 psi, slurries up to 3000 psi.
  - Design temperature—liquids from –45°F to +200°F, solids from 10°F to 175°F, slurries from –150°F to 300°F.
  - Minimum material density—liquids and slurries down to 1 lb/ft^3 (0.016g/cm^3) and solids about 0.2 SG.

Case Histories for Various Level Applications

In the following case studies, difficult measurement applications are considered. In most cases, more conventional measurement methods are used. For example, when the process characteristics and conditions, vessel configuration and other conditions that affect a measurement technology's applicability are known to be constant and stable, the choice of a measurement method may be well understood. As is true in any process measurement application, one should first consider the standard, proven and simplest approach. For clean processes, constant specific gravity, easy tank access, and dry calibration applicability, head measurement may be desired. For the same given criteria, displacers can be considered, especially for interface measurement, which is very prominent in the oil and gas industries for interface control in oil and water separators. Conventional displacer signal conditioning circuitry is being replaced by magnetostrictive and guided wave radar but it is still a displacer application. When given conditions are such that head type, displacers, and other, more conventional methods are not appropriate, more sophisticated approaches will need to be considered. The following examples will aid in the selection of such measurement methods.

> *Application 1*—To measure an acid slurry at 300 psi and 450°F with an agitator causing some surface turbulence with occasional foam. The product tends to solidify at reduced temperature and to form solids over a period of time. High temperature and pressure as well as the possibility of sensor coating eliminate the use of most measurement technologies.
>
> *Solution*—A non-contact radar transmitter was used, with a special process seal and the ability to filter noise caused by agitators and other internal objects. A 6 in. nozzle with an extended cone antenna was used to eliminate disturbances by the antenna walls. Insulation was added to prevent condensation on cooler areas along the nozzle.
>
> Another option for the above application could be radiation level technology. A detector inside the vessel could withstand the temperature and pressure, but the nature of the process could result in eventual coating over a period of time, which could cause erroneous level indications.
>
> *Application 2*—To measure the level of a diesel oil and water interface in an absorber-stripper. The specific gravity of the oil varies from 0.7 to 0.8, and the dielectric is about 2; the measurement range is 4 ft. Pressure is 250 psig and temperature is from 80° to 120°F.
>
> *Solution*—The change in specific gravity of the oil precluded the use of either a head type or displacer sensor. A GWR (guided wave radar) with an interface level transmitter and a coaxial probe was used.
>
> *Application 3*—To measure the level of ammonia in a storage tank. Ammonia level measurement can be difficult because of its tendency to change between the liquid and vapor phase. The fluid density is also subject to change depending on the phase, overall working pressure and density. The surface of the

liquid is unstable because of the boiling conditions, and the heavy ammonia vapors in the vapor space of the vessel can attenuate the signal from a non-contact radar transmitter.

Solution—The changing density will render head type and displacer technology inoperable, and the attenuation of the radar transmission in ammonia could preclude the use of non-contact radar, although for consideration of the technology, the manufacturer should be consulted for possible use. A small change in velocity could be compensated for by software in the transmitter by noting the dielectric of the ammonia vapor. A low frequency GWR has been used successfully in this application, with immunity to the vapors and turbulent surface. Since the distance from the transmitter to the liquid surface is measured, the variation fluid density does not degrade the measurement.

Application 4—To measure the level of hexane in a nonstandard-shape evaporator without access to top or bottom to mount transmitters. The material dielectric is less than two and it has changing densities. The level is always maintained in the bottom portion of the vessel.

Solution—A rigid probe, usually used for top down measurement, was mounted on the side of the vessel in the lower portion at an inclination angle of 40° which was entered into the transmitter configuration data. Compensation was made for accurate vertical level values. Immunity to density variations by radar technology did not require additional compensation.

Application 5—To measure the level of an adhesive that is viscous and sticky, and undergoes changes in temperature and density. It also tends to coat and solidify on surfaces in the vessel. The vessel is filled from the top, which causes a turbulent process surface and vapor condensation.

Solution—The nature of the substance precluded the use of many more conventional measurement technologies. This included head type because of density variations, and ultrasonic because of vapor characteristics. A non-contact radar transmitter with a 4 in. cone antenna and immunity to turbulence and condensation was used.

Application 6—To measure soda powder with a dielectric of 6 to 9 and changing surface angle in a 2-ft diameter silo. Large amounts of dust are produced during filling. High accuracy and reliability are needed for inventory and supply. Close control of level is desired.

Solution—Load cells have been used in many such situations when other technologies proved inadequate. They can require more maintenance and present greater challenges in installation. Ultrasonic technology can be considered but the heavy concentration of dust and the possible fluffy surface of the product can result in inaccurate readings. Rotating paddle and point measurement technology can be considered but generally do not provide the measurement required for continuous control; however, they can be used for point control and alarming purposes. Unless equipped with air jets for cleaning, the heavy dust can result in fouling. A sensor with no moving parts and one not affected by the heavy dust is needed for this situation. A GWR is a top mounted sensor that can be used in retrofit applications when needed and being very reliable in such applications make it suitable for small diameter silos where tighter control is required.

Application 7—To measure a high and variable temperature edible oil in a 50-ft tall tank with agitation and a low dielectric material with vapors. Since it is an edible process, any measuring device must be easy to clean, with no dead-end cavities or moving parts. Highly accurate measurement is required for material and balance inventory control and product quality. The material is viscous and can cause coating.

Solution—Generally, the hygienic requirements will preclude the use of head type and displacers, and the low dielectric conditions of the process could make radar measurement difficult. Depending on the vapor conditions, ultrasonic could be considered when sufficient accuracy is provided. The agitated surface condition may require a considerable amount of damping, which can limit the measurement sensitivity by an unacceptable amount. However, a 4-wire radar transmitter with a seal antenna that is easy to clean is accurate and has proven reliable for such applications.

Application 8—To measure the interface of diesel oil and water in an accumulator of a distillation column. The material is a clean fluid; the oil's specific gravity changes between 0.7 and 0.8, and the dielectric of the water is relatively high.

Solution—Displacers represent an ideal consideration for this application and many are used for such operations. The changes in specific gravity of the oil, however, will result in fluctuations of the interface, which can cause poor operation of the process. A GWR with a coaxial probe was used for this situation; the probe was not affected by specific gravity changes.

Application 9—To measure the level in a pressurized propane tank with a 3-in. stilling basin. The material is relatively clean, with constant process conditions.

Solution—For this application, the most probable choice is a displacer. Should the possibility of coating or changes in process density occur, a GWR probe or magnetoresistive transmitter could be used.

Application 10—To measure the level of a dusty powder with a dielectric of 2.9 to 3.5 in a 6-ft silo. Non-contact radar proved unsuccessful because of antenna coating, resulting in false echoes and a weak signal.

Solution—A low-frequency GWR with a flexible probe was used to measure the solid level through the dust in the vapor space. A very small diameter probe proved to be resistant to coating.

Application 11—To measure a mixture of methanol, water, ammonia and minerals in a distillation column with the possibility of foam and bubbles present. The mixture is not constant so the change in densities makes head type and displacer measurement inappropriate; the presence of bubbles and foam renders other non-contact methods unreliable. Turbulent surface conditions also make the situation more difficult.

Solution—A guided wave radar was used, with a coaxial probe to lessen the effects of disturbances and foam by providing a stronger signal.

Application 12—To measure a solution of 35% caustic soda in water in an isolation tank, with corrosive vapors and changes in concentration and specific gravity. Condensation of water vapor is also prevalent.

Solution—As is always the case, changing conditions in the vapor space and product in the vessel will make ultrasonic, head type and displacer technologies difficult to apply, resulting in erratic operation and inaccuracies. Liquid from condensation on objects in the vapor section can also accumulate on a capacitance probe, resulting in erratic and noisy readings. For this application, a non-contact radar transmitter was used, with a 3-in. cone antenna which was selected to match the small connection availability in the tank. The cone was not affected by condensation and had high echo tracking capabilities which resulted in strong signals.

Application 13—To measure chemicals in a heating process, with a low to moderate dielectric liquid with heavy vapors that is subject to changes in concentration, temperature and pressure. The vapors are also subject to coating objects in the vapor space.

Solution—A non-contact radar transmitter was used with a process seal to prevent coating of the antenna and contact with the process liquid. Coating and changes in process characteristics made other technologies difficult for this situation.

Application 14—To measure the level in a reactor vessel containing a mixture of solvents of various and changing densities, with agitation by mixer blades creating turbulence. The heated product produces vapors and condensation. The product is also viscous and sticky.

Solution—Level measurement in a reactor is a very difficult application because of the obstructions in the vessel and the nature of the process. Non-contact measurement is usually preferred. A radar level transmitter with a 6-in. cone proved successful in this situation. Head measurement is not possible because of density variations. The changing temperature and pressure along with the characteristics and physical arrangements of the process render other non-contact technologies unsuitable.

Application 15—To measure the liquid level in a digester with various levels of foam due to upset conditions of the process.

Varying densities of the product preclude the use of head measurement techniques, and ultrasonic measurement is affected by the presence of foam.

Solution—A radar level instrument capable of the required 30 ft measurement range with immunity to the heavy wastewater with solids proved successful. The digital format makes the echo-curve visible, and process changes can be made in response to the foam level data to correct upset conditions.

Application 16—To measure corrosive chemicals (acid and caustic) in bulk storage and day tanks. The issues of chemical compatibility, product density variations caused by temperature changes and the desire to use non-intrusive and non-contact measurement technology narrowed the selection process.

Solution—Ultrasonic technology is common in some such applications but maintenance and associated safety considerations were limiting factors. Head measurement could be considered but the high cost of exotic material for wetted parts discounted this technique. A radar instrument proved successful to measure acid, caustic, and polymer products.

Application 17—To measure the level in a lime silo that experiences build-up and coating on equipment, with the presence of dust and powder during filling.

Solution—Ultrasonic measurement has been successful in these applications but the dust during filling causes problems. Load cells can be considered but are expensive and difficult to retrofit. The use of standard non-contact radar can be considered but the heavy dust can cause problems in reflection. A guided wave radar with designed cable length was used [Ref. 4].

Applications in Solids Level Measurement

Determining the level of solids in vessels is important and often overlooked when considering the overall subject of measurement in the processing industries. Most of the level measurement technologies previously discussed can be used in solids applications; the selection process of finding the best method for a given application can be compounded when one considers the type and shape of the vessel as well as the contents. Solids containers can be silos, hoppers, bins or other types that often differ in physical configurations size and shape. Some are used for storage while others are used for intermediate buffers between inventory and production or between production and use. The materials to be measured include cement, sugar, grain, aggregate and flour, with properties that can affect the reliability and accuracy of level measurement.

These methods of measurement can range from very basic to high tech. Earlier methods that are still used are based on vessel observations or on striking and pounding on the side of a vessel to determine the material content. These techniques may still be used, perhaps as a back-up, but most applications require better and more reliable methods because of safety considerations, time constraints, and the need for increased reliability and accuracy.

Point Measurement of Solids

Most of the point and continuous level measurement principles discussed in earlier chapters can be used for solid as well as liquid level determination. Point level switches require little engineering and setup, are low in cost, and are available in many technologies that include radio frequency (RF), capacitance, rotating and vibrating paddles, and tuning forks. When moving devices contact the solids in the container, they can often be damaged by abrasive material and may fail to operate in sticky substances and heavy materials. An obvious limitation, also, is that they provide indication at one point only. They are generally used to determine level at a target point to detect a low point of material, or alarming conditions to prevent overfilling. Midpoint determination can also be made but adding more switches adds to capital and maintenance costs. Requirements for greater accuracy have created a demand for reliable continuous measurement technology that provides measurement information over the full range of the vessel.

Continuous Solids Measurement

The various technologies discussed earlier that can be applied to solids measurement will be given with applications listed for solid measurement. A review of the specific measurement technologies detailed in previous chapters may be needed as only the basic principle of operation is given here.

Lasers

Laser technology uses infrared light. This light can be formed as electromagnetic radiation or corpuscular current, with a wavelength of around one micrometer, to measure the distance in the vapor space from the instrument to the material surface by the time-of-flight principle. This technology has been used to effectively scan the surface of the piles of solids and to accurately determine the actual volume of material. Two distinct advantages of this technology are the long measurement range and very fast updates. Also, the laser beam is very narrow, which enhances its operation. A main limitation of laser solids measurement is degraded performance in extremely dusty service where the laser fails to penetrate to the surface. Also, the lens must be kept clean because of fouling by dust particles.

Nuclear

Nuclear radiation uses a radioactive source material like cesium or cobalt, located on one side of the vessel and projecting gamma rays through the vessel to a detector located on the opposite side. The radiation is attenuated by passing through the material, which indicates its presence between the source and detector. Either a long receiver section or a long source is used, and these are aligned to make certain the vessel's contents along the designed range are measured. An advantage to nuclear measurement is that it is non-invasive by being mounted outside the vessel. Higher cost is usually involved, and some radiation sources require licensing and knowledge of nuclear safety onsite. The source will deteriorate after a time of use, requiring proper disposal and replacement of the source.

Capacitance

This is a mature technology that has been used for about fifty years. The principle of operation is based on a capacitor with two parallel plates, with the material as the dielectric between the plates. As the position of the level of the material rises and falls between the plates there will be a resulting change in capacitance, which is measured and represents level. This technology can be used in a wide range of applications that can involve adverse conditions like high temperature, high pressure and corrosive chemicals. In most cases, mounting arrangements are flexible for either top or bottom mount. This is a contact technology, which can result in damage to equipment by abrasive material. Measurements may be affected by plate coating or by changes in material dielectric caused by moisture or changes in composition.

Ultrasonic

Also a mature technology, ultrasonic uses high frequency sound waves transmitted toward the material surface and reflected. The time for the sound waves to make the round trip is measured and from the known velocity of the sound, the distance can be calculated. Frequencies ranging from 5 to 30 kHz are used for measurement ranges to as high as 60 m. The range and frequency are inversely related and are selected for the particular application. Ranges are limited by vapor space atmosphere. Dusty conditions will limit the range, which also is inversely related to the dust content present.

Sound wave propagation requires a carrier medium, which is usually air, and the velocity of the sound wave is a function of temperature. The vapor space material and temperature will result in measurement error if not compensated for. The material surface, such as piles when filling and cone shapes when emptying, can result in a weak echo because of the reflection angle. Objects in the vessel can produce parasitic echoes, which can be discounted by profiling the vessel with instruments capable of intelligent echo processing. Advantages are that the technology is well understood, is reliable, non-contact and relatively inexpensive. This technology can be used in open or closed tanks.

Radar

Guided wave or time domain reflectometry (TDR)

The electromagnetic pulse follows a wire anchored to the vessel below the material surface. The electromagnetic pulse travels along the wire and is reflected by the material surface and then returns. The time-of-flight principle is used to locate the surface of the material. For an acceptable application, the range is usually limited to about 30 m and the material should not be extremely abrasive. Care should be taken during installation so the cable does not touch the vessel wall, and material shifting must not cause damage to the wire. The actual level is represented at the point where the wire intersects the surface. Piles and slopes along the surface can result in level/volume relationship errors.

Non-contact radar

Once thought to be too expensive and experimental, radar technology is expanding. First used in the processing industries for liquid level measurement in large storage vessels in the early 1970s at a relatively high cost, radar technology is used in a wide range of applications that include small bulk tanks for storage and agitated process vessels. Not until recently has radar been used for solids measurement, which requires higher frequencies. The frequency spectrum for radar measurement is from 1 to 300 GHz but is generally between 6 and 26 GHz for liquid applications. The first radar devices were 10 GHz and that is still used for FMCW technology in liquid applications. For a given output signal amplitude, lower frequency radar requires a larger antenna. A 6 GHz signal needs a 16-in. diameter horn for the same signal as a 24 GHz device using a 4-in. diameter horn antenna. Low frequency radar is well suited for liquid measurement, but for solids applications a large horn up to 10-in. diameter or even a parabolic dish antenna to capture the weaker signal from a solid surface is required. To avoid the use of larger antennas which may be difficult to install in some vessels, a higher frequency design with an antenna of 3-in. or 4-in. diameter is used for most solid measurement.

The prominent advantage of radar technology for solids measurement is the same as for liquids. Radar is non-contact and is virtually immune to vapor, pressure, temperature and dust in the vapor space.

High Range Frequency Applications—24 to 26 GHz

The frequency range of 24 to 26 GHz is well suited for solid measurement by radar for long range and/or cluttered solid applications. Silos may be tall, assembled in sections, and of small diameter or have internal seams, ladders, and other internal objects and obstructions which produce false echo signals. In such cases, narrow conical beam angles are preferred to reduce side-wall path interference and to reduce false echoes. The higher frequency radar will produce a lower beam angle. For example, 24 GHz radar instrument will produce a 9° conical beam angle compared to 36° for a 6 GHz signal with the same size antenna; therefore, the smaller antenna, easier installation and narrow beam angle are combinations that are important characteristics for solids measurement because of vessel configuration. Also, the reflection properties of shaped surfaces relating to wavelength yield better echo signals. A wave striking a sloped surface may reflect directly back to the receiver or it may bounce from the sloped surface, causing the pulse to split and resulting in the detection of two signals. High frequency signals will result in the greatest amount of signals reflecting directly from a sloped surface and being captured by the receiver.

Radar Signal Processing for Solids Application

The processing of echo signals for solids measurement is much more challenging than for liquid measurement. The flat, smooth surface of liquids produces higher amplitude echo signals with relatively simple echo characteristics and profile. For example, it is easier to see the reflection in a calm surface of a liquid than the irregular surface of a solid. Reflected signals from the two surfaces vary significantly. A radar system designed for liquid will likely not function for solids. Also, as a silo is filled or emptied, the characteristic profile changes and the signal processor must be able to respond to these differences. The reflection from a conical surface in the vessel that is there to facilitate contents discharge will cause problems because of the highly reflective sloped surface. Multiple reflections, indirect signal paths, weak signals, and false signals from internal obstructions will also contribute to problems in measurement.

Other challenges to solids measurement are tensile forces caused by changing levels in silos that can reach several tons. This can cause severe damage to guide-wires and moving parts in the vessel. The surface slope of solids may be flat and smooth for materials like cement and sharply sloped for aggregate materials. Particle size can range from fine powders like flour to very coarse, like coal. All of these characteristics can change the surface dynamics and must be considered in equipment design.

Some manufacturers of non-contact radar devices offer protective covers for antennas to protect them from moisture, dust and other material that can cause excessive build-up and erratic operation. The cover is usually constructed of PTFE (Teflon®) with a low dielectric that enables the microwave to propagate without interference by the cover. The cover must be machined to close tolerances so the thickness optimally matches the frequency. Antenna lens covers can also be used for materials such as flour, soda and others that can compact over time inside the antenna.

10.4 SUMMARY OF SOLIDS LEVEL MEASUREMENT

As is the case for every situation, a system must be designed that is compatible with the service and application. Many variables must be considered when making a choice for solids level measurement. A manufacturer can be very helpful in the selection process but first must have pertinent information relative to the operating conditions. The user should have the following information:

- Vessel Characteristics
  - Height
  - Diameter
  - Shape
- Material Properties
  - Bulk density
  - Dielectric properties (capacitance, radar or guided wave)
  - Color (laser)
  - Consistency
    - Build-up
    - Sticky
  - Temperature
  - Abrasiveness
- Filling method, location
- Obstructions (internal)
- Required accuracy
- Mounting details
- Cost
  - Purchase
  - Maintenance/installation
  - Operation

Table 10–5 will help in the selection of instruments for solids level measurement.

Demands on industry to make a better product with reduced resources have resulted in improvements in measurement technology. This is mostly a result of the need to reduce material and labor costs and conserve energy—to make a better product with less.

As is true with other measurement techniques, it is no longer acceptable to guess or estimate the quantity of solid material in a vessel nor to rely on manual measurement methods. In the last several decades, solid measurement has evolved into a high tech arena.

Point level technology may still be a favorite because of its simplicity but it is no longer adequate. The development of continuous solid level technology has resulted in this application being on even par

Table 10–5 Measurement Technology Characteristics for Solids Applications

	Ultrasonic	**Load Cell**	**Radar (non-contact)**	**Guided Wave Radar**	**Weight & Cable**	**Capacitance**	**Nuclear**	**Laser**
Range	**200 ft**		**328 ft**	**80 ft**	**60 ft**	**50 ft**	**90 ft**	**1000 ft**
Heavy dust	Caution	OK	OK	Caution buildup	Caution	NO	OK	NO
Vacuum	NO	OK	OK	OK	OK	OK	OK	OK
Tensile forces	NO	OK	OK	Caution	Caution	Caution	OK	OK
Effects from buildup	Caution	OK	Caution	Caution	NO	NO	OK	NO
Temperature above 300°F	NO	OK	OK	OK	OK	OK	OK	OK
Effects from external elements	NO	YES	NO	NO	NO	NO	YES	NO
Moving parts	NO	OK	NO	NO	YES	NO	NO	NO

with other highly technical measurement means. In the last decade, radar seems to be the method of choice for solid as well as liquid level measurement [Ref. 5].

10.5 RADAR APPLICATIONS IN STILLING WELLS AND BYPASS PIPES

Many process vessels are equipped with stilling wells or basins and bypass pipes. Radar transmitters can be used in such applications for level measurement, and function differently in pipes than in normal vessel applications. Pipes offer a calmer, less turbulent, and generally cleaner surface. This increases the reliability and robustness of level measurement for most types of sensors. Pipes isolate the transmitter from disturbances such as other pipes, agitation, fluid flow, foam and other objects. The pipe can be located anywhere in the vessel that allows access to the instrument. This can be especially important for interface measurement near the bottom of a taller vessel or a distillation column. Valves are usually provided for instrument calibration or removal.

There are limitations in the use of bypass pipes and stilling wells, however. They should be used with clean fluids that do not leave deposits and are not highly viscous or adhesive in nature. Additional cost may also be a consideration. Following are guidelines for use in radar applications.

Non-contact radar works well in pipe applications but may need to meet certain requirements such as antenna selection, design and installation. In many cases GWR may be a better choice in shorter range applications where rigid probes can be used. The probes are available in a variety of materials for corrosive fluids. For taller applications (usually greater than 10 ft), non-contact applications are generally more advantageous. Non-contact is preferred for fluids that are highly viscous, sticky, or that leave heavy deposits. GWRs with rigid probes are preferred for pipe installations in metal pipes smaller than 3 or 4 in. They offer a stronger reflected signal in pipes than in open applications, making them suitable for interface measurement and fluids with low dielectric. Flexible probes can be used in larger pipes, usually greater than 4 inches to ensure that the probe does not touch the pipe.

When using non-contact radar in a pipe application, the microwave signal is guided and contained in the pipe. This results in a stronger signal on the surface, which is advantageous for low dielectric fluids, longer distances and turbulent conditions. Radar signals inside a pipe have more than one generated microwave mode and each mode has a unique speed of propagation. The number of generated modes depends on the frequency of the radar signal in the pipe and the pipe diameter. A 2- or 3-in. pipe diameter will minimize the number of modes, and higher frequency radar should be limited to these small diameter pipes. Low frequency radar performs well in pipes up to about 8 in. in diameter and is better suited for dirty pipes and heavy vapors with condensation.

Cone antennas are suitable for pipe installations and are generally available in SST (stainless steel), Hastelloy and Monel. Antennas are selected to match the pipe size and are designed for pipes up to schedule 80 wall thickness. The instrument accuracy can be degraded as the clearance between the pipe and antenna increases.

The following pipe requirements for non-contact measurement in stilling basins and bypass applications should be considered:

- All metal material—non-metallic pipes do not shield the signal from outside disturbances.
- Uniform diameter.
- Smooth inside pipe surface—pipe joints may reduce accuracy.
- Only one hole above the product surface.
- Holes should be deburred.
- Isolation ball valves should be full port and fully open.
- The inlet pipe in a bypass arrangement should not protrude into the measuring pipe.

10.6 HEAD LEVEL MEASUREMENT WITH DENSITY COMPENSATION

Even with the advancement of various level measurement technologies, head level measurement methods are still used extensively. In fact, the number of differential pressure transmitters used for level measurement is second only to flow applications. The accuracy of head type level measurement is affected by process and wet leg density variations, hence many limitations exist. Recall that the pressure exerted by a column of fluid is a function of height and density, or specific gravity:

$$P = 0.433\text{psi/ft (Hft)(sp.gr.)}$$

where P is the head pressure caused by a column of fluid H feet in height.

In most head level procedures, the specific gravity is considered to be constant. This is not always the case because of variations in process material and temperature or both. While the formula (above equation) and this method of level measurement are fairly simple, most head level applications include one or more of the following issues:

- Location of transmitter other than at the point of zero reference level.
- Transmitter location above or below the pressure tap connection.
- Transmitter has an isolation flange or seal fluid.
- Pressure in the vapor space of a closed tank application not constant.
- Vapor above the liquid may be the same composition as the liquid or it may be another gas as in a blanket pressure application.
- Dry leg on low side.
- Wet leg on low side or wet leg on high side.
- Wet leg on both sides.
- Environmental conditions different on the high and low side wet legs.
- Wet leg environmental conditions different than process conditions.
- Changes in liquid and vapor densities.

Any or all of these situations may exist and all will result in measurement degradation; some will be difficult to compensate for and will be neglected. It will be shown that density variations will result in significant error and can be compensated for.

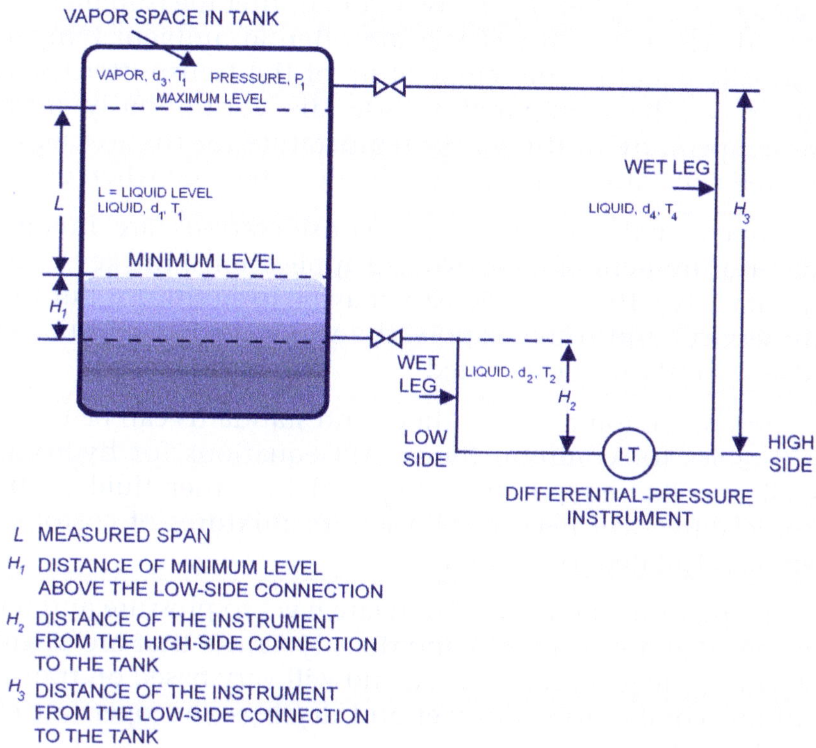

Figure 10–4 Head Level Measurement with Density Compensation

In this example, four density values are required by the transmitter based on measured pressure and temperature. Figure 10–4 shows the application and the following parameters:

- Density of the tank liquid is d_1 at T_1 and pressure P_1 on the liquid.
- Density of the liquid in the high side wet leg is d_2 at T_2 and P_1.
- Density of the vapor above the liquid is d_3 at T_1 and P_1.
- Density of the liquid in the low side wet leg is d_4 at T_4.

The temperatures of the liquid and vapor in the tank are considered to be equal. The pressure, P_1, is calculated from the pressure measured by the multivariable transmitter less the hydrostatic head of liquid in the low side wet leg $(H_3)(d_3)$. The multivariable transmitter has one RTD (resistance temperature device) for temperature measurement, and it can be assigned for either. T_1, T_2, or T_4 or another measurement source can be used. The external wet legs are often at ambient temperature that is usually equal to the temperature of the transmitter itself. The transmitter can be configured to use the continuously measured internal temperature or the sensor temperature for the wet legs. Also, user-entered constant temperature values can be used when desired.

As mentioned, many variables and considerations are involved in accurate measurement of level by head applications; the key values are density data. Also, the vapor density may be insignificant but for high pressure services and heavy vapors, the vapor density will need to be entered as a constant or calculated.

Many different computation methods and standards can be used, such as ASME tables for common fluids, API equations for hydrocarbons and ANSI and ISO equations as required for other fluid properties. Other procedures will need to be used for mixtures of custom fluids that can be calculated for user entry.

Finally, an equation can be used to relate level to pressure and temperature values that are measured and density values that are calculated. The following equation can be used and will vary based on transmitter orientations, combinations of wet and dry legs and open- or closed-tank applications.

$$L = \frac{\Delta P - (d_3)(H_3 - H_1 - H_2) - (d_1 H_1) - (d_2 H_2) + (d_4 H_3)}{d_1 - d_3} \tag{10.1}$$

Density-compensated level measurement eliminates the greatest source of error resulting from a single head measurement with a single variable transmitter. Multivariable transmitters can provide a continuous and dynamic calculation for various densities such as

- Tank liquid
- Tank vapor
- Wet leg on high side
- Wet or dry leg on low side

The error resulting from the other listed considerations is minimal and usually contributed by the inherent error of the transmitter and other instruments. With accuracies of digital multivariable transmitters and other measuring devices as great as ±0.1% of span, it can be seen that head level measurements with density compensation can be very accurate.

10.7 SUMMARY OF DENSITY COMPENSATION FOR HEAD MEASUREMENT

The application of density compensation in head level measurement is but one example of improved accuracy with density compensation. Most level measurement for steam drums in boiler and heat recovery steam generators (HRSG) are with a differential pressure transmitter that has been referred to as head type. Other applications can include various hydrocarbon storage and particular LPG (liquified petroleum gas) storage and transfer units. The principle applied and discussed here can also be used in the various head level measurements mentioned. The accuracy improvement with density compensation will depend on temperature variations, process configuration, pressure, measurement span, product, operating point and other considerations. In worst-case considerations, density compensation can reduce the error from as great as ±35% or more to less than ±0.5% of measured span; therefore, opportunities abound for improved measurement by density compensation. This statement and the material presented here may appear to be in contrast to the solutions given for many of the case history applications in the section "Case Histories of Various Level Applications." One should consider, however, that level measurement by radar is a growing option, especially for unconventional and difficult applications. Head level measurement is still commonly used and can yield improved results with density compensation [Ref. 6].

10.8 CONCLUSION—LEVEL MEASUREMENT

In specifying level measuring instruments, one usually examines the data furnished by the manufacturer or supplier. This data, when used to evaluate equipment from various manufacturers, can be difficult to use for the comparison of actual performance. Such issues as instrument location and mounting, differences in the expression of terms, the amount of detail and completeness given, and the methods used to obtain instrument data should be considered. Another issue of great importance is the fact that some manufacturers are conservative in publishing statements about their product, while others publish data that really has not or cannot be qualified or verified.

Items to be considered in evaluating instrument performance are

1. *Ambient limits*—These parameters are usually temperature, pressure, humidity, environment, vibration, clearance for mounting and mechanical vibration.
2. *Process limits*—Material compatibility, the pressure and temperature of the process material, the relationship between process conditions and those at the measuring point, and process composition changes.
3. *Sensor installation and maintenance*—The physical clearance and location should enable sensor removal, maintenance and calibration. The instrument should be mounted securely in accordance with the manufacturer's recommendations, which should be stated. When the sensor is not integrally mounted with the signal conditioner or transmitter, wiring connections, grounding and wire types should be considered. Devices are two wire, three wire, or four wire types. Two wire systems are generally more common and acceptable.
4. *Transmitter requirements and specifications*—Some are loop powered and others need external power—120, 240 VAC, 50 or 60 Hz. Power wires should run separately from signal cables to reduce interference caused by electromagnetic induction. Digital transmitters can communicate by a bus network and be bus powered. Protocol compatibility—HART, FOUNDATION Fieldbus, Profibus or Ethernet—must be ensured.

The performance criteria provided by the manufacturer usually describe a reference performance based on conditions established in a laboratory and for a specified measurement range. Reference accuracy can provide a means of comparison between measurement systems operated at or near ideal conditions, but does not quantify performance based on operating situations apart from those stated for laboratory conditions. These differences can cause a degradation of performance. It is common, however, for the stated laboratory conditions to be wide enough to cover most normal operating conditions.

Specifications from manufacturers of several different types of non-contact level measuring instruments were gathered and analyzed. Reference accuracy calculations were tabulated by model and significant differences were noted between instruments within a given range.

Also, discrepancies were found between measurement systems from different manufacturers.

Installation considerations and material characteristics are very important in level measurement applications and especially for non-contact devices. The following recommendations are listed to help the overall performance of measuring systems:

- Material characteristics can have adverse affects on the measurement quality of the measurement system.
- Turbulent surface conditions should be avoided and stilling basins should be considered in severe turbulence situations; more energy is needed in turbulence situations and the effective measurement range can be limited, caused by a reduced energy level.
- Foam on the liquid surface should be avoided.
- Dust and dirt build-up at the sensor can reduce the transmitted and/or received signal strength.
- The conditions at the liquid surface that affect reflectivity of the energy beams can result in error. For radar systems, this is dependent on the dielectric constant, which should be above 10. Levels below 1.5 can preclude the use of radar applications.

Vapor characteristics also have adverse affects on non-contact level measurement systems.

- The vapor space is considered to be transparent to the energy of the transmitted and reflected energy beam. Materials in the vapor portion of the vessel that are translucent to the traveling wave can affect the energy level and the speed of the wave.
- The velocity of an ultrasonic wave is dependent on the pressure, temperature and composition of the vapor. Compensation for changes in the vapor is necessary at times. The condition of the vapor should be constant. For example, when the conditions change during start-up, shut-down and normal operation, and appropriate compensation is not made, error can result.

The vessel's geometric design and other conditions can affect the performance of a measuring system. The configuration of the transponders and material entering or leaving the vessel can result in measurement error. The formation of solids at various points along the beam path should be avoided. Parasitic echoes and erratic energy reflection caused by internal tank structures should be discounted by the measuring system. Moving parts like agitators can produce false reflections that can be sensed as level measurement. The wetted portion of the sensor should be mounted perpendicular to the liquid surface. Condensation forming on the sensor should be avoided. Multiple reflections caused by cones, nozzles and other geometric shapes of the vessel should be avoided by proper mounting and compensation of false echoes [Ref. 7].

Manufacturers have developed and tested instruments that can successfully measure level for nearly all applications with the required accuracy. It is incumbent on the engineer and designer to research a given situation, explore the possibilities and find the best and most acceptable approach to a measurement application.

EXERCISES

10.1 List a primary limitation to the following measurement methods:

A. Radar
B. Head type
C. Displacer
D. Ultrasonic

10.2 True or False?

The quality statement for a pressure or level instrument is generally as good for ideal conditions as for installed conditions.

10.3 Which of the following is best for level measurement of a slurry?

A. Displacer
B. Ultrasonic
C. Radiation

10.4 Which measurement technology is most dependent on the reflectivity of the surface of the product?

A. Capacitance
B. Radiation
C. Radar

10.5 Match the following for prominent sources of error and criteria for application:

1. Coating	A. Ultrasonic
2. Changes in specific gravity	B. Radar
3. Material in vapor space	C. Head type
4. Dielectric of process	D. Displacer
5. Foam on liquid surface	E. Capacitance

REFERENCES

1. Lipták, Béla, *Instrument Engineers Handbook: Process Measurement and Analysis*, 4th Edition, Volume 1, CRC Press, 2003.

2. Omega Handbook and Encyclopedia, Volume 29, 1995.

3. Rosemount technical literature "Transactions" Volume 4

4. www.Emersonprocess.com/rosemount/documentation/notes/00830

5. Little, Tim and Boisvert, Jerry. "Level Measurement Trends in the Solids Industry" ISA EXPO 2007.

6. Harrelson, Dudley and Rowe, Jonathan. "Multivariable Transmitters, A New Approach to Liquid Level Measurement" ISA EXPO 2004.

7. Spitzer and Boyes, *The Consumer Guide to Non-Contact Level Gauges*, Copperhill and Pointer, Inc., 2005.

A

Definition of Terms

accuracy—(l) The ratio of the *error* to the *full-scale output,* or the ratio of the *error* to the *output,* expressed in percent. (2) The departure of the measured value of a quantity from the accepted standard for that quantity. Accuracy may be expressed as a fractional part of some quantity and may be divided into parts having different origins.

> Note: It is common practice in the scientific community to add the systematic errors of a measurement process to three standard deviations of the process variability. For a satisfactory number of observations, this policy results in 99.7 successes in 100 trials. Some manufacturers adopt a policy of adding two standard deviations of the variability to the systematic error. The result is a process having probability of 95 successes in 100 trials.

accuracy, measured—The maximum positive and negative *deviation* observed in testing a *device* under specified conditions and by a specified procedure.

> Note 1: Measured accuracy is usually measured as an inaccuracy and expressed as accuracy.
>
> Note 2: Measured accuracy is typically expressed in terms of the measured variable, percent of span, percent of upper range-value, percent of scale length, or percent of actual output reading.

accuracy rating—In process instrumentation, a number or quantity that defines a limit that errors will not exceed when a *device* is used under specified *operating conditions.*

> Note 1: When operating conditions are not specified, *reference operating conditions* shall be assumed.
>
> Note 2: As a performance specification, *accuracy* (or reference accuracy) shall be assumed to mean accuracy rating of a *device* when used at *reference operating conditions.*
>
> Note 3: Accuracy rating includes the combined effects of *conformity, hysteresis, dead band,* and *repeatability* errors. The units being used are to be stated explicitly. It is preferred that a ± sign precede the number or quantity. The absence of a sign indicates a plus and a minus sign.
>
> Note 4: Accuracy rating can be expressed in a number of forms. The following examples are typical:
>
> (a) Expressed in terms of the *measured variable.* Example: The accuracy rating is ±1°C, or ±2°F.
>
> (b) Expressed in percent of *span.* Example: The accuracy rating is ±0.5% of span. (This percentage is calculated using scale units such as °F, psig, etc.)
>
> (c) Expressed in percent of the upper range-value. Example: The accuracy rating is ±0.5% of upper range-value. (This percentage is calculated using scale units such as kPa, °F, etc.)
>
> (d) Expressed in percent of scale length. Example: The accuracy rating is ±0.5% of scale length.
>
> (e) Expressed in percent of actual output reading. Example: The accuracy rating is ±1% of actual output reading.

adjustment, zero—Means provided in an instrument to produce a parallel shift of the input/output curve. See also *zero shift.*

ambient temperature—See *temperature, ambient.*

calibrate—To ascertain outputs of a *device* corresponding to a series of values of the quantity which the device is to measure, receive, or transmit. Data so obtained are used to: (1) determine the locations at which scale graduations are to be placed; (2) adjust the *output,* to bring it to the desired value, within a specified tolerance; (3) ascertain the *error* by comparing the device output reading against a standard.

calibration—A test during which known values of *measurand* are applied to the *transducer* and corresponding *output* readings are recorded under specified conditions.

calibration curve—A graphical representation of the *calibration report.*

calibration cycle—The application of known values of the *measured variable* and the recording of corresponding values of *output* readings over the *range* of the instrument, in ascending and descending directions.

calibration report—A table or graph of the measured relationship of an instrument as compared over its *range* against a standard.

calibration traceability—The relationship of the calibration of an instrument through a step-by-step process to an instrument or group of instruments calibrated and certified by a national standardizing laboratory.

conformity—Of a curve, the closeness to which it approximates a specified curve (e.g., logarithmic, parabolic, cubic, etc.). See also *linearity.*

> Note 1: It is usually measured in terms of nonconformity and expressed as conformity—for example, the maximum deviation between an average curve and a specified curve. The average curve is determined after making two or more full-range traverses in each direction. The value of conformity is referred to the *output* unless otherwise stated.
>
> Note 2: As a performance specification, conformity should be expressed as independent conformity, *terminal-based conformity,* or *zero-based conformity.* When expressed simply as conformity, it is assumed to be independent conformity.

conformity, terminal-based—The maximum deviation of the calibration curve (average of upscale and downscale readings) from a specified characteristic curve so positioned as to coincide with the actual characteristic curve at *upper* and *lower range-values.*

conformity, zero-based—The maximum *deviation* of the *calibration curve* (average of upscale and downscale readings) from a specified *characteristic curve* positioned so as to coincide with the actual *characteristic curve* at the lower range-value.

correction—In process instrumentation, the algebraic difference between the *ideal value* and the indication of the *measured signal.* It is the quantity which when added algebraically to the indication gives the ideal value. See also *error.*

> Note: A positive correction denotes that the indication of the instrument is less than the *ideal value.*

cycling life—The specified minimum number of full-scale excursions or specified partial-range excursions over which a *device* will operate as specified without changing its performance beyond specified tolerances.

dead band—In process instrumentation, the *range* through which an *input signal* may be varied, upon reversal of direction, without initiating an observable change in *output signal.* See also *zone, dead.*

> Note 1: There are separate and distinct input/output relationships for increasing and decreasing signals.
>
> Note 2: Dead band produces phase lag between input and output.
>
> Note 3: Dead band is usually expressed in percent of span.

dead time—See *time, dead.*

dead zone—See *zone, dead.*

delay—The interval of time between a changing *signal* and its repetition for some specified duration at a downstream point of the signal path; the value L in the transform factor exp ($-L_S$). See also *time, dead.*

deviation—Any departure from a desired value or expected value or pattern.

device—An apparatus for performing a prescribed function.

drift—An undesired change in *output* over a period of time, unrelated to the input, environment, or load.

drift, point—The change in *output* over a specified period of time for a constant input under specified *reference operating conditions.*

> Note: Point drift is frequently determined at more than one input, e.g., at 0, 50, and 100% of *range.* Thus, any *drift* of zero or *span* may be calculated. Example: The drift at midscale for *ambient temperature* (70 ±2°F) for a period of 48 h was within 0.1% of output span.

element—A component of a *device* or system.

element, primary—The system element that quantitatively converts the *measured variable* energy into a form suitable for measurement.

> Note: For a *transmitter* without an external primary element, the sensing portion is the primary element.

element, reference-input—The portion of the controlling system that changes the *reference-input signal* in response to the set point.

element, sensing—The element directly responsive to the value of the *measured variable.* It may include the case protecting the sensitive portion.

elevated range—See *range, suppressed-zero.*

elevated span—*See range, suppressed-zero.*

elevated-zero range—See *range, elevated-zero.*

elevation—*See range, suppressed-zero.*

error—See also correction. (1) In process instrumentation, the algebraic difference between the indication and the *ideal value* of the *measured signal.* It is the quantity which algebraically subtracted from the indication gives the ideal value. (2) The algebraic difference between the indicated value and the true value of the *measurand.* (3) The difference between an assumed value and the true value of any quantity. In some instances, the term may be used interchangeably with "uncertainty."

> Note: A positive error denotes that the indication of the instrument is greater than the *ideal value.*

error band—The band of maximum deviation of *output* values from a specified reference line or curve.

error curve—See *calibration curve.*

error, environmental—Error caused by a change in a specified *operating condition* from a *reference operating condition.* See also *operating influence.*

error, frictional—Error of a *device* due to the resistance to motion presented by contacting surfaces.

error, hysteresis—See *hysteresis.*

error, hysteretic—See *hysteresis.*

error, inclination—The change in *output* caused solely by an inclination of a *device* from its normal operating position.

error, mounting strain—Error resulting from mechanical deformation of an instrument caused by mounting the instrument and making all connections. See also *error, inclination.*

error, position—The change in *output* resulting from mounting or setting an instrument in a position different from that at which it was calibrated. See also *error, inclination.*

error signal—See *signal, error.*

error, span—The difference between the actual *span* and the ideal span. Usually expressed as a percent of ideal *span.*

error, systematic—Error that in the course of a number of measurements made under the same conditions of the same value of a given quantity, either remains constant in absolute value and sign or varies according to a definite law when the conditions change.

error, zero—In process instrumentation, error of a *device* operating under specified conditions of use, when the input is at the *lower range-value. Usually expressed as percent of ideal span.*

full-scale output—The algebraic difference between the *output* at the specified upper and lower limits of the *range.*

hysteresis—(1) That property of an *element* evidenced by the dependence of the value of the *output,* for a given excursion of the input, upon the history of prior excursions and the direction of the current traverse. (2) The maximum difference in *output* at any *measurand* value within the specified *range,* when the value is approached first with increasing and then with decreasing *measurand.*

> Note 1: Hysteresis is usually determined by subtracting the value of *dead band* from the maximum measured separation between upscale going and downscale going indications of the *measured variable* (during a full-range traverse, unless otherwise specified) after *transients* have decayed. This measurement is sometimes called hysteresis error or hysteretic error.
>
> Note 2: *Some reversal* of *output* may be *expected* for any small reversal of input; this distinguishes hysteresis from *dead band.*
>
> Note 3: Hysteresis is expressed in percent of full-scale output during any one *calibration cycle.*

ideal value—See *value, ideal.*

indicating instrument—See *instrument, indicating.*

indicator travel—The length of the path described by the indicating means or the tip of the pointer in moving from one end of the scale to the other.

> Note 1: The path may be an arc or a straight line.
>
> Note 2: In the case of knife-edge pointers and others extending beyond the scale division marks, the pointer shall be considered as ending at the outer end of the shortest scale division marks.

instrumentation—A collection of instruments or their application for the purpose of observation, measurement, or control.

instrument, indicating—A *measuring instrument* in which only the present value of the *measured variable* is visually indicated.

instrument, measuring—A *device* for ascertaining the magnitude of a quantity or condition presented to it.

instrument, recording—A *measuring instrument* in which the values of the *measured variable* are recorded.

> Note: The record may be either analog or digital and mayor may not be visually indicated.

linearity—See also conformity. (1) The closeness to which a curve approximates a straight line. (2) The closeness of a *calibration curve* to a specified straight line.

> Note 1: It is usually measured as a nonlinearity and expressed as linearity, e.g., a maximum *deviation* between an average curve and a straight line. The average curve is determined after making two or more full-range traverses in each direction. The value of linearity is referred to the *output* unless otherwise stated.)
>
> Note 2: As a performance specification, linearity should be expressed as independent linearity, terminal-based linearity, or zero-based linearity. When expressed simply as linearity, it is assumed to be independent linearity.
>
> Note 3: Linearity is expressed as the maximum deviation of any *calibration* point from a specified straight line, during anyone *calibration cycle.* Linearity is expressed in percent of *full-scale output.*

linearity, independent—The maximum *deviation* of the *calibration curve* (average of upscale and downscale readings) from a straight line positioned so as to minimize the maximum deviation.

linearity, terminal-based—The maximum *deviation* of the *calibration curve* (average of upscale and downscale readings) from a straight line coinciding with the calibration curve at *upper* and *lower range-values.*

linearity, zero-based—The maximum *deviation* of the *calibration curve* (average of upscale and downscale readings) from a straight line positioned so as to coincide with the calibration curve at the *lower range-value* and to minimize the maximum deviation.

lower range-limit—See *range-limit, lower.*

lower range-value—See *range-value, lower.*

maximum working pressure—See *pressure, maximum working.*

measurand—A physical quantity, property, *or* condition to be measured. See also *variable, measured.*

> Note: This term is preferred to input parameter or stimulus.

measured accuracy—See *accuracy, measured.*

measured signal—*See signal, measured.*

measured value—*See value, measured.*

measured variable—See *variable, measured.*

measurement process—A method of comparing an unknown quantity with a known one. The method involves operator observations of the performance of the apparatus used in the comparison.

measuring instrument—See *instrument, measuring.*

mounting position—The position of a *device* relative to its physical surroundings.

noise—In process instrumentation, an unwanted component of a *signal* or variable. Noise can be expressed in units of the output or in percent of *output span.*

operating conditions—Conditions to which a *device* is subjected, not including the variable measured by the device.

> Note: Examples of operating conditions include ambient pressure, ambient temperature, electromagnetic fields, gravitational force, inclination, power supply variation (voltage, frequency, harmonies), radiation, shock, and vibration. Both static and dynamic variations in these conditions should be considered.

operating conditions, normal—The range of operating conditions within which a *device* is designed to operate and for which operating influences are stated.

operating conditions, reference—The range (usually narrow) of operating conditions of a device within which *operating influences* are negligible.

> Note: These are the conditions under which *reference performance* is stated and the base from which the values of operating influences are determined.

operating influence—The change in a performance characteristic caused by a change in a specified operating condition from a reference operating condition, all other conditions being held within the limits of reference operating conditions.

> Note 1: The specified operating conditions are usually the limits of the normal operating conditions.
>
> Note 2: Operating influence may be stated in either of two ways:
>
> (a) As the total change in performance characteristics from reference operating condition to another specified operating condition. Example: Voltage influence on accuracy can be expressed as: 2% of span based on a change in voltage from a reference value of 120 V to a value of 130 V.
>
> (b) As a coefficient expressing the change in a performance characteristic corresponding to a unit change of the operating condition from reference operating condition to another specified operating condition. Example: Voltage influence on accuracy can be expressed as:
>
> $$2\% \text{ of span}/(130\text{ V} - 120\text{ V}) = 0.2\% \text{ of span per volt}$$
>
> If the relation between operating influence and change in operating condition is linear, one coefficient will suffice. If it is nonlinear, it may be desirable to state more than one coefficient, such as 0.05% per volt from 120 to 125 V, and 0.15% from 125 to 130 V

operating pressure—See *pressure, operating*.

operative limits—The range of *operating conditions* to which a *device* may be subjected without permanent impairment of operating characteristics.

> Note 1: In general, performance characteristics are not stated for the region between the limits of normal operating conditions and the operative limits.
>
> Note 2: Upon returning within the limits of normal operating conditions, a device may require adjustments to restore normal performance.

output—The (electrical) quantity, produced by a *transducer,* which is a function of the applied *measurand.*

overrange—In process instrumentation of a system or *element,* any excess value of the *input signal* above its *upper range-value* or below its *lower range-value.*

overrange limit—The maximum input that can be applied to a *device* without causing damage or permanent change in performance.

pneumatic delivery capability—The rate at which a pneumatic *device* can deliver air (or gas) relative to a specified output pressure change.

> Note: It is usually determined, at a specified level of input *signal,* by measuring the output flow rate for a specified change in output pressure. The results are expressed in cubic feet per minute (ft^3/min) or cubic meters per hour (m^3/h), corrected to standard (normal) conditions of pressure and temperature.

pneumatic exhaust capability—The rate at which a pneumatic *device* can exhaust air (or gas) relative to a specified output pressure change.

> Note: It is usually determined, at a specified level of input *signal,* by measuring the output flow rate for a specified change in output pressure. The results are expressed in cubic feet per minute (ft^3/min) or cubic meters per hour (m^3/h), corrected to standard (normal) conditions of pressure and temperature.

precision—Essentially the same as *process variability;* the *reproducibility* of the measurement.

pressure, ambient—The pressure of the medium surrounding a *device.*

pressure, design—The pressure used in the design of a vessel or *device* for the purpose of determining the minimum permissible thickness or physical characteristics of the parts for a given *maximum working pressure —(MWP)* at a given temperature.

pressure, leak—The pressure at which some discernible leakage first occurs in a *device.*

pressure, maximum working (MWP)—The maximum total pressure permissible in a *device* under any circumstances during operation at a specified temperature. It is the highest pressure to which it will be subjected in the *process*. It is a designed safe limit for regular use. See also *pressure, design.*

> Note: MWP can be arrived at by two methods: (1) designed—by adequate design analysis, with a safety factor; (2) tested—by rupture testing of typical samples.

pressure, operating—The actual pressure at which a *device* operates under normal conditions. This pressure may be positive or negative with respect to atmospheric pressure.

pressure, process—The pressure at a specified point in the *process* medium.

pressure, rupture—The pressure, determined by test, at which a *device will burst.*

> Note: This is an alternate to the design procedure for establishing *maximum working pressure (MWP).* The rupture pressure test consists of causing the device to burst.

pressure, static—The *steady-state* pressure applied to a *device;* in the case of a differential-pressure device, the *process pressure* applied equally to both connections.

pressure, supply—The pressure at the supply port of a *device.*

pressure, surge—*Operating pressure* plus the increment above operating pressure to which a *device* may be subjected for a very short time during pump starts, valve closings, etc.

primary element—See *element, primary.*

process—Physical or chemical change of matter or conversion of energy (e.g., change in pressure, temperature, speed, electrical potential, etc.).

process control—The regulation or manipulation of variables influencing the conduct of a *process* in such a way as to obtain a product of desired quality and quantity in an efficient manner.

process measurement—The acquisition of information that establishes the magnitude of *process* quantities.

process pressure—See *pressure, process.*

process variability—The reproducibility of the measurement process. The degree by which the process fails for repeated measurements of the same quantity.

random error—The *error* arising from the variability of the process. The value is computed according to the laws of probability; it is stated as a statistical quantity to which may be assigned a degree of confidence.

range—(l) The region between the limits within which a quantity is measured, received, or transmitted, expressed by stating the *lower* and *upper range-values* (e.g., 0 to 150°F, –20 to +200°F, 20 to 150°C. (2) The *measurand* values over which a *transducer* is intended to measure (specified by their upper and lower limits).

> Note 1: Unless otherwise modified, input range is implied.
>
> Note 2: The following compound terms are used with suitable modifications in the units: measured variable range, measured signal range, indicating scale range, chart scale range, etc.
>
> Note 3: For multirange devices, this definition applies to the particular range that the device is set to measure.

range, elevated-zero—A *range* in which the zero value of the *measured variable, measured signal,* etc., is greater than the *lower range-value.*

> Note 1: The zero may be between the *lower* and *upper range-values,* at the upper range-value, or above the upper range-value.
>
> Note 2: The terms "suppression," "suppressed range," and "suppressed span" are frequently used to express the condition in which the zero of the measured variable is greater than the lower range-value. The term "elevated-zero range" is preferred.

range-limit, lower—The lowest value of the *measured variable* that a *device can be adjusted to measure.*

> Note: The following compound terms are used with suitable modifications to the units: measured variable lower range-limit, measured signal lower range-limit, etc.

range-limit, upper—The highest value of the *measured variable* that a *device* can be adjusted to measure.

> Note: The following compound terms are used with suitable modifications to the units: measured variable upper range-limit, measured signal upper range-limit, etc.

range, suppressed-zero—A *range* in which the zero value of the *measured variable* is less than the *lower range-value* (e.g., 20 to 100). (Zero does not appear on the scale.)

> Note: Terms such as "elevation," "elevated range," and "elevated span" are frequently used to express the condition in which the zero of the *measured variable* is less than the *lower range-value.* The term, "suppressed-zero range" is preferred.

range-value, lower—The lowest value of the measured variable that a *device* is adjusted to measure.

> Note: The following compound terms are used with suitable modifications to the units: measured variable lower range-value, measured signal lower range-value, etc.

range-value, upper—The highest value of the *measured variable* that a *device* is adjusted to measure.

> Note: The following compound terms are used with suitable modifications to the units: measured variable upper range-value, measured signal upper range-value, etc.

recording instrument—See *instrument, recording.*

reference-input element—See *element, reference-input.*

reference-input signal—*See signal, reference-input.*

reference operating conditions—See *operating conditions, reference.*

reference performance—Performance attained under *reference operating conditions.*

> Note: Performance includes such factors as *accuracy, dead band, hysteresis, linearity, and repeatability.*

reliability—The probability that a *device* will perform its objective adequately, for the period of time specified, under the *operating conditions* specified.

repeatability—(l) The closeness of agreement among a number of consecutive measurements of the *output* for the same value of the input under the same *operating conditions,* approaching from the same direction, for full-range traverses. (2) The ability of a *transducer* to reproduce readings when the same *measurand* value is applied consecutively, under the same conditions and in the same direction.

> Note 1: It is usually measured as a nonrepeatability and expressed as repeatability in percent of *span.* It does not include *hysteresis.*
>
> Note 2: Repeatability is expressed as the maximum difference between output readings. Two calibration cycles, taken one right after the other, are used to determine repeatability.

reproducibility—In process instrumentation, the closeness of agreement among repeated measurements of the *output* for the same value of input made under the same *operating conditions* over a period of time, approaching from both directions.

> Note 1: It is usually measured as a nonreproducibility and expressed as reproducibility in percent of *span* for a specified time period. Normally, this implies a long period of time, but under certain conditions the period may be a short time during which *drift* may not be included.
>
> Note 2: Reproducibility includes *hysteresis, dead band, drift,* and *repeatability.*
>
> Note 3: Between repeated measurements the input may vary over the range, and operating conditions may vary within normal operating conditions.

resolution—The smallest change in the *measurand* that will result in a measurable change in *transducer* output.

response, step—The total *(transient* plus *steady-state) time response* resulting from a sudden change from one constant level of input to another.

response, time—An *output* expressed as a function of time, resulting from the application of a specified input under specified *operating conditions.*

response time—The length of time required for the *output* of a *transducer* to rise to a specified percentage of its final value as the result of a step change of *measurand.*

rise time—See *time, rise.*

rupture pressure—See *pressure, rupture.*

scale factor—The factor by which the number of scale divisions indicated or recorded by an instrument should be multiplied to compute the value of the *measured variable.*

> Note: "Deflection factor" is a more general term than "scale factor" in that the instrument response may be expressed alternatively in units other than scale divisions.

sensing element—See *element, sensing.*

sensing element elevation—The difference in elevation between the *sensing element* and the instrument.

> Note: The elevation is considered positive when the sensing element is above the instrument.

sensitivity—(1) The ratio of the change in *output* magnitude to the change of the input that causes it after the *steady-state* has been reached. (2) The ratio of the change in *transducer* output to a change in the value of the *measurand* (i.e., the slope of the transducer output versus measurand curve).

> Note 1: Sensitivity is expressed as a ratio with the units of measurement of the two quantities stated. (The ratio is constant over the range of a linear device. For a nonlinear device, the applicable input level must be stated.)
>
> Note 2: Sensitivity has frequently been used to denote the dead band. However, its usage in this sense is depreciated since it is not in accord with accepted standard definitions of the term.

sensor—See *transducer.*

servomechanism—An automatic feedback control device in which the controlled variable is mechanical position or any of its time derivatives.

settling time—See *time, settling.*

signal—In process instrumentation, a physical variable, one or more parameters of which carry information about another variable (which the signal represents).

signal, actuating error—In process instrumentation, the *reference-input signal* minus the feedback signal.

signal converter—See *signal transducer.*

signal, error—In a closed loop, the *signal* resulting from subtracting a particular return signal from its corresponding input signal. See also *signal, actuating error.*

signal, measured—The electrical, mechanical, pneumatic or other variable applied to the input of a *device.* It is the analog of the *measured variable* produced by a *transducer* (when such is used).

> Note: In a thermocouple thermometer, the measured signal is an emf which is the electrical analog of the temperature applied to the thermocouple. In a flowmeter, the measured signal may be a differential pressure which is the analog of the rate of flow through the orifice. In an electric tachometer system, the measured signal may be a voltage which is the electrical analog of the speed of rotation of the part coupled to the tachometer generator.

signal, output—A *signal* delivered by a device, element, or system.

signal, reference-input—One external to a control loop, serving as the standard of comparison for the directly controlled variable.

signal-to-noise ratio—Ratio of *signal* amplitude to *noise* amplitude.

> Note: For sinusoidal and nonsinusoidal signals, the amplitude may be peak or rms and should be so specified.

signal transducer (signal converter)—A *transducer* which converts one standardized transmission *signal* to another.

span—The algebraic difference between the *upper* and *lower range-values.*

> Note 1: For example, (a) range 0 to 150°F, span 150°F; (b) range –20 to 200°F, span 220°F; (c) range 20 to 150°C, span 130°C.
>
> Note 2: The following compound terms are used with suitable modifications to the units: measured variable span, measured signal span, etc.
>
> Note 3: For multirange devices, this definition applies to the particular range that the device is set to measure.

span error—See *error, span.*

span shift—Any change in slope of the input/output curve.

stability—The ability of a *transducer* to retain its performance characteristics for a relatively long period of time.

> Note: Unless otherwise stated, stability is the ability of a transducer to reproduce *output* readings obtained during its original *calibration* (expressed in percent *of full-scale output)* for a specified period of time. As defined, stability does not include effects of temperature.

static characteristics—Those characteristics of a *transducer* that relate to its response to constant (or quite slowly varying) values of the *measurand.*

static pressure—See *pressure, static.*

steady-state—A characteristic of a condition, such as value, rate, periodicity, or amplitude, exhibiting only negligible change over an arbitrary long period of time.

> Note: Steady-state may describe a condition in which some characteristics are static, others dynamic.

supply pressure—See *pressure, supply.*

suppressed range—See *range, elevated-zero.*

suppressed span—*See range, elevated-zero.*

suppressed-zero range—See *range, suppressed zero.*

suppression—*See range, elevated-zero.*

suppression ratio (of a *suppressed-zero range)—The* ratio of the *lower range-value* to the *span.*

> Note: For example, range 20 to 100; suppression ratio = 20/80 = 0.25.

surge pressure—See *pressure, surge.*

systematic error—See *error, systematic.*

temperature, ambient—The temperature of the medium surrounding a *device.*

> Note 1: For devices that do not generate heat, this temperature is the same as the temperature of the medium at the point of device location when the device is not present.
>
> Note 2: For devices that do generate heat, this temperature is the temperature of the medium surrounding the device when it is present and dissipating heat.
>
> Note 3: Allowable ambient temperature limits are based on the assumption that the device in question is not exposed to significant radiant energy sources.

time constant—(1) In process instrumentation, the value T in an exponential response term A exp (–t/T) or in one of the transform factors 1 + *s*T, 1 +*jw*T, 1/(1 + *s*T), 1/(1 +*jw*T), where *s* is a complex variable, *t* is time in seconds, T is a time constant, *j* = (square root of (–1)), and *w* is angular velocity in radians per second. (2) The time required for the *output* of a *transducer* to rise to 63% of its final value as a result of a step change of *measurand.*

> Note: For the output of a first-order system forced by a step or an impulse, T is the time required to complete 63.2% of the total rise or decay; at any instant during the process, T is the quotient of the instantaneous rate of change divided into the change still to be completed. In higher order systems, there is a time constant for each of the first-order components of the process. In a Bode diagram, break points occur at *w* = 1/T.

time, dead—The interval of time between initiation of an input change or stimulus and the start of the resulting observable response.

time, ramp response—The time interval by which an *output* lags an input, when both are varying at a constant rate.

time response—See *response, time.*

time, rise—The time required for the *output* of a system (other than first order) to change from a small specified percentage (often 5 or

10%) of the *steady-state* increment to a large specified percentage (often 90 to 95%), either before or in the absence of overshoot.

> Note: If the term is unqualified, response to a unit step stimulus is understood; otherwise, the pattern and magnitude of the stimulus should be specified.

time, settling—The time required, following the initiation of a specified stimulus to a system, for the *output* to enter and remain within a specified narrow band centered on its *steady-state* value.

> Note: The stimulus may be a step impulse, ramp, parabola, or sinusoid. For a step or impulse, the band is often specified as ±2%. For nonlinear behavior, both magnitude and pattern of the stimulus should be specified.

transducer—See also *primary element, signal transducer,* and *transmitter.* An *element* or *device* that receives information in the form of one quantity and converts it to information in the form of the same or another quantity. (2) A device that provides a usable *output* in response to a specified *measurand.*

> Note: The term "transducer" is preferred to "sensor," "detector" or "transmitter."

transduction element—The (electrical) portion of a *transducer* in which the *output* originates.

transfer function—A mathematical, graphical, or tabular statement of the influence which a system or *element* has on a *signal* or action compared at input and at output terminals.

transient—In process instrumentation, the behavior of a variable during transition between two *steady-states.*

transient deviation—See *deviation, transient.*

transient overshoot—The maximum excursion beyond the final *steady-state* value of *output* as the result of an input change.

transient overvoltage—A momentary excursion in voltage occurring in a *signal* or supply line of a *device* that exceeds the maximum rated conditions specified for the device.

transmitter—A *transducer* that responds to a *measured variable* by means of a *sensing element* and converts it to a standardized transmission *signal* that is a function only of the *measured variable.*

upper range-limit—See *range-limit, upper.*

upper range-value—See *range-value, upper.*

value, desired—In process instrumentation, the value of the controlled variable wanted or chosen.

> Note: The desired value equals the *ideal value* in an idealized system.

value, ideal—In process instrumentation, the value of the indication, *output,* or ultimately controlled variable of an idealized *device* or system.

> Note: It is assumed that an ideal value can always be defined even though it may be impossible to achieve.

value, measured—The numerical quantity resulting, at the instant under consideration, from the information obtained by a measuring *device.*

variable, measured—A quantity, property, or condition that is measured. Sometimes referred to as the *measurand.* Commonly measured variables include temperature, pressure, rate of flow, thickness, and speed.

warm-up period—The time required after energizing a device before its rated performance characteristics apply.

zero adjustment—See *adjustment, zero.*

zero-based conformity—See *conformity, zero-based.*

zero-based linearity—*See linearity, zero-based.*

zero elevation—For an *elevated-zero range,* the amount the *measured variable* zero is above the *lower range-value.* It may be expressed either in units of the measured variable or in percent of span.

zero error—See *error, zero.*

zero shift—In process instrumentation, any parallel shift of the input/output curve.

zero suppression—For a *suppressed-zero range,* the amount the *measured variable* zero is below the *lower range-value.* It may be expressed either in units of the measured variable or in percent of *span.*

zone, dead—(1) For a multiposition controller, a zone of input in which no value of *output* exists. It is usually intentional and adjustable. (2) A predetermined range of input through which the output remains unchanged, regardless of the direction of change of the input signal.

> Note: For definition 2, there is but one input/output relationship, and dead zone produces no phase lag between input and output.

B

Deadweight Gage Calibration

B.1 INTRODUCTION

A deadweight gage is calibrated using another deadweight gage in a process known as crossfloating. When one gage is crossfloated against another, the two are connected together and brought to a common balance at various pressures. The balancing operation is identical to that employed on an equal-arm weight balance where the mass of one weight is compared with another. In each instance, the operator must decide when the balance is complete. In a crossfloat, the two gages are considered to be in balance when the sink rate of each is normal for that particular pressure. At this condition there is no pressure drop in the connecting line and, consequently, no movement of the fluid. The condition can be difficult to recognize, particularly if there is no means of amplification in the observation method. The precision of the comparison will depend directly on the ability of the operator to judge the degree to which the balance is complete. This procedure is repeated for several pressures, and the values of areas obtained are plotted against the nominal pressure for each point. A least-squares line is fitted to the plots as the most probable value of the area at any pressure.

Of the different methods used to amplify the signals that are generated by the crossfloat process, one is particularly rapid and convenient. An electronic sensor, which indicates the floating position of the piston, is placed beneath the weights of each gage. The output signal from the sensor is processed and fed to an analog meter that has a vertical scale, the value of which is adjusted to indicate units of displacement of the piston. Two meters—one for each instrument—are placed contiguously for simultaneous viewing. A constant-volume value, inserted between the gages, supplements the sensors.

Other less precise methods of estimating the true balance include:

- optional amplification of the sinking stack of weights of one of the gages while timing the descent with a stopwatch
- interposition of a sensitive null-pressure transducer that displays small pressure differences directly

When using a suitable amplifying device, the scatter in the plotted areas from a good-quality piston gage should not exceed one or two parts in 10^5.

B.2 CALIBRATION OF PISTON GAGES BY CROSSFLOAT

Two procedures can be followed when calibrating a deadweight gage against a standard. The procedure for manipulation of the apparatus in each instance is approximately the same, but the results differ somewhat in accuracy:

In one procedure, all known variables influencing the results of the test are controlled or are corrected to the best of the operator's ability. Observations are made using currently established techniques with estimated corrections for transient or recurring disturbances. The accumulated data are recorded in a way that will permit an error analysis of the completed experiment. Constants obtained for the test gage are reported such that the necessary corrections can be applied for the proper use of the gage. At the conclusion, the results of the test can be considered to be the best obtainable from the equipment and techniques administered. This procedure is used when it is desired to transmit the accuracy of the standard to another instrument with the least loss.

In the second procedure, an instrument is crossfloated against a standard in the manner of an acceptance test. A typical application is a transfer standard that must be periodically recalibrated against a primary one. It is not necessary that the total error of the experiment be determined, but only that the error of the test instrument not exceed a certain value. In this test, the corrections of certain variables of the test gage are intentionally disregarded because they are not applied during normal use of the instrument.

This type of test usually falls in the category of the one-in-ten rule for calibration of production equipment. The standard must have an uncertainty of one-tenth that allowed for the test instrument. It should be noted that the calibration of an instrument under the one-in-ten system does not permit the operator to disregard any of the corrections that should be applied to the standard gage. The standard is capable of measuring a pressure to the manufacturer's claimed accuracy only when it is operated in the manner prescribed by the manufacturer. Since the accuracy of a deadweight gage is largely dependent on the operating technique, some operators tend to avoid applying the displeasing corrections with the false justification that they are not apparent in the test.

A frequent practice involves use of one of the less accurate deadweight testers on a production line for the calibration of dial-type pressure gages or transducers. In some instances, it is claimed that without corrections, the gage is capable of making pressure measurements with an error that will not exceed 0.1% of the observed pressure. This claim may be considered somewhat liberal when it can be shown that variations in the force of gravity (which is a multiplier in the determination of the total force exerted by the weights) may exceed the claim by almost 100% within continental United States. The most elementary error analysis indicates that the uncertainty in the observed pressure is the sum of the uncertainties of the area of the piston, of the force acting on the piston, and of the failure of the piston to reproduce the pressure faithfully. If all the error is assigned to the force acting on the piston, there is nothing left for the uncertainty of the remaining contributors. It is possible and highly probable that these gages will perform according to the claims if only the one correction is made for the effects of gravity on the weights.

Some caution should be exercised in the use of these gages after an acceptance test or periodic recalibration. It frequently happens that errors in the adjustment of the weights and in the area of the piston will compensate for the local value of gravity. Under these conditions, a gage will measure a pressure within the claimed limits. The weights, however, may not be pooled with other sets of weights of the same make because, when used on another piston, the observed pressure may be out of tolerance. If a number of such pistons and weights are to be used interchangeably in a production process, an arrangement must be devised that will ensure that all the measurements fall within the specified limits. Such an arrangement could involve the actual readjustment of all weights to produce the proper force at the local value of gravity and on the piston of mean area of all those in use. The tolerances for adjustment of the weights would, of course, make allowance for the total variation in piston, areas, temperature effects, resolution of the gages, and effect of air buoyancy.

B.3 INSPECTION OF WEIGHTS

The weights for a deadweight gage must be thoroughly clean and in good condition before the gage can be properly calibrated. If the weights are constructed so that axial alignment is obtained by nesting, it is possible that, by careless handling or such, the pilots and recesses may have become upset and no longer mesh properly. In reconditioning the weights, the upset portions should not be cut away to restore the mesh, because the loss of metal would certainly be undesirable. Instead, the metal should be forged back into place as completely as possible to conserve the metal of the weight. The history of the mass will then be more reliable.

Any severe hicks or edge crushes should be remedied in the same way. Craters or knots of metal formed by nicks or crushes are subject to rapid wear and subsequent loss of metal; also, the projecting metal causes the weight to be misaligned in the stack.

B.4 CALIBRATION OF WEIGHTS

The weights, thoroughly clean and in good condition, must now be calibrated. A distinction is made between the terms "calibration" and "adjustment." Adjustment of a weight implies removal of metal in controlled manner such that the mass of the weight is changed to a specified value. An adjusting tolerance allows the mechanic to accomplish this assignment economically.

In a controlled environment, and with more expensive equipment, the actual mass can be determined with an accuracy that is only a fraction of the adjusting tolerance. It is this determination and subsequent reporting of the results that is defined as calibration of the weight.

No adjustment is normally made on the weights of a set unless some of the smaller weights, of two-piece construction, have obviously been tampered with. Adjustment must then be made by adding brass slugs to the adjusting cavity, and the adjustment must be of a precision comparable to that of the remaining weights of the set.

The method of weight calibration is that of direct substitution of a known mass for the gage weights. That is, after a gage weight has been balanced by some mass on a suitable instrument, the weight is replaced and the original balance restored by mass of known quantity. With this method, some of the instrument errors are canceled. The precision to which these measurements must be made depends, of course, on the class of equipment being calibrated.

Values of mass are reported in units of apparent mass versus brass standards. In the table of reported masses, values under the heading "Apparent mass versus brass" are those which the weights appear to have when compared in air under "normal conditions" against "normal brass standards"—*no* correction being made for the buoyant effect of air. Normal conditions are 25°C and air density of 1.2 mg/cm^3. Normal conditions have a density of 8.4 g/cm^3 at 0°C and a coefficient of cubical expansion of 0.000054/°C. True mass values are those which might be observed in air of zero density (i.e., in a vacuum).

B.5 DEADWEIGHT GAGE INSPECTION AND PREPARATION FOR CALIBRATION

The piston gage must be in good mechanical condition before a calibration is started. Any damage that has resulted from normal wear, misuse, or shipping must be repaired to the extent that is possible by the calibrating facility: When the gage is in good mechanical condition, it must be cleaned before installation on the crossfloat bench.

Cleaning procedures must be adequate for elimination of all substances that may affect the performance of the gage. These substances vary from particulate contamination of the liquid to deposits of oil varnish on essential parts of the mechanism. It is left to the operator to decide what cleaning procedures will satisfy particular requirements.

Certain physical measurements must be made on the piston assembly. These measurements may be made before the assembly is reinstalled in the gage and are described in the next section.

B.6 PRELIMINARY CALIBRATION OPERATIONS

In order to simplify the instructions, procedures are given in command form. The instrument being calibrated will be referred to as the test gage.

First, set the test gage on the test bench and connect it to the standard gage, making certain that the connecting tube is in good alignment and free of stresses. Remove the piston assembly and rinse it with a solvent in preparation for the measurements that follow.

Next the position of the reference plane of measurement of the piston must be determined with respect to some external fiducial mark on the gage. The dimensions of the piston determine the position of the reference plane. If the measurements are made carefully, the computed position of the reference plane will probably be within 0.01 in. of the true position. All measurements should be recorded in a manner that will cause the least ambiguity. If a form is not available, one should be designed. The geometry or profile of the piston assembly and loading table should be sketched with dimension lines.

1. Measure and record the dimensions shown for the piston assembly, weight-loading table, and sleeve weight.
2. Remove the piston from the cylinder, clean it with a mild solvent, and determine the mass of the complete piston assembly, including the enlarged portion at the bottom. Thrust bearings and their retainers usually are not a part of the tare mass. When the piston is arrested from overpressure by an enlargement that acts against a bearing, the retainer is captured by the pressure housing and is not permitted to ride on the enlargement during a measurement.
3. Clean the piston/cylinder assembly thoroughly and reassemble it to the gage. Remove any trapped air from beneath the cylinder.
4. Clean the weight-loading table, measure its overall length, and determine its mass. Record the dimensions.
5. Measure the distance from the index line on the sleeve (or hanger) weight to the seat where the weight rests on the weight-loading table. Record this value.

6. Compute the volume of the enlarged (submerged) portion of the piston and divide by the area of the piston. The quotient is the length that the enlargement would have if all the material were contained in a cylinder of the diameter of the piston. The increase in length of the enlarged portion is added to the overall length of the piston with its enlargement.
7. The various dimensions are summed as follows:
   (A) Add the new length of the piston to the length of the weight-loading table.

   (B) Subtract the combined lengths from the depth of the sleeve of the sleeve weight as measured with respect to the index line. The difference obtained is the relative position of the reference pressure plane to the index line on the sleeve weight. In order to associate the reference plane of the deadweight gage to the remainder of the system, readings are made when the weights are floating at a particular height with respect to some common plane of the system. This plane is usually chosen at the position of the index line on the post attached to the base of the gage.

   (C) Measure the difference between the chosen plane (which is the index line on the index post) of the test gage and that of the standard gage. This measurement may be made with a cathetometer, a water-tube level, or a bar on which a sensitive level vial is mounted.

   (D) Add this value with its appropriate sign to the difference obtained in (B).

   (E) Add the difference in the reference plane of the standard gage and its index plane to the value obtained in (D). The result is the difference between the two reference planes of both gages when each is floating with the line of the sleeve weight coincident with the mark of the index post. The total difference between these two planes is referred to as the hydraulic head, oil head, or Dh. The equivalent pressure is Dh × liquid density = DP.
8. The approximate value of the hydraulic head of the oil used is 0.031 psi per inch of height. Therefore, if the difference in height (in inches) of the reference planes of the two gages is multiplied by 0.031, the product will be the correction in

pounds per square inch that one gage will have with respect to the other. If the reference plane of the test gage is below that of the standard, the pressure seen by the test gage is the pressure of the standard plus the oil head (DP).

B.7 CALIBRATION OF PISTON GAGES

In some instances, a piston gage can be calibrated by direct measurement of the piston and cylinder. The measurement can be made to a high degree of accuracy and will satisfy the requirement for determination of the effective area of the piston, except when a pressure coefficient must also be determined. A piston gage can be directly compared with a standard instrument, the pressure coefficient and piston area of which are known. The comparison—or crossfloat, as it is called by pressure technicians—can be accomplished with a precision of only a few parts per million. The following paragraphs describe the crossfloat procedure and tell the reader what to expect.

It is assumed that the instrument to be calibrated is one that has been in service and that certain inspection and preparation procedures have been followed before the calibration is started. It is further assumed that the instrument normally operates with a liquid as the pressurizing medium and that it will operate properly with the same liquid used in the standard. Allowances must be made in the interpretation of the suggested procedure for instruments that employ gases or liquids other than those of the standard.

The plan is to determine the constants of the piston being calibrated in terms of those which are known for the standard. The area of the piston and the pressure coefficient of the area are the quantities of primary interest. If a given pressure is established with the standard and the test gage brought to balance, a value of pressure can be assigned to the test instrument after suitable corrections have been made. The masses on the test gage are known; they can be summed and converted to force units. When the force on the test piston is divided by its pressure, the quotient represents the piston area. These observations are conducted at various pressures throughout the range of the instrument. The reduced data result in three values of interest:

- the area of the piston at no-load pressure
- the coefficient of elastic distortion of the cylinder
- a coefficient of precision for the test

A sufficient number of observations must be conducted in order to provide the redundancy needed for determination of the precision coefficient. The quantities of *a* and *b* are related in Equation B.1:

$$A_e = A_0 \,(1 + bp) \qquad \text{(B.1)}$$

Where A_e is the mean area of the piston and of the cylinder at any pressure greater than no-load conditions, A_0 is the value of A_e at no-load, *b* is the fractional change in area per unit of pressure with dimensions A/F/A pressure unit, and *p* is the pressure beneath the piston.

Data for a number of observations at different pressures are reduced such that values of A_e are of the same population. These areas are plotted on coordinates having scales that expose the scatter in the observed values. A best-fit (regression) line is fitted to the plot. The intercept of the line with the zero pressure coordinate represents the value of A_0 for the test piston; the loci of all other points of the line are values of A_e. A particular A_e and its corresponding pressure, *p*, are selected near the maximum pressure and are substituted in the relation:

$$A_e - A_0 / b = (A_e - A_0)/pA_0 \qquad \text{(B.2)}$$

After evaluation of *b*, the computed A_e is determined for each of the pressures in the observation above. The differences between the computed A_e and the observed A_e are the deviations, or residuals, from which the standard deviation of the variability of the test is calculated.

Forms and work sheets should be designed for convenient tabulation and reduction of the observed data. The completed forms and work sheets serve as the basis for documentation of the measurement process. They must be retained as evidence of the procedures followed in the experiment and of the continuous pedigree of the test instrument.

B.8 CROSSFLOAT BALANCING WITH THE PROXIMITY INDICATOR

When comparing one deadweight gage to another, the two gages are connected together and brought to balance at various pressures. A description of the crossfloat process follows.

At a given pressure, and with the valve open, the two gages are brought to an approximate balance—either by observing the weights directly or by using the two proximity indicators. The degree of balance must be reasonably complete so that the floating positions will remain constant for a few seconds after being manually adjusted. The relative sink rates of the two gages must be determined before the balance can be completed. The criterion for a complete balance is the condition in which each gage sinks at its normal rate when both are connected.

From the arrangement of the components illustrated in the accompanying plan, it can be seen that, with the isolation valve open, the entire system can be pressurized with the manually operated pump. A procedure for completing the balance—given in command form—will be described in this section.

When possible, the calibration should be conducted without the use of electric drive motors. The motors add heat to the instrument and increase the uncertainty in the measurement of the piston temperature. The drive belt and piston-driving sleeve should be removed from the gage so that the weights will spin freely, with only the friction of the weight-loading table spindle and piston acting to decelerate the motion. When the stack of weights becomes small enough that there is insufficient momentum to carry the rotation through an observation, the driving mechanism should be reinstalled.

It is probably only a matter of preference whether the test gage is calibrated with increasing pressures or with decreasing pressures. Since the test gage has been disassembled for the cleaning operations and for the preliminary measurements, the newly assembled gage should be subjected to its maximum pressure before the test is started. This pressure seats the components together with the maximum force and presumably eliminates the possibility of a change occurring during calibration as a result of improperly seated components. The pressure serves further to assist in exclusion of residual air by forcing it out around the piston and by driving it into solution. It is of some advantage, therefore, in starting the calibration at the greatest pressure.

The following procedures are based on the practice of beginning the calibration at the maximum pressure.

1. Load both gages with weights equivalent to the approximate highest test pressure. The weights should be applied in numerical order for reasons that will be explained later. It is not

important for either of the gages to have weights equivalent to an exact value of pressure, because disturbances in the conditions that occur during the observations will result in variations in the pressure. The true pressure will be computed from observations at the time each balance is completed. Check the level vials for a change in the instrument that may have resulted from the applied load.

2. Open the isolation valve and pressurize the system until one of the gages rises to a floating position. It is convenient to have a particular group of weights on the standard instrument. If the group is used repeatedly for calibrations, the calculations for this group will be reusable and time will be saved in the measurement process. This practice requires the balancing or trimming to be performed on the stack of weights on the test gage.

   In the beginning, the sink rates of the gages will be found to be unusually large. The unusual rate is a result of the shrinkage that occurs when liquid loses its heat. The heat is produced by the work of pressurization. A period of at least 10 min should be allowed for the heat to become distributed and for the system to become quiet. During this time, some preliminary balancing operations can be started.

   With the pump, raise the floating piston to a position near the top of its stroke and set the weights in motion. Test the state of the initial balance by pressing down on the floating weights. Adjust the weights of the test gage so that the balance is improved. Continue the adjustment until it is apparent that a systematic procedure should be started.

3. While the valve is open, adjust the floating position of the left gage piston with the pump (it may be the test gage or the standard) to a value between 24 and 28 on the position meter. The starting figure for the sink rate measurement is 20; it is purely arbitrary. As the left piston sinks to the 20 mark, there will be time to adjust the right piston to match it.

4. Close the isolation valve.

5. The float position of the right piston must be adjusted for arrival at the 20 mark at the same time as the left one. If it happens that the right piston sinks more slowly than the left, it

is likely that the final adjustment of the hand pump will be made by withdrawing the plunger. This situation is to be avoided because the slack in the pump spindle nut will be in a direction that allows the plunger to creep out of the cylinder by the action of the pressure. A false sink rate will be observed. The final adjustment with the pump *must* be made with an advancing motion of the plunger.

It is not necessary for each piston to be exactly at the 20 mark at the beginning of the measurement. A mental correction may be applied for one that is high or low.

6. Allow the two pistons to sink undisturbed. When the faster-sinking piston arrives at the zero mark, make a note of the position of the other.
7. Open the valve and adjust the position of the left piston to 4 to 8 units above its intended starting point. If the left piston was the faster-sinking one, the starting point will be 20. If it was the slower one, the starting position will be $20 - x$, where x is the number of units noted in step 6. Close the valve.
8. Adjust the right piston to a position that will allow both pistons to arrive at their starting points at the same time. Again, the final pump plunger motion must be in an advancing direction.
9. Open the valve when each piston is at its intended starting position.
10. Allow the two pistons to sink undisturbed. If they do not arrive at zero at the same time, make a correction in the trim weights on the test gage and repeat 7 to 10.

Wind currents, temperature ripple, and body convection currents may disturb the process. Indeed, expansion of the oil in the connecting tube from an air current of different temperature can seriously affect the apparent sink rate of the piston. The operator should stand away from the table as the final test is observed.

An advantage of the float-position balancing indicator results from the type of information displayed. The operator can complete the balancing operation without looking away from the meters except when changing weights and adjusting positions for the next observation. When a satisfactory balance is obtained, the decision is made at the moment the two pistons are floating at their correct measuring positions. This condition is essential for good precision.

The value for relative sink rates at one pressure may not necessarily be valid at other pressures. Cylinders of different designs and of different materials will have characteristic sink rates. The best work requires a determination of relative sink rates at each pressure level.

The quality of a piston gage is judged by its sink rate and by its ability to detect small changes in pressure. The latter characteristic is called resolution and is measured and recorded at each pressure point. Small weights, in decreasing size, are added to or subtracted from the group of weights until a weight added or removed no longer causes a perceptible change in the balance. The smallest increment that will produce a detectable change in the state of equilibrium represents the resolution of the entire system.

With the completion of the test for sensitivity, the work of obtaining the first test point is finished except for releasing the pressure. Operation of the equipment should be carried out so as to ensure maximum safety at all times. When raising or lowering the pressure of the gage, the hand pump should be turned slowly to prevent damage to the piston or its stops. If the pressure is changed too quickly, the piston is likely to be damaged by the sudden shock as the piston is stopped at the end of its travel. The isolation valve should be opened after each observation.

The pressure is reduced to a value somewhat less than the next test point and the rotation stopped so that the weights can be counted and recorded.

11. The total weight load contributing to the pressure measured must be recorded with care. Each weight is identified by a number or a symbol, and its individual mass is tabulated in the mass-section report. The symbols representing the weights are recorded on the data recording sheet, and the number of individual weights in the stack is determined by count; the number of entries on the data sheet must agree with the counted number of weights. The tare mass must also be included in the count and tabulation. The process of tabulating, counting, and

verifying must be repeated to make certain that no mistake has occurred.

For convenience of reporting, the large platters should always be kept in numerical order and placed on the stack in that order. The values of the masses are listed as cumulative sums, and it is necessary to list only the first and last platters of the stack when tabulating on the work sheet. It is agreed that the first (1) and last (5) weights will be listed as '1-5,' which means that the first and last weights are included in the group.

A symbol is entered in the column under "Tare" so that this tare value is not overlooked in summing all the weights of the stack. The tare is a constant mass that contributes to pressure and must be added to every measurement taken with the deadweight gage. In addition to the mass, the surface tension effects of the oil against the piston have some influence on the tare pressure. This value can be computed from the surface tension of the liquid and the circumference of the piston. Record the temperature of each gage.

12. The gages can now be prepared for the next lower pressure. Certain precautions must be taken to ensure that the change is made safely. A rule that must be followed during the operation of any deadweight gage concerns the act of releasing the pressure after a measurement. The pressure must never be released by opening a valve or bleed screw, because if the piston is in a floating position, it will fall on the limiting stop and may be damaged. The pressure must be reduced by withdrawing the hand pump plunger until the piston is resting on the thrust plate or whatever stop is provided. Then, if the pressure is to be removed entirely, a valve can be opened.

    In changing the weights, care must be taken to prevent damage to the piston or thrust bearing as the weight is added or removed. If the weights are floating and one is removed quickly, the entire mass will be forced upward against the limiting stop, and the resulting shock may cause damage. A safe procedure is to reduce the pressure comfortably below the next lower test point before removing any weights.

    When the weights are added to the stack, they must be set down accurately and deliberately so that the pilot of the lower weight enters the recess of the weight being applied. If the weights are carelessly stacked, after years of service the edges of the nesting surfaces will become upset and the weights will no longer fit together. The resulting misalignment may cause the weights to hang with uneven axial load on the weight table.

B.9 TEST REPORT

Upon completion of the calculations and determination of the effective area at 0 psig, a report must be compiled. The individuals for whom the gage was calibrated must be supplied with all the data that were obtained in the calibration. The accompanying format is an example of a report that satisfies these conditions for a particular gage.

The report must identify the gage and/or piston calibrated and the master to which it was compared. The identifying number and calibrating facility of the master gage are shown, in order that complete traceability to the National Institute of Science and Technology (NIST) can be made at any time.

Values of mass for all the weights must be listed and the values qualified as to their unit of measure (e.g., grams, apparent mass versus brass standards, etc.).

The effective area of the piston for 0 psig must be given with its corrections for change with pressure and temperature. The reference plane of the piston, to which all measurements must be reduced, is noted.

Finally, a statement concerning the accuracy and precision of the determination must be made in order that the technician using the gage may evaluate subsequent measurements. The usual manner of denoting the accuracy of a piston pressure gage is an expression of the expected error of a single measurement as a percentage of the quantity being measured or as some minimum value of pressure-whichever is greater. Some level of confidence must be chosen that will give the accuracy figure a realistic meaning. If it is reported that the figure is based on the sum of the estimated, systematic errors and two standard deviations of the variability of the calibrating process, the user may infer that, for 95 measurements out of 100, error may not exceed the stated value. All of the uncertainties preceding the test are accumulated and shown as the systematic error for the area of the reference piston and for the mass of the weights used in the test. In estimating the uncertainty of the area of the test piston, the uncertainties of each of the following quantities contribute:

- area of the reference piston
- mass of weights on the reference piston
- mass of weights on the test piston
- precision of the comparison
- piston temperatures and relative axial tilt

Although the last two items are likely to be of random character, their contribution to the uncertainty of the test piston area is small. The errors arising from these effects are added numerically to those of the remaining quantities rather than being added vectorially.

B.10 ADJUSTMENTS OF PISTON PRESSURE GAGE WEIGHTS FOR A SPECIFIC ENVIRONMENT

It has been the practice of some manufacturers to mark the weights of a piston pressure gage in units of pressure. The value indicates the pressure that the weights are assumed to represent when they are placed on the piston of the gage. Unfortunately, this practice has survived the long period of piston-gage development. The variables that influence piston-gage pressure measurements are so numerous and of such magnitude that pressure designations on the weights are, at best, only an approximation.

Occasionally, a manufacturer is requested to provide a piston gage that is intended to operate in a given set of controlled environmental conditions. It is supposed that such a gage could be operated without application of corrections for the usual variables. This supposition is correct, except that the adjustments that are made by the manufacturer for this set of conditions are imperfect and result in a reduction of overall gage accuracy: The factors that are of importance in the decision to adjust a piston gage for a particular environment are listed here, along with a general discussion of each. The list is intended to inform the user of what he can be expected from the performance of a piston gage that has been adjusted for specified conditions.

1. Because the relationship between the pressure inside the chamber of the piston gage and the force on the carrier caused by the weights is a function of the area of the chamber, $P = F/A$,

it is apparent that the piston area should be constant or compensated when the area changes. It is easier to adjust the weights for an error in area than to actually change the area itself. Once the area of a piston has been established, the weights are adjusted as required for the piston. A routine calibration will include a procedure to make certain that the force/area relationship is correct.

2. Again for economical reasons, the weights are adjusted to a nominal value of mass having an uncertainty of 1 part in 20,000. Normally, the actual value of each mass is determined and reported with an uncertainty of 1 part in 100,000. Clearly, if one uses the adjusted values of mass rather than the actual values, the pressure may be in error (from the weights themselves) by 0.005% rather than 0.001%. The additional 0.004% uncertainty cannot be tolerated in the manufacturer's standard claims.

3. The claims for accuracy are based on a knowledge of the force of the earth's gravity: If the uncertainty in the force of gravity is greater than 1 milligal (mGal) (approximately 1 in 10^6), the error cannot be tolerated. The force representing 1 mGal is roughly equivalent to a vertical displacement of the gage of 10 ft. If the gage is moved to another location, even within the physical boundaries of a single organization, the adjusted value of the weights may be out of tolerance.

4. Upon recalibration of both the gage and the weights, it is probable that new numbers will be determined. The old adjusting figures for the weights will probably no longer be correct and readjustment may be necessary. In the case of one-piece weights, it is expensive to make them heavier.

5. Although the ambient temperature can be controlled within a few tenths of a degree, the system temperature may vary by several degrees. The work of raising and lowering the pressure and of rotating the weights causes rather large temperature changes within the system and usually results in a gradual increase in the temperature of the piston. These changes are normally accounted for by addition of extra weights, the values of which are obtained from temperature/pressure curves. If these facts are disregarded, the variations in temperature may create more error than was allowed in the claims for accuracy.

6. Where a pressure coefficient has been observed for a piston/cylinder assembly, the effective area of the piston changes throughout the range of pressures. Either the claims for accuracy must include the variation in area, or the large weights must be calculated and adjusted individually and placed on the piston in a particular sequence. The small weights that are used in various combinations throughout the range can be adjusted for only one area. The accuracy claim must at least include the range of pressure covered by the small weights.

7. The buoyant effect of the normal atmospheric air is approximately 0.015% of the mass. When weights are adjusted for a given locality and are intended to be used for measurement of pressures above atmospheric, the buoyancy of the average air for that locality must be included in the adjustment. For measurement of absolute pressures, the buoyancy is not included in the adjustment; the unit of adjustment is true mass to (true) weight. Also, for absolute measurements, the piston tare must be adjusted to include the residual bell-jar pressure. This pressure must, of course, be maintained at the stipulated value.

8. When a set of weights is adjusted for a particular set of conditions, the markings on the weights are seldom changed to indicate that a special adjustment has been made. As a rule, the only clue to such an adjustment is a statement in the documents supporting the equipment. There is always the possibility that absolute control of the various types of pressure measurements will be lost through the chain of command as a result of different methods of equipment operation.

9. When calculating the pressures for a piston gage employing a liquid pressurizing medium, the pressure gradients that result from the heads within the measurement system are significant and must not be disregarded. This statement is not to imply that gradients within a gas system are not significant; they are, only less so. All pistons, therefore, have a unique horizontal plane to which pressure measurements must be referenced.

   This plane is that which satisfies the relation $P = F/A$. The point at which the actual pressure is to be determined seldom lies on this horizontal plane, and a correction must be made for the difference. Since the difference in the two values of pressure varies with the experiment, it is usually more convenient to make a correction at the time of the measurement. This situation exists whether the weights are, or are not, adjusted to match the piston area. It is mentioned only as a reminder that variability in the measurement process does include the effects of the hydraulic and gaseous heads within the system.

This list is incomplete, but the major factors have been covered. It appears that many metrologists are discouraged in the use of a piston gage because of the need to consider the variables listed. When the gage is used as a standard of pressure for the calibration of other devices, the problem is not a serious one as a single set of calculations will usually satisfy the requirements. Small corrections are then made for the variables that cannot be accurately predicted. The effort required to consider and compensate for these various variables seems worthwhile because of the high degree of accuracy available with piston gages. Although listing the many conditions relating to gage accuracy is probably unnecessary in most calibration situations, it seems important to understand this when maximum accuracy is required.

B.11 RECALIBRATION INTERVAL FOR HYDRAULIC DEADWEIGHT GAGES

Any calibration laboratory must maintain reliable test instruments. Thus, the question often arises, "For what period of time is the calibration of a deadweight gage reliable?" Considerable expense could be avoided if a dependable answer were known.

The period of time for which the calibrated values of a deadweight gage are valid depends on two factors: the rate of decay or deterioration of the original accuracy and the total deterioration that can be tolerated. Deterioration of accuracy results from unpredictable changes in the piston area and in the mass of the weights. These changes occur as a result of normal and abnormal wear of the critical parts and, to some extent, from dimensional aging effects of the construction materials. The changes begin immediately after calibration. Estimates of the magnitude of the changes are obtained from the results of recalibrations of a number of gages that have been in service for a period of time.

Tests have shown that the recalibrated values for a group of 15 pistons indicated an average annual change in area of less than 3.5 ppm. A similar study of a group of 90 weights indicated that the average annual change in mass would not be expected to exceed 1.3 ppm. These figures were not obtained from a statistical treatment of the data. They represent the numerical average of changes in the nominal values reported to the customer. No allowance was made for the care with which the gages had been operated during the period in question.

Assuming a change in overall accuracy of 0.0025% (25 ppm) could be tolerated, the expected life of a calibration would be approximately 4 to 5 years. Some allowance must be made for the fact that the figures quoted are average ones and would be greater in some instruments. Also, the figures would be valid only in instances where the instrument was operated properly and was not subject to abnormal wear.

C

Pressure Instruments Form ISA-20.40a1

ISA	PRESSURE INSTRUMENTS				SHEET	OF
					SPEC. NO.	REV.
	NO	BY	DATE	REVISION	CONTRACT	DATE
					REQ. - P.O.	
					BY / CHK'D	APPR.

	1	Tag Number ______	Service ______
GENERAL	2	Function	Record ☐ Indicate ☐ Control ☐ Blind ☐ Transmit ☐ Other ______
	3	Case	Mfr Std ☐ Nom Size ______ Color: Mfr Std ☐ Other ______
	4	Mounting	Flush ☐ Surface ☐ Yoke ☐ Other ______
	5	Enclosure Class	General Purpose ☐ Weather Proof ☐ Explosion Proof ☐ Class ______ For Use in Intrinsically Safe System ☐ Other ______
	6	Power Supply	117V 60Hz ☐ Other ac ______ dc ☐ ______ Volts ______
	7	Chart	______ Strip ☐ ______ Roll ☐ ______ Fold ☐ Circular ______ Time Marks ______ Range ______ Number ______
	8	Chart Drive	Speed ______ Power ______
	9	Scales	Type ______ ______ ______ ______ Range 1 ______ 2 ______ 3 ______ 4 ______
XMTR	10	Transmitter Output	4-20 mA ☐ 10-50 mA ☐ 21-103 kPa (3-15 psig) ☐ Other ______ For Receiver, See Spec Sheet ______
CONTROLLER	11	Control Modes	P = Prop (Gain), I = Integral (Auto Reset), D = Derivative (Rate), Sub: s = Slow, f = Fast P ☐ PI ☐ PD ☐ PID ☐ I_f ☐ D_f ☐ I_s ☐ D_s ☐ Other ______
	12	Action	On Meas. Increase Output: Increases ☐ Decreases ☐
	13	Auto-Man Switch	None ☐ Mfr Std ☐ Other ______
	14	Set Point Adj.	Manual ☐ External ☐ Remote ☐ Other ______
	15	Manual Reg.	None ☐ Mfr Std ☐ Other ______
	16	Output	4-20 mA ☐ 10-50 mA ☐ 21-103 kPa (3-15 psig) ☐ Other ______
ELEMENT	17	Service	Gage Press. ☐ Vacuum ☐ Absolute ☐ Compound ☐
	18	Element Type	Diaphragm ☐ Helix ☐ Bourdon ☐ Bellows ☐ Other ______
	19	Material	316 SS ☐ Ber. Copper ☐ Other ______
	20	Range	Fixed ☐ Adj. Range ______ Set At ______ Overrange Protection To ______
	21	Process Data	Press: Normal ______ Max ______ Element Range ______
	22	Process Conn.	1/4 in. NPT ☐ 1/2 in. NPT ☐ Other ______ Location: Bottom ☐ Back ☐ Other ______
	23	Alarm Switches	Quantity ______ Form ______ Rating ______
	24	Function	Press ☐ Deviation ☐ Contacts To ______ on Inc. Press.
OPTIONS	25	Options	Filt Reg ☐ Sup Gage ☐ Output Gage ☐ ______ Charts Diaph Seal ☐ Type ______ Diaph ______ Bot Bowl ______ Conn ______ Capillary: Length ______ Mtl. ______ Other ______
	26	Mfr & Model No.	______

NOTES:

ISA FORM S20.40a

ANSWERS TO EXERCISES

Chapter 1

1.1 A. Field-mounted controllers

B. Transmitters and panel-mounted controllers in control rooms

C. Direct digital control

D. Distributed Control Systems (DCSs)

E. Digital Transmitters

F. Programmable Logic Controllers (PLCs)

G. Single Loop Digital Controllers (SLDCs)

(Note: D, E, F, G were all developed about the same time)

H. HART

I. FOUNDATION™ Fieldbus

1.2 The flyball governor for the steam engine

1.3 Flexible automation—configuration for different control strategies

Economics

Greater accuracy and dependability

Networking of instruments

Communication via a bi-directional data highway

Field control systems

Asset Management

Intelligent field devices or digital transmitters

1.4 Pneumatic transmitter usage has greatly declined in recent years, but some may still be used in hazardous atmospheres, and for remote locations where electricity may not be available.

1.5 The interconnection and configuration of all instruments to monitor, control and retain data in control systems.

Chapter 2

2.1 The area of the wall is: (10ft)(12ft) = $(120ft^2)(144in^2/ft^2)$ = 17,280 in^2 (.5 $\#/in^2$) = 8,640 pounds

2.2 Pressure is proportional to force and inversely related to the area over which the force is applied. For liquids at rest, the pressure exerted by the fluid at any point will be perpendicular to the boundary point, when an external pressure is applied to any confined fluid at rest, the pressure is increased at every point in the fluid by the amount of the external pressure.

As a force is applied to one of the pistons, usually the smaller one, the result in pressure created in the fluid is applied to the area of the larger piston. This creates a force on the larger piston which is proportional to the areas of the two pistons.

2.3 Gage pressure is referenced to atmospheric pressure. Absolute pressure is referenced to a vacuum.

Absolute pressure = gage pressure + atmospheric pressure

2.4 $$\text{Atmospheric pressure} = \frac{14.7\#/in^2}{0.433\#/in^2/ft} = 33.9 \text{ ft } H_2O$$

$$\frac{33.9 \text{ ft } H_2O}{13.6} = 2.5 \text{ ft mercury} \times 12 \text{ in. ft} = 29.91 \text{ in. mercury}$$

2.5 An accuracy statement expressed as a percent of full scale is reference to that constant value. An accuracy statement expressed as percent of measurement is referenced to the value of actual measurement. Accuracy expressed as a percent of measurement is more accurate.

2.6 Accuracy—The departure of the measured value of a quantity from the accepted standard for that quantity.

Precession—The ability of an instrument to reproduce a given measurement.

Hysteresis—The error resulting in a change in the direction of measurement. This can be caused by friction and worn or damaged components and is usually more prominent in mechanical instruments.

Linearity—The closeness of a calibration curve to a specified straight line, which represents a device with a constant gain.

Dead band—The range through which an input signal may be varied, upon reversal of direction without initiating an observable change in output signal.

2.7 The purpose of calibration is to verify the accuracy of a device. A calibration procedure generally consists of simulating and measuring, with a test instrument, the measured variable, observing and verifying the corresponding instrument response or output.

2.8 Zero, span, nonlinearity

2.9 Primary standards are test instruments against which the accuracy of all other calibration devices is compared. They are maintained by NIST.

Secondary standards are test instruments where accuracy is directly traceable to a primary standard.

2.10 Test instruments are generally a higher quality, are more accurate and have greater precision than a production instrument.

Chapter 3

3.1 C-tube—less expensive, less free-end travel, rugged and reliable—used mostly in process gages.

Spiral—greater free-end travel, higher torque at free-end—more sensitive than c-tube—used in some pressure gages and transmitters.

Bellows—greater movement to pressure ratio, increased sensitivity and accuracy, low to intermediate pressure range—can be used in differential pressure applications.

Diaphragms—sensitive to low pressure ranges because of larger wetted surface area—used in differential pressure instruments.

Helical—high free-end travel and torque, high range applications—can be used in direct reading indicators.

3.2 In motion balance instruments, the relative position of the rotation of the pointer or movement of the sector is a direct result of the amount of deformation of the pressure element caused by the applied pressure.

3.3 Zero elevation is an adjustment made to the instrument to raise the zero point to compensate a net result of a wet leg in the low side of a differential pressure instrument. Zero suppression is an adjustment made to lower the zero point to compensate a wet leg on the high side of a differential pressure transmitter or the measuring port of a pressure transmitter.

3.4 Pressure elements:

Stainless steel

Bronze

Beryllium copper

Monel

3.5 Separators are used to keep process fluid from contacting the pressure element.

3.6 AAAA

AAA

AA

A

3.7 Classification by case design and arrangement

Classification by field and special use

3.8 Generally, for liquid and vapor service, the instrument and impulse line should be gas free, for gas service, the instrument and impulse line should be liquid-free.

3.9 If the process fluid is likely to freeze in the environment in which it is to be used, the instrument and impulse line should be insulated and/or heat traced or filled with a fluid that will not freeze. This is called sealing the instrument.

3.10 Pulsation dampers can be used to prevent sudden and large changes in process pressure variations causing undesirable spurious responses of the instrument. Static pressure measurement on the discharge of a positive displacement pump.

Chapter 4

4.1 A transducer converts one form of energy to another, more useful form, of energy. A transmitter converts a process measurement to a scaled electronic or pneumatic signal to be transmitted to a location for measurement. A transmitter has an input transducer to convert the process variable to a motion, strain or force that the output transducer uses through signal generation or conditioning to develop the output signal.

4.2

Type	Input	Output
potentiometer	motion	resistance
inductor	motion	inductance
linear variable differential transformer (LVDT)	motion	voltage
capacitance	force movement	capacitance
strain gage	force movement	resistance
resonate frequency	force movement	frequency
Piezo-resistive	force movement	resistance
Reluctance	force movement	inductance

4.3 A DC Wheatstone Bridge circuit is an electrical transducer to convert a small change in resistance to voltage or current. By using equivalent circuit theorems, exact relationships between resistance, voltage, and currents can be established.

4.4 A differential pressure transmitter measures small changes in pressure, usually expressed in inches of water, applied to each side of a pressure element. The pressure element, usually a capsule or diaphragm, does not respond to static process pressure.

4.5 Differential pressure instruments are used to infer a flow valve by measuring a pressure drop across a restrictor in a flowing process fluid.

Differential pressure instruments are used to infer a process level by measuring a hydrostatic head pressure exerted by a column of liquid differential pressure instruments are used to negate the affects of static pressure in the vessel.

4.6 Analog electronic transmitters produce 4.20 mA current and 3-15 psi. Process connection is 1.2″ NPT. Three valve manifold arrangements and connections are standard for isolation and equalization for differential pressure applications. Body is stainless steel or other exotic materials used for wetted parts as required for more corrosive process fluids. Later standards such as HART, FOUNDATION Fieldbus (FF) and Profibus PA are used to set standards for communication and field control applications

4.7 Force balance principle for measurement and transmission imply that a measurement produces a force on the input transducer which is applied to the signal conditioner or secondary transducer which produces an output. The output is applied to the input to produce a negative feedback response that provides a stabilizing affect for the measurement system.

Motion balance systems provide stabilization to the measurement system whereby the measurement force produces a motion and output measurement. This motion continues to provide a negative feedback.

4.8 Digital transmitters enable two-way communications between field instruments and operator control stations. This has made possible the following features that are not available with conventional transmitters:

- Remote calibration and reranging
- Self-diagnostics
- Calculations for linearization and characterization
- Configuration for specific applications
- Maintenance records and asset management

Other advantages are:

- Less maintenance required
- Price compatibility with conventional transmitters
- Less susceptible to error caused by environmental and ambient conditions (temperature and pressure)
- Greater accuracy, procession and turndown

4.9 Sensor trim is a means to position the factory characterization curve to optimize transmitter performance over a specified pressure range or to adjust for mounting affects.

4.10 HHC is Hand Held Communicator

HHT is Hand Held Terminal

It is used as a human machine interface (HMI) unit for digital field devices, normally digital transmitters and digital value positioners (controllers).

4.11 Configuration is the process of setting the parameters (input and output range, engineering units, etc.) for a specific application.

4.12 4–20 mA transmission is still used when there is a lack of compatibility for digital communication between transmitters and receiving instruments. The use of 4–20 mA will no doubt decline as FOUNDATION Fieldbus and all other digital systems become more common.

4.13 In HART technology, a digital signal is superimposed on a standard 4–20 mA DC signal.

4.14 It uses both digital and analog signals for transmission. No, HART usage is expanding.

4.15 An all-digital serial two-way communications system used to link field devices.

4.16 By using a frequency hopping spread spectrum

4.17 FCS uses a programming language available with FOUNDATION Fieldbus that enables field instruments (transmitters and valve positioners) to perform control functions.

4.18 Star, Mesh, and Cluster-tree

Chapter 5

5.1

Type	Application
Dipstick	engine oil, tank inventory
Lead Lines	tank inventory, lake and reservoir depth
Sight Glasses	open and closed tank inventory
Float Gages (automatic tank gages—ATG)	open and closed tank inventory (greater level distances than others

5.2 A float moves on the surface of liquid to be measured which causes an indicator to move along the tank edge or on a take-up reel.

5.3 Gagehead, Float, Tape or Cable

5.4 A displacer is immersed in the liquid to be measured. As the liquid level changes, the buoyancy of the liquid causes the weight to change which results in a small movement. This movement through a signal conditioner generates a pneumatic or electronic signal (3-15 psi, 4–20 mA).

5.5 The amount (volume) of the liquid displaced and the specific gravity of the liquid.

5.6 Volume of displacer $= \frac{\pi D^2}{4}(L) = 21.98 \text{ in}^3$

Weight of Water Displaced = $(21.98 \text{ in}^3)(.036 \text{ \#/in}^3) = 0.7195$ pounds

Weight of Oil Displaced = (.7915 pounds)(.6) = 0.4749 pounds

Weight Difference = 0.7915 pounds – 0.4749 pounds = 0.3166 pounds

5.7 Because the length/weight ratio determines how far the liquid level must rise on the displacer, the actual change in level, to create a resulting weight change and instrument response.

5.8 The function of the torque supports the weight of the displacer in the lighter fluid and recoils as the weight of the displacer changes. It also provides a seal between the process and atmosphere.

5.9 Fill the displacer volume with the lower specific gravity fluid and zero the instruments output. Fill the displacer volume with the heavier fluid and check for 100% instrument output. There will probably not be a span adjustment as this is determined by the displacer volume—length/weight relationship—and the specific gravity of the fluid. When full calibration is not desired, the process level can be established a mid-point as can normally be detected by a reference mark on the displacer gage. The output of the instrument is set to 50 percent by the zero adjustment.

5.10 For direct measurement, the instrument's response is measured as the process variable. For inferential measurement, the process variable is calculated or by some means related to the instrument's response. Most measurements are inferred values.

Chapter 6

6.1 Hydrostatic head level measurement is based on measuring the pressure exerted perpendicular to a surface area by a vertical column of liquid. A level value can be inferred by using the value of 0.433 pounds per square inch per foot of water and correcting for specific gravity of a process fluid.

6.2 Open tank measurement has atmospheric pressure exerted on the surface of the fluid to be measured. Atmospheric pressure is also applied to the low side of the differential pressure measuring instrument.

6.3 A. 10.39 ft.

B. 10.39 psi

6.4 For closed-tank level measurements, the pressure inside the vessel must be compensated by applying it to the low side of the differential pressure transmitter.

6.5 A diaphragm box operates by measuring the pressure inside a closed system. The box is located on the bottom of the liquid in the vessel. As the level changes a flexible diaphragm on the bottom of the box flexes causing a change in the volume of the closed measuring system resulting in a change in pressure which is used to infer a level.

6.6 This is another name for the diaphragm box.

6.7 In the bubble system, airflow is established in an open tube located in the vessel. The back pressure on the tube is measured and is equal to the hydrostatic pressure exerted on the open end of the bottom of the tube. The location of the open end of the tube in the vessel is the zero reference level.

6.8 Span = 30 in. H_2O

Suppression = 7.2 in. H_2O

6.9 Span = 63.2 in. H_2O

Elevation = 75.05 in. H_2O

6.10 A diaphragm and flange assembly is mounted to the process and connected to the instrument body by way of capillary tubing and filled with a noncompressible seal fluid. Pressure on the measuring diaphragm is transferred to the instrument measuring element by the seal fluid.

Diaphragm seals are used to isolate the process from the otherwise wetted portion of the instrument and to eliminate wet legs.

Chapter 7

7.1 As the level changes on the resistance tape, the head pressure of the liquid causes a resistance helix or slidewire to touch a shorting base strip. This changes the measured resistance of the slide wire that is proportional to length of the unshorted portion and the level of fluid.

A process fluid forms the dielectric between two plates of a capacitor. The vapor above the liquid forms another dielectric as the level of the process changes between the plates, the resulting change in dielectric causes a change in capacitance between the plates and is calibrated as level.

A conductance cell to measure level operates much like the capacitor except the measured electrical quantity is conductance that is calibrated as level.

In ultrasonic level measurement, a waveform is transmitted toward the level in a vessel and is reflected by the surface of the liquid. The time of travel of the transmitted and reflected wave is measured. From the known velocity of the waveform and time of travel, the distance is calculated and is proportional to level.

Pulse radar—A radar signal is transmitted through the tank atmosphere, travels to the liquid level and travels back to the receiver. The total time of the transmitted and reflected waveform is measured. Knowing the velocity and time of the traveling waveform, the distance is calculated and divided by two to determine the position of the liquid level.

Frequency modulated continuous wave (FMCW) – A radar signal is transmitted through the vapor portion of the vessel and reflected by the liquid surface. There is a frequency shift between the transmitted and reflected signals that is a function of the distance traveled.

7.2 Point level measurement is used to determine the location of liquid in a vessel.

Continuous level measurement is used to give a measurement that indicates instantaneous tracking of level in a vessel. Point can be used for alarming, shut-down, and on-off control. Continuous is used to determine the level at any and all points on the measurement range and continuous control applications.

7.3

Contact

vibrating switch	float switch	thermal detector	nuclear
rotating paddle	pressure switch	ultrasonic	

Non Contact

radar	ultrasonic	capacitance	fiber optic
head	resistance tape	nuclear	magnetostrictive
weight	magnetostrictive		

7.4 Cobalt 60 and cesium 137

Alpha, beta, and gamma particles are emitted by the source.

Gamma particles are used in nuclear level measurement.

Nuclear level measurement is based on transmitting gamma particles through the level process and detecting the signal after passing through the process along the axis of measurement. The difference in strength between the transmitted and received signal is a function of the path length distance of the signal.

7.5 Nuclear measurement is immune to adverse process conditions such as temperature, pressure, abrasive and corrosive materials and vessel geometric design.

7.6

weight	radar
nuclear	rotating paddle
ultrasonic	vibrating switch

7.7 The speed of the waveform changes with temperature.

7.8 Liquid-liquid interface can be measured by guided wave technology where electromagnetic pulses are transmitted along a conductor. When the wave form contacts an interface—liquid-vapor and liquid-liquid—a portion of the wave is reflected. By measuring the time between transmission and reflection waveforms, the distance can be calculated.

7.9 In end-of-the-probe applications, the transmitted waveform along a guide travels to the end of the probe and is reflected. The time delay between the signal traveling to the end of the probe through the vapor portion of the tank and through the liquid is proportional to the level and the dielectric of the material. It measures the dielectric or the material and can operate with lower dielectric, also, it will operate with a high noise to signal ratio.

7.10 A. multiple reflections

B. multiple propagation

C. other microwave transmitters

7.11 An inductive sensor/encoder

7.12 A bottom-based system eliminates possible error caused by tank-roof movement.

7.13 Interphase measurement is provided by installing a float along the measuring element that is positioned by the liquid-liquid interphase surface.

7.14 High temperature, corrosive product, and slurry applications. All may require noncontact level measurement. Installed agitators can also result in a non-constant interface point.

7.15 Because the speed of light is nearly constant with regard to tank atmosphere, a medium is not required for the propagation of light waves.

7.16 The nature and amount of laser beam reflection is a function of the characteristics of the liquid surface.

7.17 Because of the very short wavelength and nearly parallel path of laser beams.

Chapter 8

8.1 Both head level and density measurement operate on the principle expressed in the following equation:

$$\text{From equation 7.4,}\ P = \frac{0.433\ \#/\text{in}^2}{\text{ft}}(G)(H)$$

For level applications, the G term is constant and pressure reflects level. For density applications, the H term is constant and P changes with density (gravity).

In displacement applications for level measurement, the specific gravity of the process fluid is constant and the weight of the displacer is a function of the amount of the displacer submerged in the process fluid as the level changes. For density applications, the displacer is always fully submerged and the weight is a function of the density of the process fluid.

In radiation level measurement, the strength of the gamma rays at the detector is a function of the path length through the process fluid which changes with level. The density of the process fluid is constant. For density measurement, the path length of the gamma rays is constant and the strength of the gamma rays at the detector is a function of the density of the process fluid.

8.2 For density applications, the static pressure in the process vessel must be referenced to the low side of the transmitter. When the low pressure tap is in the vapor section of the vessel and the vapor condenses, a wet leg results. Also, when the low pressure tap is in the liquid portion of the tank, there will be a wet leg (Figure 8-3A). A repeater can be used as shown in Figure 8-3B to eliminate the wet leg.

8.3 Oscillating coriolis—The end of the tube is forced to oscillate at the free end and the frequency of oscillations is a function of the tube mass which changes with the density of the material in the tube which has a fixed volume.

Ball type density meter—Silica balls with different densities are immersed in the process enclosed in a glass tube. A light source projects light into the tube which contacts photodetectors on the side of the tube opposite the light source. The height of the highest ball is detected and relates to the density of the product.

Vibrating Spool—A cylindrical spool is immersed in a process fluid and forced to oscillate along its longitudinal axis. The oscillation frequency is a function of the stiffness of the spool and the oscillating mass. The entire system consists of the spool and fluid around the spool which also oscillates. Since the spool has a fixed mass and the frequency of oscillations is a function of the total system mass, fluxion in oscillation frequency is a result of the mass (density) of the process fluid.

U-Tube—Process fluid flows through the U-tube section of the pipe supported by flexures at the open end. As the mass of the tube changes, the closed, free end of the tube will experience a small defection. This defection is measured by an appropriate proximity detector and the resulting displacement is a function of the total mass of the tube (tube mass + mass of process fluid).

Tuning Fork—A tuning fork oscillating at resonant frequency is submerged in a solution whose density is to be measured. The frequency is constantly monitored and relates to the density of the fluid.

Chapter 9

9.1 Automatic Tank Gaging (ATG) defines an instrument or system which automatically measures and displays liquid level or ullage in one or more tanks either continuously, periodically, or on demand.

Hydrostatic Tang Gaging (HTG) incorporates a means by which the true volume or mass of a material in a tank can be determined.

Also, for HTG applications, the exact height/volume relationship of a vessel must be known. This is determined by a tank-calibration (strapping) method.

In most terminology, HTG is considered an application of ATG.

9.2 Manual strapping

Liquid calibration

Optical reference

Optical triangulation

Electro-optical distance ringing

Tank calibration is required to determine the height/volume relationship of the vessel.

9.3 Tank recomputation is to be considered when:

- Tank bottom causes inside diameter changes
- Tank bottom causes plate thickness changes
- Tank tilt requires volume/height correction
- There are variations in tank shell temperature
- Uncompensated process specific gravity

9.4 Level

Temperature

Pressure (less critical and may not be necessary in some applications)

Density (mass)

9.5 Calculations to determine:

- The tank height/volume relationship
- Density unless directly measured
- Level from hydrostatic head unless directly measured
- Volumetric units as desired/required
- Mass unless directly measured
- Equivalent tank area
- Net volume

9.6 Stratification of:

- Temperature and density values in the tank between measuring points

- Changes in tank geometry that are not compensated resulting in inaccurate level measurement/calculation—this error can result from an inaccurate tank strapping table data

9.7 HTG can provide volume measurement with sufficient accuracy for custody transfer.

Chapter 10

10.1
A. Low dielectric
B. Changes in fluid specific gravity
C. Change in process specific gravity, coating on displacer
D. Material in vapor space, fluffy process material

10.2 False

10.3 C. Radiation

10.4 C. Radar

10.5
1. D
2. C & D
3. A
4. B
5. E

Index

D

E

I

L

N

O

P

S

U

V

W

Printed and bound by CPI Group (UK) Ltd, Croydon, CR0 4YY
13/09/2026
14961057-0004